MARS CITY STATES
New Societies for a New World

Edited by
Dr. Frank Crossman
The Mars Society

Published by Polaris Books
Lakewood, Colorado

 **THE MARS SOCIETY**

TABLE OF CONTENTS

PREFACE: WHAT WILL WE CREATE ON MARS?

Robert Zubrin

Mars appears barren to many people today, just as ice age Europe and Asia must have appeared to early humans migrating out of our original tropical African natural habitat. Yet, by developing new technologies, new attitudes, and new customs, our ancestors were able to create the resources to not only sustain themselves, but flourish with ever increasing prosperity across the entire planet. In doing so, they transformed humanity from a local biological curiosity of the Kenyan Rift Valley to a global family, hundreds of nations strong, sporting innumerable contributions to thought, literature, art, science and technology. Having achieved such success, the task before us is take it further, by developing the technologies and ideas to create the resources that will enable the birth and sustain the growth of new vibrant branches of human civilization on the many worlds that surround us.

Of the worlds currently within our reach, Mars possesses by far the richest assortment of raw materials for transformation into resources. It thus presents the best prospect for settlement. Furthermore, Mars offers not just one new world, but an open frontier where we will learn what we need to know to take on the rest. It is there that we will develop not only many of the key technologies, but more importantly the attitudes needed to go, settle, and create life beyond.

The possibility of not merely exploring, but beginning the settlement of Mars within our time, has recently been brought into view by the advances in space launch technology demonstrated by Elon Musk's SpaceX company. After a 40-year period from 1970 to 2010 during which the cost of space launch remained static, the remarkable SpaceX team's introduction of mostly reusable launch vehicles has cut launch costs by a factor of five – from \$10,000/kg to \$2,000/kg - over the past decade. As I write these words, they are taking steps to slash it by a further order of magnitude through the rapid development of a fully reusable two-stage to orbit methane/oxygen propelled heavy lift launch system they call Starship. With a 100,000 kg to orbit capability, no hardware expended, and propellant costs of about \$1 million per launch – in addition to being a high-performance propellant readily manufacturable on Mars, methane/oxygen is the cheapest of all propellants producible on Earth – Starship could potentially cut Earth to orbit launch costs to less than \$100/kg– two orders of magnitude lower than those prevailing a decade ago. Once reaching low Earth orbit, each Mars-bound Starship will need to be refueled by six more tanker Starships, putting the propellant cost of sending a Starship to Mars with 100 tons of payload, including 100 passengers, at around \$7 million. SpaceX's Boca Chica facility currently employs about 2,000 workers and is turning out Starship prototypes at a rate of 1 per month. Assuming an average cost, including benefits and taxes, of \$10,000/month per employee, that works out to a labor cost of \$20 million for each Starship prototype. If they can increase the rate to one per week, and cover the cost of materials and overheads with an amount equal to their labor costs, they can probably cut the cost of building a Starship to something in the $10 million range. (The very large booster first stage, called the Superheavy, would cost more, but, as these always return to the launch site the same day they take flight, they would only need a few of these to support a large Starship fleet.) So, if passengers were charged $400,000 each – or $40 million total for the hundred – that would be enough to pay for the Starship (the immigrants would be wise to keep it for housing), its cargo, and the launch operations and propellant to get it to Mars, with some profit for SpaceX to spare.

Now $400,000 for a ticket to Mars, plus tools, provisions, and a starter apartment in a landed Starship, is an interesting figure. It is the equivalent, in modern terms, of what it cost to travel from England to colonial America in the 1600s. At that time a middle-class person could pay for his family's one-way passage by selling his house and farm, while a working man could get his ticket in exchange for seven years working for room and board. Roughly speaking, that's about what $400,000/ticket represents today. It's a price that a sizable number of people could muster, if they were willing to cash in their chips, pull up stakes, and take a chance on a new life in a new world.

So people will soon be able to go to the Red Planet. But that very possibility opens a still more interesting – indeed truly grand - question: *What will we create on Mars?*

It was to answer this question that the Mars Society sponsored its Mars City State Design Competition in early 2020. The challenge: Design a city state for 1 million people on Mars. The prizes: $10,000 and a grand trophy for the best design, $5,000 and an excellent trophy for Second place, with lesser prizes and trophies on down to Fifth.

Each team entering the contest could win up to 100 points for their design. The design for a Mars city must clearly be technically feasible, so 30 points were awarded for the technical merit of the design. But unlike an exploration mission, or a scientific base, which could potentially be supported by government, non-profit society, foundation, or corporate largess, a Martian city state will need to pay its own way. 30 more points were awarded based on the soundness of the economic plan for the city. If a city is to succeed and grow, it will need to be a place that people will want to move to. Few will want to move to Mars to live in dingy circumstances. A Mars base may only need to be functional, but a Mars city needs to be beautiful. 20 points were awarded for the aesthetic merit of the design. But it is not only physical beauty of its surroundings that make a city joyful, it is the opportunity it offers its citizens to develop and exercise their full human potential. 20 more points were offered for the social, political, and cultural foundation of the city.

The response to this challenge was extraordinary. Some 176 teams from all over the world entered the contest, each submitting 20-page design studies. This was far more than the Mars Society leadership could read through and properly assess, so we put out a call for volunteers, and were delighted when fifty well-qualified individuals stepped forward to offer their services as judges. With this team available, we were then able to divide the 176 entrants into ten groups of 17 or 18 teams, with each group judged by a panel of five judges. Those panels then selected the two best from their group, to forward to all the judges for consideration. This group of twenty semifinalists were then assessed by all the judges, with the ten strongest teams chosen to be the finalists. These final ten then presented their designs to the judges live at the international Mars Society teleconvention in October 2020, resulting in the selection of the top five as winners.

All twenty of the semifinalist, finalist, and top five winning designs are presented in this volume. The range of creative ideas is extraordinary, and while in a contest of this kind, there perforce can only be a few winners, there is much of merit in all of them. Indeed, what we have here is an intellectual banquet, a feast for thought, that will be of enduring value for all those who will consider how to initiate human civilization on Mars in the future.

I designed the contest and served as one of the fifty judges. I thus had the pleasure of reading through one of the initial groups of 18 teams chosen at random, as well as all the 20 semifinalists, and witnessed the presentations all ten finalists at the convention. I was struck by the ingenuity displayed by the teams in coming with extremely clever technical, economic, and aesthetic solutions to the problems of designing a practical and beautiful Mars City state. But what moved me most of all was the glowing idealism of the teams in striving to define a new and better way for humans to live together in a new world. To be sure, the teams did not agree on the specifics, and their ideas ran the gamut from concepts that could be broadly described as social democratic to libertarian. But what they all had in common was a passionate commitment to the search for something better. What could be more important?

The ideas of 18th Century enlightenment liberalism that America's Founders used to initiate their new republic were not original with them. On the contrary, they were well known to educated circles in Europe at the time, but those who believed in such concepts as government of the people, by the people, for the people, freedom of speech, freedom of worship, and inalienable individual rights that stand above King, state, or church, were dismissed as impractical dreamers, perhaps pleasantly insane. It took a new start in a new world – a place where the rules hadn't been written yet – to give those ideas a chance to show what fruit they might yield. The Founders called their project a "noble experiment," as indeed it was. To be sure it did not produce perfect results, but it was the best the world had to offer at the time and did well enough that millions of people voted with their feet to come to America to be part of it and yo help to build it into a beacon of hope and a shining example to be developed further by, and for, all humankind.

As the varied ideas in this volume indicate, there are likely to be many noble experiments on Mars. Some will fail, but others will succeed, leading their cities to grow, and by their example, set a new standard for the further progress of humanity everywhere.

As Thomas Paine said at the time of the American Revolution; "We hold it in our power to begin the world anew."

What a grand thought. What a grand opportunity. Let us prove worthy of our moment.

Golden, Colorado, January 2021.

EDITOR'S NOTES

Frank Crossman

As outlined in Robert Zubrin's Preface, in 2020 The Mars Society sponsored a contest to create the best plan for a Mars city state of 1,000,000 people, focusing on making a sizable human urban settlement on the Red Planet as self-supporting as possible and developing the city state's economy, politics, society, and culture in an aesthetically pleasing environment. Of the 176 teams that submitted twenty-page proposals, judges picked twenty semifinalists, all of which have a chapter in this 580-page book.

To help you develop a reading strategy (front to back or skimming by topic(s) of most interest to you), I established a rather arbitrary 6- line set of Editor's Notes for each chapter that provide some highlights of the breadth and depth of the plans. The top ten finalists who all presented at the TMS October Convention and the First through Fifth place finishers are identified below.

Chapter 1 - "The Mars Union," has a mystical, philosophical, sociological and theological focus in addition to a detailed economic one, and it is written in a future-history style. It offers a detailed radar chart strategy for where to establish cities, and a pros vs. cons tradeoff as to habitat architecture. The external trade among Mars, Earth, asteroids, and moons is detailed. Access to the detailed charts of the original report are linked to QR codes and URLs.

Chapter 2 -"Aurora," establishes ice filled Korolev Crater as the spider webbed, city location with Earth sourced micro 10KWe nuclear reactors providing power, and transportation utilizing underground tunnels and maglev trains. Perchlorate eating bacteria cleanse the Martian regolith and release oxygen as a byproduct. Attention is paid to community and aesthetics to inspire the populace. Exports are focused on the exporting of ideas and arts.

Chapter 3 Finalist - "Mareekh," describes crater habs - domed craters upwards of 500 m. diam. The domes are a translucent silicone inflatable with high tensile strength cables in a hexagonal grid and with another cable set serving to anchor the domes to the concrete blocks buried in the crater floor. Small modular reactors 100-300 MWe supply most of the power requirement. Mareekh is envisioned as geographically disperse with 50 settlements averaging 20,000 individuals each.

Chapter 4 foresees a **"Mars City State"** that requires 130 years to reach one million inhabitants from its founding in a phased economic development starting with Science and Tourism, then Real Estate, and finally Industrial goals. A network of cities linked to the Mars space elevators give Mars the advantage over Earth of building a large industrial complex in space. A modular vault-like habitat architecture in a honeycomb-like tiling arrangement is detailed.

Chapter 5 - "Brahmavarta," details the initial founding of the city and its hexagonally shaped city building block. Its outer wall has an 11-layer outer thermomechanical coating over regolith brick. Brahamvarta foresees the long term terraforming of the planet. Profitable operations discussed include; asteroid mining, research & tourism, deuterium sales, and broadcasting. A 62-year investment and economic breakeven is projected.

Chapter 6 - "Hive City," is built into the slopes of Hellas Planitia in a interlocking hexagonal structures resembling honeycomb and giving the city its name. Attention is paid to the individual apartment designs. The education process is focused on education of truly creative individuals, providing them with everything they need for self-realization, happiness, as they become the pride and hope of the city. Pets are important to the aesthetics of the social structure.

Chapter 7 Finalist - "The Republic City State of Tharsis ," is a small network of population centers located in near-surface caves in the region of Tharsis Mons. Robotic dogs and cyber trucks are important to mining operations. A rotating hydroponics module, food printing, and a cultured meat process are described. Energy is supplied by tethered He balloons covered with solar panels and by KW size nuclear reactors. A blockchained cryptocurrency economy model is detailed.

Chapter 8 - "Tolma City," has a 5 ringed city architecture. The city-to-people connection is established by a narrative (light gray background in chapter) of a typical day for three inhabitants of the city. Each worker is paid the same, since each job is considered vital to the city. The importance of the Rodriguez well in extracting water from buried ice is detailed as is a scheme for aeroponic tunnels and a network chart for ISRU.

Chapter 9 - "Pax Ares," reaches one million persons nearly 500 years after its founding at Meridiani Planum, which has already been identified as a good source of raw materials and is near the equator. An expanding "snowflake" hexagonal city design is described. Net worth analysis is applied to funding of the city state. The importance of cross training in medicine and AI diagnostics is detailed in the Health section.

Chapter 10 Finalist - "The Nüwa Concept," has a self-sustainable development focus based on Energy Return on Investment. Each of five cities envisioned utilize architectures appropriate to the mesa, wall, and valley of craters, and many details of these excavation and tunnel systems are presented. A hybrid energy distribution system of photovoltaics, concentrated solar, and nuclear power is used. Details of early, intermediate, and university levels of education are discussed.

Chapter 11 - "Phronesium," establishes isolatable pods of 30-50 people with a dozen or so pods specializing in a specific technology activity (e.g., U-enrichment) to insure a diversity of experience and a simultaneous exploration of many options. The concept of a democratized decentralized economy based on small-scale firms is proposed, as is the need to bind the populace together with some shared understanding of values, identity, and morals.

Chapter 12 - "Cosmocity" is sited in Gusev Crater and a 1000-year terraforming project paralleling Kim Stanley Robin's Red-Green-Blue Mars trilogy is envisioned. There is a focus on phased urban planning architectures and detailed analysis of the supporting facility requirements for the sub cities in the crater. The design includes a space elevator and displays a city cross section showing the elevated mag-lev train system and intercity utility lines.

Chapter 13 Finalist - "Nerio," describes an underground city architecture. The tunnel habitat is envisioned as a European style metropolis with multi-story skylighted alleyways and gardens. Energy is provided by small- and medium- scale, molten-salt, nuclear reactors providing heat and electricity that is stored in a compressed air "peaker plant.". A chart detailing technology development on Mars and the terrestrial applications of that IP gives the export strategy of Nerio.

Chapter 14 Finalist-4th Place "Phlegra Prime," is comprised of ten spatially distinct districts located in the Phlegra Montes region with its abundant shallow ice deposits. The 260 years taken to reach a population of one million is predicated on a detailed plan of bootstrapping the industrial base and its expansion as the population grows. The utilization of Martian thorium deposits is described as key to a self-sustaining power production using thorium bed nuclear reactors.

Chapter 15 Finalist-3rd Place - "Foundation," has a settlement in the water-rich Arcadian Planitia linked to a one near the sun-rich equator. The populace is organized into 10,000-person autonomous districts. Real estate and IP are projected to be the primary sources of income and solar energy is emphasized over nuclear in order to ensure energy independence from Earth. A personal view of life is richly described by the narrative of Lucie's diary which runs throughout the chapter.

Chapter 16 - "Vanaheim," is 5 towns and 2 spaceports on the equator in Libya Montes just south of Isidis Planitia. Parabolic solar thermal trenches are build to produce superheated water for power. Sugarcane is the primary energy storage media to be refined to sugar, biochemicals, flammable gases, mulch, and charcoal. A detailed description of the infrastructural requirements of the distributed city state is presented, as is a scheme for file encryption and trade of IP to Earth.

Chapter 17 Finalist - "the SGAC city state design," starts with a discussion of aesthetics and living a happy life on Mars and uses a hybrid system of governance based on direct democracy assisted by an AI administration. The population is skewed toward engineers, scientists, and innovators. The settlements are located within lava tubes in Hellas Planitia and within craters in Terra Cimmeria and Terra Sirenum. Geothermal plants provide three quarters of the power for the cities.

Chapter 18 Finalist-2nd Place - "the Southern Cross Innovations City State," describes a manufactorum for the production of basic materials from generalized Martian regolith and extraction of rarer materials like thorium (for molten salt nuclear reactors) from specialized mining sites. 1000-person colosseum shaped habitats are the basic building block of the hexagonally arranged city, which is located near Chryse Planitia. Asteroid mining services are the prime income source.

Chapter 19 Finalist-5th Place - "Korolev Crater Special Administrative Region" of the USA is a city built in domes under the ice filled crater. "It operates as an essential autarky, producing the vast majority of its economic resources in situ" and importing only a few in trade. Power is provided by thorium reactors and stored in LN2 cryobatteries. Many chemicals and materials are produced in bioreactors. Attaining a steady age demographic pyramid is recommended.

Chapter 20 Finalist-1st Place - "Nexus Aurora," starts with a narrative of a typical family's day to illustrate that happiness and tranquility can be found on the red planet. The location is Dao Vallis on the eastern rim of Hellas Planitia. Food is grown in basalt fiber reinforced plastic farming modules with artificial lighting. The basic housing block holds 570 people. Medical facilities utilize tele-triage systems for emergencies. Several 3D models of urban planning designs are illustrated.

I hope these notes help to get you, the reader, started on this topic of how to design for a million-person settlement on Mars. I will remind you that the design teams all have email addresses, and many have website addresses linking to much more information on their designs than can be expressed in this book format. I encourage you to learn more about these exciting projects!

Palo Alto, California. January 2021

1: THE MARS UNION IN 149 MARTIAN YEAR
(2251 A.D. on Earth)
General Report on the Activities of the Mars Union

Zilu Wan
The Ohio State University
zwan.001@avaento.com

Angela Cui
Mars Society China
cyanctt@gmail.com

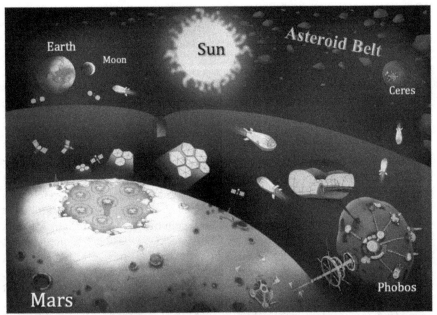

Figure.1 The EMMA (Earth, Moon, Mars, Asteroid Belt) space economy zone in 149 Martian Year. (*Image Credit: King Animation Inc.*)

Prologue: The Monolith Code behind EMMA Atlas

Back to 2034 A.D., when human being first landed on Mars, "Pioneer Martian" found a Monolith. On its surface, it was painted the EMMA Atlas from Martian prospective, in which the solar and planetary configurations of Sun, Earth, Mars and Ceres are shown above in **Figure.1**. Below the Atlas, there were four numbers engraved on the Monolith (**2034, 2108, 2251, 2674**, respectively). Scientists from Planetary Institution soon discovered that the solar and planetary configurations at these given Years precisely match the EMMA Atlas above (See **Appendix 1**). The

2034 A.D. is the year of one small step for mankind on Mars. Then, what do other three Calendar Years signify?

Prophets believe these given Years are milestones for Mars Civilization. Exactly as what they claimed, in 2108 A.D. (72 Martian Year), the population of Mars had reached to 1,000. Today, in 2251 A.D. (149 Martian Year), Mars Union is able to be a planetary independent civilization, because it has formally welcomed its 1,000,000 citizens. At this historical moment, Mars terraforming process has begun. Martians firmly believe the planetary scale environmental transformation will complete "ON TIME" in 2674 A.D. (378 Martian Year), simply because Monolith has already known EVERYTHING.

Chapter 1 Executive Summary

Part 1: A Review of Mars Union History
Rome was not built in a day, to outline a million-city state on Mars, the knowledge and experiences of thousand people town are critical, because these know-hows provide us a clear path to follow. Such knowledge and experiences mainly reflect in three fundamental aspects: technical feasibility, economic profitability and society harmonization.

1.1 Technical feasibility
Throughout the history of Mars immigration, the large-scale commercial transportation between Earth and Mars is fundamental. The use of plasma rocket propulsion engine helps in reducing the travel time to 42 days during Hohmann transfer period (Steve N., 2014; Levchenko L., et al., 2018). Such a significant decrease in travel time has prompted a reduction in transportation cost to $50,000, making the trip affordable to everyone. The maturity of the Giant Biological Life Support System (GBLSS) has enabled more than 1,000 people to live on Mars permanently, which indicates its safety and sustainability in terms of food, water, thermal control, radiation protection and air supply for respiration (Demidov O., et al., 2019). Moreover, the use of In Situ Resource Utilization (ISRU) technology has guaranteed the uninterrupted industrial and agricultural activities on Mars by sourcing food and construction materials locally rather than importing them from Earth as they used to. Lastly, the use of artificial wombs technology enables Martian babies to be bred and delivered in Earth Gravity Station located at the Mars Phobos Lagrange point. Given all physiological indexes have been well controlled during all breeding period, medical tests have confirmed no major differences between a child born on Mars and that born on Earth (Greenblatt J., et al., 2019).

1.2 Economy profitability
According to Greenblatt and Rao's analysis, the gross investment of the first thousand people town is about 150 billion Maroken since 2034 A.D. (32 Martian Year). After ticking the 1,000 population in 2108 A.D. (72 Martian Year), the Mars economy has been growing steadily at a rate of 5-10 % annually. The rate of inflation has remained stable and low (2-3%), indicating a positive and performing

economy (Greenblatt J., et al., 2019). Moreover, for many decades, the overall rate of employment has been at 70%, with slight fluctuations in the unemployment rate (3% to 5%). Due to the initiation of terraforming project and large-scale infrastructure construction, Mars economy has suffered from a high deficit rate of 5% over several decades, and may continue to increase by some percentage in the next few decades. Despite such economic challenges being anticipated, it is believed that successful investments on Mars will bring Martians with a tremendous advance in both Martian environment and their living conditions.

1.3 Society Harmonization

Martians with different religious, ideologies, nationalities and interests constitute various political parties that reflect social activities on Mars (Demidov O., et al., 2019). **Rainbow Seven** (See **Chapter 5**) is the ultimate goal for Mars Union government, since it guides Martians and Mars society to maintain sustainable development and continuous terraforming process. Focus has also been given on ensuring the Martians equality in the aspects of education and working opportunity---to guarantee our human beings stay at the forefront of any technologies such as artificial intelligence (AI) robots. As for religious, Marsianism becomes the primary choice for local Martians---the life found on Mars gives them new explanations for life and reasons for being (Rabb H., et al., 2018). Regarding this, Marsianism is well-developed and takes the form of faith within the colonization of Mars for the advancement of humanity and its survival. All the key components of Marsianism are found in the Marsian Creed, a revelation constituting the new and dominant religion on Mars (Tom K., 2018).

Part 2: The Mars Union has welcomed its one million citizens

A virtuous cycle of economic development, together with continuous technological advance, institutional reform, and environmental improvement have been nourished Mars community constantly. After 76 Martian Years, the Mars Union has welcomed its one million citizens. Throughout the developmental history, asteroid mining is one of the major industrial sectors. Without the supports from Mars, such as food, methane, consumable goods supply, asteroid mining mission would not be possible. As the indispensable part of EMMA Trade, such logistical supply needs facilitate the development of relevant industries on Mars. Today, in 149 Martian Year, the exports of agri-products, energy & fuel, rocket part as well as industrial services generate a staggering 31.44% of total Mars Union exports (See **Chapter 3, 3.9**), and provide more than 30,000 job opportunities.

1.4 Considerations for Multiple City Site (MCS) Strategy

As Mars Union population grows, citizens need new place. It seems that expanding current habitat is much easier than making another one at different place due to the facts of resource and energy convenience, environmental similarity, BLSS system expansion convenience, developmental cost saving, etc. However, Multiple City Sites (MCS) strategy shall always be the priority of Mars Union. Concrete analyses are as follows:

1.4.1 Preventing of serious contagious diseases
Humans have limited knowledge regard to Mars environment so far, thus any mishandling of biochemical materials from Earth could be catastrophic to our health. There is highly possible that, micro-organisms, pathogens, and other unknown biological things that, when taken to extra-terrestrial environment, are prone to cause biological mutation (Rabb H., et al., 2018). For Martians, epidemics like COVID-19 could have devastating outcomes due to the unknown influence of Mars environmental conditions on the pathogenesis of this virus.

1.4.2 Resource proximity
The significance of MCS strategy also reflects in the effective utilization of resources on Mars. Likewise on Earth, where the Middle East region is rich in oil, and South Africa is rich in diamond, Mars Tharsis regions are endowed with volcanic debris and ferric oxide. Valles Marineris has low levels of radiation and high thermal inertia property, which makes it an ideal place to build the City No.1. The temperature of Elysium Planitia is warm, and resources such as volcanic debris, sunlight, and water ice are abundance, thus having great potentials for agriculture. Isidis Planitia's substantial deposits of silica make it the most promising place for technology development (See **Chapter 2, 2.2-2.3**).

1.4.3 Culture diversity
Cultural diversity is important since it offers the possibility of variations in lifestyle and thinking, which is believed to be necessary for long-term human settlements on Mars. Such variations reflect in food cuisines, art and music, sport and entertainments, holidays and memorial days (Lordos G., et al., 2019), climate and habitat modular style (See **Chapter 2, 2.4.2**). Moreover, immigrants from Earth have diverse cultural backgrounds and religions. As a result, one super city would be challenging for them to be unified. Such a destabilizing risk will further build up, and finally result in ethnic conflicts among immigrants.

1.4.4 Terraforming process
MCS strategy also contributes to the terraforming process. Apparently, the planetary engineering project won't be able to complete in one place. Enormous evidences have shown that Hellas Planitia is prone to be the origin of global dust storms. Thus, city located in this region should prioritize dust storm monitoring and controlling. Modified green algae, together with silica aerogel, are promising solutions. Another place for terraforming is Boreum Vastitas. The combined uses of Orbital Reflectors Array (ORA) and the Ocean Transfer Energy Conversion (OTEC) machine enable the ice cap melting, thus further moistening the atmosphere and rising surface temperature (See **Chapter 4, 4.6-4.8**).

Chapter 2 Climate, Geology and City Layout analysis (CGCL analysis)

2.1 Overview
The growing and expansion of human settlement depends on the ability of Martians to explore and utilize the natural resources nearby. In this chapter, both geological

and climatic conditions determine how human settlements on specific site would be (Dodd K.C., et al., 2016). Fundamental parameters include ice water abundance, landforms, radiation level, annual min/max temperature, annual surface solar flux level, rocks, minerals abundance, etc. Once the settlement site has been selected, the community will first grow to a thousand people town, and a hundred thousand thereafter. It is to be noted that the "right place" doesn't mean the parameters are all look good. Such a potential place needs the knowledge of climate and geology, decision makers' wisdom and perspective, and all citizens determinant and efforts. With respect to terraforming process, climatic and geological data are significant. Identifying the origin of Global Dust Storm (GDS), tracing its route, then by using the combination of Genetically Modified Green Algae and Silica Aerogel, the GDS will be gradually under our human control.

2.2 Climate

According to Marohasy (Marohasy J., 2017), the climate variation depends on critical parameters. Such parameters include annual min/max surface temperature, solar flux to surface, thermal inertia, dry ice distribution. The analysis of GDS formation is also significant since it strikes at the very foundation of our community. Climate data is a decisive factor that pilots development of agriculture, energy consumption, and terraforming strategy.

2.2.1 Annual Min/Max Surface Temperature

The Mars Climate Database is a repository of information on the various climate elements of Mars (Millour E., et al., 2018). It provides engineers and scientists with vital information that articulates the model of the Martian climatological system. As for the annual min/max surface temperature, 4 critical time points for each global daily maximum and minimum temperature have been selected to simulate the temperature change throughout the whole year. These time points are Northern Hemisphere spring equinox (Ls=0), Northern summer solstice (Ls=90); Northern autumn equinox (Ls=180) and Northern winter solstice (Ls=270). In total (As shown in **Appendix 2**, **Column A & B**), 8 figures clearly unfold the surface temperature change throughout the whole year. As a summary, it can be concluded that Isidis Planitia, Elysium Planitia, Central Meridiani, Kasei Valles belong to tropical region, while Valles Marineris is Tropical Lowland; Olympus and Tharsis regions are subtropical highland; Hellas Planitia is subpolar lowland, etc.

2.2.2 Annual Solar Flux to Surface

The annual maximum and minimum solar flux to the surface is essential in understanding the type of agricultural activities that could be initiated on Mars. Similar to Annual Min/max surface temperature, solar flux has a seasonal change, thus 4 critical time points (same as temperature) for global daily solar flux to surface have been selected to simulate the flux value change throughout the whole year. In summary (as shown in **Appendix 2**, **Column C**), 4 figures clearly indicate its annual value change. Regions such as Valles Marineris, Isidis Planitia, Elysium Planitia, Meridiani Planum, Terra Sabaea receive more solar flux energy than others, which could potentially be utilized by large-scale farming activities.

2.2.3 Thermal Inertia (TI)

The term 'inertia' refers to the tendency of something to resist change, thus the word 'thermal inertia' refers to the tendency of something to resist changes in temperature (OSIRIS-REx Mission). The Mars Climate Database outlines the heat map of the planet and therefore provides a brief understanding of the possibilities of habitation on Mars. According to the thermal inertia map (**Appendix 3**), the red region tends to have high inertia capacity. These regions include Valles Marineris, Isidis Planitia, south rim of Hellas Planitia, Argyre Planitia, Acidalia Planitia, Utopia Planitia, Chryse Planitia. Considering Acidalia, Utopia, and Chryse Planitia will be submerged when the ice cap begins melting, these areas should not be selected as potential settlements.

2.2.4 Global Dust Storm

The original regions of Dust Storm sequences have been summarized in **Figure.2** according to Dr. Wang's research (Wang H., et al., 2015). Blue bars indicate that the regions will highly possible be submerged in case of terraforming. If so, ground terraforming methods like green algae plantations won't be feasible. As a consequence, though Acidalia Planitia is obviously leading the pack, Hellas Planitia, Solis Bosporus, Cimmeria Sirenum are the critical regions to control GDS. Next, among the fore mentioned three regions, Hellas Planitia is the best candidate for ground terraforming, as the terraforming plant will highly possible to survive considering its lowest elevation (high air pressure), low radiation level, and mild temperature comparing with the others.

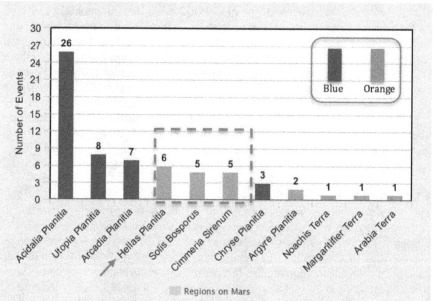

Figure.2 The original regions of dust storm sequences during Martian Year 24-30

2.3 Geology

An understanding of geology should precede considerations on human habitation of Mars. Topographic information, as it reveals the surface landform and elevation, should be taken as first priority for settlements and city planning. Besides the fundamental topographic, resource distribution such as water ice, POPF resource, Bouguer gravity, and volcanic debris foresees the region's potential for developing industries and cities. It shall be noted that Bouguer Gravity (**Appendix 4, figure D**) shows gravitational abnormality in Olympus and Amazonis region. High gravitational index at such a high elevation region suggests that the denser mantle is likely closer to the surface than average, which further tells the abundance of geothermal resource. **Appendix 4** summarized all the geological information as discussed in this section.

2.3.1 Water Ice Distribution

Water ice is of the most significant resource for habitation and industry. The water ice distribution map outlines essential information regarding the possibilities of settlement, agriculture, industry manufacturing, and fuel production on Mars. As shown in **Appendix 4, figure C**---*Distribution of water ice on Mars*, the water-rich areas are Valles Marineris, meridian equatorial regions, and south part of Elysium Planitia. These regions have great potentials for developing large scale agriculture and metropolitans.

2.3.2 POPF Mineral Resource Distribution

Plagioclase (silicon, aluminum, calcium oxygen, potassium, sodium): Plagioclase is some of the fundamental minerals that can be found on the surface of Mars. They are the type of minerals that hold value to human settlement. Their distribution is, however, dependent on climatic conditions that define various parts of Mars. As such, there are only specific places where plagioclase can be found. This mineral is vital in making various artifacts for human settlement.

Olivine (mineral magnesium iron silicate): The other fundamental mineral resource found on Mars is the Olivine group of minerals. Physical, mechanical and thermal properties enable the material relevant as a slag conditioner. The mineral is also used for blasting sand, ballast, and shield material. The properties of the mineral are ideal for building construction.

Pyroxene: This mineral is to be found mostly on metamorphic rocks. It is rock-forming inosilicate mineral that rich in calcium, and therefore, the primary source for building materials. The properties of the mineral also make it ideal for various aspects of human life. When it applies to tissue engineering and healthcare, the artificial bones as well as calcium-rich diet solve the bone loss issue effectively under the microgravity condition.

Ferric oxide (steel material): The steel material has been a defining symbol of human culture as it processes from one stage to another. The mineral is used in the production of steel materials, which have been useful in manufacturing and construction. The mineral is also useful in polishing jewelry. In medicine, ferric oxide used to cure itchiness, and to irritate human skin.

2.3.3 Potential City Zones

Based on Climatic and Geological data discussed above, parameters that are used for urban development research have been determined---temperatures, radiation level, water ice abundance, POPF Abundance, Soil Fertility, and ISRU Energy. Each parameter in spider chart has level from 1 to 7, thereby visualizing potentials for each region in agriculture, manufacturing industry, terraforming process, and urban liveability, etc. As summary, 8 candidate regions on Mars surface have been selected --- they are Valles Marineris, Tharsis Montes, Chryse Planitia, Elysium Planitia, Isidis Bay, Boreum Planum, Meridiani Planum, and Hellas Planitia, as shown in **Figure.3a&3b**

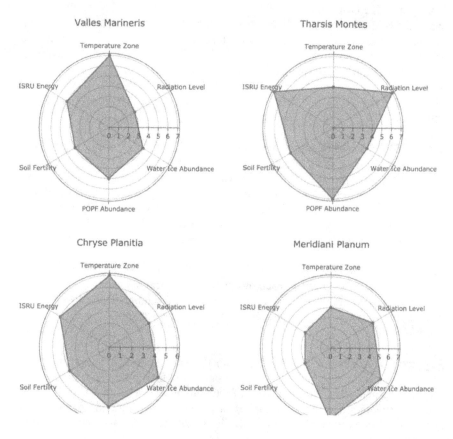

Figure.3a 8 potential regions for city and industrial development

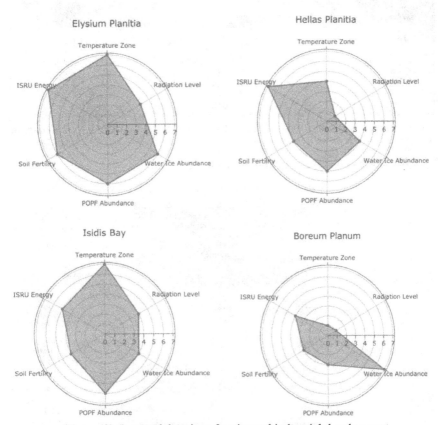

Figure.3b 8 potential regions for city and industrial development

2.4 City Layout
2.4.1 Fundamental Land Resource Law
Back to 2015 A.D., there was a strong consensus that in an circle area with diameter of 100 km---typically refer to a single Exploration Zone (EZ), multiple places within it are both sufficient to sustain multiple scientific investigations (Science Regions Of Interest---ROIs), habitat development, and potential industrial activities (Resource ROIs) by using In-Situ Resource Utilization (ISRU) (Ben B., et al., 2016). Over time, the concept of EZ first proposed by NASA has further developed into Mars Land Resource Law, therein defined the Exclusive Economic Zone (EEZ) for Human Settlement on Mars. Specifically, when the population of a settlement is under 10,000, it has EEZ circle with diameter of 100 km. The Mars Union Environmental & Sustainability Law states that the maximum population of a single settlement is 100,000. A list regard to population size and EEZ circle area has been summarized in **Table.1**.

Population of Settlement	Diameter of EEZ circle	Average land area per person	Total EEZ Area (km²)
1001-10,000	100 km²	0.785-7.85 km²	7,850
10,001-30,000	150 km²	0.588-1.762 km²	17,662.5
30,001-50,000	180 km²	0.508-0.848 km²	25,434
50,001-100,000	240 km²	0.452-0.904 km²	45,216

Table.1 Populations and city Exclusive Economic Zones, enforce by Fundamental Land Resource Law of Mars Union

2.4.2 Human Settlement Modular Analysis

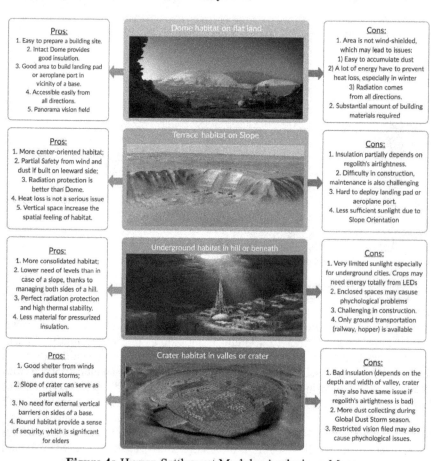

Figure.4a Human Settlement Modular Analysis on Mars
Scan QR code in the end for Full Resolution

*Information source: Joanna K., **2008**, Architectural problems of a Martian base Design as a habitat in extreme conditions. Doctoral Thesis, Gdańsk University of Technology.*

Pros:
1. Ideal shield of radiation, wind and global dust storms;
2. OTEC system speeds up vertical water circulation, thus reinforce ice cap melting process.
3. Spatial inner space, sufficient sunlight by ORA, these provide Mars residents a Mars Heaven to live

---*Credit by Zilu Wan*

Cons:
1. Extremely difficult in construction, huge investments upfront.
2. Maintenance is challenging, as the sphere is constantly moving.
3. Transportation is not convenient, only ground hopper and air ship are available
4. Primary energy inputs come from ORA, when it goes wrong, cities may face tight odds of survival.

Figure.4b Human Settlement Modular Analysis on Mars
Scan QR code in the end for Full Resolution
Information source: Joanna K., 2008, Architectural problems of a Martian base Design as a habitat in extreme conditions. Doctoral Thesis, Gdańsk University of Technology.

2.4.3 The Atlas of Mars City Layout

At the end of **Chapter 2**, there is a Mars city layout map made by Zilu's team, of which 8 primary city zones as well as Stanley Space Port & Earth Gravity Station have been visualized. Primary information for each zone such as center city, notable tour sites, GDP, Global Liveability Index, industry activities, and influential companies have been summarized in the map. Due to the page limit, the whole map will be listed as **Appendix 5.**

Chapter 3 Mars Union Economy

3.1 Overview

The economy of Mars Union is a highly advanced free-market, primarily knowledge-based economy. The prosperity of Mars Union advanced economy allows the country to have a sophisticated welfare state, a post-industrial infrastructure, and a high-technology sector competitively on par with Earth.

The major economic sectors of the Mars Union are energy, rare metal industry, space engineering, big data, 3D printing and smart factory manufacturing, agriculture, terraforming, biological science, pharmaceutical industry, and tourism. The Meridiani Planum rare metal and diamond industry is one of the Earth Moon Mars Asteroid (EMMA)'s centers for rare metal refinery and diamond cutting, amounting to 15.4% of all exports. Relatively poor in complex industrial manufacturing capability, Mars Union depends on imports of Robotic, Complex 3D printer, and other heavy machines for industry use. Its prosperity in high technology and rapid economic development is attributed to the presence of high-quality education and an extraordinary quantity of population. Mars Union has a reliable educational infrastructure that has enabled it to create value for goods and services. Also, with the presence of a high-quality incubation system, the country can create value for products and services, which has contributed to the establishment of high-tech firms in abundance. Moreover, all these factors are supported by the presence of a reliable venture capital sector. Its primary high technology station, referred to as 'Silicon Isidis,' is regarded as the second leading, as compared to its Californian equivalent on Earth. The Silicon Isidis Hi-tech Park (SIHP) has significantly acquired a majority of the Mars Union companies for their quality and consistent human resources.

Since Mars Union has already showcased a remarkable track record for generating lucrative technologies, it has emerged as a top priority for several of EMMA's leading shareholders, industrial echelons, and entrepreneurs. Therefore, EMMA business leaders have been signaled by the economic dynamism of the Mars Union. Examples of such business leaders are the pioneers of Heaven Garden Jewelry Auction, Starry Space Construction, EMMA Space Company, Hermes Data, and Stefan Genes. Each of them has acclaimed the economy of the Mars Union and invested substantially across several industries of exceeding their traditional business tasks and portfolios.

3.2 General Economic Data

GDP	1,065 billion (nominal, 149 MY est.); 1,015 billion (nominal, 150 MY est.) 1,098 billion (nominal, 151 MY est.)	Export	155 billion (149 MY est.); 135 billion (150 MY est.) 151 billion (151 MY est.)
GDP growth	7.8% (nominal, 149 MY est.); 4.7% (nominal, 150 MY est.); 8.2% (nominal, 151 MY est.);	Export Goods	Agriculture products, Consumer Goods, Rocket and Space Part, Industrial Services, Big Data, Pharmaceuticals, Fuel and Energy, License, Tourism;
GDP/capital	1,070,351 (nominal, 149 MY est.)	Import	148 billion (149 MY est.); 113 billion (150 MY est.) 141 billion (151 MY est.)
GDP by sector	Agri 2.5%; Industry 18.8%; Services 78.7% (nominal, 149 MY, est.)	Import Goods	Nuclear Reactor, AI Robotics, Aero Engines, 3D Printer, Rare Metals
Inflation rate	2.3% (nominal, 149 MY est.)	FDI Stock	210 billion (149 MY est.)
Human Development Index (HDI)	0.991 very high (149 MY est.)	Gross External Debt	292 billion (149 MY est.)
Populations	995,000 (149 MY est.)	Public Debt	52% of GDP (149 MY est.)
Official Languages	English & Chinese	Budge balance	-5% of GDP (149 MY est.)
Immigrants from earth	823,000 (149 MY est.)	Revenues	172 billion (149 MY est.)
Life expectancy	65 Martian Years	Revenues structure	Tax revenues: 165 billion Income and property tax: 93 billion Customs, Duties & VAT: 71 billion Fees: 997 million Other revenues: 7 billion
Currency	Maroken (MRK, MK)		
Exchange rate MRK per USD/RMB	0.83/0.22 (149 MY est.)		
Labor Force	580,750 (149 MY est.)		
Unemployment	3.7% (149 MY est.)	Expenses	187 billion (149 MY est.)
AI/Labor rate	38.6% (149 MY est.)	Expenses Structure	Terraforming: 39 billion; Education and R&D: 51 billion; Social Welfare and Medical: 22 billion; Administrative:16 billion; Interest on Debt: 13 billion; Repayment of loans: 14 billion
Labor Force Structure	Primary Industry 2.6% Secondary Industry 22.4% Tertiary Industry 75%		
Ease-of-doing business	1st (very ease, 149 MY est.)	Foreign Reserves	35 billion (149 MY est.)

Table.2 Gross Summary of Mars Union Economy Activities in 149 Martian Year

3.3 Energy and Electricity

Mars Union was depending on Earth's imports to satisfy its energy needs. Solar energy was once the Union's primary energy source; however, the seasonal Global

Dust Storm (GDS) contaminated the atmosphere with tiny dust particles (Brophy J.R., et al., 2013). As such, the Mars energy potential was reduced remarkably because sunlight was prevented from reaching the solar panels. Fortunately, energy import demand is gradually weakening thanks to the application of methane and geothermal resource in Tharsis Montes, Amazonia Planitia, Hellas Planitia, and Elysium Planitia. In 149 Martian Year, total energy generation of Mars Union amounts to 1.18 quad (equals to 345.8TWh). Its total consumption amounts to 1.05 quad (which equals to 307.7TWh) accompanied by a remainder of exports that amounts to 38.1TWh. Production of electricity amounts to 172TWh, whereas consumption equals 159TWh. The fixed generating capacity equals to approximately 44 GW, primarily from Geothermal (39.1%), Solar panel (16.6%), Nuclear power plant (34.4%), and Methane fuel plants of 9.9%. The energy sector has employed over 50,000 individuals.

3.4 Transportation and Logistics

It is quite challenging to develop global transportation system on Mars. Issues such as Global Dust Storm, radiation, cost of vehicle-based life support system, vehicles endurance mileage and backup support greatly limit ground transportation development. Currently, global ground transportation is primarily dominated by highspeed railway, aero plane, and ground hopper. Airship used to be an option, however, due to the thin atmosphere of Mars, its carrying capacity is very limited. Today, airship rides are promoted as sightseeing tour vehicles by traveling agents. In 149 Martian Year, statistics show that total railway operation mileage on Mars is about 35,300 kilometers, with approximately 396 railway stations already established and over 30 million passenger rides recorded. It also includes about 45 million tons of freight transported. As for groundhopper system, 1,205 stations have constructed, with 2 million travelers assisted, and 4 million tons of goods transported. The space system has accommodated about 100,000 traveler rides, and transported 120,000 tons of goods. Of 100,000 individuals who have reached the Stanley Space Port, 95% of them are Mars & Earth round-trip travelers, while 3% of them are staffs working at Earth Gravity Station and Stanley Space Port. The rest 2% of these individuals are scientific researchers, and resource explorers whose destinations are Enceladus, Titan, and Asteroid Belt.

3.5 Agriculture

Agricultural sector of Mars Union is one of the most developed industries. Regardless of the climate and geography of Mars Union, the industry is leading in the EMMA space region of agricultural technologies. The success of Mars Union Greenhouse Farm Units (GFUs) to nurture and grow a variety of crops is attributed to Giant Biological Life Support System, and hydroponic, aeroponic technology, LED lighting, RNAi, CRISPR, and clean soil technology (Lakkireddy K.K.R., et al., 2018).

Today, thanks to fore mentioned technologies, Mars Union is able to produce 100% of its food necessities (Latruffe L., 2014). There are 3,500 GFUs in 149 Martian Year, with a standard size of 15 acres per unit. At least 15,000 staff members work in GFUs. Out of total 52,000 acres under farming, approximately 28% is allocated for long-lasting cultivation. Rotating cultivations takes about 72%. GFUs are primarily situated along the equatorial region, particularly Elysium Planitia, Valles Marineris, Hellas Planitia, and Isidis Planitia. Field crops took approximately 38% of the total agriculture produce, whereas vegetables and fruits were attributed to a maximum of 35% in 149 Martian Year. Moreover, plants like rubber, mulberry, and cotton took 16%, as Cannabis and other specialties were reported to have taken 11%. As summary, approximately 41% of agricultural produce is apportioned for consumption, manufacturing has taken 32%, and direct exports accounted for 7.06% (See **3.9 External Trade**).

3.5.1 Meat Substitutes Industry
Some of the meat alternatives in the meat industry are insect, fungus, and cellular protein. Insect provides significant doses of calories per unit but utilizing less water and feed (Bonny Sarah P.F., et al., 2017). A comparison of insects and other plants and farm animals finds that insects offer similar macronutrients but substantially produce higher yields per unit. In other words, insects have a significant food conversion ratio than other protein sources. Regards to Cellular Agriculture, it allows people to have a familiar diet on Mars with constrained environment. Clean meat, cow-less milk, algae, fish, and chicken-less egg are utilized for generating food on a cellular platform by utilizing tissue engineering and cell culture methods (Elzerman J.E., et al., 2013). Quorn is also regarded as an alternative to meat. The producers of this product have claimed it has the same taste as chicken. In Mars Union, Quorn was once considered as the best source of protein (Whittaker J.A., et al., 2020).

3.5.2 Silkworm industry
The silkworm industry is associated with the farming of silkworms to generate silk. The marvelous natural material was a magnificent export of Chinese dynasties for several centuries, and it is still regarded as an exceptional commodity. The record of Chinese silk extends back to at least 2,000 years. Chinese have developed several applications for this precious material. For instance, it was used to make high-end apparels, bowstrings, and canvas for painting.

Silk can also be used to make protein drink. Its natural sweet taste, as well as its skin delicate functional property making it an ideal ingredient for sugar substitutes. Additionally, silkworm is also making a good insect food. Some Martians have recommended it as a potential candidate food for asteroid space exploration mission. Besides, feces from silkworm can be used for fertilizers in several ways. This fertilizer can be made through a fermentation process to generate a product referred to as the silkworm excrement organic fertilizer (Yuming F., et al., 2016). As such, silkworms form a crucial part of the Biological Life Support System of ground habitat and space station.

3.5.3 Liquor and Cannabis
Similar to Liquor & Cannabis Industry on Earth, these sections are critical origins of governmental tax. Generally, Liquor in Mars Union contains beer, Saka, brandy and whisky, Chinese Liquor, and grape wine. Longping wine specialists are working hard in an attempt to discover a variety of grape that might survive in the harsh conditions on Mars. Christians on Mars may use wine as a representation of the blood of Jesus Christ in the holy communion.

In Mars Union, Cannabis may present alternatives for the treatment of a wide range of medical conditions and diseases. Mars has a different surface condition as compared to that of Earth as scientists claim that it has different atmospheric pressures, and 0.38G in its gravitational force. Therefore, Cannabis is regarded as a significant contender for the treatment of chronic illnesses, inflammatory infections, and mood disorders. Martians may also use Cannabis for spiritual purposes.

3.6 Industrial manufacturing (Made in Mars, advanced manufacturing).
The first settlement of the red planet triggered the need for industrial expansion and development to cater to the rising needs of the inhabitance. In particular, the need for industrial expansion was triggered by the rising population and also a rise in living standards. Examples of the industries that emerged on the red planet are steel, mining and refinery, chemical, AI smart factory, and 3D printing. Mars Union has generally benefited from the emergence and expansion of the industries above, as they have facilitated the development of human settlements globally. The manufacturing capabilities of these industrial sectors have also made terraforming processes feasible. In 149 Martian Year, 30.5% of Mars Union economic gross value was attributed to its manufacturing sectors (Brophy J.R., et al., 2013). Approximately 80,000 staff members have participated in 792 industrial businesses.

3.6.1 Resource mining industry (also refer to POPF industry)
Plagioclase, olivine, pyroxene and ferric oxide are critical resources on Mars, and they serve as upstream industries for Mars Union manufacturing. Olympus & Tharsis, as well as Isidis Planitia are primary regions for resource mining industry, as geological data indicates these regions are rich in POPF. The geological data also shows that Meridiani Planum, Schiaparelli Crater, and Noachis Terra are regarded as sources of rare metals. Numerous craters in these places suggest that asteroid collisions are of frequent occurrence. Therefore, the abundance of heavy and rare metals is naturally higher than other regions on Mars. The extraction of rare metals and processing involves a significant part of exports, possibly around a quarter or more. In 149 Martian Year, total POPF production is 592 million tons, with 5.1% of the production belong to heavy and rare metals.

3.6.2 CGMP (Concrete, Glass, Metal, and Polymer) Industry
The emergence of Mars concrete---by using ISRU technology, enables large scale construction on Mars. One of its advantages is compressive strength--- approximated to be above 50 MPa. It is also capable of protecting Martians from gamma rays, X-rays, alpha rays, and other forms of radiation. A comparison with

other shielding substances finds that it is relatively low-priced and easier to mold into several shapes. As for glass, it is essential for industrial activities and life support on Mars. Production of photovoltaic arrays, glass domes, silica aerogels, as well as chips for robotics require substantial silica material and thereof. Metals are essential for infrastructure and manufacturing industry itself. Primary customers for metal are Global Highspeed Train Group, Space EMMA Rockets and hundreds of manufacturing factories. The applications of polymers are almost everywhere. Synthetic polymers are critical for construction of Phobos Space Elevator, Blue Origin Pearl City. The demand of polymers in Medical devices and pharma industry is also strong, as health issues are the first priority for Martians. In 149 Martian Year, total CGMP production is 298 million tons. To be specific, Concrete is 109 million tons, Glass 32 million, Metal 65 million, Polymer 92 million tons, respectively.

3.6.3 Life support necessities industries (water, oxygen)
Water and oxygen are basic elements for Mars citizens. As the living capsules grow to permanent bases, and further develop to urban cities, life support necessities are no longer critical. Terraforming process of ice cap melting further releases significant amount of water, moisturizes the atmosphere, which enables the global atmospheric circulation. In 149 Martian Year, production of clean water is $3.3*10^9$ cubic meters, of which 77% is reclaimed water. As for consumption, propellant plant usage is 5.7%, POPF and CGMP industry usage is 16.1%, Agriculture usage is 29%, human consumption is 41%, respectively.

3.6.4 3D Robotics Printer and AI Smart Factory
As for downstream of industrial manufacturing, 3D printing and Artificial Intelligence Smart Factory are as indispensable as water and oxygen on Mars. The flexibility is what makes it a valuable tool for manufacturing everything includes but not limit to industrial machines, major part of robotics and vehicles, building and infrastructural construction, medical devices, tissue plant, living goods, and even for food. By the end of 149 Martian Year, 5,500 smart factories have been built. More than 38,000 personal 3D printer workshops have been widespread used across the Mars Union.

3.7 University and Education systems
The red planet boasts some of the most developed and highly industrialized segments. Science and technology are at the forefront of the Martian industries. In 149 Martian Year, the percentage of Mars citizens engaged in scientific and technological inquiry, and the amount spent on R&D in relation to GDP, are among the highest in the EMMA space region. Mars Union education & research institute, together with kinder garden, primary school, middle and high school, attracted more than 200,000 people from Earth, Mars, and Asteroid Belt. Also, in the same year, the planet has recorded total employment of close to 100,000 individuals in the education sector. Statistic data shows that this sector has contributed to the creation of a significant number of job opportunities. As per the planet's per capita income, it has the most significant numbers of scientists and technicians in the EMMA space region, with 350 of these professionals in every 10,000 staff members. Such High-

Tech Talents have played a role in the development of natural science, agricultural science, space science and engineering, medicine, AI technologies, and other forms of technologies.

3.7.1 Universities, Majors
Mars Union has some leading universities and competitive courses offered. An example of such a learning institution is Valles Marineris University that offers programs in Finance, Space Science, Agriculture, Artificial Intelligence, and other related courses. Isidis Institute of Technology, located at central Silicon Isidis, is leading the development of Space Rocket Engineering, Big data, and Telecommunication. The Elysium University of Agriculture and Terraforming specializes in offering programs associated with logistics, agriculture, energy, food, and terraforming. Olympus Industrial University is a specialist in the provision of machine and energy courses. Moreover, Hellas Environmental Science Institute has also specialized in providing Martians with environmental science and terraforming programs. Kasei Life Science Institute offers hotel management, media, pharmaceutical, and life science courses. Also, Borealis Ocean College has specialized in marine-related courses like Ocean biology and hospitality management. Schiaparelli Diamond College is a specialist institute in the provision of diamond cutting and political and law-related programs. It is to be noted that there are two academies in Mars Union have the highest rejection rate among the colleges and universities. They are Lass Witz Fashion School and Isidis Film & Art School, both accepted rates are below 5%.

3.7.2 Intellectual Properties and Technology Transfer
Mars Union has made a rigorous effort to enhance the ability of the economy to benefit from a reinforced system of intellectual property rights. Such an initiative involves increasing the resources of Mars Union Patent Office (MUPO). It also involves reconstructing enforcement activities and incorporating programs to generate ideas sponsored by government research to the market. All Mars Union research universities have technology transfer offices. Their share of patent applications encompasses about 15-20% of the planet's inventive activity of Mars Union patent applicants .

3.8 Healthcare and Pharmaceutical
The availability of advanced and modernized technologies has enabled the economy of Mars Union to enjoy the privileges of having an innovative and high-quality healthcare system in the EMMA space region. Healthcare facilities on the red planet are fitted with cutting-edge facilities and state-of-the-art medical technology. Moreover, medical practitioners are highly qualified and trained. The population density of these medical doctors finds that there are six of them in every 1,000 individuals on Mars. As per the last stages of 149 Martian Year, there are 22 hospitals, with 12 of them being general hospitals, 6 being specialized medical centers, and 4 being mental health hospitals. They are accompanied by 86 long-term residential facilities. The government owns 4 of the general medical centers, whereas other private health centers are owned by non-profit charitable

organizations such as EMMA Charity Federation. NGOs also have partial management of the public clinics and pharmacies with other hospitals, roughly 580 treatment centers, and 225 drug stores globally. Also, by the end of this Martian Year, Mars Union has about 6,800 registered doctors, with an extra 5,000 who are licensed but not practicing (working in other fields or retired).

3.8.1 Diseases and Treatment

In 149 Martian Year, a round trip for Earth & Mars travel is 84 sols. Therefore, diseases affecting Martians' health are not primarily from travel. Radiation, contagious bacteria and virus, biosystem failure due to accidental events are the primary factors contribute to health problem. In addition, Mar 0.38g gravity may have long term side effects on health, of which include bone loss, muscle atrophy, visual damage, metabolically inactive or inert (Clément G., 2011). Treatments for those diseases include drug, diet therapy, tissue engineering, CRISPR therapy, etc. In 149 Martian Year, Mars Union hospitals treated about 100,000 patients, with 12,000 hospitality admissions and 6,500 emergency visits.

3.8.2 Artificial womb (AWB) industry

With the development of artificial womb technologies, several ethical issues are raised. On Earth, it is a general consensus that the technology may limit or eliminate the interaction between parents and infant. Hence such a technology may be unacceptable. In the case of Mars, 38% gravity force could significantly slow down embryotic development, and cell differentiation may severely be affected, which may extend pregnancy period to 15-18 months (David W., 2016). As a solution, by simulating Earth gravity in Earth Gravity Station, fetal development could be back to normal. Moreover, it is quite challenging for women living in EGS during pregnant period due to the unaffordable living cost and limited space, thus artificial womb has been allowed in EGS. Facts prove that babies born in EGS using AWB are generally safe. Thus, it become the most preferred alternative for having babies on the red planet. In 149 Martian Year, Living Force Medical Center in Earth Gravity Station received 5,200 applications from Mars Union's expectant mothers, 4,850 babies were delivered on time with good health, 280 were born prematurely, 70 were died before delivery or in early infancy.

3.8.3 Pharmaceutical

The pharmaceutical industry of Mars Union encompasses generic juggernaut Kasei Pharmaceutical industries and other startups---the research and development-driven firms. The industry is generally growing for several reasons. Regular drug consumption and patent licensing contributes 50% of profit for those pharma companies. Compare to Earth counterparts, Pharma companies in Mars Union have a natural sort of advantage of being a Contract Research Organization (CRO) partner due to Mars unique environment and 38% gravity. Thus, CRO business contributes another 50% profit. The continued stream of profitability through patent licensing and CRO business for those pharma company attracts hundreds of technological ventures in EMMA space region, several of whom believe pharmacy is the most prominent industry on Mars.

3.9 External trade

Exports has amounted to a total of about 155 billion MRK in 149 Martian Year. The red planet has imported approximately goods worth a whopping 148 billion MRK. However, in the following Martian Year, the estimate external trade volume is decreased by at least 15%. This is because Non-Hohmann Transfer Period (NHTP) significantly prolongs the travel distance and time. As **Table.3** shows, the significant imported products of Mars Union are Aero Plane Engines, 3D printers from Earth, Nuclear Reactor (IMSRs), AI robotics, and rare metals from an asteroid. However, its export products are 3D printed rocket, space station assemblies, liquid methane gas, rare metals, agricultural products, consumer goods, diamonds, and pharmaceuticals.

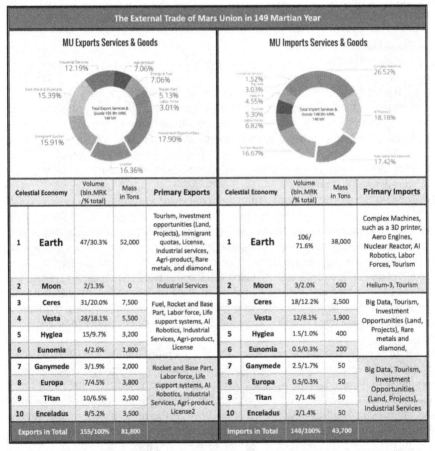

Table.3 The Summary of External Trade for Mars Union in 149 Martian Year
Scan QR code in the end for Full Resolution

A trade surplus is attributed to services from the hospitality and service sectors such as engineering services, software development, biomedical and scientific R&D.

Earth is the leading trading partner of Mars Union, and Mars Union is the second-leading trading partner of Earth. Both of the economies had entered into a contractual agreement in 72nd Martian Year, in which the economies agreed to increasingly removed tariffs on most goods traded between two planets since then. Mars Union has also been able to develop a trading relationship with Vesta, Moon, Hygiea, Eunomia, Ceres, Jupiter, and other planets. Based on regional terms, the Asteroid Belt is regarded as the leading and potential destination for Mars Union exports. In 149 Martian Year, Mars Union exported goods have amounted to about 78 billion MRK to the Asteroid Belt. These figures account for about 52.8% of the red planet's entire exports. Although there is trade surplus for Mars Union in total, Mars & Earth trade has huge trade deficit, mainly due to the high value of complex machines, AI robotics, and nuclear reactor. It is to be noted that trade deficit from Earth will exist for decades until terraforming process complete.

3.10 Banking & Finance
The currency for Mars Union is Mars Union Token, also being called Maroken (MRK or MK). It is a crypto currency using block chain technology (Heidi H., 2018). Basically, Martian crypto currency can already do four things:
1) Provide fast settlements limited only by the laws of physics regardless of the locations of the sender and the receiver in a transaction.
2) Provide a highly portable medium of exchange that doesn't rely on a middleman and doesn't care about Mars Union political regulations.
3) Provide a tamper-proof means of determining the details of any given transaction on demand.
4) Reward the production and deployment of resources.

However, the natural characteristic of crypto is deflation, which possibly could result in constrictive monetary policy and inhibit economy growth. As a solution, Mars Union Currency endorsed by governmental credit is indispensable. Mars Union Token has been issued by Schiaparelli Central Bank. Methane and potato are selected as monetary equivalents to stabilized price for Mars Union market. It is to be noted that Mars Union allow several types of tokens circulate among industries, and their exchange rates to MRK are considered as barometer of the industry.

As for financial services, Mars Union thriving venture capital industry plays an important role in funding the booming high-technology sector. The Union is now teeming with hundreds of prosperous private equity and venture capital (VC) firms looking for investment in the next potential million or billion startup. One of such a VC locates in Kunlun city, named Vien Venture Partners. Its major investment sector is Apparel and Fashion Design. By the end of 149 Martian Year, 8 fashion startups have gone IPO through its capital raisings and mergers.

Figure.5 shown below is a venture banking system that explains how collateral loans combine with unlimited insured capital by introducing Default Insurance Notes (DIN). It is expected that with appropriate DIN rate and leverage ratio, VC firms will become the most profitable capitalists of all time (Hanley B.P., 2019).

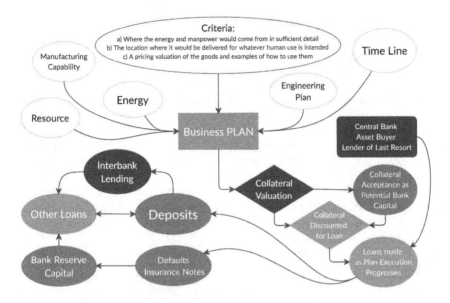

Figure.5 A Banking/Financing System for Mars Union Development

1. Mars Banking formally designates that plans based on utilization of resources are acceptable as collateral for a loan.
2. The loan collateral is discounted based on risk and capacity to execute to arrive at a loan amount. In some cases the discount could be eased if the plan is of high priority for the colony.
3. As the plan executes, the collateral value rises commensurately. In standard insured banking, the revaluation of the company only occurs at IPO or M&A.

*Information obtained by: Hanley, B.P. **2019**, Banking/Finance for Mars, 22nd*
Annual Mars Society Conference, University of South California, Los Angeles.

3.11 Comprehensive Economy Flow chart

In the end, we've made the comprehensive economy flow chart that summarizes capital, trade, technology and labor force flows of Mars Union, see **Appendix 6**. Such a flow chart exemplifies how Mars Union economy be made successful. The economy society begins with capital, Hi-tech, natural resource, external trade and skillful Mars Citizens and immigrants. Capital Entities invest real and service economy; hence job opportunities have been created. Labor force, technologies, together with resource input from ISRU or external trade, bring valuable goods, service, and revenue taxes naturally, thus investments have their return, people have salary, valuable goods and services benefit Martian life, protect Mars environment, and advance technologies. With such a virtuous circle, Mars Union economy will continue to grow, and the living standard and environment for Martians will continue to improve. In the future not too far, Mars will be our second home world (from 149 Martian Year, not from 2020 A.D.). See **Appendix 6**.

Chapter 4 Mars Union Technology and Engineering

Part 1 Technologies
4.1 Technologies Summary of Industry 4.0

Back to Earth in 2020 A.D., there was a profound industrial revolution named Industry 4.0, which also exerted significant influence on Mars society development. In essence, Industry 4.0 is the trend towards automation and data exchange in manufacturing technologies and processes which include cyber-physical systems, the internet of things, cloud computing, cognitive computing and artificial intelligence. Considering Martian harsh environment and poor living conditions, the need for Industry 4.0 for Mars exploration is even greater than on Earth. Consequently, such the advancement and improvement of technologies substantially adopted by Mars Union enables large scale industries on Mars and in EMMA space region possible and efficient (Brukner Č., et al., 2003). Their applications on Mars industry have been summarized in **Figure.6**.

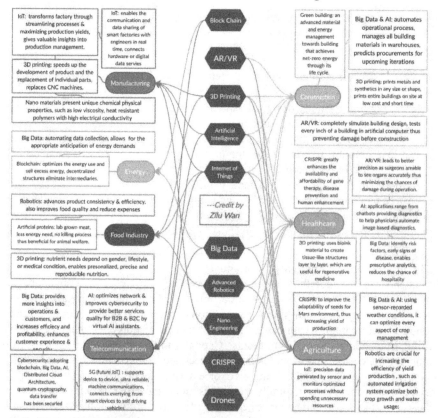

Figure.6 State-of-arts technologies and their influence on Mars Industry
Scan QR code in the end for Full Resolution

Information Source: Research Blog, 2018, Industrial 4.0 Innovation Breakdowns,
Startup Insights. https://www.startus-insights.com/innovators-guide/

4.2 Systematic Nodes and Energy Mass Information Labor (EMIL) Flows

Summary: A comprehensive systematic flow chart has been provided to visualize all mankind activities on Mars. The systematic nodes start from NGSuM (Energy input) and Labor Force. The energy and labor input enable space and ground explore facilities (Satellite, Drones, Tumble Weed, RC Explorer) to collect data and send it to mineral extractors. The raw materials are then transmitted to Concrete Glass Metal Polymers (CGMP) factories, of which transforming them into items required by the down steams. The last phase of industrial manufacture are 3D robotic printers and AI smart factory. The printer and smart factory can produce almost everything for life support units, agricultural activities, and valuable goods for EMMA trade. **See Appendix 7.**

Part Two: Engineering
4.3 Transportation

The environmental conditions of Mars create major challenges for transportation and communication with factors such as global dust storm (sand will cover the road), drastic temperature change. These challenges negatively influence the development and progress of the colonization and rehabilitation of mass for human habitation. As a result, vehicle system must have basic essentials such as life support unit maintenance (oxygen and thermal), energy supply (endurance mileage less than 100 miles) (Baratta M., et al, 2019). Consequently, only a few transportation systems have potentials to develop on a global scale. The transportation methods that can be used include high-speed railway, ground hopper, aero plane, and plasma propulsion rocket.

4.3.1 High-Speed Railway

By the end of 149 Martian Year, total mileage of the railway system opened to traffic is 35,300 kilometers, with 396 railway stations that have been built facilitating the transport services. Thanks for the thin atmosphere and low air resistance, train speed is estimated to 600-1,000 km/h (Eric A.T., 2019). To prevent the accumulation of sand and dust on the train's path, viaduct railway is widely used for majority sections. As for life support unit, a centralized system applied on board has significant superiority over single units. It is stable, reliable under the extreme conditions. Also, it ensures the maximum efficiency of the trains by using systematic design and approaches, thus greatly reduces the energy consumption.

4.3.2 Ground Hopper

The advantage of ground hopper is that Martians will have the ability to traverse more aggressive terrains but also maintain wider mobility. Therefore, ground hoppers are usually the most convenient vehicles when it comes to coverage of short distances (Geoffrey A.L., et al., 2012). Powered by a small nuclear reactor, the vehicle is a robotic exploration craft that used the power of the reactor to compress CO_2 from Martian atmosphere. When liquid CO_2 collecting is ready, a Nuclear Thermal Reactor will superheat the gas and turn it into rocket exhaust (Robert Z., et al., 2012). In 149 Martian Year, Ground Hopper Vehicles are the primary commuting tools for residents living in Blue Origin Pearl City.

4.3.3 Aero plane
Aero planes are an effective mode of travel to space since they glide like an airplane on the edge of Martian atmosphere and maneuver like space craft when in outer space. They are mainly used in vehicle travels between ground city and Phobos Space Elevator near surface terminal (60 km above the ground) (Koji F., et al., 2012). It significantly reduces the cost of Mars-Earth commuting in both cargo and passenger transport, since it replaces rocket launching on ground and save fuel consumption by 90%. Additionally, aero plane can also be used as vehicle travels among ground cities, as alternative for high-speed train. Such a long distant travel route is New Dubai (Dao Vallis in Hellas Planitia) to New Televiv (northwest of Olympus Montes).

4.3.4 Plasma Propulsion Rocket
This is one of the leading transportation modes when it comes to space travel. Due to the continuous propulsion from plasma, it significantly reduces the time travel from Earth to the red planet (Levchenko L., et al., 2018). During the Hohmann transfer period, the time spend on travel has been reduced to only 42 days. The chemicals that are embedded in the rocket engines produce continuous thrusts that increase the speed at which the rocket moves. The rocket is known to move at a very high speed of approximately 160 km/s, which makes it efficient and costs affordable for space mission travels.

4.4 Ocean City on Mars (Blue Origin Pearl City)
Melting ice cap is critical for Mars terraforming. Orbital Reflectors Array (ORA)---as the optimal melting solution, provides continuous thermal input for ice melting (Wood M., et al., 2018). The melted ice, which leads to the formation of North Sea, provides an opportunity for Martians to develop sphere city underwater. Primarily, it consists of a sphere that has an approximate diameter of 500 meters, which floats in the deep sea as a space vehicle (SHIMIZU Co., 2017). The city is known to be more habitable and even safer than some ground cities as it is not affected by GDS and Marsquakes. It happens to be a very prosperous city with minimal temperature changes, high moisture content, lowest solar and cosmic radiation level, and higher concentration of oxygen than on the ground. Construction of Blue Origin Pearl City is a crucial step in making Mars habitable by speeding up the melting process while providing an ideal place for living.

4.4.1 Ocean Thermal Energy Conversion (OTEC)
Ocean thermal energy conversion utilizes the concept of difference in temperature from the sea surface and the deep sea. This provides energy self-sufficiency. By introducing heat exchange medium such as liquid ammonia, it generates electricity through a heat engine that recognizes the temperature of both the deep sea and the sea surface (Ristiyanto A., et al., 2020). This vertical thermal circulation further speeds up the melting process---as it pumps cold water up and drops surface hot water to bottom, hence facilitating terraforming process.

4.4.2 Deuterium Water Production

Deuterium water is a form of water that contains a large amount of hydrogen isotope, which believes to be an ideal source for nuclear fusion. Geologically data indicates that the content of Deuterium in H_2O on Mars is 5 time higher than which on Earth (Robert Z., et al., 2012). Deuterium water provides a clear and efficient energy source. The demand for energy sources on Earth is high considering the continued depletion of fossil fuel sources. Therefore, the abundance of deuterium water creates a lucrative industry, and its high demand would become primary economic sector of Blue Origin Pearl City.

4.5 Algae Revive Hellas: Hellas Planitia Global Dust Control

As discussed in chapter 2, Hellas Planitia is one of the primary sites that GDS originated from. To control the GDS, plantation on Hellas surface is critical. However, the low temperature and high ultraviolet radiation levels at the surface of Mars preclude the plant survival. Recent research discovered that silica-gels are known to have capability of radiation blockage, and a certain level of thermal conductivity that are effective in mimicking the green-house effect of atmosphere on Mars (Wordsworth R., et al., 2019). Specifically, a 2-3 cm-thick layer of silica aerogel is able to transmits sufficient visible light for photosynthesis, block hazardous ultraviolet radiation and raise temperatures underneath without the need for any internal heat source. It is believe that the combined approach of silica aerogel and algae is possible to make a Green Hellas Planitia with minimal subsequent intervention, hence the GDS would be gradually under controlled.

4.6 Space Infrastructure
4.6.1 Phobos Space Elevator

Phobos Space Elevator, as in its name indicates, significantly reduces the cost of Mars-Earth commuting, both for cargo and passenger transport. The delta Velocity at near surface terminal is 0.6 km/s. However, when reaching to Phobos, the number goes up to 2.15 km/s naturally. Anchored at Phobos Stanley crater, the total length of space elevator is 6,000 km. Made from carbon-nanotube and KEVLAR fiber, the elevator is strong and flexible enough to lift as many as 100 tons for a single ride (Perek L., 2010). Terminating at the upper edge (60 km above ground) of Mars not on the ground (Leonard M.W., 2003), the elevator is free from GDS and air turbulence, thus its stability is guaranteed. Aero craft would be launched from Martian surface to rendezvous with the moving elevator tip and their payloads detached and raised with solar powered loop elevators to Phobos. Phobos, after that, can be used as a source of raw materials for conducting of space-based activities. Another space elevator would be made to extend from Phobos into space towards the Earth/Moon system or instead of the asteroid belt. The tip going outwards will also be used in monitoring and catching arriving crafts.

4.6.2 Earth Gravity Station:

Earth gravity station locates at Lagrange point between Phobos and Mars. Connected by Phobos Space Elevator, EGS is 2.5 kilometres to the surface of Phobos thus serves as the gateway to Mars ground (Hitoshi I., et al., 2017). The station is always on rotation depicting the fact that there is provision for the stability of the elevator. Gravity is quite essential in this case, as human babies can only growth normally under Earth gravity. The constant rotation of the station provides some level of linearity between Mars, Earth, Phobos, and any other space body that is being studied.

4.6.3 Stanley Space Port

Stanley Space Port locates at Stanley crater on Phobos, which is used for cargo and passenger travel from Earth, Mars, Phobos, Moon, and within the asteroid belt. Due to the tidal locking effect, Stanley crater points toward Mars permanently. The orientation of the crater provides a natural shield from radiation, which is crucial for passengers and staff health in logistic center. However, consider its orbital phase change, radiation protections are not constant. The radiation shielding duration for each period is 4 hours 50 minutes, while the radiation exposure is average 3 hours 10 minutes. Similar to Earth ground airport in 2020 A.D., Stanley Space Port has central terminal and several docking systems, which can be accessed by the space vehicles. The spaceport also provides an area for cargo & fuel storage and spaceship maintenance factory. These resources and facilities minimize the costs incurred by space travelers traveling in space and bring significant convenience for industries and commercials.

4.6.4 Orbital Reflectors Array

The Orbital Reflectors Array is one of the fundamental space terraforming infrastructures. The proposed ORA would be designed to be used in a Highly Elliptical Polar Orbit (HEPO) (Rigel W., 2007). Analyses suggest that HEPO design is the most convenient solution over others. When comparing to 6,000 km proposed by previous solution, the average 100-300 km orbital altitude of ORA dramatically increases the energy flux reaching ground. Next, HEPO allows for easy and low fuel consumption adjustment to be made at the far end of its orbit, which would allow for flexibility in controlling the surface heating as well as allowing for environmental conditions adaptation. Last, HEPO design doesn't need big reflectors, the size of a deployed reflector is 150m in diameter, which is technically feasible based on currently state-of-art. In the case of Mars Union, calculations show that an array of 16,800 reflectors is capable to have reflected solar energy that constantly melting ice cap in a $15*15km^2$ block area with average energy inputs of $1000W/m^2$.

4.6.5 Lagrange Global Artificial Magnetic Shield (LGAMS)

LGAMS is the other fundamental space terraforming infrastructure. By placing a satellite producing powerful magnetic field at Mars Lagrange point 1, Mars will have an artificial magnetic shield. Mathematic shows that such a LGAMS around Mars only needs to have a strength of roughly 11% that of Earth, which will create magneto sheath long enough to extend half million kilometers beyond Mars (Brandon W., 2018). As for LGAMS itself, the total mass of the craft is about 317 tons. The primary parts include an 830-Megawatt nuclear fission reactor, four copper solenoids with each size of 3.5m in diameter and a mass of 57 tons, four cooling panel with square dimensions of about 9 meters per side, a 25m diameter sunshield to protect LGAMS from sun radiation. Attitude control thruster and electronic & computer housing are also integrated into LGAMS. Consequently, with nothing more than 300 tons of material and current state-of-art, Mars could once again boast a firm line of defense against solar winds.

Chapter 5 Religious, Laws, Government and NGOs

5.1 Myths and Pantheon system

Myths are not just the dusty old stories of the Greeks and Romans, they are the stories we tell ourselves to explain the world surrounding us (Pamela J.S., 2018). Myths also tend to explain the world we have shaped, influenced and created to our benefits. Myths are based on traditions, being of factual or fictional origin. The First Gods of Mars may ultimately be explained by the compression of every Mars Mission involving several people into just one individual.

According to Smith, it is important to carefully select people who will be the First Gods of Mars, perhaps a thousand years from now, and the big question is how this should be done. The problem of choosing the first inhabitants of Mars is that the world is faced with several challenges that are supported by belief systems. Pantheons tend to allocate deities in covering natural events like volcanoes, water, plague, wind, as well as psychological aspects like love, marriage, hate, death, childbirth and the afterlife, as shown in **Figure.7**.

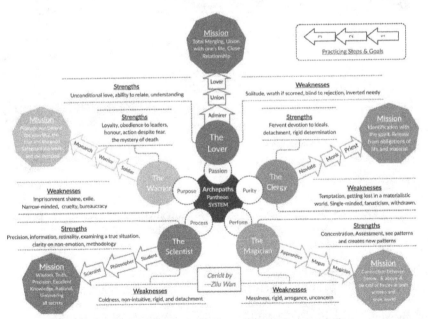

Figure.7 Five Arche paths and their characteristics
Scan QR code in the end for Full Resolution
Information Source: Pamela J.S., 2018, The first God of Mars. 21st Annual Mars Society Conference, University of South California, Los Angeles

5.2 The significance of Martian life and inspiration (Mind Flow Chart)

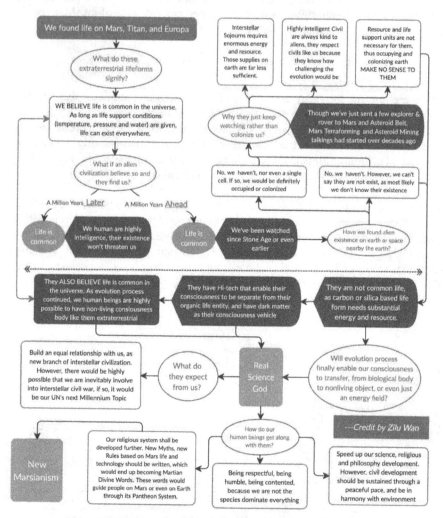

Figure.8 First Martian Life and its significance to Marsianism
Scan QR code in the end for Full Resolution

5.3 Mars Political Society

In **Appendix 8**, the political society begins with that prestige, outstanding Martians and immigrants from Earth. According to Mars Union Constitution and Election Law, these groups are 22 political parties that represent Mars Union citizens and constitute Mars Union Parliament. These political parties would enable the citizens to vote for the prime minister and the 201 seats of parliament. The electoral committee nominates Prime Minister, Chairman of Parliament, and Chief Justice.

Similar to countries on Earth, the candidates running for the Prime Minister's post by participating in elections whereby the winner will form the government, appoint Chief Justice and Chairman of Parliament. The citizens vote for their parliament members who will represent their will in the parliament by making laws. The laws created by the parliament ensure that the government practices are supervised, and rules established to maintain the societal stability and order.

Mars citizens, as well as enterprises and organizations, shall be entitled to a mandatory court ruling. The courts under the chief justice shall be funded exclusively by the government hence facilitating full and independent justice according to the Mars Union Laws. The prospective goals of the Mars Union, are listed on the two sides of the chart, which could be interpreted as **Rainbow Seven.** These prospective goals include the environment, life quality, social equality, society development, education and employment opportunities, safety, and wealth.

Chapter 6 Life on Mars

6.1 Livelihood and Social Welfare System
Thanks to the advancement of technology, the living standard of Mars Union is significantly higher than that of most others in EMMA space region, and is comparable to that of other highly developed countries on Earth. In 149 Martian Year, Mar Union Human Development Index is 0.991, indicating "very high" development. Mars Union also has a very high life expectancy at birth, with the residents' average age of 65 Martian -years old (See **Chapter 3, 3.2**).

Throughout human history, livelihoods have been secured through work. In 149 Martian Year, more than 21,000 jobs have been created in Mars Union. The average salary is 18,000 MRK monthly, however, the living cost (15,000 MRK) is significantly higher than that on Earth due to the inhabitable environment. The minimum monthly salary is about 6,500 MRK, which is stipulated by Ministry of LESS (Labor, Employment, and Social Services). Teachers and nurses should expect a median salary of 18,000-22,000 MRK, business managers, engineers, doctors and lawyers have some of the highest median salaries in Mars Union. As a doctor, 40,000-60,000 MRK shall be expected. As for engineers---in particular space engineers, salaries could be as high as 80,000 MRK or more.

Mars Union has Career Service Center (CSC) spread all over Mars. Operated by the Ministry of LESS, the joint of CSCs and Continuous Education School (CES) in universities provides career guidance, training opportunities and job information for every Mars Union citizens and legal immigrants from Earth. Through the training program, participants will have basic life skills on Mars. At the end of training program, there is a Vocational Qualification Examination (VQE) with 3 Martian Years period of validity, of which technology development results in such a short period. It is to be noted that VQE is not only for assessing the trainees' eligibility, but also for judging if the Universal Basic Income (UBI) shall be given. Mars Union won't keep lazy, such a policy promotes citizens' initiatives in learning and working will until their retirement.

Welfare in Mars Union refers to the series of social welfare schemes which are administered by Ministry of LESS, Ministry of Health, and EMMA Charity Federation (ECF). All citizens and Earth immigrants must pay insurance contributions in order to qualify for welfare. In 149 Martian Year, Mars Union public welfare expenditure amounts to 22 billion MRK (See **Chapter 3, 3.2**). The central budgets are focused on Old-age pensions, disability pensions, medical subsidiary, UBI and unemployment benefits (Doron A., et al., 1991).

ECF runs the old-age and disability pension system. Funds are paid to those who have paid insurance contributions for a minimum of 6 Martian Years. The basic old-age pension is 4,500 MRK monthly, which rises to 5,500 MRK after the pensioner has reached the age of 45 Martian-years. A seniority increment is added based on the number of years the pensioner has paid contributions, up to a maximum of 50% of the basic pension.

Individual officially recognized as disabled who do not work or homemakers whose ability to perform household tasks has dropped by at least 50% due to disability, are entitled to a disability allowance from ECF. The pension rate is determined by the level of disability, with the full pension being 6,000 MRK monthly.

Medical subsidiary is governed by Ministry of Health through EMMA Health Insurance Company. Similar to the previous two, a Mars Union health basket has been offered to insurance paid residents. It covers all costs of medical diagnosis and treatment in the areas of family medicine, hospitalization, surgery, vaccination etc. As for serious illness, patients must partially pay for these medications with copays: medications included in the basket are covered vary from 50%-90%

Unemployment benefits are paid to individuals who lost their jobs. These individuals shall first register with the CSC, and subsidiary can only be issued after CSC's assessment. Those who are involuntarily terminated or can provide evidence of having left their jobs for a justified reason receive subsidiary immediately, while those deemed to have left their jobs for no justified reason start receiving subsidiary only 90 sols after their registration. Same rule applies for those being offered suitable work or training program by CSC but refused to do so.

6.2 Entertainments, Sport

As for Entertainments, Games (VR/AR games), Bars, Films & Shows dominates Martian's daily life. Casino, theme parks, hot springs and indoor skiing are getaway choices for vacation. X-screams and EMMA studio are most prevalent theme parks in Mars Union, with annual visitor's receptions more than 2,000,000.

Sport Club: The aim for sport in a sense is not just for fun, especially for those people planning back to Earth. Fitness, Swimming are primary options for Martians. Other activities such as tennis, table tennis, golf, baseball, basketball, soccer,

American football---as long as teamwork is necessity---are popular among universities. The most influential sport event in Mars Union is ARES Cup Football Game, the registered member of athletes in 149 Martian Year is 75,000.

Space diving: A novel space sport for the earthlings. It typically requires special cabins, which must be covered with a layer of waterproof material. Under microgravity condition, the internal pressure of water ball is zero, and its structure is maintained by surface tension. Space diving is to shuttle back and forth between these water balls with different sizes, which provide participants an illusionary experience with various space effect.

Space mixed martial arts: Low gravity effect makes martial art action more challenging, thus new fighting techniques have been developed. Some of the participants become experts in this field, which is called Space Mixed Martial Arts (SMMA). SMMA attracts a large number of viewers and followers, some representative technique such as Chi-gong becomes prevalent among MU cities.

6.4 Holidays, Events, Exhibitions

Mars has a variety cultures, which means Martians could have lots of festivals. Martians holidays not only include earthlings such as Thanksgiving, Christmas, Spring Festival, Bon Festival, Valentine's Day, but also Martian unique festival such as balloon festival, graffiti festival, chef festival, etc.

The most influential Mathew Balloon Festival on Mars is held in the central city of Elysium Planitia --- Kunlun. It is set to be on the 2nd day of the summer solstice month, which is the first Monday of that month. On the day of festival, the city is full of balloon decorations like a fairy tale world. The festival is originated by an Engineer named Mathew. Early in the development period of Mars, he died unexpectedly of a scientific mission in Ma'adim Valles. To commemorate him, his friends in Kunlun city sent balloons to Ma'adim Valles every year thereafter. As growing number of balloons were received by Martian engineer in the Valles, they began to hang balloons in front of their bases. The commemorative activity finally moved tens of thousands of Martians, until Kunlun government voted to make such activity as an official balloon festival.

Fully commercialized Events such as Mars Olympic, Ares Bowl, fashion show, food show, robotics show, auto show and luxury good auction are not held every year. However, these events usually attract tourists and visitors more than any other holidays, and thus naturally become an indispensable part of Martian life.

6.5 Art, Music, Luxury Goods

Art: In the age of interplanetary era, new painting schools have emerged rapidly. Space nihilism, space realism, and other space-related painting schools have become increasingly popular. Unlike space paintings in industrial and postindustrial age, painters today are more focused on expressing the spirit of space. They often take a spaceship to experience the taste of space, and then show the inspiration from their experience in the work. Space realism painting style, among the others, has always been the most popular painting since it truly depicts the scenes of space and other celestial bodies. However, in the early days of immigration to Mars, space nihilism painting reached the peak. This group uses various painting techniques to express the confusion of human in the space. Most of paintings have a very strong philosophical meaning---that space is just a symbol of nothingness, and human being are only a minor component of this symbol almost invisible.

Music: Punk music and death metal account for majorities of Mars-themed music. Mars music takes the development of metal music as a starting point, and many of them involve nothingness in space, violent death, exploding spaceships, etc. However, Mars punk gradually gains the upper hand in pop music. Similar to the punk culture of the twentieth century, Mars punk is also new and distinguished from tradition. However, Mars punk music did not last very long, after thousand town age, the Mars music entered into blooming state.

Hit You Hard is a well-known Mars metal band. Although the name does not look like a death metal band, their music is indeed close enough to death metal. However, the band's lead singer was jailed for murder (throwing the waitress into the airlock, and blowing into the space), and thus ending the band's leading position in Mars music world.

Down the Spaceship is noteworthy because this is a space punk bank in a spaceship maintenance shop. All the members are spacecraft maintenance personnel, and their songs often use a lot of professional vocabulary for spacecraft maintenance, which is quite unique among others.

Luxury goods: It is to be noted that the most luxury goods on Mars and its nearby space are Crystal Heart launched by Heaven Garden Inc. In this masterpiece, the plant nutrient solution is suspended in a microgravity environment, and the white gypsophila centered around the nutrients grows outwards, and finally forms a three-dimensional heart shape. Heaven Garden Inc. seals the fixed Crystal Heart in glass and ships it back to Earth for sale. This space made masterpiece has been really loved by Earth consumers. However, due to its low productivity and high shipping cost, even a small piece of work is expensive (1,000 MRK). Since then, Heaven Garden has naturally become a holy thing in women's mind, and thus a nightmare in men's heart.

6.6 Gastronomy Culture
Most visitors to Mars Union agree that the food is always among the most memorable aspects of their trip. Martian cuisine is as diverse as the Mars Union citizens with recipes from North America, Europe, Eastern Asia, Middle East, and Latin America highlighted by local ingredients. The last decade has been a time of explosive growth in the Mars Union catering industry with must-go restaurants popping up in nearly every center city. There is no better way to sample all the culinary delights of Mars Union than to head to one of many food festivals held

yearly-round.

Kunlun City Food Festival: You could easily spend your entire trip visiting amazing restaurants in Kunlun alone, but you might make the most of your trip by joining 50,000 others at the annual Taste of Kunlun City Food Festival in the spring equinox. Professional exhibitors include some of the most famous chefs in Mars Union, offering their delicacies at a bargaining price.

Chef's Restaurant Week: Similar to Restaurant Week on Earth, Mars Union restaurants have been offering customers delicious two-course meals for reduced prices on a seasonal basis. If you missed the Kunlun City Food Festival, this would be a great opportunity to try some of Mars Union best restaurants at affordable prices.

6.7 Epilogue
In the end, there is a 30-sol trip brochure of Mars Union (See **QR table** in the end). Trip starts from Stanley Space Port, with the blessing of newborn Martian baby, and ends up with funeral event in Bezosphere. Visitors will have a chance to experience early settlers' work and life. Trip also provides lots of opportunities for visitors to learn science, industry, fashion, art and music, and sport activities during the stay. Being as one day volunteer offers precious experience for travelers, since it tells how challenging the terraforming process would be, and how Marsianism approaches death.

We choose to go to Mars!

Acknowledgments

Special Thanks
Longfei Chu (CEO of Link Space, Spacecraft Engineer in China Academy of Launch Vehicle Technology), for his help of data processing of Mars Climate Database using Python.

Cover Art and Drawings
"A Mars global 30-sol Trip in 149 Martian Year"
Original Sketches by Zilu Wan.
Drawing by King Animation Inc. Hongjiang Li, Yan Li

*Maps and Flowcharts are ESSENCE of the report !

URL links:
Reference: https://h5.9fm.cn/hnk7n6?url=M666G_kSh3ysG

Vocabulary & Acronyms: https://h5.9fm.cn/hnk7n6?url=M666G_7b7phLG

Figure 4a &4b: https://h5.9fm.cn/hnk7n6?url=M666G_oU8ayJ

Table.3: https://h5.9fm.cn/hnk7n6?url=M666G_Evf5du

Figure.6: https://h5.9fm.cn/hnk7n6?url=M666G_jOn1Pf

Figure.7 &8: https://h5.9fm.cn/hnk7n6?url=M666G_kyZPnZG

Appendix 1-4: https://h5.9fm.cn/hnk7n6?url=M666G_6e3gDY

Appendix 5: https://h5.9fm.cn/hnk7n6?url=M666G_2udru5

Appendix 6: https://h5.9fm.cn/hnk7n6?url=M666G_1y5V3d

Appendix 7: https://h5.9fm.cn/hnk7n6?url=M666G_f9xCX2G

Appendix 8: https://h5.9fm.cn/hnk7n6?url=M666G_ONsSuvG

Mars Global Trip: https://h5.9fm.cn/hnk7n6?url=M666G_Ted9qv

QRs (references, figures, charts, and Mars trip)

References	Vocabulary & Acronyms	Figure.4a & 4b Settlement Analysis
Table.3 External Trade	Figure.6 Industry 4.0	Figure.7 & 8
Appendix 1-4	Appendix 5, The Atlas of Mars City Layout	Appendix 6, The Mars Union Economy Chart
Appendix 7, EMIL Flow Chart	Appendix 8, The Political Society of Mars Union	A Mars global 30-sol Trip in 149 Martian Year

2: AURORA: A MARTIAN CITY DESIGN

Chris Thomas and Will Rosling
Southhampton, UK
hereloss@googlemail.com and wrosling@aol.com

INTRODUCTION

For decades, humans have fantasised about life on Mars. The red planet is arguably the most iconic and the most well-known planet (apart from Earth) in our solar system. It is a cultural symbol referenced in many works of media and entertainment in science fiction: films, literature, TV, music, artwork, video games amongst others. The most resonant theme of the planet is the speculation of life, and indeed through years of research this speculation has increasingly become more substantial. Humans have been sending probes and exploratory vehicles to Mars to learn more about the planet, and although it would seem we are unlikely to find any advanced alien civilisation there, there is plenty of evidence to suggest life may have existed on Mars in the past and possibly may even still remain on a microscopic scale. However, in more recent times a new question has arrived on the stage--will *humans* be able to live on Mars?

As the prospect of facing significant environmental challenges on Earth in the near future becomes increasingly apparent, this perhaps lends more and more sense to the idea of turning the science fiction of extra-terrestrial colonisation into reality. There is currently an aim to send the first humans to Mars within the next couple of decades, and beyond this there is also the proposition of setting up a permanent human colony on the red planet. However, there are obviously many environmental challenges that Mars poses which would need to be overcome for such a feat to be accomplished. The first obstacle will simply be being able to provide sufficient supplies for a journey which would be expected to take no less than three months; the willing astronauts will require enough water, food, air and protection from radiation for the duration of the transit--and that is before they have even landed on Mars—and, Mars, unlike Earth, does not have an effective magnetic field to shield away the harmful cosmic radiation abundant in outer space, nor does it have a breathable atmosphere. The entire planet is a barren desert, yet if such a colony was to be established, it would need to be almost completely self-sustaining due to the tremendous costs of transporting mass off the surface of Earth and landing it on Mars. Everyday necessities such as food, water, air, energy and building materials amongst other things will need to be sourced from Mars itself. This is before the financial obstacle has even been mentioned; in order to afford for such a settlement to be built, the overall cost would be at least $100 billion, based on the principle of around $140,000 per ton of mass to be transported for the initial settlement - the cost would lower over time [1].

This report illustrates how such a colony could be established in the future. The aim was to realistically design a city on Mars supporting a population of one million people--a figure comparable to Cologne, Germany. It encompasses a wide range of considerations to be taken and proposes solutions to the multitude of obstacles in the way of building the city. Choosing a suitable location is imperative to give the first settlers as good an advantage as possible in terms of accessibility to resources. The infrastructure and layout of the city will need to be planned in such a way to mitigate against the environmental hazards imposed by Mars but must also ensure sustainability and be cost-effective. Additionally, it was considered important that the city had high aesthetic value. Like any other human settlement on Earth, there will need to be a government in place to support the population and this presents the opportunity to explore new political ideas since there is no current jurisdiction on Mars. Life in the city will need to be enjoyable to attract immigrants from Earth and to keep residents entertained. Finally, we also consider the possibilities even further beyond into the future; how would the city develop and grow even after it reaches the target population of one million people?

There are a wide variety of visions of what a future on Mars would look like and be like. This is our example--a fantasy of the first ever human city on Mars, which we have named *Aurora*.

LOCATION

Mars Background

Mars is the fourth planet out from the sun in the solar system, and on average is the second closest planet to Earth (the nearest is Venus). The distance between Earth and Mars varies from as close as around 34 million miles to as far as around 250 million miles, depending on their position as they orbit around the sun. By timing a flight to Mars such as to minimise the distance required to travel, the journey time would take approximately 3 months. It is roughly half the diameter of Earth and has around a third of the gravity, as well as having a very thin atmosphere at only 1% of Earth's atmospheric pressure. Without an atmosphere to stabilise surface temperatures, it can be as hot as 35 degrees Celsius at midday at the equator but can also be as cold as -63 degrees Celsius at the poles--of all the other planets in the solar system, Mars has the closest temperature range to Earth. A day on Mars is almost the same length as on Earth (about 25 hours), although a year on Mars is approximately equal to 2 years on Earth. Similarly, to Earth, Mars rotates on a tilted axle meaning it effectively has "seasons", although naturally these have markedly different characteristics compared to Earth due to the lack of oceanic and atmospheric dynamics at play. It also has two moons, Phobos and Deimos, which are thought to be asteroids captured by Mars' gravity from the asteroid belt which divides the solar system into the smaller terrestrial worlds of the inner solar system and the magnificent gas giants of the outer solar system. Mars' surface is rich in iron oxide which gives it a reddish-brown appearance, leading to its colloquial nickname "the red planet".

Choosing the location on Mars for the city is an important decision, as it will need to give the colonists the best accessibility to resources whilst mitigating against environmental hazards as best as possible. It is important also to mention, however, that Mars itself as a whole poses significant environmental hazards that need to be taken into consideration regardless of specific location. As discussed previously, Mars is not protected against solar radiation and Galactic Cosmic Rays (GCRs), nor does it possess breathable air. A large proportion of the city will be built underground such that the topsoil acts as a barrier against the radiation, and an artificial atmosphere will need to be created. The soil is malnutritious for plant growth and contains perchlorates which are toxic to humans--therefore the soil will need to be modified in order to be optimal for growing food, and only certain foods will be able to be grown initially. Despite being the next most habitable planet to Earth in the solar system, Mars still experiences temperatures well below freezing so sufficient heat energy generation and insulation will be essential. The reduced solar flux hinders solar power harnessing, and the thin atmosphere makes wind power virtually negligible--therefore the most optimistic renewable energy source is most likely nuclear. Fitness will also be vital for humans living on Mars as the lower gravity will mean less stress on the body and increased likelihood of muscle atrophy if not adequately maintained. Furthermore, living on Mars will be testing psychologically due to being so distant from Earth--and especially for the first settlers, there will only be a very small population. The candidates who would be eventually selected to set up the colony on Mars would have to pass mental as well as physical assessments.

Specific Location

It has been suggested that the ideal location for a city on Mars would have to be one that has already been extensively explored by the Martian rovers, thereby providing much greater depth of data already available. However, other important considerations needed to be taken into account which will be covered in greater detail. Some of the leading candidates for the location of *Aurora* included:

- Phlegra Montes: a mountain range in the northern lowlands.
- Tharsis: a volcanic plateau to the east of Olympus Mons, the largest volcano in the solar system
- Amazonis Planitia: a plain to the west of Olympus Mons
- Valles Marineris: a megascale canyon forging some 4,000 km across the equator of Mars
- The poles, which are both capped by extensive ice fields

However, the chosen candidate was the Korolev Crater, situated in the northern plain at 73 degrees latitude and 165 degrees east longitude. Due to Mars' axial tilt, which is almost the same as Earth, the Korolev crater (which is named after the Soviet scientist Sergei Korolev) is situated in the region of Mars where the midnight sun is visible on the summer solstice--effectively Mars' equivalent of the Arctic

Circle. This means the area experiences long daylight hours in the summer but very short daylight hours in the winter. This is disadvantageous compared to other locations, but its assets should not be overlooked too quickly.

The crater is filled with an ice field, kept naturally stable by the crater walls. This ice field provides an obvious water source that the city could use; the crater contains approximately 2,200 cubic kilometres of ice, which if 100% recycled every day would be able to sustain a population of around 5.5 billion people. By situating the city due south of the crater, there is flat land which is easier to build on and is also free from the dust pockets scattered around Mars where vehicles and buildings would sink through [2]. It could also act as a shelter against dust-storms from some wind directions; despite the atmosphere on Mars being very thin making wind strength very weak compared to Earth, they are able to lift up vast quantities of dust which soar upwards. Dust storms have also been found to occur more often in the southern hemisphere compared to the northern [3], but nonetheless will be one of Mars' most significant environmental hazards.

The crater itself would have been formed by a meteorite impact and subsequently is likely to have a higher variety of minerals and ores in the bedrock [4] compared to flatter areas, which will be useful for construction as well as potentially for economics. Additionally, the crater would have a high aesthetic value which would help draw tourism. Cooling from the ice makes the air directly above the crater cooler and denser, which could be beneficial for generating an artificial breathable atmosphere for the city.

Being at a higher latitude compared to the other locations presents some other advantages and disadvantages. On one hand, being at a higher latitude means the gravity is very slightly stronger and so acclimatisation to microgravity would be less extreme, and there is also slightly more atmospheric protection against radiation. On the downside, the slower planetary rotational velocity at higher latitudes would make it more fuel-expensive to take off from Mars for return missions to Earth. Additionally, there is a lower solar flux at higher latitudes meaning temperatures will generally be colder and solar energy production will be more limited; however, since the primary energy source will be nuclear this should not be a significant issue. By using strong insulating materials for habitat construction, which will be discussed in the engineering section, heat loss will be minimal regardless such that the colder outside temperatures will not be overly consequential. Another disadvantage of reduced sunlight is it can have a negative effect on mood since it is associated with deficiencies in serotonin production, a hormone which increases mental well-being. Deprivation of sunlight can be addressed using phototherapeutic light screens inside the habitats, which use polychromatic polarised light to mimic sunlight, and possibly also pills containing vitamin D and drugs which help trigger production of serotonin.

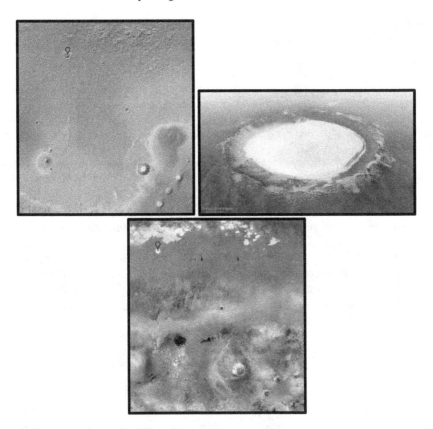

The above pictures [5] show an elevation (left) and visible (bottom middle) image of the Korolev crater (indicated by the red marker). For reference, the edge of the polar ice field can be seen at the top of the visible image and the volcanic region of Tharsis, as well as Olympus Mons, can be seen in the bottom right corner. In the top right is an aerial view of the Korolev crater taken from the Mars Express spacecraft.

FOUNDING OF THE CITY

The founding of the city will happen in a few distinct phases, and we will go into detail for each one of them in the following sections. The city will grow from an outpost, to town, to a large city, and beyond! In each section we will state exactly what is being set up, and any special methods and ideas which are required for this.

Pre-Arrival

Even before the first settlers arrive, we will wish to set up supplies and some basic utilities for them so that life is easier upon arrival. This work will be done by robots and drones and will likely take a good few launches (and so probably about 4 years due to the window on the closest approach of Mars) to set it all up.

Firstly, we will want to deploy a robotic taskforce to begin to set up the water supply and prepare the ground for the settlers. To do this, the robots will need to set up a system which will harness the frost from the ground and melt it to form drinkable water. All of the filtration machinery for this will be brought from Earth with the first robots. They will also have to bring a Nuclear Power supply, as to keep them powered up and ensure we have power to both melt the ice when required, and for the settlers when they arrive. These robots should also begin to set up some solar panels during this phase.

We will also send another ship with some basic mining and refining machinery in it, so we can begin to get our raw materials to build the base. This ship could also contain a small piece of machinery to create Martian bricks - Martian rocks are in fact very easy to turn into bricks simply using a hammer [6] - or stronger bricks can be created using high-pressure machines, which can also be imported with the first colonists [7]. We will also send with this a filtration machine, which can filter Martian soil so that it is not toxic to the touch for humans.

Finally, we will send a ship with some non-perishable food supplies (E.g. Canned foods, Nuts, Dried food etc.) and some air and oxygen tanks. These will need to be enough to last a year, to hopefully prevent any major incidents from occurring.

First Arrivals

The first settlers to arrive will be living in the housing areas of the spaceships in which they arrived, so we don't need to build any new houses for them. They will already have a robotic task force, along with water, energy, and some raw materials to help get the base up and running. The pods they will arrive in should be well insulating, radiation-proof and carry their own energy source. They will also carry a supply of non-perishable food, and a water recycling system - and a small garden to grow some food would be very beneficial. These circular pods will also be equipped with "landing legs" - which will be able to drill into Mars to provide a foundation. We will describe the engineering behind these later.

To decide on the first settlers, we will need to pick from current astronauts - they will need to have a full physical and psychological screening before we will let them go, however. For future arrivals, we will also need to check for any illnesses, along with making them take a test to ensure they have the right attitudes and skills to thrive on Mars. We will go into further detail on this process later.

The first settlers, once they have landed, should begin by using the robots (which will be much easier to command once there is no delay due to being on the same planet) to cover their "houses" with Martian soil to ensure no radiation can get in, and then dig underground tunnels between their houses, as to limit exposure to Mars. The first settlers should be mainly scientists; however, they should have a medically trained professional among them, with supplies - in case of an emergency!

The robots will then start their two major tasks for this period - the water system, and a food dome.

The water system will mainly involve drilling into the side of the crater, and then putting together and installing the heating system, filtration systems, and delivery systems for the water. All parts from this will have to come from Earth (excluding the pipes, which can be made from the raw materials harvested from Mars) and should come with the settlers.

The food dome will involve using the raw materials, along with using the method to create Martian glass (see the Engineering Section) [8], to put up the first domes. Some of the parts from the rockets could be used to make this, and the initial design may have no windows at all - but simply use lights to simulate sunlight to make the plants grow, replacing the metal with windows as the glass is created, although some extra lighting will almost certainly be needed even after the glass is installed. The machinery to create the glass will need to be brought from Earth as well initially. We also know that the soil around the Korolev crater is surprisingly good for growing food - referred to as "Garden soil" like [9] - and so, once filtered to remove any toxins, adding some human waste to this will likely allow us to create Manure, which will be great for growing our first plants. Aquaponics (using Hydroponics) [10] can also be utilized to improve this once we have created the water system, and so we should import some fish eggs to allow us to set this up. This will improve the settlers' diets and make our job of setting up the food and water easier.

Throughout this period, extra solar panels will also be being created - however, we will get our main power from Nuclear generators, similar to the kind NASA are already using [11], for which we will go into more detail in the Engineering Section. Air will be recycled and filtered from the resources brought with the settlers and using the air we stored on Mars prior to arrival. Using the Sabatier method [12] and also slightly adjusting the concentration of air in the building and it's pressure (similar to how diver's tanks are set up), we can ensure a constant, sustainable and healthy air balance.

Finally, we can use this stage as a point to set up methods of more easily travelling to Mars - such as a Skyhook on Earth, and one circling Mars, making future imports of machinery and supplies easier. This should also reduce the cost. Space elevators will probably not be created on Earth by this point - however, if they are, these would also help us out. We do not want to import too much heavy machinery; however, some will have to be imported at the beginning.

Once this stage is complete, we will have access to water, food, shelter, and power, and can move onto the next stage.

Growing the Colony

Once the necessities are in place, we can start to expand the city. We can firstly use our materials and power to improve our mining and refining operations - at this point, a 3-D printer will be imported, along with some other industrial level machinery to allow us to produce the mining tools on Mars. This will give us more materials to work with.

Next, we will begin to build houses. We have decided on a Spider-Web like design of the city (of which we will go into more detail later), with small community centres for each sub-district. Underground tunnels should be dug and reinforced for each new building that is created, and a community centre should be designated. For the initial centre, one of the already existing houses will probably be the best bet, however, we will build underground cavern-like centres for each sub-district, with larger dome-like centres, similar to the food dome, for the district centres. The caverns should be spread with soil and have plants grown in them - lighting should also be set up in each one, as to make them a bit brighter and less gloomy. All the houses of a subsection will connect to the community centres.

Whilst building the houses, the robots should be connecting each one up to the waste-filtration system - as we do not want any of our resources to go to waste. Piping should also be set up for each house, and there should be a Nuclear Reactor per person, or solar panels, since as one reactor provides 10kW of energy, one per person will be sufficient [13]. The Nuclear Reactors will still have to be sourced from Earth at this point. Research into whether Geothermal could be an option is also a good idea at this time - however, from our research from Earth, it seems Mars may be too cold.

The walls of the houses will be explained later in our Engineering section, and once we have a few sub-districts set up, we can begin to create the City Centre. The districts will be connected using a tube system - and this will be a priority once we have begun work on the City Centre. At this point, we expect to be around 100-1000 citizens, and so we will build a hospital. As we build out and get more citizens, we can also invest in a University, sports facilities, and government buildings. Of course, we will build more food domes as we add citizens - ensuring we always have food for at least 1.1-1.2x as many people as we currently have living on Mars.

The colony will be very dependent on Earth at this point - any specialist equipment will have to be shipped in, however most building and robotics equipment should be fairly easy to import and once it is on Mars, won't need to be brought again. Food and water will be sustainable at this point, as will energy - however, some necessities, such as vitamins, may need to be imported. A set of satellites - orbiting Earth, Mars and the Sun, at points in between the 2 planets, should also be set up - if enough are set up in the middle, we should have almost no points where we lose connection with Earth - and satellites are not that expensive, and will be easy to keep in orbit around the Sun and Mars. These satellites will also need to know the locations of the two planets, as to make sure they bounce signals correctly.

At this point, we would expect the city to grow fairly organically - many people have in the past shown interest in living on Mars, and if we add incentives, we should be able to easily grow our population. Screens should be installed in the subterrain community centres (or windows installed at the top of the roofs) and the larger ones should have large, domed parks under them.

Founding of Services

Once the city is at a reasonable size, we will also have to create certain services on Mars.

The Government will be the first, and main, service to be created. At first, this should simply be one of the initial settlers, voted for by the rest, and who is more like a team leader than a politician. However, as time goes on, political parties will form, and so once the city hits 10000 people, a Martian constitution should be created. All arms of government should be created at this point and expanded and amplified as time goes on. We will go into further detail on the Government in a later section - however, at this point a building should be created specifically for this purpose.

Fire and police should also be set up. Since any fires that happen could cause an atmosphere leak and be dangerous, firefighters should be highly trained and prioritised. They will need land vehicles, and a connection to the tube network (which will be more like an underground rail network) to allow them to respond quickly. We should create a police force - although they will probably not be very necessary since we will be screening all applicants before sending them to Mars.

Waste Management should be dealt with by robots, although a human overseer will be required - details of the filtration system can be found in the engineering section. The spaceport will also have to be created, and this will have a robotic crew - however, this too should have a human to oversee this. Once the university is founded, teachers and professors should also be imported, as to ensure our population is well educated, as this will allow us to gain new insights and techniques to help make the city of Mars more comfortable.

CITY PLANNING AND INFRASTRUCTURE

We will go more in depth on how the city will be laid out and work and will expand upon the points we made in the previous section.

City layout

The City will be laid out in a spider-web shape, as seen below, splitting into three whenever the web strands get too far apart. This will allow us to expand the city easily, whilst creating a central hub for commerce and socialisation, and an easy transit network. The tube will run along the "webs" and will have stations at each intersection. At each of these intersections, we will have a community centre, as previously stated in the last section. Each of these community centres will be like a town square on Earth - with a park, some shops, a sports facility, and other amenities. Each one should ideally have slightly different amenities from the ones nearby it - so for example, one has a badminton court, and another a tennis court - but this is not vital.

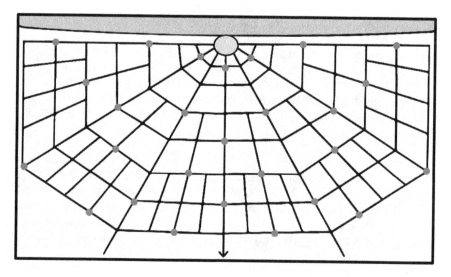

This diagram shows the spider web design of the city. The yellow circle indicates the centre, with the crater up top. Each red dot is a District centre. Every crossing of the lines will be a community centre, and the spaceport will be in the direction of the arrow at the bottom. Trains will run along these lines and will all go to the centre - merging into one line where the lines split.

These sections will also have the houses of the people who live in that community attached to it via underground tunnels - so we can foster a sense of community in these. The houses will be a mix of sizes and heights, as to make the city feel less monotonous, and people should also decorate the front doors or tunnel entrances. Each community centre will also have its own food domes connected to it, and these should be numerous enough to keep the population well fed. All electric and water supplies should come through the ground, so these won't be visible. Each of the community centres will be in a cavern - thanks to Mars' lower gravity, these won't need as much support as they would on Earth, but adding some beams to help hold up the ceiling should be done to be safe. A window could also be installed in the top, and screens around this can help simulate the feeling of being outside - without the harm it would bring.

For each of these sections, we will aim to let the people living there add their culture to it - we will get more into who we would like to have in our city in Demographics later on. Furthermore, in the centre of a group of community centres, in the place of the red dot on our diagram above, we will have a larger district centre. This should have more variety in shops, a school, a medical clinic, parks, and various other amenities. This will, instead of being underground, be in a bigger surface dome - and should therefore have larger and more aesthetically pleasing buildings nearby it. The dome will be made in the same design as the food domes - only larger. Each of these should also have a tower in the centre, as to allow for a viewing platform for the citizens of the district to see the city from on high.

Finally, the city centre will have all the really important buildings - the government offices, the university, the hospital, and some more touristy ones, such as the "Mars Palace", "Mars Tower" and the "Grand Library". The original landing site should also be preserved here - it would be a great tourist destination! In this section, we will have multiple smaller domes with connecting underground pathways, as to allow for easy walking between the centre. More shops, restaurants and the nightlife should also be in the city centre - and so should any major sport facilities. These will be in domes, or, if they do not require much space, be simply in buildings around the city centre. The city centre should be using the grandest architecture we can design (as will be discussed later). The spaceport will be directly linked to the city via the transport system, and will be south of the city, as to prevent noise pollution. We will now move on to this in the next section - transport.

Transport

The city will be connected mainly using a series of underground tunnels - as surface cars would be expensive and not very useful.

Firstly, each house will be connected to their community centre via a small underground path. This path should be at least 6 feet deep into the Martian soil and reinforced with iron and steel created from our Mining facilities and covered in a protective layer from the toxic soil. We can make Martian bricks from the soil (simply by crushing it using a hammer, once it is filtered to remove toxicity) to create the base of these tunnels, and lights should be placed in the ceiling. Since these are underground, we cannot add any windows - however, if we have money spare, screens can be put down to simulate the outside.

Next, each of the community centres will be connected together by large tunnels. These should have steel support beams and tracks, to hold onto the trains and keep them stable, and should be sealed off using airtight doors. For future far off new settlements on Mars, they may even be able to go via the surface. We should use electric, but otherwise fairly normal, subway trains. However, these should be airtight - we don't want to lose any of our precious air! - with docking stations in each community centre. The bigger district and city centres can have larger stations - as multiple train lines will run through them. Cargo train lines will also be placed down - sometimes using the passenger lines, sometimes not, like an overground rail system is on Earth. The fire, ambulance and police services will also have a link to this - as well as surface cars in case they require to complete the job on the surface.

We will also connect the community centres together by smaller tunnels made for walking. These should be deeper - the community centres are in caverns, and so we can place these at the bottom of those caverns, maybe about 10 feet down, and then reinforced and lightly covered in a protective layer. The floor should also be bricks, and have some lights along it - however, not as heavily done as for the tunnels from houses to the centres. No screens are required here, however some could be placed - like advertising boards today. We also have the raw materials on Mars to provide bicycles - which are fantastic for keeping people healthy, lowering the load on our transit system, and keeping the city connected.

Finally, we will build our spaceport south of the city. We also plan to build a skyhook, using Phobos as the point to attach this too - as this will make landing and take-off cheaper and easier. This will be resource intensive however, and so we should set this up when the city is at a fairly large size. The spaceport will be connected to the rail system for both freight and passengers, and landing ships should be able to easily dock to the spaceport. It will have airtight seals for these arriving ships - in a similar fashion to today's airports - and a team of robotic workers will be on hand to move and sort the ships. We will not have many amenities here, since it is easy to get into the city, however a couple would be useful, such as a guide booth and a medical facility.

We will build all of these early on in the colonies founding - as stated earlier - and use robotic drills and workers to create them. New districts will only be opened once all the centres are built, and each tunnel between them will have blast doors - as to allow us to shut off any section where air is escaping or we have a problem. Each will be built in a similar style, and so will expand naturally as the city does.

Demographics

We hope to encourage a sense of closeness and community in our city, and since all the colonists will need a variety of skills to help them achieve this and will naturally be brought close together by the experience, we would like our city to be as diverse as possible. This will allow us to include many different cultures and bring fantastic new ideas to the city, as well as making all of the Earth governments collectively take an interest in our city.

We had a few initial ideas on the structure and percentages we would like in the city. Firstly, we investigated using an existing city and copying this - however, since we will face many different challenges, and do not want to bring any problems from Earth cities to this, decided this was unviable. We could also have brought people from just one nation - this would mean they could all communicate, however we would lose out on the ideas from other nations - and many Earth nations would not have any strong economic interest or diplomatic interest in Aurora.

Therefore, we decided on the following method to decide who should be in our city: Each Government who wishes to take part will be allocated a set number of places - evenly split between the number of nations that take part. The initial settlers will have to be put forward by these nations and agreed upon by all of them as being the best for the initial mission - the main focus here will be scientific knowledge and engineering and mechanical knowledge, as well as psychological strength. The next wave of settlers will have to pass extensive mental, physical, and psychological tests, proving they will be excellent on Mars before they will be allowed to go. The waves after that will have to simply pass an easier version of all three tests, before they can be given a date to leave. As the city grows, we will invite more of the colonists to come along, bringing over most of a nation's people at once if possible, until the city is full.

We will then let nations decide which place they would like their citizens to live. We think it is likely they would choose a single community centre to place all of their people - which would create cultural hotspots for each nation, and so they could show off and share their culture easily in these centres. Furthermore, since we have an even mix, we can create a diverse government and bureaucracy, as well as a diverse set of teachers and professors. The community centres are also so close, and the city compact enough that the cultures and peoples will mix easily and make friends and new traditions among them. This will allow for knowledge to flow between nations easily and will help us make people feel not like the nation they're from, but as a different nation: Martian. On this identity, we move to the next section: Architecture.

Architecture

To create our architecture, we had to consider the materials available to us on Mars, and any problems different styles of Architecture would face.

Firstly, Mars has a much lower pressure than Earth. This means rounded shapes are preferable – even though we will have a lower pressure atmosphere than Earth inside our city (see Engineering for details on this). Since Mars has less gravity, building higher is also less of an issue – but since it has less soil tension, we should also build bigger bases and foundations to prevent the city from sinking. Mars does have a layer of fine dust in places – but our city is well away from these areas. We also have less materials than we have on Earth – we can create Metal, Glass, Plastic, Martian Brick, and basic concrete from these, along with most colour of paint – however, since the dust storms will degrade these, we should avoid outside paint if possible.

Hence, we will have a couple of different styles of building. We have already discussed the tunnels (Martian brick floor, steel supports, light metal coating with lighting and screens) and the Caverns (Steel supports, but sparingly used - windows and screens at the top, parks and shops on the ground, with a covering on the side to prevent toxic Martian soil from getting in), and so will go onto the housing and more prominent buildings.

Our taller buildings will have a smooth, shiny steel outside - this will prevent heat loss and will make the city look futuristic. Large, white struts will be placed along these too, in a similar fashion to our dome, and will make them look less plain. All tall buildings will have a larger underground base - this should be built in the same fashion as the tunnels, with a brick floor and steel reinforcement, with a light metal coating. The domes for both food and the centres will be large, with triangular, thick glass windows and white struts to hold them together. The more prominent buildings should have a Martian brick base, as this will give them an older feel, with some domes in gardens or outdoor sections. White beams should also be used strategically in these, and windows should be common. The windows and walls will be thick and done in the Martian way (see engineering). Houses should be a mix of these styles - with a shiny metal roof, brick walls on the outside with white metal struts keeping them together and adding a more futuristic feel. Small patches of shiny metal on this could also add some more character. We have put some concept drawings (see the references for the original picture credit), of a shiny steel tower, dome of similar style to how the domes of the city will look, and how a pure brick building would look on Mars - although all of our buildings made of brick will have white metal beams in a similar fashion to the domes, and shiny steel accents on them [14].

Parks and Leisure

The last thing to consider for the city plan and layout is the parks and leisure facilities. These will be either indoors - and so follow the architecture stated in the previous section, but with a few amendments depending on the sport or activity - or will be in a dome. If they are in a dome, they should have a soil (with grass) floor, with trees, bushes, and other small plants around - this can help make our city feel more pleasant. They can have white strut beams to keep a roof in place, however they do not need a thick roof since the dome is protecting them. Parks, outdoor sports fields, and even some other leisure, such as music concert venues and cinemas can all be built in this way. These will be in the district and city centres - and some can be built as offshoot projects from community centres, however these should be in every other (or possibly every third) centre. Small versions of these should also be in every community centre - however, these will mainly be parks and small sports fields as the community centres will not be that large. Statues can also be placed in the centres and will help us gain a bit of culture.

The indoor parks and sports facilities will instead have to always be in offshoots from city centres, as we need the space in the centre for shops and restaurants. These can be built in the normal style of the city - most will probably be built in the tower style due to their size. The shops and restaurants should be mainly placed in the community, district and city centres rather than as offshoots - since this can create a high street and a great sense of community spirit - and encourage people to spend and join clubs and meet with people. VR centres can also be set up - in either location - which will allow people to feel as if they are traveling, and so make the city seem much larger than it is. We can also create low-g sports, using Mars's gravity to our advantage, and encourage sports leagues and competitions - and once the city is large enough, we can even invite Earth athletes to this! We can also have "Surface explorer" stations - places where you can rent a car and drive across set routes on the Martian surface.

All of this city plan should help us create a spacious, pleasant, safe, and well-designed city, with a unique character - however, some difficult concepts had to be thought up to make this work. We will go through all our more complex designs now in our next main section: Engineering.

ENGINEERING

Food and Water

The ice field contained within the Korolev crater will be the primary water source for the city. To save heat energy, the ice won't be heated and liquified straight away as it doesn't need to be until it enters the main water pipeline network system. Cryobots will drill and extract ice from the crater [15] and transport it to a heating chamber, where the ice is melted before flowing into pipelines taking it to a set of complex filtration units [16] which each contain different phases. The first phase is to effectively "sieve" the water to remove any solid objects suspended in the flow. The second phase is to allow the water to settle in a large tank to let any remaining contaminants denser than the water sink to the base of the tank, forming "sludge". These are then collected and removed by scrapers which slowly move around the tank. In the third phase, the air is aerated to remove any bugs in the water by enhancing bacterial growth (this will become more important for recycling water). In the fourth treatment phase, there is a second settlement tank to allow bacteria to sink to the base and this can then be removed as well. The final filtration step [17] is a combination of carbon filtering (for which carbon can be extracted from the air and human respiration) which is effective at removing harmful contaminants such as chlorine, mercury and lead; and then finally the water is distilled to remove microorganisms and other contaminants with a high boiling point. The majority of used water can be recycled using the same filtration process, and the byproduct "sludge" removed during the cleansing phases can be used as fertiliser for soil or also as an energy source by heating it. The water should be looped through multiple filtration systems to mitigate against the eventuality of one of the filtration units malfunctioning, which would compromise the safety of the water if not accounted for. Water pipelines will have to be thermally insulated very well to prevent the water from refreezing, so they should be stored mostly underground and also use non-corrosive, strongly insulating material. Earth water pipes are often made from copper as they are corrosion-resistant; these could be incorporated with the addition of a metallic foil covering for extra insulation.

Crops will be grown in large "trays" to prevent water percolating through the soil and being lost. These trays will be stored in underground tunnels to protect against radiation but will use multi-wavelength LEDs to provide a light source for the plants to grow. Martian air already naturally contains carbon dioxide, but the atmospheric pressure can be increased using air pumps through check valves (see Atmosphere). The byproduct oxygen produced by plant respiration can be extracted for use in the air generator system. Martian soil contains toxic perchlorates, which can be removed by using perchlorate-eating bacteria which also produce oxygen as a by-product. Halophilic bacteria found in hypersaline soils have been shown to be able to achieve this and already exist on Earth [18] and would not cost much to be transported. It is estimated that when the population of the city reaches one million people, the amount of space required to grow enough food to support the entire population would be around 7 square kilometres [19]; however, these will be divided up around the city rather than having one singular agricultural centre, if only to mitigate against the risk of the environmental control systems to fail in one of the tunnels which would potentially lead to a large loss of crops. Over time, bio-waste products removed from the water filtration process as well as organic detritus material from the crops themselves would help to enrich the soil, making it more nutritious for plant growth and increasing capacity for new foods to be grown. Insects will also help by pollinating the plants as they do in Earth's ecosystems, and they will also provide a source of protein for the Martian diet. Because the food will be grown underground and in controlled environments, seasonality would not be an issue so production of different foods would not be restricted to certain times of the year.

Atmosphere

The two key issues of the atmosphere on Mars are that it is too thin to breathe, and it doesn't contain a large enough quantity of oxygen. To replicate air on Earth, the artificial atmosphere will require a pressure of at least 800 hPa (lower than this would risk symptoms associated with oxygen starvation, observed in altitude sickness on Earth) [20], and contain 21% oxygen whilst containing no more than around 1% carbon dioxide. Oxygen can be produced by passing Martian air heated up to 800 degrees Celsius through a solid oxide electrolysis unit, which has been developed as part of the NASA MOXIE project [21]. Additionally, oxygen can also be extracted from water present in the ice field, and from plant photosynthesis. Carbon dioxide will naturally be part of the atmosphere due to respiration. Nitrogen makes up 78% of Earth's atmosphere but is not essential. A combination of nitrogen and argon, which collectively make up around 4.5% of Mars' atmosphere, could be used to make up the remainder of the air. To increase the air pressure, air can be pumped into habitats through spring-loaded check valves. A multiple-valve system may need to be used if the springs cannot cope with the large pressure differential between Mars' atmosphere and the artificial city atmosphere--essentially the system would iteratively crank up the air pressure between valves, like a gearing system. The overall atmosphere system would consist of generators outside the city which extract Martian air and convert it into a breathable mix, and then the air would fill up air pumps situated next to valves on the exterior of habitats. The air pumps would then pump the breathable air into the habitat just like pumping up a football, and would be controlled by monitoring the atmospheric pressure inside the habitats which should regulate the air pressure to a minimum of around 800 hPa.

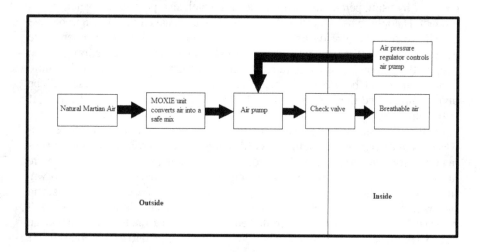

To save energy costs, all artificial air would be recycled as much as possible. The ISS is currently able to recycle around 40% of the oxygen which the astronauts breathe [22]. A Sabatier reactor would be used to break down the carbon dioxide from human respiration whilst an electrified membrane can split water molecules into hydrogen and oxygen. The oxygen can be used for breathing whilst the hydrogen would be used in the Sabatier reactor to be combined with the carbon dioxide. This reaction produces water and methane, the latter of which could be used as an extra fuel source. Additionally, plants used for food will produce oxygen as well so this would also be harnessed.

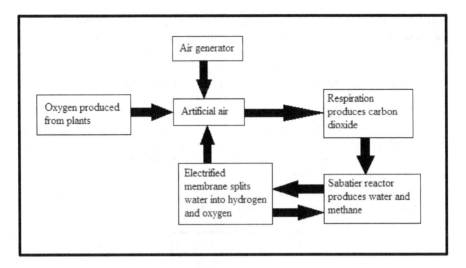

Energy

Nuclear energy would be the optimal energy source on Mars as it is efficient and a lot more reliable compared to the alternatives. Solar energy would be viable in summer with long daylight hours, but the solar flux would be lower compared to Earth and also being at a higher latitude. Contrastingly, in the winter the shorter daylight hours would make solar power almost useless unless solar panels were also built at lower latitudes and the energy was transported by pipelines to the city. The distance to such a latitude which would receive sunlight on a daily basis all year round would be around at least 1000 km. Nevertheless, it would not be harmful to make use of solar power when it can be used. Wind energy is not viable at all due to Martian air being so thin. Consequently, this probably leaves only nuclear as a feasible option for the primary energy source. Compact nuclear reactors have been in development already with future Mars missions in mind, as these will be fuel-cost-effective and power-efficient. The NASA Kilopower project [23] successfully demonstrated that a small nuclear reactor, capable of generating up to 10 kW continuously for 10 years, could be used for long duration visits to Mars. The system utilises a uranium core whilst sodium heat pipes transfer heat from the reactor to a high-efficiency engine which converts the heat energy into electricity. The presence of uranium on Mars is considered likely, and especially in places such as craters formed by impacts of meteorites. Energy consumption on Earth varies depending on the level of infrastructure; the country of Cyprus, whose population is a comparable 1.2 million, averaged around 445 MW in 2016, while most developed countries averaged in general between 500 and 1000 W per person with the whole world average being 309 W [24]. However, energy demand will likely be very different on Mars: large quantities of heat energy will be essential for melting ice from the crater, powering the MOXIE air generator unit, and keeping habitats sufficiently warm since temperatures on Mars are often extremely cold. To be as cost-effective as possible, building materials will need to be strongly insulative to conserve as much heat energy as possible. There are caveats with nuclear reactors, the most obvious being the risk of a devastating release of radiation which can contaminate the surrounding area if the reactor core overheats, the most famous example of which being the Chernobyl accident in 1986. Whilst more modern reactor designs have stronger safety measures in place and it has been suggested the majority of nuclear accidents are simply down to human error [25] a contingency plan should be put in place as a failsafe regardless. An evacuation procedure should the worst happen would make use of the underground shelters that would come with every building in the city. Keeping the reactors above the surface would allow the ground itself to be used as a shield against the radiation, protecting the city inhabitants. The reactors should also be encased in strong protective shielding, perhaps also made of Martian regolith, to contain any radioactive leaks and prevent them from contaminating the Martian environment.

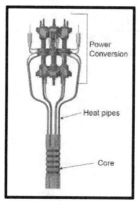

Diagram of a prototype Kilopower reactor based on Uranium core and sodium heat pipe transfer to the power converter.

A further consideration of the energy infrastructure will be to mitigate against induced fluxes caused by interactions with Coronal Mass Ejections (CMEs). Despite having a much weaker magnetic field compared to Earth, it is not completely lacking a magnetic field of sorts and there would still be potential for an incoming CME to generate an electromagnetic force if it were to hit Mars. Such an interaction would spark a sudden induced current in electric systems and risk overwhelming them. Such risks are already considered on Earth, which has a much stronger magnetic field; the National Grid in the UK will power up the current when a CME is forecasted to make the resulting induced current less marginal [26]. A similar protocol would be required on Mars, although initially there will need to be some research into the interactions of Mars' magnetic field with CMEs.

Buildings

The most important specifications of buildings exposed to the hostile Martian environment will be to protect against radiation and to insulate as much heat energy as possible. Experiments have been done to test which materials best shield against radiation, and leading candidates include lithium hydride [27]and Boron Nitride Nanotubes (BNNTs). BNNTs are around 100 times stronger than steel in tensile strength [28]. Bioplastics have also been cited as an effective construction material [29] since they can be produced from plants grown on Mars, making them renewable, and is also effective at shielding against radiation due to their low atomic weight. Foam glass [30] is an effective insulating agent which would be cheap and easy to produce on Mars, by using Martian rock combined with carbon as a chemical foaming agent. About 6 m of Martian regolith is enough to shield against normal radiation levels that can be expected to occur on the surface--hence having large portions of the city underground is beneficial. Each building, and any area in general that people can occupy in the city, should have an escape shelter deep underground in the eventuality that a CME hits. Forecasting CMEs could be done by setting up a coronagraph on a satellite orbiting at a lagrangian point between Mars and the Sun. If the spaceship transporting the first settlers separates into landing vehicles upon arrival, the satellite could be left behind with the coronagraph equipment already in place. Used in conjunction with the existing coronagraphs used for space weather monitoring on Earth, this would in fact not only help triangulate the trajectories of CMEs approaching Mars, but also Earth--so this would have significant value to Earth as well [26]. An alert system would be made using the Martian internet, depending on each resident carrying a connected device on them at all times (i.e. a Martian mobile). CMEs can normally be forecasted with around 10 hours warning on Earth, although the severity of their impact can often only be accurately estimated with half an hour of notice [26].

Overall, the building frame can be made from steel which can be produced using the iron which is prominent in Martian soil, and carbon extracted from the atmosphere. It's notable that buildings on Mars would be under less stress compared to Earth due to lower gravity and lighter wind strength, as well as not having to factor in tectonically induced stresses from earthquakes. Therefore, buildings on Mars should have greater stability naturally. Building exteriors would be designed with a multi-layer approach to maximise radiation shielding and thermal insulation. The outermost layer would be a simple metal foil coating, giving the building a shiny appearance that would be strong aesthetically. Beneath this would be a layer of compact Martian regolith, beneath which will be a layer of radiation shielding material. The final interior layer would consist of a foam glass vacuum insulated panel. Windows would be challenging as they could potentially be weak spots for radiation exposure but using double-paned lead glass [31] would mitigate this risk as they are used commonly for experiments on Earth involving harmful levels of radiation. This would be coupled with exterior sheltering around the perimeter of the window to reduce the angle of exposure to radiation, and emergency shutters can be deployed for high-risk events such as CMEs. Other see-through insulators, such as Aerogel, could also be utilized in creating the windows to improve heat retention - but this would be in conjunction with the radiation shielding, as Aerogel does not provide radiation shielding [32].Dust removal on the outer layer of the buildings can be achieved using a pulsed "plasma broom" [33] which is powered by low pressure carbon dioxide, matching the characteristics of natural Martian air.

Transport

Automotive transport will be included in the form of an underground tube network similar to those used in large cities on Earth such as London and New York. It should be noted that in order to promote an active lifestyle which will be important for mitigating against the effects of microgravity, residents will be encouraged to not rely heavily on this transport system--instead, walking or cycling for longer distances should be the primary commuting methods. The tube system will keep all parts of the city interconnected, including the spaceport, and can be designed in a similar format to existing tube networks on Earth.

Conventional trains rely on engines which require air in order to burn fuel and generate power. If a conventional train system were used, the entire tube network would need to be constantly aerated to keep the air pressure in the tunnels high enough for the engines to work. An alternative method, instead, would be to base the design of the tube system on Maglev trains [34]. Maglev trains use electromagnetic fields between the track and the train (steel is a good material for this) to propel the vehicle forwards, and hence they do not use a normal engine. They are also very efficient as the magnetic repulsion keeps the train off the track, meaning there is no rolling resistance and also much less noise. Furthermore, by not having to boost the air pressure inside the tunnels, there would be less air resistance as well meaning that the trains would require less energy to run compared to Earth Maglev trains. Breathable air would be maintained inside the vehicle and an air-lock system would keep the stations aerated for when passengers embark or disembark off the train. Because the tube system will be underground, the construction material of the trains will not require radiation shielding but will require good insulation, so a non-magnetic metallic foil coating could be used to keep temperatures sufficiently warm.

GOVERNMENT, DIPLOMACY AND ECONOMICS

We will now go into further details about the formation and setup of the Government, and the Diplomacy and actions this government will undertake. We will then go through the Economics of the city, going into detail about what the government should do to encourage this, and what outcomes we will be hoping for.

Government

Our government will be set up in democratic style. Initially, the colonists will simply elect one of them to be the "Team Leader" - who can delegate other tasks to others. Since the colony will be small at this point, this should easily suffice. Once the city is larger, we will need a more formal government. We have decided that designating a "Powerless figurehead" - a so called "King of Mars" - would be useful for tourist reasons, diplomatic reasons (an easy ambassador and head of state to visit others) and would help foster a sense of community. The rest of the government should be regularly elected - every 2 Earth years (so upon closest approach) would be good to make the government accountable and uncorrupt. We will make departments for the most vital sectors first - such as city planning, fire, medical, finance etc, and these will be more like a city government at this point. As the city expands, we will add other branches as they are required, and all department heads should be elected on the 2-year cycle.

A few branches of government will be most vital to the city, however. Initially, the scientific branch and industrial branches will be heavily required - since we need the science to get the city set up and allow us to build new things more efficiently, and the Industrial branch - since we need a lot of Raw Materials. Diplomacy with Earth will also be vital, as we need to get many of our initial machinery and resources from it, and the colony will cost money at first, rather than make it.

After some time, we will instead shift our focus to education, the arts, tourism and expanding and preserving the city. A special department for preserving the city will be vital - to keep an eye on resources, to check we are being efficient enough in recycling, and to make sure no disaster will wipe us out. Tourism, education, and the arts will be our major ways to make money, and so should be very heavily encouraged - but the details of this will be talked about in the Economy section. Each small community centre will get to elect a representative, and these together can form the Parliament of Mars - who will have final say on all of the laws to be passed, and any initiatives undertaken by the arms of the Government.

The city will also require a police force, small army, and laws to be created. The army will be very small - with good diplomacy and a large distance between the nearest nation, it should be very easy to avoid war - however, they should be around just in case, and also take care of threats, such as asteroid impacts (which will be rare) and be on call in case of an emergency at the spaceport. The police can also be more friendly and lightly armed, since we are vetting our colonists before they arrive, so no criminals should be imported. However, since we cannot fully guarantee this - and as time goes on, crime is probably going to happen - we will require a police force. This should be comparatively a little smaller than a force would be on Earth in a city of similar size.

The laws will be decided in a constitution - developed by both a council of Earth governments, and by the first Martian Colonists. On this should be provisions to ensure that the Earth governments cannot unduly influence Aurora, and to state some basic laws which all countries already abide by. We will also create a few other laws at the start - including one that, until repealed, no person may buy or sell Martian land. This will allow us to manage the expansion of the city and the layout more effectively. We will also set up laws such as that "No Person May be Denied access to Air, Water, Food or Housing in Aurora", and other similar laws. We will also initially let the government control all land, utilities, and resources - and set this out in law at the start. However, the government will have the power to later repeal these and sell the rights to these. Of course, the government can sell dug out resources - just not the right to dig them out.

Finally, on diplomacy - the Martian government will not be allowed to go to war or declare a trade ban with any nation sitting on the Martian council on Earth. Furthermore, they will also be considered as "Allied" to all of the states on this council - and neutral in any war for which they partake in. Diplomacy will take place via messages, including video ones - and there will be a Martian embassy in every capital for the nations that sit on the council. They will do the main liaison with the governments of Earth, and the "King" will also be able to visit - with a diplomatic entourage - to finalise any major deals or similar. These in person meetings should happen every 2 years - on closest approach - with other smaller meetings once a month, although these will be done over a whole day due to the time difference and will be by message only.

Imports, Exports and Tax

Since Mars is so far from Earth, any imports and exports will be costly to send and receive - and so only certain items will be profitable. The rest should be imported as little as possible and will not be any use for exporting.

For Imports, they will, of course, be required for the first settlers - machinery, tools, and even just basic supplies (including air) will have to be imported. Importing machinery that can build more of itself - such as a 3-D printer - will be a great use of resources at this point and will allow the colony to become self-sustaining. Once the main imports have been completed, the mining and refining operations should reduce the need to import anything further - apart from the colonists! We have estimated the cost per colonist (using $500 per kilo) to import them is around half a million dollars. Hence, they should be highly trained before arriving on Mars - and since they are being imported by many different governments, this cost can be shared between the nations. We could even ask the colonists for an "Investment" amount, since they do get a house, food, water, and electricity on Mars, to lower this cost further. Clothing, vitamins, and other helpful items - which do not weigh too much - will also have to be imported initially.

Once the first settlement is complete, we will be able to drastically reduce our need for imports. Vitamins will need to be imported in micro-doses until the city is completely self-sustaining - which could be achieved with a constant "recycling" of previous colonists - and most may also be able to be obtained out of the plants and diet. Most plants will be able to be grown in our city since the atmosphere and temperature will be kept steady, and so growing cotton for clothing is also not very difficult. Plastic can also be made on Mars using the soil (silicone plastics)[35] or plants [27] and so, most clothing does not need to be imported, and food will be grown solely on Mars once the city is up and running. The only other imports people will require are luxuries.

For importing and exporting luxuries, we have come up with a rule. Since most people would not be pleased with paying more than 10% of an item's value for shipping, Imports from Earth would be economically viable if the monetary value (in $) of the commodity was at least 5000x its mass in kg. To be viable to export, the value should be at least 3000x its mass in kg. This means many items are not viable - however, Jewellery, artwork, valuable documents, and some light tech (such as an iPhone) all fall in this category and so can be imported. This means factories on Mars could create these products and export them for a profit - but since they are rare and cheaper to make on Earth, this is unlikely.

For exports, we will mainly be exporting ideas and arts. Since we have an internet connection with Earth, exporting ideas, music, books (in electronic format) and digital art will be very easy - and so the government will want to heavily encourage this. We will be aiming to make all of our income in this fashion - and then the Martian people can sell the ideas and inventions to Earth people, who can develop them either with them or on behalf of them, and send some of the profits to Mars. We can even export musicians - smaller ones can play holographic, or VR sets to people on Earth and Mars, and, if they are large enough, we could also export the real ones via spaceship. Books would have to be printed on Earth but could then be easily distributed and pay the author royalties, and art would be very similar. We would aim to have a Grand library with all Martian books and academic papers in it - which would help inspire people and hence we could have even more people exporting ideas and entertainment.

The only other export would be, as stated in the last paragraph, goods which are light, but expensive, such as jewellery. This would probably be a niche industry, making items such as "Mars Rock Rings" - however, the government should give some tax cuts to encourage this business. In future, asteroid mining using a skyhook in orbit around mineral-rich asteroids may be profitable, since Mars is close to the asteroid belt and has a lower cost of space travel - however, this is not going to be a main source of income until the city is already wealthy. These minerals could then be exported to either Earth or Mars to be used by the citizens.

Taxes would be done in a similar manner to Earth. Since no-one is paying rent or utilities, we would simply charge a sliding-scale income tax - meaning you're always better off earning more. Since you don't have to pay if you don't earn - but need money if you want to buy any luxuries, this would encourage people to try to earn through their passions - music, art, knowledge and writing - and means people will not feel the need to work just for the sake of earning money. However, since we do require some lower-skilled workers (e.g. restaurants, bars, overseers etc.) these jobs will come with slight tax cuts, to make them pay a slightly better wage. Tax on purchasing any items would also apply - since this will give the government a regular, easy source of income, and basic food would be exempt from this - but not more expensive food, as to keep the cost of living low. This would give people a good reason to work - but, since they don't have to pay many costs, they can work where they wish to work.

Economy

We have already covered most of the points in the previous sections, however we will now cover the final few points on the economy of Aurora.

Employment will be, due to the nature of our exports and large robotic workforce, mainly focused on Arts and Knowledge - for which the tax structure and free housing and utilities (and cheap food) will encourage. However, at the start of the colony, we will need to hire people to simply help us to build the city. This will be still highly skilled - however, we will hire contractors for this role, as to encourage people to come to Mars, since the city will not be built at all at this point. These should be long term contracts - e.g. two or four years - as to match up to the closest approach and encourage people to stay, and, of course, the job interviews for these will have to be done on Earth. As the city grows, we will need less contractors, since we will have a big enough population, and can instead encourage locals to take on these jobs by giving them good salaries and incentives. Robots will end up doing many of these jobs in the long run - and so very few people will be required for these roles, so they will be highly sought after. This will give the Martian job market a very different feel to Earth, where many people have these roles and relatively few are artists or scholars. We may also have some people start up construction companies - however, these need to be highly vetted and checked due to the hazardous nature of the work, and the risks involved if it is done incorrectly.

Finance can also become a major Martian market. We will set up a Martian stock exchange, and so this will function in the exact same way. They will be taxed on earnings as if they were income taxes, and we will encourage companies to open a Mars branch and list themselves on the Martian stock exchange with some tax incentives - which should help boost the economy initially. We will also give incentives and loans to new businesses starting on Mars, and businesses will be allowed to be multi-planetary - and just need to pay some tax if they are registered on Mars. Mars will also have a central Bank, which will provide loans in a similar fashion to Earth. Mars will also therefore print its own currency - not pegged to any Earth currency - for which it will have all the normal measures to keep the economy afloat.

New Martian ideas and inventions, along with standard Earth industries, will be the main businesses on Mars. We will therefore allow people to patent ideas and inventions easily, and make sure we have strict copyright laws - since we will lose money if it is stolen, meaning legal practices can also start up and be successful on Aurora. Web companies and businesses will need to be registered on Mars - but will have free and easy trade with Earth. The knowledge can also be sold via VR and holographic conferences, for which an entry fee can be charged.

Tourism will be our final method of keeping our economy going. We will make sure that Earth has many spaceports with travel to Mars and advertise this fact - and once the city is large enough, people will want to visit! Souvenirs and keepsakes will also give a slight income, and since tourists would have to stay for long periods of time at once due to the planetary cycles and travel time, hotels and bars should have a guaranteed income during this season. Since we can travel when the planets are not perfectly aligned - it is just a bit longer and more expensive - we will get tourists about half of the year, and people will be encouraged to spend a "gap year" on Mars, which can provide more workforce, and more money in the economy. Using VR travel, we can also charge people to visit Mars from Earth, and Earth from Mars - hence earning from tourism in both directions. To make sure tourism is as profitable as possible, we will need to make the city attractive to visit - which brings us onto our next section, City Life and Culture.

CITY LIFE AND CULTURE

The aim of the project is not just to build a city that is habitable for human settlement, but also enjoyable to live in. Without plenty of scope for leisure activities, entertainment and places of interest, it will be difficult to attract people to visit or move to the city and this will be detrimental to economic and population growth--and it can be expected that the novelty sensation of being on a new planet, as grand as that may be, will not be everlasting. There is the additional factor that the choice of places to travel once on Mars will virtually be limited to only within the city itself, apart from journeys back to Earth. Therefore, it will be important to ensure that there are plenty of things to do in the city itself.

Community and Aesthetic

The overall theme of the city should be of inspiration and curiosity, to reflect the magnitude of the occasion of settling on a new planet and representing the start of a new chapter in human history. The name *Aurora* was chosen on this basis; to most people, the word *Aurora* resonates with the spectacular phenomena of the northern lights--*Aurora Borealis*--and the southern lights, *Aurora Australis*. These colourful light shows in the night skies seen at high latitudes on Earth, caused by the magnetic deflection of the solar wind, is an inspiring scene which many travellers aspire to witness. It is also a visualisation of the large-scale forces of work between Earth and outer space. In a more literal sense, *Aurora* is also the Latin word for "Dawn", and hence this name can be used to emphasise the beginning of a new era. Following the trend of inspiration, the city should be a symbol of inspiration to people on Earth, in order to encourage people to visit. The architecture should have a distinct and iconic design to make it instantly resonant and identifiable with people on Earth, in the same fashion as famous buildings such as the Antoni Gaudí designed architecture in Barcelona. Buildings on *Aurora* would typically be dome or tower shaped, with a glimmering metallic exterior for a futuristic look coupled with Martian masonry to blend in more naturally with the surroundings and give the architecture a closer sense of "belonging" to its world. Caverns would also feature heavily in the designs, but these would not be visible from the outside - and seem very similar to domes (apart from the walls) on the inside. There could also be a singular magnificent structure whose only purpose is for touristic value with a cityscape viewing platform, similar to the iconic Eiffel Tower in Paris. In *Aurora*, the iconic buildings would be the Palace, the Grand Library, the Tower of Mars, and the Government buildings. The big central dome of the city centre would also be an iconic site of the city. Science fiction culture has always come up with fantastical visualisations of what a futuristic, extraterrestrial civilisation would look like and this should be a reference guide to artistically aim for and even go beyond.

Maintaining good mental health within the population will be important in order to create the enjoyable, vibrant community which this project aims for. There will be several obstacles to overcome in order to achieve this. Firstly, by having large parts of the city underground and also by using a lot of shielding from the sun, this would have a negative impact on the mental health of the population. Such a phenomena is also observed on Earth, where human settlements located in the Arctic circle which is deprived of sunlight for long periods of time throughout the year tend to generally less happy residents compared to cities and towns at lower latitudes. This will also be the case in *Aurora* as Mars has an axial tilt of 25 degrees (almost the same as Earth), meaning it will have its own "Arctic circle" type region situated above the latitude of 65 degrees, which the Korolev crater is well within. Due to the high latitude, *Aurora* would have large seasonal variations in diurnal sunlight hours; in the summer, the days will be very long, but in the winter the days will be very short. A comparable settlement on Earth at the same latitude is the Greenlandic town of Ittoqqortoormiit. To mitigate against the deprivation of sunlight, phototherapeutic light screens would be used inside the habitats to create artificial sunlight--however, residents should also be able to spend a sufficient amount of time on the surface and small windows can allow a limited amount of natural sunlight in. A second obstacle would be the actual remoteness of the city, being millions of miles from the nearest human civilisation on Earth. Although this can be mitigated by using a satellite link to keep residents on Mars in contact with their friends and families on Earth, it will be psychologically challenging to cope with living so far away--especially for the early settlers who will only have a small number of people to interact with on a daily basis. Similar requirements are already needed for jobs on Earth in certain remote locations, such as scientists working in Antarctica.

The city should strive for a socially well-knitted community, with most people living in communal accommodation and regularly participating in group leisure activities such as sports clubs. Ironically, one hypothesis is that the most likely people who would volunteer to move to Mars initially would be fairly asocial; it would be sensible to predict that more social people would find it more difficult to leave friends and family behind, and would therefore be less inclined to volunteer for the colonisation missions. Additionally, there is also the risk that after spending months living together in a small spacecraft, the first colonists may already be fed up with living with each-other by the time they actually reach Mars and so would maybe prefer their own space in terms of accommodation. Since the spaceships used to transport the first colonists will also be used as the first habitats, they should be designed to be able to provide for this. Robots will also be very commonplace around Mars and people will likely have frequent interactions with them. Not only will this mean a requirement for residents to be comfortable with being around and depending on robots (the robots themselves should be designed and programmed to facilitate this as best as possible), but it may also revolutionise the relationship between humans and technology--and this could become a unique attribute of Martian city culture.

Since a day on Mars is very close to the same length as it is on Earth (only about 1 hour longer), normal business hours would not need to be changed much; however, it would be complicated to stick with the Gregorian calendar since this would lead to many issues in resolving time conversions between Earth and Mars due to the different length of days and the fact a year on Mars is about twice as long as a year on Earth. This would be a significant issue for business trading with Earth and also for seasonal planning-- for example, solar power usage would depend heavily on the time of year. For this reason, a new "Martian calendar" is proposed. A year on Mars is about 669.79 Mars days. This could be rounded up to 670 days, divided into 10 months of 31 days and 12 months of 30. Every 5 years, there would be a "leap" year, except that instead of being 1 day longer it would be 1 day shorter. Earth holidays which fall on regular dates (e.g. Christmas on 25th December) could be kept but would simply have to change day each year--much in the same way that some holidays (e.g. Easter) already do so. The Mars-Earth close-by date could also become a celebrated day on the Martian calendar.

Tourism, Sports and Cuisine

Tourism will be a key part of growing the Martian economy as it is the catalyst for generating interest in the city and encouraging new people to visit or permanently relocate there, so it is important to maximise the experience for tourists. Places of interest may be realised in natural course--for example, the spaceships which land and habituate the first colonists could be preserved after they are disused and serve as a tourist site. A future project could also be to build museums showcasing exhibitions which teach about the history and culture of the city--for example, a "Martian History" museum. Tourism on Mars should be advertised strongly on Earth and managed via a designated tourism agency in charge of organising flights, accommodation, and guided tours. Like many cities on Earth, there should also be a tourist information point located in the city that tourists can visit to learn more about places of interest.

Fitness will be important in order to combat the negative effects of microgravity as well as having an overall positive effect on mental health, and therefore sports and other physical activities should be encouraged. Houses would be fitted with small gyms and fitness equipment, and city planning would ensure all residents have easy access (target of 10 minutes travel time or less) to a park where playgrounds, sports pitches and leisure facilities can be built. Sports would have to be adapted due to the lower gravity--football, for instance, could be played with a heavier ball and inside a cage to prevent the ball being kicked too far away. Ultimately, new "Martian sports" could be invented tailored to the microgravity environment. Clubs could be set up allowing for residents to regularly practice sports as well as being social gatherings. In the long term, competitions and leagues could be organised which could generate revenue from attending fans or even broadcasting.

Martian cuisine will initially be heavily dependent on what can and cannot be grown in Martian soil. It is economically strenuous to import large quantities of food from Earth, due to their low value per unit mass--although this may not be the case for some more luxurious foods such as lobster. Over time, as the soil becomes more modified and more nutritious as well as with increasing amounts of scientific research, there will be increasing availability of new ingredients. The Martian diet will at first be largely plant-based, as meat will not be available in large amounts and would mostly have to be grown in laboratories. Insects would be an efficient substitute to supplement protein for a balanced diet, as they have high quantities of protein per unit mass and would also be beneficial for plant growth. Due to low diversity of foods that can be grown especially in the early phase of colonisation, herbs and spices would form a significant part of Martian cuisine to add interesting flavours and prevent Martian food from being too bland. Theoretically, in the long-term plants may adapt to Martian soil in such a way that they genetically become a new species, meaning new foods would be created and be endemic to Mars.

Media and Entertainment

Theatres, cinemas, and music venues would be built to showcase Martian media productions. A TV network would also be set up via a satellite (which would be geostationary since it would only need to broadcast to one location) and present not only Martian news and entertainment programmes, but also those from Earth. Due to the multicultural demographic that would be expected for the Martian population, it would be preferable to show programmes in a choice of languages; although it may be logistically impossible to cater for every existing language, it would certainly be beneficial to allow for the most spoken languages such as English, Spanish and Chinese. Alternatively, there could be a subtitle system to translate for other languages. Other entertainment venues such as bars, nightclubs and casinos would be built--we hypothesise hard liquor may become a cultural normality on Mars as it requires less plants to grow.

It will be necessary to build a "Martian internet" allowing for easy communication of information, particularly for keeping in contact with Earth but it will also be essential for warning people of environmental hazards such as incoming solar storms or dust storms. The Martian internet would also be used as a platform for other online networking services, including social media applications. These could both be imported from Earth (e.g. Facebook, twitter) as well as including new ones invented on Mars.

Education and Other Culture

Knowledge and education will be an important part of Martian culture, especially as the first settlers will have to be highly skilled scientists who are able to engineer the infrastructure of the city. As the younger population grows, there will be a need for an education system; to start there would be a university to take up Earth students as well as Martian residents. The "University of Mars" could be a tourist attraction in the same way that famous currently-existing universities are such as Oxford and Cambridge, and for the purpose of the university there would be a grand library storing both Martian and English works. After the university, schools would be built for younger students--it would be important to build the university before the schools as if the university was still under construction by the end of the first school term, school leavers would have nowhere to go unless they could afford to move back to Earth--or they would simply have to find work, but being likely a lot less qualified compared to the highly-skilled majority of the population. Teaching in Martian schools would focus on Mars-specific science. For example, learning about the hydrological cycle--a common topic in geography syllabi--would not be useful in a world without surface liquid water. Tectonics may have some relevance due to the formation of the vast volcanoes in the Tharsis region and Olympus Mons, although Mars is not currently a tectonically active planet. Academic subjects such as maths and science will be the most important subjects due to the high skill requirements of many jobs that will exist on Mars. Education would also emphasise the multiculturality of the city, so it should be considered important to learn about various Earth cultures, religions and also languages. An additional factor to consider is that optimal transit times (when Earth and Mars are closest together) will also be the windows for new students to relocate to Mars, and since these are only every 2 Earth years that means student take-ups will only be able to happen every other year. One logical solution to this dilemma would be to effectively double the length of academic terms on Mars such that each "year" of education is actually the equivalent of 2 Earth years, before students transition to the next stage of education.

Due to differences in materials available (although some luxuries would be economically viable to import, as discussed in the economy section), as well as the multiculturality of the population, clothing styles may possibly end up being different on Mars. They might seem odd to newcomers at first, but this could possibly be eased by producing online magazines to promote Martian clothing and make them seem more normal. Magazines and newspapers will most likely have to be exclusively online applications due to the lack of trees to produce them in paper format.

Overall, building a city on Mars allows a window of opportunity to experiment with new lifestyles as we would effectively be starting on a new blank canvas of society. The most important target should be to recognise the disadvantages of lifestyle on Earth which are hardwired into societal structure and carefully plan to avoid them, whilst simultaneously keeping and perhaps enhancing the advantages to ensure a high quality of life and also keep some familiarity to balance out the novelty of living in a completely new world.

THE FUTURE OF AURORA

Our final section of the report is on the future of the City. Once the city has been completed, is completely self-sustaining and has one million citizens, we have a few plans for what comes next.

Firstly, offshoot cities and even cities on other planets can be founded using Aurora as a base. This will allow for cheaper missions than from Earth and will allow for even more Martian settlements to be founded - providing new knowledge, materials and making Aurora rich and prosperous, and a place to go to pursue a brighter future. Once we get to this stage, bringing big, tourist heavy events from Earth - such as the Olympics and big festivals - can happen on Mars. We will already have sent teams and entertainers to Earth, but we can now welcome them to the city!

Secondly, we can make Mars travel even cheaper and more pleasant, allowing us opportunities to explore and visit other planets and set up grand projects. Supply ships in orbit around Earth and Mars could also help people travel easier and with more luxury, and this could be put into practice reasonably early on. A Dyson Swarm could be put in place around the sun, providing a clean, easy and renewable source of energy for the whole Solar System – and due to the lower gravity, this would be cheaper to put in place from Mars, or a Mineral rich Asteroid. If Cryo-Sleep becomes possible, along with faster rockets, and structures such as space elevators are set up on both planets, travel between them will be cheap and easy - like simply visiting another country. This would also make tourism very highly profitable, and asteroid mining could become commonplace and help the space-economy - and spaceports in the middle of space or on small asteroids could even become common too.

Finally, and most ambitiously, we can attempt to terraform the entire planet. Mars has a very weak magnetic field due to a lack of a Dynamo - so we would have to either restart its dynamo or simulate it using magnetic fields originating from under the cities of Mars. This is currently not known how to achieve - however, if it was eventually created, then it would allow Mars to hold an atmosphere - and so, it could be thickened to warm it and create running water, detoxify the soil and prevent radiation. After this, we could build cities outside, with forests, fields, and nature, in the same manner as Earth - and then, Mars would truly be a new dawn for humanity, with a profitable and unique capital city - Aurora.

REFERENCES:

[1]. https://futurism.com/the-byte/elon-musk-mars-city-cost-10-trillion

[2]. https://www.nasa.gov/feature/jpl/nasas-treasure-map-for-water-ice-on-mars

[3]. https://www.nasa.gov/feature/goddard/2019/martian-dust-could-help-explain-planet-s-water-loss-plus-other-learnings-from-recent-global

[4]. www.spaceref.com/news/viewpr.html?pid=37633

[5]. https://www.google.com/mars/

[6]. https://www.forbes.com/sites/bridaineparnell/2017/04/27/mars-colonists-can-build-red-brick-houses/#198799d26712

[7]. https://www.sciencealert.com/it-turns-out-martian-soil-can-probably-be-compressed-into-bricks-stronger-than-concrete#:~:text=Scientists%20from%20the%20University%20of,%22similar%20to%20dense%20rock%22.

[8]. https://www.sciencedaily.com/releases/2016/01/160120122756.htm

[9]. http://news.bbc.co.uk/1/hi/sci/tech/7477310.stm

[10]. https://en.wikipedia.org/wiki/Aquaponics#:~:text=Aquaponics%20(%2F%CB%88%C3%A6kw,water)%20in%20a%20symbiotic%20environment.

[11]. https://www.engadget.com/2018/05/02/nasa-completes-full-power-tests-small-nuclear
reactor/?sr_source=Facebook&guccounter=1&guce_referrer=aHR0cHM6Ly93d3cuZmVjZGl0LmNvbS8&guce_referrer_sig=AQAAAAhL0HvTZ5zkbt7Gzh-kWHcWf-WodKPm6yOjPkiywTK23awfEbS6Z9AhN4n3M6ngJk50M9nsdUZiAkdEOH1xpPwlBgtekmT08nazSvtk5XpuKSTNo77ZiBA-RE5jt0FRs3bux6DkJCyyFVlpbBmN52uPtgWBFDiE_AiB6Qz5MJCP

[12]. https://www.nasa.gov/mission_pages/station/research/news/sabatier.html

[13]. https://www.nasa.gov/directorates/spacetech/kilopower

[14]. https://www.stainless-steel-world.net/blogs/100/shiny-facades-and-the-issues-of-glare-and-solar-energy.html, www.depositphotos.com Image ID: 169372884 by author: lookaround

[15]. https://www.nasa.gov/sites/default/files/atoms/files/mars_ice_drilling_assessment_v6_for_public_release.pdf

[16]. https://cycles.thameswater.co.uk/Accessible/The-sewage-treatment-process

[17]. https://learn.allergyandair.com/water-filters/

[18]. https://www.hindawi.com/journals/ijmicro/2019/6981865/

[19]. https://www.space.com/how-feed-one-million-mars-colonists.html

[20]. https://www.nhs.uk/conditions/altitude-sickness/

[21]. https://mars.nasa.gov/mars2020/spacecraft/instruments/moxie/

[22]. https://www.popsci.com/how-iss-recycles-air-and-water/

[23]. https://www.nasa.gov/directorates/spacetech/kilopower

[24]. https://en.wikipedia.org/wiki/List_of_countries_by_electricity_consump
tion

[25]. https://www.world-nuclear.org/information-library/safety-and-
security/safety-of-plants/safety-of-nuclear-power-reactors.aspx

[26]. Met Office Space Weather

[27]. https://pubmed.ncbi.nlm.nih.gov/30499384/

[28]. https://www.nasa.gov/feature/langley/team-discovers-new-method-of-
making-promising-material

[29]. https://www.aispacefactory.com/marsha

[30]. https://marspedia.org/Insulation

[31]. https://www.raybar.com/xray-glass/lead-glassh

[32]. https://www.chemistryworld.com/news/better-windows-with-
aerogels/1010158.article

[33]. https://iopscience.iop.org/article/10.1088/1367-2630/aa60e5/pdf

[34]. https://en.wikipedia.org/wiki/Maglev

[35]. https://space.stackexchange.com/questions/2569/can-materials-be-
found-to-make-plastic-on-mars

3: MAREEKH
An introduction to Engineering, Economic, and Social Perspective of Establishing a Human Civilization on Mars

Dr Muhammad Akbar Hussain
Founder and Director
Southern Cross Outreach Observatory Project, Australia
akbar.astronomer@gmail.com

PREFACE

Humanity's centuries old dream of a human civilization on another world is as fanciful as it is inevitable. With the advent of 21st Century, humanity seems to have acquired bare minimum potential of colonizing other worlds. We now find ourselves at crossroads with a choice to make; either leapfrog into exploring new worlds beyond our home planet, or wait for an unforeseen future on the only world known to harbour life.

I am going to share my vision of how a human civilization can be established on Mars with existing technology and expertise. I will briefly discuss the engineering, economic and social aspects of establishing such a civilization that can achieve self-sustainability and independence as a parallel human civilization on a different world.

OBJECTIVES

Following the establishment of the first group of bases on Mars demonstrating the ability to sustain a permanent human presence on the red planet, humanity will need to overcome barriers in laying foundations of a permanent human civilization there. Creation of a million strong human settlement will be a good start in the long-term goal of making Mars our true second home, and ultimately terraforming it in distant future.

To recap, following major challenges need to be addressed in every step of the way.

Main challenges
1. Low atmospheric pressure: Mere 0.6% of that of Earth, essentially vacuum.
2. Low temperature: Roughly -65 degrees Celsius, ranging between -153 degrees Celsius at poles at night to +20 degrees Celsius near midday at equator.
3. Radiation: Up to 2.5 times higher than experienced by astronauts on International Space Station.

Other modifiable problems
These problems can either be modified, adapted to, or be overcomed through technology.
1. Lack of breathable oxygen
2. Lack of surface water
3. Low gravity
4. Isolation

Attainment of self-sustainability and self-sufficiency with only minimum ongoing support from Earth will mark the birth of human civilization on Mars.
The main objectives that need to be accomplished are:

1. Ability to manufacture
-   Construction material: Concrete, bricks, cement, fibreglass, plastic and steel.
-   Hardware: Machinery for construction, agriculture, and travel.
-   Fuel: Important for transit between colonies and also to lift off from surface of Mars for journey to Earth for people and trade, travel to asteroids for mining and exploration.

2. Ability to grow food
-   In form of traditional agriculture of crops and breeding animals, fish etc.
-   Novel ways of preparing food using bacteria and algae to reduce reliance on mainstream agriculture and breeding of animals.

3. Healthcare
-   Ability to address common human pathologies.
-   Ability to identify and manage illnesses resulting from unique characteristics of life on Mars, including low gravity, radiation exposure, isolation etc.

4. Economy
- Ability to achieve not only self-sufficiency but also production of materials in surplus that can be exported to Earth in exchange of goods that may still be hard to procure or create in Martian environment.
- Tourism: For Earth residents visiting Mars for recreation, holidays, adventure, and exploration.

5. Self-governance
- A human colony on Mars should be able to govern itself independent of Earth.

MAREEKH

I would like to introduce my model of Mareekh Civilization. **Mareekh** is an Urdu word for planet Mars. This civilization comprising of one million individuals would be living across dozens of individual settlements (or cities) on the red planet.

In this model, individual Mareekh settlements would consist of surface colonies inside inflatable structures over Martian craters anywhere from 150m to 500m in diameter called **Craterhabs**, and underground habitable spaces in tunnels roughly 8m – 10m in diameter, and several hundred meters to few kilometers in length connecting spacious multipurpose atriums. The settlements will be connected through dirt tracks for rovers which will later be modified into rail tracks and eventually overground tunnels and hyperloop systems. The purpose of building these inflatable domes over craters is to take advantage of nearly circular crater rim acting as reinforcing wall and the depth of the crater will provide additional volume, saving precious structural material for a similar volume on a flat surface.

The main purpose of large inflatable structures is to provide sufficiently large pressurized space for humans to comfortably move around between buildings and structures, undertake farming, mining and construction of underground tunnels and spaces, without the cumbersome ritual of going through airlocks and changing spacesuits frequently; essentially a shirt-sleeve environment. These inflatable domes will house buildings for different purposes, farms, warehouses, factories and mining units.

The main problem in erecting and maintaining such large volumes of space is the need to maintain nearly 1 bar of internal pressure against 0.006 bar outside; practically vacuum. Such a structure has to withstand an outward force of nearly 1kg per square cm or 10,000 kg per square meter. Multiplying it with the total area of, say, a 300m diameter dome or roughly 70000 square meters, results in a force equivalent of $7 \times 10e8$ Kg or 0.7 million tonnes trying to blow apart the structure constantly. Flimsy looking sci-fi glass domes will not stand a chance and a radical engineering approach to build such structure is required.

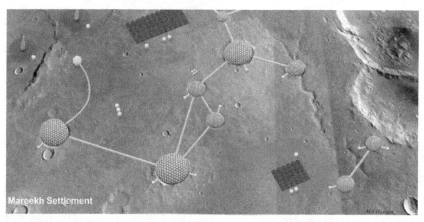

STRUCTURE AND DESIGN

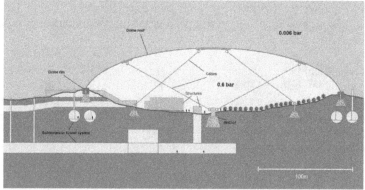

The Craterhabs will be made from fabric that can withstand a pressure of at least 1 bar.

Materials like certain varieties of Silicone e.g. RTV Silicone[1] used in various space applications can withstand temperature range of -115°C to over 200°C. Silicone also has an excellent tensile strength[2] and can withstand up to 60 kg per square cm. It is also fairly translucent. Another even stronger fabric which is tough and can withstand tension, temperature differences and vacuum is **Dyneema®** which has a tensile strength[3] of nearly 20000 kg per square cm; far stronger than silicone but it is opaque. Both these materials offer many times the minimum safety factor required to maintain structural integrity of the dome. The periphery of dome will be anchored into a system of concrete blocks buried under a reinforced trench encircling the entire crater along its periphery. These concrete anchors or blocks will be assembled from Martian regolith and water from subsurface permafrost. To further augment the support for the anchoring rim of dome, the main body of the dome will be anchored to the floor of crater through **Dyneema® cable systems**, tens of them, connected to concrete blocks buried deep into the crater floor. Each cable will be a set of three, with tension in one or two cables and third acting as a backup, should any cable in the system snaps or experiences too much tension.

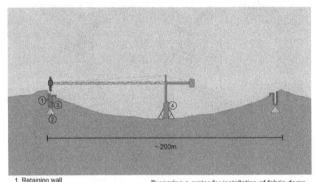

1. Retaining wall
2. Concrete anchor system
3. Reinforced dome rim
4. Crane shaft

Preparing a crater for installation of fabric dome

A Craterhab dome will be made of a hexagonal lattice of Dyneema® with patchwork of Silicone, 5-10 m in size. The Dyneema® skeleton will be made of three layers and the silicone patches will be a bi-layer. This design will offer an additional protection against micrometeors that have the potential of punching millimeter-size holes in the silicone layer. In such an event, one silicone layer in the patch can be moved by just a centimeter or so, fixing the hole immediately, while further repairs can be carried out. Drones can be used to glue small repair patches from inside, which will simply be held in place by the internal pressure until the glue takes hold.

These fabric domes will be inflated with mixture of nitrogen (either extracted through atmosphere or minerals), oxygen and carbon dioxide using gas turbines. Due to translucency of silicone material, these habs will act as greenhouses and trap heat, warming up the whole structure, further increasing the pressure without additional expenditure of energy.

The pressure inside a Craterhab will be maintained to nearly 0.65 bar which is roughly equivalent to the atmospheric pressure at 4100 m altitude above sea level, analogous to the pressure at El Alto district of La Paz, the capital of Bolivia (Happiness index 5.5. Not bad!)[4]

The temperature inside the Craterhab will be maintained between 15-20 degrees Celsius. The amount of air scavenged by the Martian soil at the base of the crater may be balanced by the outgassing from Martian soil when it is subject to over 10 degrees Celsius temperature, further balanced by the gas turbines that can buffer the amount of air and pressure inside the crater hab.

One of the major advantages of using Silicone material is its inherent UV reflectivity; reflecting over 95% of all UV radiation yet letting sufficient visible and infrared rays to create a naturally illuminated greenhouse and creating a relatively safer open environment for survival of its inhabitants and crops.

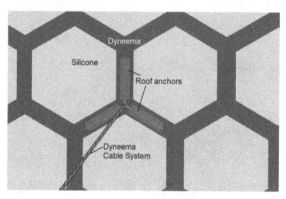

Contents of Craterhabs

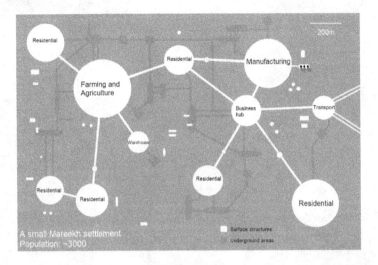

1. Living quarters where residents can roam between the tunnels and Craterhabs without spacesuits, giving them tremendous flexibility and freedom to perform their work in the most relaxed manner possible.
2. Greenhouses to grow crops. The semitransparent domes will act as greenhouses. The water to grow crops can be tapped from the permafrost from walls of the tunnels and regolith from tunnel boring process.
3. Warehouses where essential construction of 3d printed materials using raw material from Earth combined with regolith to construct building and structures, and create silicone using silicone manufacturing units for construction of more domes. These warehouses can also be used to manufacture other types of fabric, concrete, plastics, tools, machinery and parts.
4. Open water bodies for recreation, fish farming, and reservoir of water for drinking and farming etc.

With expansion of the settlements, more and more Craterhabs will be added to create more open spaces for establishment of residential areas, shopping, education, business and economic activities, and open recreational areas.

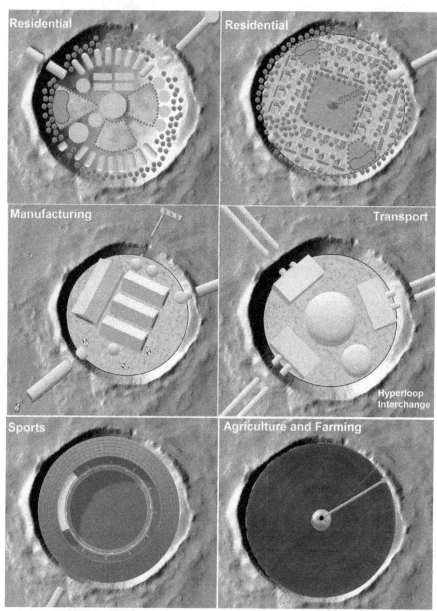

Interiors of Craterhabs can be customized for various applications and purposes

TIMELINE

10 years after the first successful landing of Mars Direct, a small network of human bases on Mars has taken shape, with a permanent presence of roughly a hundred humans. This is year 2040, and the stage is set to establish the first Mareekh settlement using the Craterhab technology to vastly expand the potential of human settlement on Mars. First fleet of twenty Starships arrives with material for the construction of first three Craterhabs. This includes prefabricated Dyneema® patches, construction and manufacturing equipment, power production units, life support equipment, fuel production units, and team of 200 engineers, mining and construction experts, mechanics, scientists, biologists, geologists, medics, and architects. The landing occurs in close proximity to one of the major Mars Direct bases established earlier, in order to utilize the pre-existing infrastructure as well. The total volume transported would be nearly equivalent of 240 TEUs of cargo (about 8000 cubic meters) weighing 3000 tonnes.

The weight and cost breakdown of payload on first Mareekh settlement mission (year 2040) consisting of 20 starships may look something like this:

Payload	No. of units	Weight per unit (tonnes)	Total weight (tonnes)	Cost per unit (Million USD)	Total cost (Million USD)
Dome fabric (tri-layer) and cable system	3	2.5	~ 8	3	~10
Construction cranes	2	50	100	1	2
Robotic modules for cranes	6	5	30	1	6
Tunnel boring moles	2	100	200	20	40
Exploration Rovers	10	8	80	3	30

General purpose cranes and forklifts	8	5	40	0.06	0.5
Mobile heavy-duty cranes	3	40	120	0.5	1.5
Excavators	6	8	50	0.3	~2
Construction and Industrial 3D Printers	10	2	20	0.2	2
Mini-dumpers	6	7	~42	0.25	~1.5
Radioisotope Thermoelectric Generators (RTGs)	10	0.5	5	50	500
Small Modular Reactors (SMRs) (300MW)	2	50	100	7000	14000
Solar Panels (2500sqm)	-	-	20	-	2
Hydroponic Modules	2	50	100	50	100
Concrete Manufacturing Unit	1	100	100	20	20
Food	-	-	150	-	3
Water production/waste management facility	1	50	50	10	10
Communication Units	2	10	20	5	10
Fuel Production Modules	2	100	200	50	100
Fuel Storage Tanks	5	4	20	0.4	2
Gas Turbines	6	1.2	~8	2	12
Crew	200	0.08	16		
Personal effects	200	0.15	30		
Cables, electronics and tools	-	-	50	-	5
			Grand Total: 1559 tonnes		Grand Total: $14,860 million
			Approx: 1600 tonnes		Approx: $15 billion
			Surplus capacity: 1400 tonnes		

Calculated cost of transport of payload weighing 3000 tonnes on 20 starships: $500/kg = $1.5 billion

Cost of payload: $15 billion

Cost of 20 star-ships: $100 million x 20 = $2 billion

Total cost: Approx. $19 - 20 billion (excluding cost of ground infrastructure and manpower on Earth, and crew training)

It seems like there is a surplus cargo capacity of 10 starships to Mars. However, it should be noted that the Starships will be home to a crew of 200 individuals for several months during transit through space, and will be used as habitats on Mars itself for nearly 2 years until construction of Craterhabs and underground spaces can be completed. The crew will be travelling in 10 starships with living spaces for 20 members each and smaller cargo, sparing 10 starships purely for carrying heavier, larger, or dangerous payload including nuclear reactors, drilling and construction equipment, and Hydroponic Modules.

Construction of first Mareekh Settlement consisting of three Craterhabs, of roughly 200-300m in diameter will start, which will take one Earth-Mars orbital cycle to complete. The construction will involve preparation of base of craters, construction of reinforced sandwich-wall along the crater edge, drilling of tunnels, construction and placement of concrete anchors inside the sandwich-wall and under the base of the crater. Assembly of the Craterhab will start soon after this, by joining together prefabricated individual hexagonal pieces of the Craterhab dome and installation of supporting Dyneema® cables system. The first 3 Craterhabs will be entirely made up of Dyneema® fabric, weighing 1000 – 2000 kg each and will be opaque, requiring artificial lighting. Silicone is heavy and bringing it from Earth with incur huge cost. The raw material for silicone is found in Martian regolith and can be synthesized on Mars during the course of progress of Martian settlement. Once the domes are laid down, sealed on the edges, they will be filled with Martian atmosphere using gas turbines to achieve a pressure of 0.6 atmospheres, composition of which can then be altered to suit human needs. After the construction of first three Craterhabs, next most important step will be to start construction of underground tunnels, 8-10 meters wide. Tunnel walls will be reinforced with 3d-printed fibre-reinforced concrete rings joined and sealed together to ensure the ability of tunnels to hold the internal pressure and prevent leakage. This fibre-reinforced concrete rings will also be used to create air-lock tunnel systems for outside access. The subsurface regolith dug out in this process will be the most important mineral produce at that time that will provide necessary substrate for manufacture of concrete and provision of water for the colony.

The next mission of 20-40 starships by 2045 will carry another thousand or so astronauts from different walks of life, and other equipment including basic mining and processing units, basic steel fabrication units, basic fabric and plastic manufacturing units, and modular silicone manufacturing units employing Muller-Rochow Process for construction of translucent domes of Dyneema®-Silicone hexagonal lattice as discussed under Structure and Design section earlier. With the construction of next 5-6 Craterhabs, creating safe internal pressurized environments for farming and construction and residential units, the Martian Settlement will achieve its first baby steps in self-sustainability. Mining and processing of ore will be the next major step in establishing the Mareekh Civilization. With each orbital cycle, starships from Mars will return to Earth, using Martian synthesized fuel (Liquid hydrogen and oxygen) as more starships from Earth arrive with people and supplies. A trade and exchange can start in due process making best use of starships returning to Earth, with cargo of Martian produce including processed minerals and ores, and novel agricultural produce like Martian grown vegetables and fruits. This will be vital in generating and sustaining cashflow to Mars for further development of the settlements and attaining sustainability and self-sufficiency.

With every half a decade or so, more and more starships will return, carrying trained crew in increasing numbers, as the requirement to haul supplies from Earth to Mars will reduce with increasing sustainability of a growing human population on Mars. By year 2060, small cities on multiple locations on Mars hundreds of kilometres apart will have taken shape, connected by over-ground railway network and hyperloop system, which will continue to expand individually, until by year 2080 to 2090, a million strong, independent and self-sufficient human civilization on Mars will have taken shape. Mareekh Civilization will be born, marking new age in human evolution; the Interplanetary Age.

Humanity will triumph against the elements by becoming an interplanetary race.

ECONOMY

In order to achieve independence and self-sufficiency, a Mareekh colony needs to be able to produce all of its food and raw material for construction of structures and machinery. However, there will still be dependence on import of Earth-based material and finished products in form of medications, certain food items, machinery, advanced electronics and skilled manpower in certain areas. In exchange, the Mareekh civilization would be able to sell its produce and raw materials like minerals and certain novel synthetic materials and drugs that may require low gravity for their manufacture.

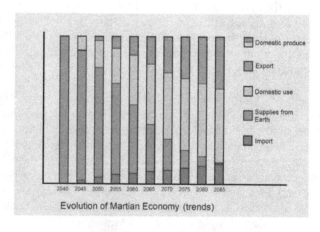

Evolution of Martian Economy (trends)

Domestic usage:
The main products for domestic consumption would be steel, fabrics like graphene and silicone, and plastics that can be manufactured from raw materials existing on Mars (Zubrin R., The Case for Mars. p306). These materials will be utilized in building structures and machinery for further exploration, mining and manufacture; a process that will become more and more self-sustained as the civilization progresses.

Export:
Once Martian industry and mining is self-sustained, there is bound to be a surplus. However, it may not be economically feasible to export these materials to Earth as it won't be cost effective. Only those minerals that are scarce on Earth and relatively abundant on Mars will be the main mining products that can be exported to Earth. One such substance is Deuterium which is far more abundant on Mars than it is on Earth and is useful in nuclear power generation. Future entrepreneurs may find a way to export it back to Earth to start a cash flow or purchase of import items from Earth in exchange.

Decades later, with further progress of heavy machinery construction capability and expertise of Mareekh Civilization, asteroid mining can be seen as a potential sector that can further boost economic growth of Mars. Not only it will provide necessary raw material for construction and industry on Mars, extracted metals mainly iron and nickel, and possibly gold and platinum may become a useful export for their smaller mass per unit volumes and high returns. These metal ores can be processed on mining outposts on asteroids themselves and semi-processed product can be directly shipped back to Earth.

Tourism:
Tourism from Earth will contribute significantly in early Martian economy. A Martian colony needs to keep this in mind and aim to build infrastructure that is feasible and attractive for Earth based tourists, including spacious Craterhabs and underground facilities, and efficient means of transport between settlements and points of geological and research interests.

Main imports:
The dynamics of imports will rapidly change with the progress of Mareekh Civilization. In early decades, a Mareekh civilization will be heavily dependent on import of most of the hardware, food and medicine, but this will rapidly change with increasing self-sustainability of a Mareekh civilization. Few items will always need to be imported, including nuclear fuel, and certain electronic items. A Mareekh civilization will also be heavily dependent on nuclear power, and radioactive elements are much harder to find on Martian surface as compared to Earth. The cost of imports needs to be balanced with exports. Some form of barter agreement with Earth might be useful in early decades to create a balance that should favour Martian economy and not otherwise!

Other venues
Martian economy can also be fuelled through more unique ways of income generation and unconventional methods can be experimented. These will require support from Earth. Since the population of Mareekh Civilization will be roughly a million, these methods can actually prove useful as they would contribute to Martian economy without diverting any sizable fraction of Mareekh population from essential tasks of building Martian infrastructure.

1. Advertisement and sponsorship:
Several large companies and corporations on Earth can be given incentives of using their support of Mareekh civilization for the purpose of advertisement and promotion of their products on Earth. This may become a substantial economic resource for an early Martain Civilization.

2. Agreements:
Right at the beginning of the establishment of a Martian colony of roughly a few hundred to few thousand individuals, agreements can be reached on Earth to ensure sustenance of future Martian Civilization. One such agreement may be to lease out mining rights on Earth's Moon to Mareekh Civilization for Helium-3 extraction for power generation not only for Earth, but also for Mars. Since Mars does not have an abundance of resources for power, including weak solar power, and lack of available radioactive heavy metals for nuclear power generation, it will heavily depend upon nuclear fuel from Earth. Regular launches of nuclear fuel cargo from Earth is rife with potential dangers, and it will take only one or two accidents of such a cargo, to halt or end shipment of nuclear fuel from Earth; essentially an end of line for the prospective Martian Civilization. Mining of Helium-3 for fusion energy, directly extracted from Lunar regolith will be safe and feasible. In fact, it may be cheaper to extract Helium-3 from Moon using Mars as a base for launching mining equipment and cargo ships to the Moon, than from Earth. It will be essential to give mining rights on Earth's Moon to Martian Colony which will not only be vital for economic sustenance of Martian Civilization, but may also provide relatively cheaper source of fusion fuel for Earth, which, combined with relatively cheaper Deuterium from Mars due to its abundance there, may eventually benefit future generation of cleaner fusion energy on Earth itself. A Lunar mining lease agreement of 100 to 200 years can be done giving exclusive rights to Martian Civilization to extract Helium-3 on Moon, until an alternate fuel source or power generation method can be discovered that would reduce or eliminate the need to depend on Lunar mining for powering the Mareekh Civilization.

A mining outpost on Moon
ESA/Foster + Partners
www.esa.int

POWER ON MARS

A Radioisotope Thermoelectric Generator (RTG)

Image source:
http://voyager.jpl.nasa.gov/gallery/assembly.html

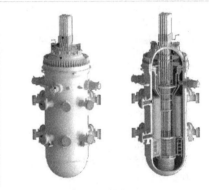

Small Modular Reactor <300 MWe

Image source: ansnuclearcafe.org

The initial power source for a Martian colony will be solar, with limited nuclear power backup in case of lack of sunlight amid global dust storm on Mars that can last for weeks or months at a time.

As the civilization progresses, Mars will become a predominantly nuclear-powered nation; relying on fission reaction using nuclear fuel imported from Earth, and later, possible generation of cleaner fusion power using local Martian Deuterium and Helium-3 mined from Moon, as discussed above.

Mareekh will be a power-hungry civilization, with very high per capita usage. The main expenditure of power will be in ore processing, maintaining of habitability of living spaces, fuel production for ascent vehicles, rovers, mining operations, construction, manufacturing of machinery, and terraforming. Personal power usage will be very similar to a modern industrial nation on Earth but combined power consumption per capita can be compared to that of Iceland[5], where over 80% of the power is consumed by industrial sector, especially mining and processing of Aluminum[6]. In a Martian context, this figure may be over 90%, with only 1 – 2 % used for domestic consumption inside the habitats and living spaces.

In fact, it will be a good comparison with Iceland to extrapolate the power consumption by a Mareekh Civilization. Per capita energy consumption of Iceland is roughly 5800 watts per person; over 80% of which is used by industry, and 15% (870 watts) is domestic consumption. Keeping this figure same, except that this would be only 2% of the per capita usage of power, the total per capita power requirement can be extrapolated to nearly 44,000 watts per person. Multiply it by a million and this gives the average power requirement by a million strong Mareekh Civililzation; a whopping 44 billion watts (44000 megawatts) at any given time.

The area required for solar panels for production of this much amount of electricity on a fine sunny day with minimal dust or haze in atmosphere will be equivalent to at least 325 square kilometres, provided 20% efficiency of solar cells. In order to provide relatively reliable power throughout the daylight hours, it may need to be three times larger, roughly 1000 – 1200 square kilometres. The main issue will be cost of construction of such a huge power array and the unreliability of it in Martian dust storm. Couple of methods can be utilized to ensure seamless supply of power from such solar arrays including storing energy by charging huge battery arrays similar to Tesla Battery in Jamestown Australia, and production of liquid hydrogen fuel to run generators during dust storms to ensure supply of electricity to meet the minimum needs. These panels will age with time and will incur huge cost and human hours for maintenance and replacement. Though promising, the solar power generation wouldn't be a long-term solution to meet energy needs of Mareekh Civilization.

This leaves us with nuclear power as the only practical long-term option for its relatively cheaper infrastructure and reliability of power output. Until a safe and practical way of producing fusion power is discovered, fission reactors will be mainstay of power generation.

Two kinds of nuclear-powered reactors can be considered:

1. Radio-isotope Thermoelectric Generators (RTG)
These relatively low cost and long-lasting power sources would be useful early in the establishment of a Martian settlements, but due to their low output and efficiency (in kilowatts only, and 3-7% efficiency), these must soon be replaced with larger and more efficient designs like boiling water reactors.

2. Small Modular Reactors (SMR)[7]
More efficient and powerful, these reactors would utilize water extracted from permafrost through a process similar to Hydraulic Fracturing or fracking, to

generate steam. In addition to power generation, this steam can also be released into the atmosphere contributing to an early process of terraforming. The steam can be also liquified relatively easily and utilized for domestic and farming usage as well. Each compact unit will be capable of generating up to 100 to 300 megawatts of power. These reactors will be dependent upon procurement of nuclear fuel from Earth, which may increase the cost of power production several folds as compared to that on Earth.

Fusion reactors are still a novel idea but if successful, these reactors will become the mainstay of power generation, fuelled by Helium-3 mined from Earth's Moon and Deuterium from Mars itself.

Fuel production

Liquid hydrogen and oxygen fuel can be synthesized relatively easily on Mars. With sufficient nuclear power and ability to extract water from permafrost, enormous quantities of liquid hydrogen and oxygen can be produced through electrolysis and used as fuel for ascent vehicles, rovers, mining machinery, transport, and for domestic use as breathable oxygen for living and work spaces, and other gear.

Roughly 40 megawatts of power is required to produce hydrogen and oxygen through electrolysis, equivalent to fill 100 space shuttle external tanks per year. Another 40 – 60 megawatts of power will be required to produce this fuel for domestic and industrial usage.

Other usage
3000 megawatts of power required to run 1000 mining trucks similar to Caterpillar 797F (3000kw), and another 3000 megawatts for underground drilling gear.

McMurdo station in Antarctica requires 72kW for heating for a population that drops to 200 individuals in winter. One can extrapolate this need to 300 to 500 megawatts for heating the internal living and work spaces on a Mareekh civilization. A breakdown of power usage in a heavily industrialized and mining oriented Mareekh Civilization might look something like this.

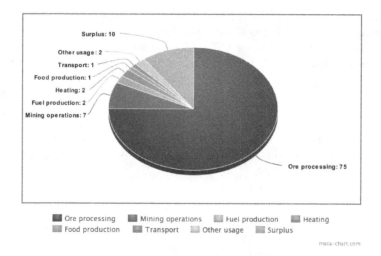

Surplus: 10
Other usage: 2
Transport: 1
Food production: 1
Heating: 2
Fuel production: 2
Mining operations: 7
Ore processing: 75

- ▣ Ore processing   ▣ Mining operations   ▣ Fuel production   ▣ Heating
- ▣ Food production   ▣ Transport   ▣ Other usage   ▣ Surplus

meta-chart.com

TRANSPORT

The main transport between colonies on Mars would be railroad, both overground and underground in form of tunnels. The rovers can be utilized for exploration missions away from the colonies. Due to extremely thin atmosphere, air travel will be almost out of question, except for short distance recreational uses. Even low Martian gravity may not compensate for the lack of lift from extremely thin atmosphere. Transport in form of hyperloop system between settlements may work even more efficiently on Mars than on Earth due to low gravity and thin atmosphere causing less drag and hence requiring less power.

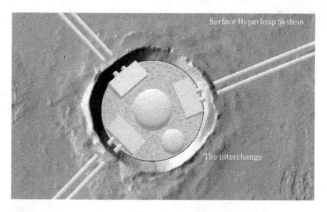

Surface Hyperloop System

The interchange

Transit

If humanity cannot come up with an alternate source of propulsion to replace the rocket technology by then, rockets will remain the mainstay of transport between Earth and Mars.

Less power will be required to lift off from Mars than it is from Earth, due to less Martian gravity and thinner atmosphere. Therefore, liquid hydrogen and oxygen fuel may be sufficient. Liquid hydrogen is difficult to store, but due to its constant usage and production, this may not be a major issue. A network of transit shuttles in Earth's and Martian orbits can serve as a useful means of frequent travel between the two worlds for cargo and humans.

AGRICULTURE AND FARMING

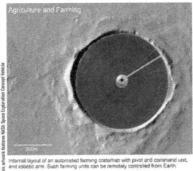

Internal layout of an automated farming craterhab with pivot and command unit, and robotic arm. Such farming units can be remotely controlled from Earth.

The types of plants need to be picked up from Earth initially will have proven ability to survive and grow in low gravity, and low-pressure environments.

Farming large cattle like cow and sheep may not be feasible on Mars, but growing chicken and fish in waterbodies in Craterhabs may provide a sustainable source of animal proteins. Mushroom farms and production of synthetic proteins using bacteria may also be seen as a viable option for alternate sources of food, should any one type of food production not being successful or face catastrophic failure.

Growing a combination of potatoes, sweet potatoes, and wheat will yield an average of 55,000 kcal/ha/day[8]. In order to meet the basic minimum for a population of one million with an average calorie requirement of say, 2000 kcal per day (men 2500, women 2000, and considering reduced calorie requirement due to low gravity, 1800 to 2000 kcal average per day may be a reasonable estimate), a total cultivated area of at least 360 square km will be required.

Craterhabs dedicated for purpose of farming can be much larger than the residential ones since plants can survive in relatively lower pressure environments. So larger craterhabs, say 500m across, may be utilized for purpose of farming with lesser requirement for pressure maintenance, hence less stress on the dome. Also, any catastrophic failure of the dome pressurized environment will be less hazardous to human life. These domes can also be utilized for experimenting ever-larger domes, perhaps several kilometers across, and to develop understanding of the dynamics and engineering requirements of constructing and maintaining such domes for future human habitation. To cover an area of 360 square km, around 1800 craterhabs of nearly 500m diameter will be required. Some of the farming may be in underground hydroponic facilities. Growing animals like chicken and fish, and novel food production methods including fungi and bacteria may also be utilized for obtaining proteins. A significant portion of grown food will be vacuum-sealed and frozen, and stored in food silo facilities as backup in any event of global catastrophe, to last for many months, even years. So, a production of surplus food will be a necessity. Certain types of food may be imported from Earth that require huge area for production but pack large amount of calories in smaller volumes, like sugar, powdered milk, dry meat and fish, and protein or nutritional supplements.

Farming will mainly be through automated methods, and the by-products of cultivation will be churned into manufacture of fertilizer in addition to fertilizers and soil supplements regularly imported from Earth. Since farming is a slow process, in addition to automation, farming equipment can also be remotely supervised and operated from Earth if a time delay of 7 - 20 minutes can be dealt with. This will spare a significant number of people on Mars to remain available for other tasks.

HEALTHCARE

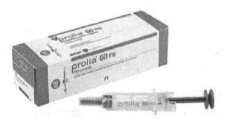

Medications such as Denosumab may be required to stregthen bones in low gravity Martian environment

Image source: https://melioguide.com/

Each Martian Colony will have primary to tertiary healthcare facilities addressing the healthcare needs of the residents from simple ailments to advanced conditions requiring complex surgical procedures. However, a number of ailments, like the need for organ transplants may require patients to be sent back to Earth.

Living in closed and isolated environments may lead to a number of mental health issues including home sickness, claustrophobia, anxiety and depression, among many others. Provision of mental health services will play a huge role in healthcare in a Martian Society.

Living on Mars may also lead to increased risk of illnesses from effects of low gravity resulting in thinning of bones, and from exposure to radiations especially in field workers working long hours outdoors in space-suits, in rovers, during surface transport, and in other less protected environments. Utilization of Artificial Intelligence and remotely controlled robots in high radiation environments may be a way around this problem.

SOCIETY

Structure

The Mareekh civilization will be radically different from early human civilization on Earth, with specific traditional gender roles obviated by the use of technology. Men and women will find use of technology allowing them to contribute to more productive roles in sustenance and progress of the society. Farming and domestic chores can be done through robotic technology, sparing humans for different, more productive roles to help consolidate the Martian civilization. The gender difference in roles may only be limited to the needs for procreation, births and raising of young ones, but that would also be heavily facilitated by the use of technology. Most of the men and women will be employed in mining, exploration and construction. A minority will contribute to "in-house" duties including running of hospitals, educational institutions, cafes and trades, maintenance of internal environment, IT, farming and food production. One major difference from Earth will be lack of defence forces and only minimal number of people employed in law enforcement. This will have an added advantage of saving economy from draining into defence and manufacture of military hardware, and will have a very positive impact on the economy and social mindset.

We can imagine a one million strong Mareekh civilization spread across 50 settlements throughout the planet, each with a population of 15000 to 25000 individuals on average. Analogies can be derived from spread of small regional towns in Australia with similar populations and distances. In an average settlement of say, 20,000 individuals, a population breakdown may be similar to that on Earth once the civilization has been completely established and achieved self-sustainability. The gender distribution will be close to 50 percent males and 50 percent females. Roughly 20 percent of the population of that particular settlement (4000) will be children and adolescents and 10 percent people (2000) may be senior citizens; most of whom will be early settlers with mountains of knowledge and experience of starting a unique and independent civilization. Through their ideas, they will contribute to the growth of Mareekh civilization with their knowledge and experience. 70 percent people (14,000) will be the main powerhouse of economy;

nearly 10,000 of them will be working in mining and construction, and that is a huge number.

15000 to 25000 population of individual Mareekh settlements will be an ample population achieving sustainability of an individual city. Movement of people between settlements will play a vital role in further strengthening of bond, and transfer of technology and assistance when required.

Governance
A Mareekh society will have an average higher level of education and community sense, and it will be a perfect model for democracy with room for experimenting on inclusion of novel ways of governance suited for the Martian way of life based on isolation, dependence on each other, and best use of limited resources.

Language
As the first million strong settlement on Mars will have members from all over the world, a good communication would be paramount to success of a new civilization on a new planet, and would require a unified language. With all its weaknesses and terrible spellings, English still has proven itself to be a powerful tool of communication that has brought the world together on Earth, and may do the same on Mars, with modifications to suit the new world. It would be fun to imagine a future Martian accent.

Education
Education will not be limited to theoretical studies and a significant amount of learning will be performed through activities of sustenance including farming, and internal maintenance of the habitats. Youth can be utilized for this purpose, sparing the adults for riskier jobs like exploration, mining and construction. Schools may be local and community based, but universities may be virtual, combining the students from different cities and settlements on one huge virtual platform. Education will also be uplinked to Earth. Educational tours between Earth and Mars may be commonplace.

Religion
People may choose to take up any personal religion based on their traditional Earth-based background. Many people may follow religion to find internal peace and maintenance of mental health and wellbeing in a new and treacherous environment. In fact, it may be promoted in order to allay anxieties and fears in an unexplored world. On the other hand, algorithms of survival on Mars and understanding of science and technology may be the new Code of life that everyone will have to submit to and follow in order to ensure safety of oneself and others on the planet, and will be subject to continuous improvements based on learning through experience. Calculation of odds may take the place of beliefs and prayers, but there will always be room for faith.

Demography

Analogies can be drawn with smaller, newly established industrialized nations on Earth to understand the dynamics of population of Mareekh civilization. Countries like UAE, Qatar and Bahrain have predominantly expat populations of workers also contributing to the development of their rapidly progressing infrastructure. While in a Martian context, analogies can be drawn in certain areas of demography with these countries due to predominantly mining and industrial infrastructure of the Mareekh civilization with limited farming and agriculture on a desert planet where most of the activities are carried out indoors, the expat population may be a minority only (10-15% of the population, in contrast to over 70% in the above mentioned countries), limited due to the cost and time involved in travel between Earth and Mars. Those expat workers from Earth who choose not to make Mars their permanent home, may have limited role in legislation as any major challenges to the economy and sustainability of a Mareekh civilization will have lesser impact on them. These expat workers will mostly be young adults from Earth with families back home. Many would choose to stay in smaller, underground accommodations to save on cost of living and would reduce the requirement of larger residential spaces which will be utilized by permanent settlers. Majority of the residents in Mareekh society will be permanent residents of Mars with families and children born on Mars; the true Martian race.

CONCLUSION

A human settlement on Mars is an exciting subject and holds many promises for the success of humans as a space-faring specie. Mars is our closest destination due to its relative proximity and certain similarities with Earth. Mars was once a dynamic, wet world with every potential of harbouring life. It is seemingly dead now, but some of us may find it a moral duty of humanity to sow the seeds of life on it, initially in the form of creating a human civilization there, and in the long run, terraforming the entire planet to make it a living world once again where rivers flow and life gets to write new chapters in evolution.

Acknowledgement

I am especially thankful to my wife Asma for her immense support in completion of this project.

REFERENCES

1. Silicones: the material of choice for airplane and outer-space uses. Space Tech Expol USA
    Link:www.spacetechexpo.com/resources/news-and-editorial/news-container/2019/05/13/silicones-the-material-of-choice-for-airplane-and-outer-space-uses/

2. Silicone 40 Transparent:
    Link:www.steinbach-ag.de/en/forming-solutions/products/product-group/silicone-40-transparent.html?sword_list%5B0%5D=%2A%2Aprovide%2A%2A
3. Dyneema® weight and strength: zpacks.com/pages/materials
4. World Happiness index
    https://upload.wikimedia.org/wikipedia/commons/4/4f/World_map_of_countries_by_World_Happiness_Report_score_%282017%29.svg
5. Energy Statistics in Iceland 2018
    Link: https://orkustofnun.is/gogn/os-onnur-rit/Orkutolur-2018-enska.pdf
6. Fossil Fuel Support Country Note (Iceland), OECD, April 2019
7. Small Modular Reactors: https://www.energy.gov/ne/nuclear-reactor-technologies/small-modular-nuclear-reactors
8. Food yield per hectare:
    Link: Fossil Fuel Support Country Note, OECD, April 2019
9. Modified images from Google Earth resources have been used in artwork in certain places in this report.

4: MARS CITY STATE

Stephan Zlatarev
stephan@SpaceColony.One
stephan.zlatarev@gmail.com

SUMMARY

The following is a design of an envisioned city on Mars capable of reaching population of 1 million people by mid-22nd century. Its design focuses on characteristics that will make it practical, vital and sustainable. As a pioneering large settlement, the design of this city will greatly differ from possible designs of future cities created after mankind has well established its presence in space. It will develop through three major phases defined by the achievement of four milestones:

- *City founded milestone* - City is founded as a small colony
- *Science and tourism phase* - Period of developing science and tourism projects
- *First space elevator milestone* - The first space elevator is built. It will boost the transport channels to the surface of the planet
- *Real estate phase* - Period of city expansion and the development of attractive marketplace for real estate on Mars
- *First industrial factory milestone* - The first industrial zone in orbit is built. It will boost the industrial potential of the city
- *Industrial phase* - Period of developing manufacturing power. It will make the city self-sustaining and autonomous

- *Population of 1 million milestone*

The design of the city is based on four key characteristics:

- *Economic attractiveness* - while Earth in 21st century is seeing the limit to its population and fierce competition over natural resources, the city on Mars will offer businesses healthy economic growth through access to fast-growing population and untapped resources
- *Development roadmap* - the identity of the city will evolve around a sequence of major milestones in its development - like building the first space elevator and the first industrial factory in space. This identity will attract the right residents to join the city and be part of its development
- *Service platform* - the city will provide a platform for services for the needs of its residents and businesses. It will make it easy for businesses to quickly start and scale their operations. It will also enforce regulation over all services to ensure they are safe to be consumed by the residents
- *Modular infrastructure* - the infrastructure provides basis for the logistics in the city through an assembly of standardized reusable modules. These modules are easy to replace when damaged and easy to relocate when the city needs to adapt and evolve

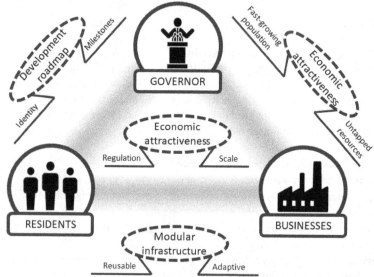

We will describe these through the interaction of three participating groups:

- *Governor* - the organization that initially founds the city and then runs it

- ***Residents*** - the people who live and work in the city
- ***Businesses*** - the service companies that supply the residents with everything needed to sustain and improve the quality of their lives

ECONOMIC ATTRACTIVENESS

The first city on Mars and its design will be shaped by the forces that lead humans settle on the planet. This design assumes the major force to be economics and we will start its description by looking at the projections of the relevant drivers of world economy in the near future.

Natural resources and population growth on Earth

The world has been experiencing exponential economic growth in the last decades with advances in technology and high standard of living. Despite the many threats to the economy (like financial crises, climate changes, depleting natural resources, social inequalities, etc.) businesses and consumers are used to this growth and are expecting it to continue. The economic growth, as measured by world GDP per capita, is projected to continue to grow but many analyses on the future of economy describe the increased effort required to maintain it, partially due to the effects of over-population. The increasing population on Earth and its increasingly demanding way of living will burden the access to the limited sources of energy, water and food.

Cost of living will increase in the following decades and will contribute to the slowdown of world population's growth rate. Projections show that by year 2100 the growth rate of human population will get back to its levels around year 1900.

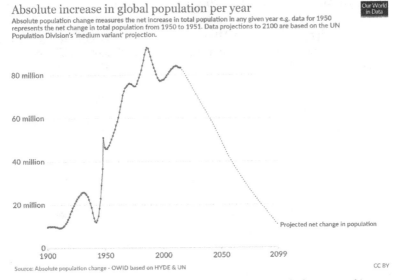

Absolute increase in global population per year

Absolute population change measures the net increase in total population in any given year e.g. data for 1950 represents the net change in total population from 1950 to 1951. Data projections to 2100 are based on the UN Population Division's 'medium variant' projection.

Source: Absolute population change - OWID based on HYDE & UN

(Source: https://ourworldindata.org/world-population-growth)

Since population growth is a key driver of economic growth, the limited growth of population will also limit the economic growth on Earth. The world economy will bear the characteristics of a saturated market with high competition for the same base of consumers resulting into limited profitability and limited growth potential for the businesses.

Natural resources and population growth on Mars
While the economy on Earth will face over-population and saturated markets, the economy on Mars will have different challenges and opportunities. Mars is neither populated, nor developed. However, it has huge potential in land to populate, natural resources to extract and better access to Space and its resources. The major barriers to untapping this potential for population and natural resources can be removed with advances in technology. History shows high tech advances follow Moore's law which reflects the observation of exponential advancements of certain technologies. Therefore, it is reasonable to assume that our abilities to expand our economy on Mars will grow exponentially, doubling every few years or decades.

Founding a city on Mars with even mediocre standard of living will unlock population growth rate like that of average Earth cities. Our proposed design will aim for relatively good standard of living with growth rate following the envisioned four milestones of development of the city:

- Annual growth rate of 20% from 2030 to 2050 with science and tourism as main source of income. The population will grow from 10 to 1000 residents. This period will bear the highest population growth rate. We will later describe the drive behind this growth that makes it realistic
- Annual growth rate of 10% from 2050 to 2100 with real estate and services as main source of income. The population will grow from 1K to 100K residents
- Annual growth rate of 5% from 2100 to 2150 with services still being the major source of income but now supplemented with manufacturing. The population will grow from 100K to 1M residents

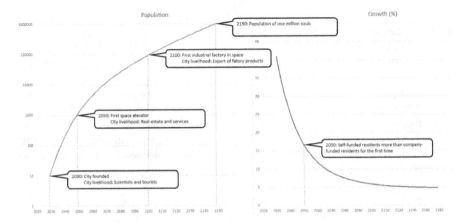

As a result, the first city, or network of cities, will reach a population of 1 million by year 2150. This population will participate in the economy and will, by itself, attract businesses.

Natural resources on the surface of Mars and in space near the planet will also become easier to harvest. Although in the beginning heavy machinery will be expensive to transport to the planet, advances in technology - lighter materials and cheaper transportation - will gradually lower the costs for opening and operating extraction sites for natural resources. In contrast, on Earth the cost of running similar sites will continue to grow due to competition over resources and environmental concerns such as pollution hazards.

In addition, the gravity of Mars is only about 38% of that of Earth and it is situated 50% closer to the edge of the asteroid belt which makes it an excellent candidate for an asteroid mining hub with competitive advantage over Earth.

We can conclude that, during the following century, there will be a great contrast between Earth and Mars considering population growth and access to natural resources. Population growth on Earth will slow down, while that on Mars will grow exponentially. Access to natural resources on Earth will slow down, while it will grow exponentially on Mars. This conclusion suggests that at a given point in time Mars will become a very interesting investment opportunity and that early investors will gain advantage over laggers.

Economic development of a city on Mars
The design of the city proposed here will follow an interesting path of economic development. According to the three-sector model of economics the main focus of societies will shift from the primary sector (extraction of raw materials) through the secondary sector (manufacturing) to the tertiary sector (services). The model classifies intellectual activities and knowledge-based activities in the quaternary sector as an extension to the model.

The city on Mars will, however, follow a reversed path by starting almost entirely in the quaternary and tertiary sectors. Within the first few decades after its foundation its main livelihood and source of income will be from exporting science and tourism. Only after the first milestone - introducing the first space elevator - will it become sufficiently affordable for various businesses and individuals to self-fund their journey to Mars so that the critical mass of residents and services can reach the point for the tertiary sector to dominate the economic activities. Later, after the second milestone of building the first industrial factory in space, the secondary sector will gain momentum as the city will become capable of producing a rich variety of goods. Finally, the city will build the needed industrial power to develop the extraction sites for it to become net exporter of raw materials.

The share of economic sectors will evolve in the following way:

| 2030 | 2050 | 2100 | 2150 |
| City founded | First space elevator | First industrial factory in space | Population 1M |

The design of the city on Mars deliberately supports this path of economic evolution and the city founder will have the responsibility to maintain it because it will ensure that GDP per capita for the city will start relatively high and will remain high, which will be an important factor for attracting businesses and residents.

Return on investment for the city founder
The city will be driven by three groups - the governor, residents and businesses. The city founder will possibly also be the first governor to run the city. The governor organization will have a few responsibilities that we will describe further. A major responsibility will be to maintain the economic viability of the city. This will include building the city by the design and managing the different phases, most importantly the first phase "Science and tourism" which will require a lot of financial investment and entrepreneurial skills until the city gains momentum to transition into the next phase.

The city founder will need to either provide or contract:

- Pipeline of science projects - for revenue
- Pipeline of tourists - for revenue
- The minimum set of services for life support and basic needs - shelter, air, water, food and waste, energy and heating, communication - through the service platform we will describe later - for supporting the science and tourism activities

The following calculation presents the scenario where the city founder provides and funds all minimum services. The calculation spreads all investments starting from a 10-year period before the city is founded. Capital and operating expenses are split in categories research and development (R&D), production, transportation, maintenance and operations with the following assumptions:

- Research and development activities start in 2020 (which is 10 years before the planned date of city founding)
- The population starts with 10 residents in 2030 and grows to 1000 residents in 2050
- The city founder provides the six life support services - the set of minimum services in the service platform
- Yearly spending of 200 million USD per service per year for research and development
- Service equipment is produced on Earth with progressively increased capacity amounting to 120% of actual population in the city
- Cost of service equipment starts at 50 million USD per resident per service and is reduced by 10% per year due to technological advances from the output of research and development
- Weight of service equipment is 10 000 kg per resident per service and is transported to Mars at 500 USD per kg
- Yearly maintenance expenses of service equipment are 20% of its production cost
- Yearly operating expenses and supplies for the service equipment is 100 000 USD per resident per service
- Scientists in city are 20% of population
- 3 active projects per scientist per year
- Yearly revenue of 25 million USD per active scientific project
- Tourists in city are 20% of population
- Revenue from one tourist year starts at 25 million USD in 2030 and falls by 20% year-by-year

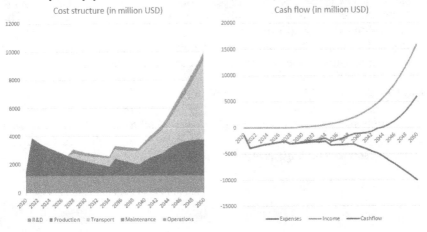

This high-level calculation shows that the city founder will reach positive cash flow around year 2045 and will break even 10 years later. Net expenses before the point of positive cash flow top at 4 billion USD per year which could be a manageable investment for a joint venture by a group of dozens interested large enterprises that wish to capture the first mover advantage in a new Mars economy.

DEVELOPMENT ROADMAP

The design of the city will follow a roadmap of four major milestones that will define three distinctive periods of economic development:

- City founded milestone - The city will be founded as a small colony with 10 residents
- Science and tourism phase - The period following the foundation of the city. During this period the city will develop unique opportunities for scientists and tourists to support its growth. The population will grow to 1 thousand residents.
- First space elevator milestone - The city will build its first space elevator as a highway connecting the surface of the planet and planet orbit
- Real estate phase - The period following the opening of the space elevator. During this period the city will develop real estate and services market attracting people to its unique lifestyle. The population will grow to 100 thousand residents.
- First industrial factory milestone - The city will build an industrial zone in orbit using its network of space elevators to access space and planet surface
- Industrial phase - The period following the opening of the industrial factory in space. In this period the city will specialize in manufacturing exploiting the characteristics of space and will support its growth through exports. The population will grow to 1 million.
- Population of 1 million milestone - The city will become a true city of our civilization with unique cultural, social, economic and political footprint

2030: City founded. Science and tourism phase
After 10 years of preparation, the city will be founded in around 2030 with the infrastructure and staff to support science projects and tourism. It will start as a small colony with 10 residents and will gradually grow adding infrastructure, habitats and service buildings as more residents move in.

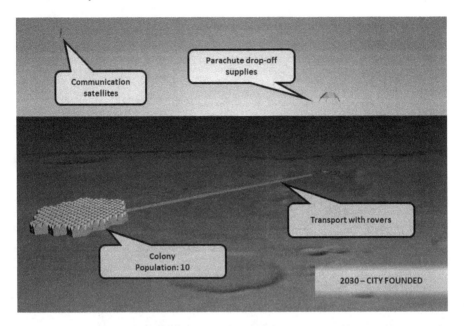

The governor will play a critical role in this period. The governor organization will be responsible for balancing revenue generation, rate of immigration and city construction activities.

The governor organization will sell seats for scientist and tourists in advance allowing enough time for building the city to accommodate them. Scientists will be limited to 20% of the total population. Tourists will be limited to another 20% of the total population. The remaining 60% of the population will be providing infrastructure maintenance and services to the scientists and tourists. During this period, the governor organization will carefully select these people such that they are likely to settle permanently. It will look for young people, couples, with complementing skills.

The governor organization will also lay the foundations of the service platform that will ensure residents can meet their needs and businesses can easily start and grow. However, the city will be still too small as a market to attract businesses. The governor organization will develop it by contracting the businesses needed for the basic needs of the residents - for shelter, air, food, water, heating, electricity, communication.

The income of the city will rely on the science projects and tourists. The city will re-invest part of the income in scaling the infrastructure and staff but will still need additional capital investments to reach population of 1 thousand. A portion of the investment will be used to build the first space elevator.

2050: First Space elevator. Real-estate phase

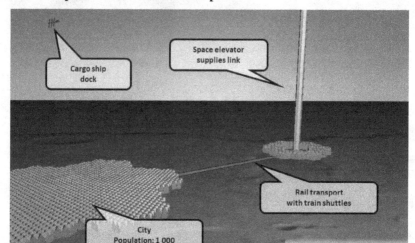

The city will build its first space elevator in around 2050. A space elevator is a type of planet-to-space transportation system that permits vehicles to travel along a cable from the surface directly into space or orbit without the use of large rockets, where the cable is held up from above by centrifugal forces. We do not yet have material for such a long cable that can withstand the gravitational forces of the Earth, but a space elevator is feasible on Mars because we have such material for its weaker gravitational forces.

The elevator will be a "wonder of the world" with great economic and social impact. It will mark a turning point for the city by offering greater mobility. With the growing city and the technology of space elevator there will be more opportunities for people to make living on Mars. A certain profile of people on Earth will choose Mars as their new home. These may include entrepreneurs who will prefer the opportunities of Mars over the saturated markets of Earth, or family people who will prefer to raise their children on Mars as a then more viable alternative than on Earth. Even if a tiny minority on our home planet, they will be sufficient in numbers to drive growth of a large city on Mars.

As the costs of migrating to Mars lowers with technology advances, one such advancement being the space elevator, there will be a point in time when the number of people who are able and willing to self-fund their migration to Mars, possibly using bank loans, will surpass the residents that are funded by the governor and businesses operating on Mars.

The growing population will support the market and the opportunities for new businesses. The governor organization will use this period to open the service platform to many businesses. It will become easier for new businesses to start operations and to introduce new products and services. The residents will enjoy higher variety and quality with lower prices as time passes.

As in the previous period, the city will use a portion of its finances for the next milestone - building the first industrial factory in space.

2100: First industrial factory in space. Industrial phase

Building on the first space elevator, a network of space elevators will speed up and improve the logistics between the surface of Mars, its orbits, the asteroid belt and Earth. The city will have a clear advantage over Earth in building a large industrial complex in space. The city will build the first factory in around 2100 and gradually form an entire industrial zone. This complex will have lower cost for transportation and construction as compared to the competition on Earth. It will employ high levels of automation with operations in space premises performed by robots and remotely controlled from Mars surface.

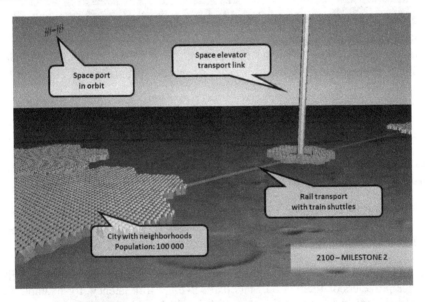

The factory will have strong competitive advantages over the industrial sites on the Earth in producing certain products, supporting special technologies and developing certain economies. For example, it will have easier access to asteroid-mined resources, will be able to use technologies relying on zero-gravity or vacuum, will be less exposed to risks of pollution, etc.

The industrial factory in space will define the identity of the city for many decades. It will direct its economy by developing its specialization in exported goods and services. Equally importantly, it will support the growth of the population of the city to a size that becomes recognizable and significant in our civilization.

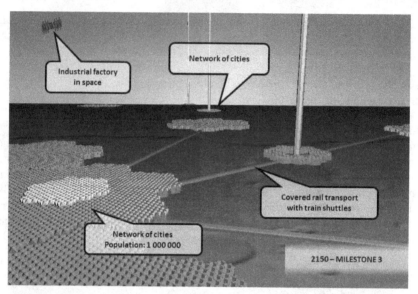

2150: Population of 1 million

After building its first industrial factory in space in around 2100, the city will maintain average yearly growth of 4-5% to reach a population of 1 million by around 2150.

The city will become fully developed and self-sustaining, but still a partner to Earth. The service platform will have a wealth of active businesses and new Mars-native businesses will appear. The city will continue to be an attractive location for businesses but will no longer be dependent on entrepreneurs from Earth as local businesses will now lead its growth.

By covering the lower level needs of the residents, the service platform will free room for higher aspects of life to develop in a natural way. The people will use their free time for socializing, for parenting, for hobbies, for culture, for business, etc.

All aspects of life will still be well connected to those on Earth. For example, with service providers of Internet on Mars that solve network latency, for example by mirroring content locally, residents on Mars will be able to enjoy the same TV shows, sports events, etc. that their friends from Earth enjoy.

Some aspects of life may, however, have diverging development on Mars. For example, with the high degree of automation and artificial intelligence of the service platform and the infrastructure, there may be no need for taxes for business activity (like VAT on Earth) and the economy on Mars may become far more fluid than that on Earth.

The city on Mars will become a unique extension to human civilization. It will participate in the "global" (now interplanetary) system but will build its own lifestyle and add to mankind's diversity.

SERVICE PLATFORM

The city growth and the development path we described are driven by the interaction of three groups - the governor, the businesses and the residents.

The governor organization provides businesses with a platform for services and the infrastructure to access consumers and suppliers. In return the businesses pay for using the platform and infrastructure. The governor will maintain the infrastructure to ensure the businesses and people can reliably exchange goods and services.

The governor organization regulates the services and provides city residents with secure services as well as housing. In return the residents pay for the housing they rent. The rent covers access to the services, other amenities and the capacity of the governor organization to ensure (through regulation) that the services are available and do not put lives at risk.

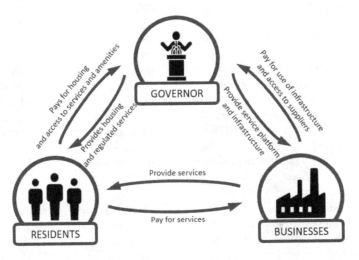

All needs of the residents are met by the services the businesses provide, for which the residents pay the businesses.

Minimum set of services

The primary purpose of the service platform is to meet the biological and physiological needs of the residents. The governor ensures the availability of a minimum set of services on the platform that provide enough supply for all residents. The governor regulates these services to ensure they are accessible, safe to consume and with guaranteed availability.

In the beginning, the governor will contract the service providers for the minimum set of services. Later, as the city grows and becomes more attractive to new businesses, the governor will ensure there is sufficient supply for the minimum set of services only by running the service platform and will only rarely directly contract service providers.

The design of the city positions the service platform to be the medium for the interactions among the governor, residents and businesses to achieve the milestones and grow the city. The design is not bound to any specific implementation of the different services. Instead it is open to accommodating the innovation businesses can bring to solve the challenges of life on Mars.

We will only look at examples for possible implementations of the different services in the minimum set:

- **Shelter** - Basic protection from the harsh environment on Mars, from radiation and bad weather, will be provided by the infrastructure, part of the service platform. A cover made of rammed soil at least one meter thick will form a cave that will provide sufficient space for living and protection from the natural environment on Mars. The cover will block the radiation from the Sun and Space. The cave will be intentionally larger than the room needed for the residents and will maintain better environment in terms of temperature and air pressure although not enough for living without additional protection. The infrastructure of the service platform will include inflatable habitat as a second level of protection. The habitat will maintain fresh air and constant room temperature. For the cases of emergencies, the infrastructure will also offer light protective suits and emergency spaces to hide. These three levels of protection will be constantly monitored and maintained in an automated way. The shelter will have an in-built communication system that will ensure connection to the control center and the rest of the world. We will look in more detail into this infrastructure in the next section.

- **Air** - Breathable air will be constantly recycled from imported and locally produced gases. Service providers will use a variety of technologies to produce and recycle air and will be able to choose the appropriate mix to optimize financial returns. For example, they can import oxygen from Earth, extract it from the water reserves of the city or recycle the used air. Regulations by the governor will include provisions for the quality of air, production capacities, redundancies and emergency procedures. Service providers will have the choice of specializing in supplying air or combining air supply with other services, such as air conditioning and heating.

- **Water** - Drinkable water will similarly be constantly recycled from imported and locally produced water reserves. Again, service providers will use a variety of technologies to produce and recycle water, optimizing their financial returns within the constraints of regulation and customer satisfaction. Since water is very heavy and costly to import from planet Earth, service providers will possibly use in-situ resources while the demand is small and may later supplement with asteroid mining if local extraction cannot keep up with the growth rate of the city. Service providers will have the choice of specializing in supplying water or combining water supply with other services such as those related to air and energy.

- *Food* - Residents require varied diet that may require a multitude of service providers. Possibly some service providers will specialize in imported dehydrated food while others will specialize in local fresh produce for example in automated vertical farms. Regulations by the governor will include provisions for the quality of food, capacity for production, import and reserve storage.

- *Electricity and heating* - Electricity and heating will be possible to produce from different energy sources - nuclear, solar, wind - and the city will have to maintain a resilient mix of those. Regulations by the governor will include provisions for the capacity and redundancy of supply.

- *Waste management* - Service providers that deal with waste management will possibly work with suppliers of food, water, air to optimize the recycling capabilities of the city.

Amenities

The secondary purpose of the service platform is to meet the higher-level needs of the residents. The governor will not necessarily enforce strict regulations on these since people's lives do not directly depend on them, as the case with the minimum set of services is. Nevertheless, the governor may contract service providers for under-supplied services to maintain better lifestyle of the residents and overall attractiveness of the city.

Some examples of services that will use the same service platform are:

- *Security* - as the city grows there will be growing need for keeping the peace and resolving conflicts. Because the city will be equipped with highly monitored infrastructure that is since day one capable of and responsible for tracking people with the purpose of protecting them in case of emergencies, police and bodyguard services will be collaborating with the governor organization and the infrastructure.

- *Health* - similarly, the health of the residents will be a concern of the governor and there will be basic but well-integrated health monitoring capabilities in the infrastructure of the city. Additional services will be further supplied by independent service providers possibly with close collaboration with the infrastructure of the city. They can start with individual doctors and as the city grows build large hospitals.

- *Education* - the needs for education will rely on connection to Earth at least in the very beginning. Virtual classrooms, individual practices, coaches - these can all be done online even considering the latency between Mars and Earth. As the needs grow and local education facilities become viable, they will appear in the service platform. The city will be founded on exporting science related to Space research and it will be just a matter of time until education in more spheres of knowledge will be exported too.

- *Entertainment* - the need for entertainment and socializing will attract a variety of venues - cinemas, theaters, restaurants, cafes, pubs.

- And many more like *sports*, *leisure*, *travel*, *shopping*, etc.

All the services will be supported by the service platform and will in return support the city with amenities. The mix of services drawn by the evolving needs of the residents will shape the lifestyle of living in the Mars city and its unique culture.

Service platform features
The service platform will be instrumental to the development of the city and its design will need to stay focused on a few simple concepts that will allow it to evolve together with the city and at the same time stay reliable. These three concepts and their high-level design are:

- *Modular infrastructure* - The purpose of the infrastructure will be to connect service providers to service consumers. The entire city infrastructure will be composed of standardized modular building blocks that can transport goods from service providers to service consumers. They will also be easy to add, maintain and recycle. These modules will be connected and have common monitoring and control aided by artificial intelligence and robots. In the simplest implementations the individual modules will be controlled by a central control center but as the technology advances they will be autonomously performing basic procedures, too. We will look into the design of the infrastructure in the next section

- *Regulation for services* - Some services will be essential to life and other services will be highly valued by residents or businesses. For those critical services the governor will impose regulation so that the entire community can rely on their availability and quality. The regulation will be in the form of rules, service providers must comply to, and will be coded into the infrastructure, meaning that the monitoring and control capabilities of the infrastructure will extend to tracking the availability and quality of the critical services. The governor will have the power to enforce the regulation by blocking non-compliant service providers. The governor will also be monitoring the risks of future non-compliance and mitigate the risks by, for example, contracting additional service providers.

- ***Marketplace for services*** - The platform will be open to businesses to provide various services to the residents and to other businesses. The marketplace will be an open IT system serving as an interface between service providers and residents. It will provide a network of consumers and suppliers building on the automation, monitoring and control of the modular infrastructure. It will offer the possibility to make direct purchase transactions between consumers and producers but also build and run entire supply chains. It will also be open for extensions so that interested businesses can specialize in extending it with value-add services over the interactions on the marketplace.

The service platform will start small with the foundation of the city and will grow with it. It will also influence the three milestones of the development roadmap.

When the city is founded and enters its *Science and tourism phase*, the governor will be hiring the companies and staff who will accommodate the scientists and tourists. It will also hire service companies to provide the minimal set of services to support life in the city. As the economy in this phase is dependent on the capability of the governor to attract science and tourist projects, the number of operating services will be small, they will provide low variety of services and most of these services, if not all, will be considered essential so the governor will be contracting and regulating them.

Already at this point the governor will provide the common *modular infrastructure* to support delivery of the services. The governor will also *regulate the services* to ensure residents receive adequate life support and that relevant risks concerning service outages and people health are mitigated. The service platform will have developed two of its three key characteristics but will still be small and funded by the governor.

After the city passes the *First space elevator milestone* and enters its *Real-estate phase* the service platform will need to expand to support more than a thousand residents. It will add its third key characteristic - the *marketplace for services*. The platform will continue providing the common infrastructure and regulation over services and will now also be driven by competition based on the free market forces of supply and demand. At this point it will be common for more than one service company to provide the same or similar services, since they are now attracted by the market and not hired by the governor.

After the city passes the *First industrial factory milestone* and enters its *Industrial phase* the service platform will need to be prepared for hundreds of thousands of residents. The expanded economy will allow for high variety of services. The service platform will additionally boost variety and innovation by opening the marketplace and the underlying infrastructure to extensions by businesses. As a result, the overall level of uniformity of the infrastructure may get slightly reduced because some residents will be using extensions that others are not using, but as long as extensions do not compromise the compatibility of the basic functions of the infrastructure modules the businesses will be able to add a lot to the evolution of the infrastructure and increase the speed of development of the city.

MODULAR INFRASTRUCTURE

Finally, we will describe the modular infrastructure - the physical support of the service platform. It consists of rammed soil protective cover, connected modules for flooring and logistics, robotic drones for maintenance and control center for monitoring and control.

Cover
The entire city is covered with a structure made of bricks of rammed soil. The cover provides space for the residents to live in and a first-level protection from radiation and bad weather. There are also the second-level protection by the module inflatables and the third level protection by the emergency bunker we will describe later.

Experiments with Mars soil simulant suggest that the abundant soil of Mars may be used to create strong bricks with little additives if any at all. The weaker gravity of Mars will allow the structures made of rammed soil to be erected taller than equivalent structures made of rammed soil on Earth. This material will not require expenses for transportation from Earth allowing for large structures to be built at lower cost.

Our design will use honeycomb tiling structure with hexagons with diagonal of 4 meters, where every hexagon borders with exactly 3 columns. The columns will be 1 meter in diameter and 6 meters tall. The columns will support 60-degree roofs that top 12 meters above ground.

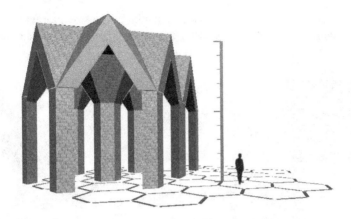

The cover will provide isolation in terms of air pressure and temperature. It will be possible to maintain constant air pressure in the range between 20% and 50% of that on Earth and temperature ranges above zero degree Celsius. Although this will not be enough for humans without extra protection it will allow for the modules inside the cover to be more light-weight and more flexible. It will also allow residents some freedom for short activities outside the modules.

The city will gradually build its cover to meet its population growth. The first phase of the development of the city will require cover for 1000 residents, which will be made of 3 neighborhoods, each being 180 meters in diameter and consisting of 6000 cells (for private, social and industrial use) arranged in honeycomb tiling in 45 rounds around a center cell. The city will use many such neighborhoods in clusters around points of interest on the surface of Mars. A total of 3000 such neighborhoods will be required for the city to reach population of 1 million. One neighborhood will look like this from the outside:

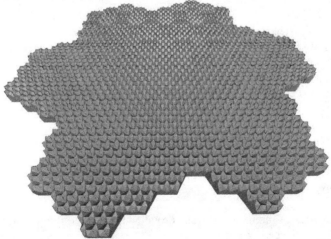

The habitat should be good from psychological point of view, too. Therefore, it should provide sufficient space and light. Artificial light sources will be attached to the bottom side of the cover to substitute sunlight. The structure will allow for a range of sight of a few hundred meters. Colored sculptures will imitate trees in the distance. One neighborhood will look like this from the inside of the deployed habitat modules we will describe later:

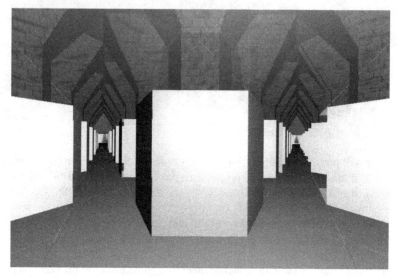

Drones

The cover is constructed and maintained by robotic drones. The drones also lay the modules inside the cover. These drones will be designed to be simple and cheap. They will be controlled centrally by the control center and will be most effective when swarming in large numbers over a task.

A single drone is configured to perform one or more of the following tasks:

- Collect soil - The drone will be able to dig into the Mars surface, extract small chunks of soil and store them in an in-built container. The drone will operate like an ant, rather than like a bulldozer, meaning that it will be equipped with small rippers - claw-like devices - and will be able to dig soil in small chunks to collect a desired amount of soil rather than collect a large quantity of soil at once

- Ram soil into a brick - The drone will be then able to press the collected soil in its in-built container into a brick. It will be equipped with the mechanism needed to mix the soil with necessary additives and exercise the necessary pressure.

- Move a brick of rammed soil - The drone will be able to move a brick - either a brick it produced or a brick it collected - to the destination pointed by the control center. It will be equipped with communication device to connect to the control center. It will also be equipped with six or eight legs to be able to move itself (together with the brick it carries) on the surface of Mars as well as climb on lain bricks. When it needs to climb on top of a tall structure, for example to lay a new brick or to collect a previously lain brick, it will swarm with other drones to first build a ramp of bricks to make path to the target position.

- Move a module (together with other drones) - The drone will be able to also move larger objects - obstacles, debris, other drones, and very importantly the modules we will describe later. Because those objects may be larger than a single drone, the control center will co-ordinate a few drones to work together and move the object as a group.

In addition, the drone will be able to perform supporting tasks, such as going to a re-charging or a cleaning station.

The following sequence shows a simulation of drones swarming to build the cover structure:

We see in the sequence how a few drones will prepare bricks, others will simultaneously build a ramp, while others will lay out the column. The drones will keep producing more bricks and constructing the ramp and the column similarly to how a 3D printer creates a print layer by layer. Once the cover is complete then a few drones will reclaim the bricks from the ramp for reuse in a next section of the cover.

The design of the drones implies that by working in groups and consuming more time to build and maintain things they will compensate the fact that they are not individually strong. If we consider the cover structure as described and scaled to support 1000 residents, which will be enough for the entire first phase, it will require 90 million bricks. We will assume 1000 drones, one per resident, and average time to build a single brick of 10 minutes and average time to move and lay a single brick of 10 minutes. The time required to build the cover structure will be 3.5 years. Since the first phase is projected to 20 years, until 2050, it will be possible to build the cover with these 1000 drones, leaving sufficient time for dealing with malfunctions, time for recharging and other tasks.

Modules

Within the cave of the cover, the city will be built on top of a layer of connected modules. The modules will be lain on the ground arranged in a layout that completely covers the surface around the columns of the cover. These modules will be the flooring of all areas in the city - private, social and industrial. The modules will be standardized and therefore they will be possible to be reused, repurposed or replaced as needed.

The modules and their accompanying maintenance drones will be produced on Earth and need to be transported to Mars. We assumed a weight of 1500 kg for a set of one module and one drone in the calculations presented in the previous sections.

The basic module is a hollow hexagonal prism of 2 meters in length for a side and 1 meter in height when deflated and 4 meters in height when inflated.

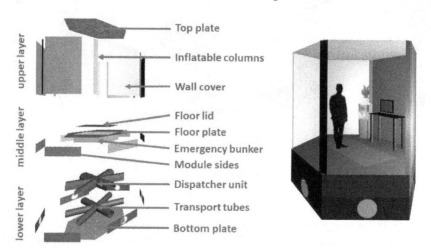

The module has three layers:

- The lower layer is for the logistics in the city - for transportation. It contains 3 tubes connecting its opposite sides with an electro-mechanical dispatcher unit that controls the movement of capsules in the tubes. The dispatcher unit will maintain the speed of the capsules in transit through the module, will pull the capsules that are sent to this module out of the transport tubes and will store them in the middle layer, and will put the capsules that are sent from this module into the transport tubes.

- The middle layer serves as a storage for resources and as a highly protected environment. The capsules that are stored in the middle layer of the module can be used by appliances placed in the upper layer of the module or can be just stored for future use.
- The upper layer is an inflatable cover that provides the living spaces of the city. When inflated it is 3 meters high with a volume of 30 cubic meters. The top plate is supported by six inflated columns. These support columns can be attached to the neighboring modules. Any of the six sides can be covered to form a wall between the modules or can otherwise be used as a door to the next module. When the module is deflated, the inflatable supporting columns and the wall covers get inside pockets in the middle layer while the top plate lays securely on top of the middle layer.

The most visible part of the module is the inflatable upper layer. This is where people live and work. The same standardized structure of the module will be used for all needs - private, social and industrial areas - so that they can interconnect and work together as the transport backbone of the city. The modules can be equipped with different instruments and will store capsules with different resources to fit the purpose of the area they are located in. For example, modules in private areas will have home furniture and appliances, while industrial areas will have office furniture, manufacturing machinery or may be even deflated if used for robotic activities.

The middle layer of the module is usually used for storage. But in private areas, its highly protected environment is also used by residents as a healthy place to sleep in and as an emergency bunker. This layer can be equipped with instruments for health monitoring and treatment.

The module is always participating in the logistics system of the city as a transport relay. It is the lower layer of the module that is responsible for that. The module is connected to other modules tiling in a honeycomb structure. The tubes of adjacent modules get connected to form long pipes for transportation of goods in the capsules.

This example city plan shows the modules of a section of a neighborhood with the private areas of residences in yellow; the social areas of gathering places in blue, streets in white and parks in green; the commercial areas of shops, offices in red; and the industrial areas of factories and labs in gray. The lower layer of the modules connects all spaces in a continuous logistics system:

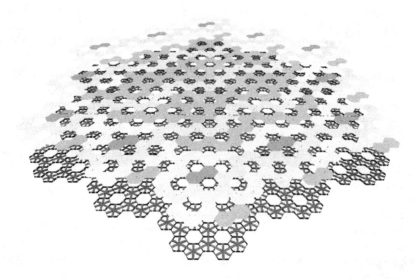

Control

All infrastructure components are controlled centrally, heavily aided by IT systems and artificial intelligence. The control center manages the transactions in the service platform, optimizes the path for deliveries, mitigates risks of failures, navigates the drones to build new housings and repair existing housings, conditions living spaces and assists the residents.

The control center also serves as the main interface in communication between the residents, the businesses, the governor, the service platform. For example, when a resident wants to report an incident they will first talk to the AI.

An important role of the control center is managing the logistics of the city. All essential goods and resources travel through the city inside the transport tubes of the lower layer of the infrastructure modules. Standardized reusable capsules are used to contain the resources. The capsules can contain and transport one resource at a time. A basic container can be used to store resources that do not require special handling, such as water, food, waste, but specialized units with the same capsule shape can be used in the transport tubes to transport and store other resources. For example, a battery capsule can store and transport energy for electricity; or high-pressure tanks can store and transport oxygen or air.

Businesses, in their role of service providers, will fill in capsules with the goods they supply and put them into the transportation system. Residents and other businesses, in their role of consumers, will receive the capsules, store them locally, consume the goods and eventually return the capsules. The control center will manage the transactions and will navigate the capsules from supplier to consumer.

The infrastructure for 1000 residents - with 3 neighborhoods each consisting of 6000 cells - will have a total capacity of 622,080 cubic meters of deliveries per day, assuming 50 cells as average distance between supplier and consumer, capsule volume of 0.02 cubic meters, a limit of one capsule at a time in a cell and average velocity of a capsule of 4 m/s. Additionally, assuming a total of 6 cubic meters of supplies necessary for the basic needs of one person (air, water, food, energy, waste), then 1% of the capacity of the transport system of the infrastructure will be required for life support leaving a lot of room for mitigation of risks of local malfunctions, congestions and other possible disruptions. The remaining 99% of the capacity is available for additional commercial use on the marketplace of the service platform.

Finally, the control center is responsible for city's urban planning. All components of the infrastructure are managed by the control center and work together to move entire neighborhoods from locations of declining attractiveness to locations with better potential. The city will be able to get closer to newly discovered deposits of natural resources, move away from locations near depleted resources or where disasters stroke, or simply move quarters around points of interest where people prefer to live.

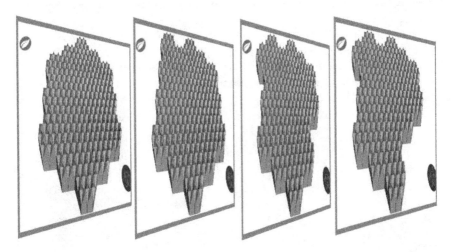

This sequence of diagrams, illustrating the move of a block that is close to an area with pollution (marked with a biohazard symbol) to a site of a new resource (marked with a green leaf symbol), demonstrates the power and flexibility of the infrastructure in the evolution of the city.

REFERENCES

1. https://SpaceColonyOne.github.io/MarsCityState/ - Web location of this document
2. https://github.com/SpaceColonyOne/MarsCityState - Source of this document
3. https://ourworldindata.org/world-population-growth - Data on world's population growth
4. https://www.researchgate.net/publication/222382174_Population_Growth_in _the_World's_Largest_Cities - Data on expected population growth for cities
5. https://en.wikipedia.org/wiki/Moore%27s_law - Information on Moore's law and empirical data on exponential growth of technology advancement
6. https://en.wikipedia.org/wiki/Asteroid_mining - Information on asteroid mining and distance from Mars to the asteroid belt
7. https://en.wikipedia.org/wiki/Three-sector_model - Description of the three-sector model in economics
8. https://ntrs.nasa.gov/archive/nasa/casi.ntrs.nasa.gov/20040075697.pdf - Comparison between the working cost estimating methodologies of the cost engineering functions in NASA Marshall Space Flight Center (MSFC) and ESA European Space Research and Technology Centre (ESTEC)
9. https://en.wikipedia.org/wiki/Space_elevator - Information on the technology of space elevators
10. https://en.wikipedia.org/wiki/Earth_structure - Information on the technology of rammed soil bricks
11. https://www.theverge.com/2017/4/27/15436154/mars-soil-simulant-study-building-human-missions - Research on use of Mars soil for bricks

5: THE BRAHMAVARTA INITIATIVE:
The First Self-Sustaining City State on Mars

Arvind Mukundan
National Chung Cheng University, Taiwan
Department of Mechanical Engineering
Institute of Optomechatronics
arvindmukund96@gmail.com

Abstract

The vast universe, from its unfathomable ends to our very own Milky Way galaxy, is comprised of numerous celestial bodies – disparate yet each having their own uniqueness. Amongst these bodies exist only a handful which have an environment that can nurture and sustain life. The Homo sapiens race has inhabited the planet which is positioned in a precise way – Earth. It is an irrefutable truth that the planet Earth has provided us with all necessities for survival – for the human race to flourish and prosper, and make scientific and technological advancements. Humans have always had an innate ardor for exploration – and now, since they have explored every nook and corner of this planet, inhabiting and utilizing its resources, the time has come to alleviate the burden we have placed upon Earth to be the sole life sustaining planet. With limited resources in our grasp and an ever-proliferating population, it is the need of the hour that we take a leap, and go beyond the planet for inhabitation – explore the other celestial objects in our galaxy. But then arises a confounding conundrum – where do we go? The answer is right next to our home – the Red Planet, Mars. Space scientists have confirmed that Mars has conditions to support life and is the closest candidate for human inhabitation. The planet has certain similarities as Earth and its proximity provides us with convenient contact. This paper will be dealing with the conceptual design for the first city state on Mars. Aggregating assumptions, research and estimations, this first settlement project shall propose the most optimal means to explore, inhabit and colonize our sister planet, Mars.

Introduction

About

This paper presents team Brahmavarta's (The term refers to the area as the place where the "good" people are born. The name can be translated to "holy land", "sacred land", "abode of gods" and "the scene of creation) response to the challenge posed by The Mars Society to design a working model for first self-sustaining human settlement on Mars. The city state should be self-supporting to the maximum extent possible – i.e. relying on a minimum mass of imports from Earth. The goal is to have the city state be able to produce all the food, clothing, shelter, power, common consumer products, vehicles and machines for a population of 1,000,000 Brahmavartans (The Citizens of Brahmavarta), with only the minimum number of key components, such as advanced electronics, needing to be imported from Earth. This paper will explore both, technical and non-technical aspects, involved in planning first self-sustaining human settlement on Mars.

Phases of Brahmavarta Initiative

Despite their obvious differences the red planet and blue planet do have somethings in common. The rotation speed for one day on Mars is only about 44 minutes longer than a day on Earth. Mars's axial tilt is also 25° compared to our planet's 23°, therefore Mars will undergo similar seasonal and temperature variations. Both the planets are made up of metallic cores and similar mineral composition. They have similar surface structures including mountains canyons and deserts. But the differences between the two planets are much more significant. Martian atmosphere unlike on Earth is very thin only measuring about 1% of Earth's atmospheric pressure and completely unbreathable for humans. It's composed of 96% of CO_2, 2% argon, 2% nitrogen with trace amount of oxygen and water vapor. Mars is drastically colder than Earth averaging -46°C with -143°C in the winter and 35°C in summer on the equator. Mars is also very dry and dusty and is buffeted by frequent sand storms. Mars lacks any reasonably sized magnetosphere measuring between 16 and 40 times weaker than Earth's which leaves more susceptible to harmful cosmic rays. Gravity clocks in at about 37% that of Earth which would be a small challenge to overcome but nothing compared to the temperature and lack of atmosphere. Any successful Martian colony would have to contend with these big problems to fight to establish any kind of hospitable environment on what should be considered a world hostile to life. To counter all the obstacles various literature were reviewed and assumptions were created around which The Brahmavarta Initiative has been developed. First, it is assumed that technology of carrying 100 tons of payload to Mars with 1100 tons of propellant, is tested and ready to use by 2026 [1]. Liquid methane and liquid oxygen ($CH4/O2$) will be used as propellant, operating in mixture ratios between 3:1 and 3.5:1 (oxygen: methane) [2]. Secondly, full support of Brahmavarta is assumed for the entirety of the mission. In order to better understand the challenges involved, the whole design for Brahmavarta is conceptually divided into following phases: Pre-Initiative phase, Settlement phase, Self-Sustaining phase. Finally, the cost of shipping goods from Earth to Mars will be 500/kg and the cost of shipping goods from Mars to Earth will be 200/kg.

Pre-Initiative Phase

Martian ADministration on Earth (MADE)

In the year 2020, The Mars Society will announce the holding of an international contest for the best design plan of a Mars city state of 1,000,000 people. The best proposals will be selected and the funding plan will be authorized. For two more years research will be conducted to check the feasibility of the project. By the year 2024, MADE will come into existence when governments and space agencies will join hands to sign Mars treaty who will become the investors on this project. The number of Brahmavartans and the authorities on MADE will be based on the investments, with highest being the greatest number of Brahmavartans. MADE is split into six different ministries with the Martian Ambassador overlooking the operations with a flat hierarchy. Head of each ministries will act as a board of directors in making a decision. In case of even votes, the Martian ambassador will have another vote to decide. The board of directors will be responsible for managing the functionality of sub departments. Goods arriving from Brahmavarta will be handled by Ministry of External Affairs in exchange of resources with governments on Earth. Ministry of Finance will handle the investments and the budgets for funding of missions and research projects on Mars. Ministry of Human Resource and Development will look after recruitment process and the training of the Brahmavartans. Ministry of Tourism will take care of the future tourists and their needs. Ministry of Planning will build the infrastructure of MADE and take care of the mission control.

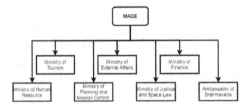

Figure 1: Structure of MADE project

Settlement Site

The goal of this design is to build sustainable and enjoyable spaces for Brahmavartans living in extreme conditions, as well as developing a reasonable community within the realistic engineering constraints. Six levels of criteria are used in the settlement site selection.

Selection Criteria

Identification of suitable atmospheric conditions such as temperature, pressure, sunlight, water and topography. Figure 5(a) shows climate zones based on temperature, modified by topography, albedo, actual solar radiation with A=Glacial (permanent ice cap); B=Polar (covered by frost during the winter which sublimates during the summer); C=North (mild) Transitional (Ca) and C South (extreme) Transitional (Cb); D= Tropical; E= Low albedo tropical; F= Subpolar Lowland (Basins); G=Tropical Lowland (Chasmata); H=Subtropical Highland (Mountain). Figure 5(b) shows color perspective as blue colors indicate high potentials, with the darkest blue as the best sites; red colors indicate fewer good sites with dark red as the worst [3]. Search for past and present life on Mars. Basic science research to gain new knowledge about the solar system's origin and history. Applied science research on how to use Mars resources to augment life-sustaining systems and how to live on other planets. Identification of specific regions of interest – local magnetic field for radiation mitigation, minerals for in-situ resource utilization, and scientific importance.

 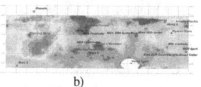

a)   b)

Figure 2: (a) Mars Global Climate Zones, (b) Map of the ideal sites from a plant perspective

Proposed settlement sites on Mars

Considering all the six criteria, four sites chosen to build our states on Mars. Arcadia Planitia, Jezero Crater, Utopia Planitia and Valles Marineris. Four different states are built with not more than 250,000 Brahmavartans in each states. Two important reasons for building four states is to utilize the resources on Mars to full extent and the safety of the Brahmavartans. These four states are named Dvaraka, Hastinapura, Kiskintha, Indraprastha.

Characteristics of selected site

1. Arcadia Planitia is flat and smooth which allows the construction of the base easier. The subsoil terrain and excess of ice are important resources for the base. Its low elevation provides good thermal conditions and solar power. This state is named as Hastinapura.
2. Valles Marineris is an ideal location because of the temperatures (203 K in winters and 313K in summers) and it is believed that there may be spring-like deposits running beneath the deep canyon, where groundwater could burst through onto the surface. It has lower radiation level (15 rem/ year) and low altitude. This state is named as Kiskintha.
3. Utopia Planitia is a huge, 3300 km diameter basin that formed by impact early in Mars' history. This city is named as Indraprasta.
4. Jezero is located close to the Martian equator, providing warmer temperatures than at the poles. There is also evidence of hydrated minerals. The availability of water is high and loss of water from the surface can be minimal due to low elevation as well as flat ground surface for easy infrastructure development. This city is named as Dvaraka.

Dawn of Brahmavarta

In the year 2024, pre-initiative phase will begin with the selection of the Brahmavartans. In 2026, experiments on Soil, communication grid, planning of the missions and ISRU experiments will begin. The Mars Sample Return mission which will bring Martians Soil will be tested in the labs. This will be followed by sets of cargo missions from 2032. Initial cargo missions will include transporting of Nasa Habitat 3D printer, Sabatier Reactor and surviving equipments followed by robots, essential raw materials, equipments for air, water facilities will be launched ISRU equipment, spacesuits and Agricultural products. Further cargo mission will include nuclear generators, Rovers and solar panels followed by medical equipment, Raw materials and manufacturing equipment. Initial cargo spacecrafts will be on round trip to Mars, by carrying few tonnes of hydrogen from Earth and converting it into return trip . By the year 2040, 410 tonnes of cargo would be sent to Mars to initially start building Hasthinapur.

Brahmavartans training phase The training program for Brahmavartans on Earth will start in 2024. Approved space organizations and governments will have to apply and at full capacity 3750 trainees will be selected by MADE for each round. The selection will be based on resilient, adaptability, flexibility, English language, medical, age between 20-45 and no criminal background. Training will be given on Martian knowledge, software skills and various group activities. At the end of cut-off round, trainee group will be left with 3350 trainees. At full capacity every year 6 rounds of training will be considered and 3350 trainees will go through basic and advanced Martian survival training as shown in Figure 3. At end, trainees will be offered to join MADE, for better understanding of Martian governance. Basic level activities will include technical, physical and social training. Every candidate will be trained with multiple skills and the points will be awards to mark their progress. Therefore, the whole training program has been designed for 4 years.

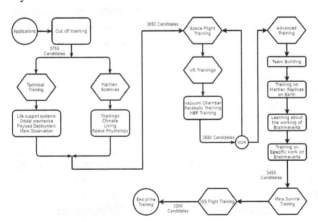

Figure 3: Brahmavartans Training

Settlement phase

Brahmavartans arrival

In the year 2032, 10 Brahmavartans will arrive at Hastinapur. A series of temporary modular pods will be printed from Martian concrete using 3D printing rovers for their stay. In the following years more Brahmavartans will start to populate the planet. Only engineers of various field and agronomists will be sent in first ship followed by 50 people for the next two every year. Between the years 2034 and 2039, 450 Brahmavartans will populate the Planet. From 2040, 10 ships carrying 1346 in total will arrive each year. Considering an improvement in the technology, Mortality and Birth rate, it will take approximately 186 years to have a 1,000,000 people colony on Mars. In early phase, department representatives will be in contact with MADE, once Brahmavarta's population reaches 60%, implementation of Martian government will be initiated. In order to avoid abnormal sex ratio in future, male to female ratio will be 50%: 49% respectively. Initially only one state, Hastinapur will be populated. But later, the population will be equally divided into 250,000 in each state.

Brahmavarta's architectural concept

Brahmavarta's Layout

The primary engineering requirements for the design are safety, efficiency and expandability. Safety requires that there be atleast two interconnected and individually pressurizable segments. In case of an accidental loss of pressure, a fire, or other failure, there must be atleast one means of backup space. Furthermore, the loss of one space must not cut off the functioning portions of Brahmavarta from each other. Hence as mentioned in the section 2.2.2 the state of Brahmavarta is divided into four cities and each city is divided into two sectors. Each sector can accommodate upto 130,000 people including the tourist, the visiting scientists and the Brahmavartans. Each block is connected physically by a road that interlinks them. All three parameters were kept in mind while developing Brahmavarta's layout.

Sector Layouts

The total area of a sector is 16km^2. All measurements and calculations to build all the components of this sector was based on [24]. The layout of each sector is designed in an octagonal shape with eight layers as represented in Figure 4. Each of these layers consists of an essential component required for the genesis and survival of a city, and are described in the order from the outermost to the innermost. The first layer of the city consists of apartments which consists of two-bedroom flats as mentioned in section 3.2.4. These apartments are placed at uniform intervals and are separated from the rest of the layers by a canal which forms the second layer. The next layer is a necessity for the sustainability of the sector – the food production belt. This layer is interspersed with industrial buildings which are constructed in a manner as to expel the least amount of particulate matter to avoid pollution in the apartment zone. Half of the size of the food production belt is made up of structures to support Vertical Farming in artificial light. The next layer is the energy belt- which contains arrays of back up solar panels affixed to the ground to fulfil the emergency requirements of the sector; along with a few apartment buildings. Beyond this layer of energy generation is the layer which consists of districts with individual housings which are single bedroom flats. These housings are structurally and aesthetically identical and are designed in accordance with the Green Building Construction concepts. The next layer consists of areas of recreation – theatres, pubs, gaming centers, aquariums, gymnasiums, a multi-purpose sports stadium and the like. They are placed strategically close to the individual housings so as to encourage congregating masses. Going further inside the sector, the next layer also houses industrial buildings manufacturing other essential requirements of the city. The outer edge of this layer consists of studio type apartments where people working in the industries would live for emergency situations. The studios are built in such a way that two studios can be combined into one forming a double bedroom apartment if necessary. The penultimate layer consists of buildings dedicated to the fine arts and educational institutions. It also consists of the research labs for various engineering and medical fields. The Brahmavarta Congress is located at the central dome of the sector. The innermost layer also contains a transportation hub, where a public transport vehicle commences its journey before travelling throughout all the layers linearly and radially, plying across all the roads connecting the different layers of the sector. The black lines represent the roads. Amalgamating with another identical sector, the city is constructed.

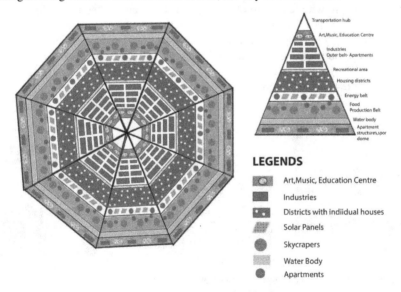

Figure 4: Layout of a block

Brahmavarta's Vernacular Architecture

Mars has a thin atmosphere and no global magnetic field, so there's little protection from harmful radiation. The thin atmosphere also means there's little air pressure, so liquids quickly evaporate into gas; despite freezing temperatures, an unprotected human's blood would boil on Mars. However, it wouldn't feel like home if Brahmavartans are walking in space suit in and around the sector. Thus, a local terraforming is important for them to feel like Earth. To maintain a comfortable temperature and habitable air pressure, each layer of the sector would be made up of pressurized and interconnected bio-domes each covered with a multilayer transparent polyethylene. The layers in the order from the outermost to innermost are Acrylonitrile butadine styrene, Fluorinated ethylene propylene, Crosslinked Ethylene Tetrafluoroethylene, Polytetrafluoroethylene. These domes are built by tying a pressuried inflatables on to the ground and then the building can be built using the martian regoliths. These layers will protect Brahmavartans from the harmful radiations and inside these domes the process of paraterraforming can be done.

Living Quarters

The living quarters is divided into three different styles depending upon each Brahmavartan's preference; Studio, One Bedroom and Two-bedroom homes as shown in Figure 5. The size of a studio 76.03 m^2 in which living room, bedroom, and kitchen are combined into a single room. The studios serve as a home for the bachelors who work in the industries. Since the studios are built next to the industries, people who take care of the emergency needs of the industries will be staying here. The size of a one-bedroom flat is 136 m^2. A one-bedroom flat is little bigger than the studio which has an extra washroom and a separate bedroom. These types of flats will be built in the second phase of the layout as shown in the Figure 4. The size of a two-bedroom flat is 163 m^2. These will be built on the outer zone of the colony in a four storey building.

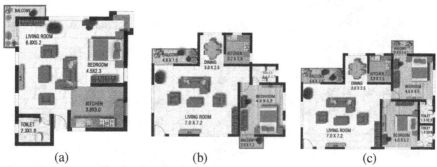

Figure 5: (a) A studio, (b) One Bedroom Apartment, (c) Two Bedroom Apartment

Infrastructure materials

Transporting all the necessities for the infrastructure from Earth is not feasible. The regolith available in Mars is rich in oxides of iron, silicon, magnesium and aluminum which can be synthesized and it is possible to extract essential materials for building infrastructure on Mars. With the abundance of CO_2 in Mars atmosphere, polyethylene, polypropylene and polycarbonate can be easily manufactured on Mars. A thermomechanical coating (TMC), which can be extracted from Mars regolith along with the bricks and sulpher based concrete [4].

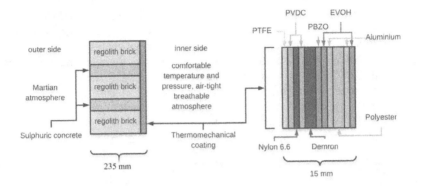

The thickness of outside walls will be ~ 250mm, of which 235mm will be brick thickness and remaining will be TMC which can be seen in Figure 6. TMC provides protection from excessively cold temperatures, impact, fire, heat, wear, abrasion, chemical degradation, thermal conduction and radiation heat losses and also flexible enough to allow an easy and repeatable folding and unfolding [5]. The characteristics of TMC is shown in Figure 7. For roof of greenhouse and inner living quarters, polymethyl methacrylate glass will be used. PMMA is an economical alternative to polycarbonate when tensile strength, flexural strength, transparency, polishability, and UV tolerance are more important than impact strength, chemical resistance, and heat resistance [6]. By extracting and synthesizing the Martian Regolith the PMMA glass can be manufactured. Also, PMMA glass provides protection from UV radiation for wavelength around 300nm which is within the range in case of Mars. Apart from manufacturing of primary materials, ethylene can also be used for manufacturing of thermoplastic elastomers which is the primary material for space suits. Carbon, nickel, manganese, aluminum, steel and other metals required for manufacturing colony use equipment can also be extracted and processed.

Material	Layer
iylene vinyl alcohol copolymer)	Pressure layer
n	Insulation layer
	Structural layer
yphenylene benzobioxazole)	Impact, puncture, fire, heat protection layer
	Energetic radiation layer
yvinyldiene chloride)	Moisture protection layer

Figure 7: Characteristics of TMC

In the Odds of Dust Storm

Martian winds can frequently generate large dust storms. When Mars is at its perihelion, the southern hemisphere is suddenly heated, and a large temperature difference relative to the northern hemisphere is generated. This drives strong winds and dust from the southern hemisphere to the northern hemisphere. Most of the storms move with velocities of 14m/s-32m/s. Taking into consideration curves and contours of the design, a 3D flow analysis is done on Brahmavarta's sector for the scenario mentioned, which is shown in Figure 8. Results obtained shows the maximum surface pressure of 1160pa. It is also found in the structural analysis; displacement has maximum value of 15mm. In the thermal analysis heat flux has a maximum value of 123.8 W/m². From structural and thermal analysis, it is concluded that Brahmavarta's sector is able to withstand the thermal displacement caused by pressure and temperature difference in case of dust storm while the pressure and temperature inside the sector is maintained constant. The infrastructure is designed in such a way that it withstands against all the odds of dust storm and Martian temperature changes with a great safety factor of 4.

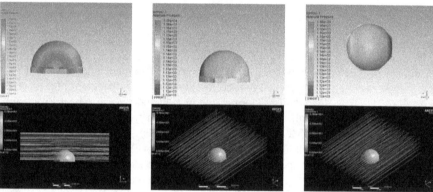

Figure 8: 3D flow analysis of Brahmavarta's block design

Transportation

The four cities of Brahmavarta are located miles apart and various transportation options for connecting the cities needs are considered and discussed below.

Long and Short Distance Ground Travel

For long distance travel, NASA's Space Exploration Vehicle can be used, the pressurized cabin can be used for space missions or for surface exploration. SEV has the cabin mounted on a chassis, with wheels that can pivot 360 degrees and drive about 10 kilometers per hour in any direction. It can house two astronauts for up to 14 days with sleeping and sanitary facilities and thus in two weeks around 3360 Kilometers can be travelled [9]. For a short distance travel, with not much to carry, an electric ATVs can be used. However, the ATVs can also work on the extra methane produced from the Sabatier reactor or better a hybrid.

Unmanned Aerial Vehicles

NASA has been testing a flying wing prototype called Preliminary Research Aerodynamic Design to Lower Drag, or Prandtl-D for short which is capable of flight in the thin Martian atmosphere [18]. Ingenuity is a robotic helicopter designed specifically to operate in the low-pressure atmosphere of Mars which could potentially cover a distance of up to 300 meters per flight. The key to its operation, though, seems to be in the shape and size of the counter-rotating rotors [18]. Balloons have been flying for decades in Earth's stratosphere which, at its upper altitudes, has an atmosphere as thin as that on the surface of Mars. Conventional stratospheric balloons, though, only stay aloft a few days because of the daily heating and cooling of the balloon. Helium super pressure balloons, on the other hand, could fly for more than 100 day and the same technology toward a balloon for Mars [18].

Martian Hopper

Planet hopping is a long-distance travel method that is unlike anything we have on Earth. Basically, a 'hopper' takes off like a rocket and begins a ballistic trajectory that shoots it into space, but not into orbit. The planet's gravity pulls the vehicle back down to the surface where it can perform a powered landing, refuel and do it again. These can be extremely tricky in a heavy gravity environment like Earth but, as both SpaceX and Blue Origin have demonstrated, clearly doable. On Mars, with only 38% of Earth's gravity, it should be much easier. The Martian Hopper will be powered by a radioisotope thermal generator for the rocket engine which could pull CO_2 from the Martian atmosphere and compress it until it is liquefied. When ready to 'hop', pump the gas into a chamber and expose it to the intense heat from the RTG. A research team from Leicester University and the Astrium space company have proposed some calculations suggest the thrust achieved could enable a one metric tonne craft to leap a distance of up to 900m at a time [21].

Brahmavarta's Life Support System

The schematic of Brahmavarta's life support system (BLSS) consists of a Sabatier reactor from which drinkable water, oxygen, propellant, Iron and steel production is shown in Figure 9. The reactor is fed with the Atmospheric CO_2 and hydrogen electrolyzed from ground water which in-turn produces oxygen and methane at a ratio of 4: 1 and the excess oxygen will be supplied to life support. Considering, consumption of ~ 9.6kg of water per person per day and ~ 2.8kg of oxygen per person per day, thus 222 tons amount of water will be required for 130000 (including temporary stays) people per day and 364tonnes of total oxygen per day. The BLSS water extraction subsystem will be able to extract 68.2kg/hr. [6], hence 132 facilities will produce 110635 tons/day. By doing electrolysis, 198 tons/day amount of O_2 is produce, out of which 20,341 kg/day for propellant generation and remaining 3,032 kg/day is sent to air processing plant. From electrolysis, 2,922 kg/day amount of H_2 is sent to Sabatier reactor for producing 5,812 kg/day CH_4, which will be used for propellant generation and 13,051 kg/day water is produced, which is sent to water processing plant. Hence, 9.54 x 10^6 kg/year of propellant is produced, which satisfies the need for incoming and outgoing spaceships. Air processing plant will convert pure O_2 into a breathable air ratio. This breathable air will be supplied to living quarters, kitchen and labs. This air with more CO_2 proportion, released from mentioned areas, will be send to greenhouse chambers. The plants in greenhouses will produce O_2 using photosynthesis process. This air with more O_2 proportion, will be input for air processing plant.

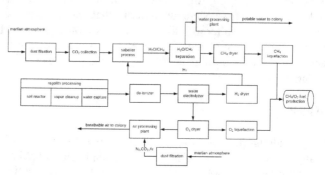

Figure 9: Breathable air circulation system

Electricity production

For countries that are highly industrialized and have high energy consumption, the amount of energy needed for a city of 1 million is about 1500 MW. The consumption of electricity in Brahmavarta will be for domestic and personal uses, scientific and office uses, BLSS and agriculture. Average electricity consumption for commercial activities is 3.5kWe per person. Also, power consumption is 17kWe per processing facility [2]. Adding all the needs together with a factor of safety the power consumption of one sector in a city will be ~ 600MWe. Electric power generation will be a combination of uranium based nuclear reactors, retracting common and the solar panels inside the dome and system of nickel-hydrogen batteries. The evolution of NASA's Kilo-power Reactor will result in a 100kWe reactor from the maximum of the currently envisioned design of 10kWe [11]. With 200 nuclear reactors, total power generated is 200MWe. The roof side of outer living quarters, passage, lab, kitchen etc, buildings will be covered with solar panels. The area covered by solar arrays on one sector in buildings will be approximately 75% the size of a sector i.e., with 33% efficiency and considering 100W/m^2, solar power generated will 400MWe. Hence, total power generation will be 600MWe satisfying the total electricity need of a sector. Other than these, a series of Mars Autonomous and foldable Solar Array will also be deployed outside the sector to charge the nickel-hydrogen batteries to prepare for a worst-case scenario and to power the transportation systems of Brahmavarta [26].

Food production and waste management

A healthy body requires 2kg of food every day for nutrition and energy which includes 1000mg of calcium, 400μg of folic acid, and 56g of protein each day [7]. But the intake varies with scenarios like the iron intake for females is double while the whole intakes doubles in Figure 10: Layout of manure and water recycle plant case of pregnancies. Considering these factors, poppy seeds, winged and soya beans, lentils, potatoes, tomatoes, onions, green beans, spinach, beets, orange, lemons will be harvested. Also, every year a cargo consisting of 100 tons of Meat will be imported from Earth to Brahmavarta. These crops require only 7 hours of sunlight and 14 hours of artificial light, manures, proper drainage system. These plants only require sprinkling of water from time to time. Soil is required to be rich in potassium, phosphorous and nitrogen nutrients which can be easily achieved by adding composites, lime and Sulphur to maintain the pH balance for respective crops. Rhizobium bacteria's will be used to make nitrogenous rich soil. For first two years, 100kg of manure's will be imported from earth, which cost $1000 for a ton and simultaneously the Martian soil will be prepared. The food for the first year will be brought by the Brahmavartans in their cargo. The layout of the water Processing and the Manure Processing is represented in Figure 10. Water treatment facilities will purify urine into struvite, a very good fertilizer that is guaranteed free from diseases. Human waste will be collected and sent to for processing which will be used as manure for Agriculture along with Struvite. It has also been found that the holes that worms dig in the soil aerate the mixture and improve the soil's structure, making it easier for water to penetrate the soil and nourish plants. By their activity in the soil, earthworms offer many benefits: increased nutrient availability, better drainage, and a more stable soil structure, all of which help improve farm productivity. It has been found that Earthworms not only can survive but also reproduce in Martian soil which can be used to aerate and improve the soil structure [10].

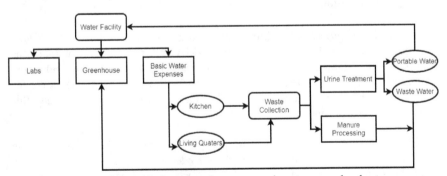

Figure 10: Layout of manure and water recycle plant

Internet

Internet consists of two big components, first servers that store information and their related infrastructure. Proving internet to Mars initially will be difficult and slow. Distance between Earth and Mars at closest is 54.6 million kilometers and the farthest is 400 million kilometers. Thus, for a piece of information to be transferred it takes around 3-22 minutes to transfer from earth to mars. A Laser wave is 100,000 times shorter than radio wave; which means more room to carry data 5 times more reliable connection. Smaller wave means better signal and fortunately the technology we need exists which is been tested by NASA Lunar Laser Communication Demonstration [12]. These Lasers will be able to handle HD Videos and more. NASA is focused on upgrading existing satellites in order to build Laser Powered Space Internet Network. With the help of Lasers and Satellites Internet works as same on Earth. However, once the settlement has been completed, an entire separate set of infrastructures including servers will be built on mars which will be working independently.

Manufacturing Industries

Iron The MER-B rover, landing at the Meridiani Planum site, found abundant hematite in the form of small (4–6 mm) spherules. These are approximately 50% hematite by weight. Similar to Earth ore deposits, the ore on Mars was concentrated and produced by an aqueous diagenesis, at a time early in Mars' history. This is an easily accessible ore; the hematite concretions exist both in the soil, where it can be simply scooped up; and embedded in soft sedimentary rock, from which it can be easily extracted. The standard terrestrial carbothermic process can be used to reduce hematite to metallic iron, since carbon monoxide is easily available at all locations on Mars from the carbon dioxide atmosphere.

Magnesium In the process of making cement, large amount of MgO are produced. Magnesium metal is made via the Pidgeon process whereby MgO is reacted with silicon metal at high temperature to produce magnesium and SiO_2 [21]. The normally unfavorable thermodynamic equilibrium is driven toward completion by distilling away magnesium vapor as it forms.

Aluminum In the process of making cement, large amount of Al_2O_3 and MgO are produced. Aluminum is produced via the FFC Cambridge process without the need for cryolite (Na_3AlF_6) that is difficult to obtain on Mars due to the relatively scarcity of fluorine [21]. Instead, calcium chloride ($CaCl_2$) is used to dissolve the alumina, and due to large quantities of chlorine available from perchlorate, this material is readily produced.

Steel Production of steel, rather than simply iron, requires an addition of carbon in controlled quantities. The amount of carbon needed is small, and carbon is easily available from the Martian atmosphere. Carbonyl processing uses carbon monoxide, which is easily manufactured from the Martian atmosphere, to transport the metals in the form of volatile iron- and nickelcarbonyl. Distillation can then separate the iron and nickel, allowing the composition to be modified as desired.

Organic Chemicals and Plastics Ethylene can be produced from CO_2 and water produced from BLSS using Intermittent Sunlight [20]. Once Ethylene is created a wide variety of more complex molecules can be created. Polyethylene, terephthalate, polyethylene, polypropylene, polyvinyl chloride, polystyrene, polycarbonate, polymethyl methacrylate, polyoxymethylene, acrylonitrile butadiene styrene and nylon which can synthesized from about three dozen reagents made from simple organic precursors. Together, these polymers enable a wide array of uses and significantly reduce the colony's earth dependence.

Basalt Basalt is very common on Mars. It is an igneous rock that is commonly formed by magma extrusions during a lava flow. Basalt is more elastic than steel and has similar elasticity to aluminum. Fused Deposition Modelling can be performed by using computer numerical control, to precisely force molten material through a die to specific places in layers. The molten material flows onto existing material thereby the new and existing material fuse together as the material freezes or solidifies. This approach is commonly used to print thermoplastics such as polylactic acid or acrylonitrile butadiene styrene [27].

Concrete and Cement Cement is produced through the combination of lime (CaO), silica (SiO_2) and alumina (Al_2O_3), with smaller amounts of magnesia (MgO). All of these compounds are found in Martian regolith, but with much lower amounts of CaO and Al_2O_3 than required. By first heating the regolith to separate the lower-melting FeO and to lesser extent Al2O3, MgO and SiO2, a mixture with composition of approximately 52% CaO, 32% Al2O3, 15% SiO2 and 2% MgO is obtained [23]. This cement is then mixed with coarsely crushed regolith and water to produce concrete.

Mirrors and Glass Using lunar regolith simulant and heating it within a susceptor-assisted microwave oven, it is possible to manufacture a variety of basaltic glasses. Furthermore, it is also possible to shape these glasses by grinding and polishing the surface flat and smooth. Glasses manufactured from different lunar regolith simulants can be coated with aluminium or silver, to create the reflective properties. With a porous and/or smooth surface finish, mirrors can be made to reflect the incident solar light (400 nm–1250 nm) in-between 30% for the worst and 85% for the best samples [25].

Communication

Communication between Earth Mars is conducted via communication channels that orbits Mars. For cases when the Sun eclipses the straight communication the kind of antennas may be placed in Lagrangian points of the solar system which provides robust communication channels with the Earth. Communication within the colony is arranged based on the G5 network with Ethernet and MQTT protocol that is used for IoT and smart cities ensuring all the devices are connected to the web and provide real-time information about its state. IoT also gives a chance to various kinds and use AI algorithms to optimize the intake of water, food and reduce the other important factors. For better GNSS navigation kind of communication, posts are printed from the 3D printers in order to improve navigation for Mars exploration. Brahmavartans will use localized cell networks for the majority of communications. Rovers and EVA suits will be fitted with Bluetooth or wired connectivity. Every colonist will carry a smart phone and wear smart watch which will provide instant communication between colonists, tracking capabilities and health monitoring. Multiple surface towers provide redundancy and geolocation with their own independent solar fields. Mars has sufficient ionosphere to bounce a signal beyond line of sight and hence self-tuning HF radios will be used for backup and emergency communications.

Terraforming Mars

There are 3 sources for CO_2 on Mars. There's the south polar ice cap which consists of water ice, interspersed with thick layer of CO_2 ice. The next accessible source is absorbed into the surface dust- the regolith up to 100 m deep. The final source is carbonate minerals in the crust. To Terraform Mars there are three requirements, Warm the planet to 290k, Increase the atmospheric pressure, supplying 240 mbar of breathable oxygen and provide sufficient water for a water cycle. A magnifying soletta and mirror lens can be used to vaporize the Martian regolith by placing a soletta 1000 kms from Mars, between the planet and the Sun, held in position by light reflected from the annular support mirror. Sunlight from the soletta must be further focused by a lens ranging some 400km above the Martian surface. Since this is within the upper reaches of the martian atmosphere, an aerial lens is used instead of arbital mirrors. If the oxygen is liberated through the pyrolysis of regolith oxides or nitrates no further action will be required. However, any excess carbon dioxide will have to be converted into oxygen by photosynthesis [17]. In addition to this technique, albedo modification, the impact of ice asteroids, and the release of artificial greenhouse agents will be considered and researched after the full colonization of Mars. Once Mars has been terraformed, we need to protect the new atmosphere. We cannot restart Mars's magnetic field so we can try to build an external magnetic shield. The easiest would be to do that in space by placing an orbiting field generator placed between Mars and the sun.

Self-Sustaining Phase

To create a self-sustainable environment, a trade system similar to Earth has to be introduced which must be profitable in the long run. The important objective of the economic model of Brahmavarta is to check the feasibility of the project which starts with generation of the investments. After the initial investments are brought in for the project the Pre-Initiative phase starts followed by the settlement phase. The cash inflow-outflow of the project and the consumables in Brahmavarta will be calculated in US dollars, since the initial investment is generated on Earth. The important constraint to the case is the transport of goods from Earth to Mars costs $500/kg and from Mars to Earth cost $200/kg. By using this constrain the economic model of Brahmavarta was developed.

Initial Investment

Generation of Initial Investment The foundation of Brahmavarta will be laid through series of investments from across several government bodies on Earth. $4 billion from the investors will be drawn as loan every year. Furthermore, the profits generated for training of Brahmavartans will also add to the initial account which is $4 million per person. Hence the total initial investment for the project is $240 billion.

Inflow and Outflow during the Pre-Initiative Phase The major outflow in the pre-initiative phase will involve thirty three cargo missions where in each of them comprises of NASA Habitat 3D printer, Sabatier Reactor, Food for the first few Brahmavartans, ISRU equipment, spacesuits, agriculture materials including seeds and fertilizers, nuclear generators, Rovers, solar panels and medical equipment and various other items required for a settlement.

Inflow and Outflow during the Settlement Phase Settlement phase will comprise a total of 156 missions. A total of 20,176 tonnes of cargo would be sent at the end of the settlement phase. The first Brahmavartans will be sent with cargo having food and various utilities for their survival. These values are rough estimations with reference to weight distribution. During this phase a few of the profitable operation mentioned in the subsection 4.2 will begin thereby bringing in inflow of cash to the economic model of Brahmavarta.

Profitable Operations

The economic model of Brahmavarta is based on the model of trade system as on earth and to keep the economy self-sustaining, this economy model discusses various profitable operations and estimates the profit generated upon their installation in a fully functioning settlement. The proposed methods of sustenance on Mars are as follows:

Research Visits Every year 85 selected scientists from countries/organizations who has sponsored for the initiative would be sent to Brahmavarta for an year of extensive research. They would be hosted at the city near- est to their area of research. The revenue generated by these research visits would be 891.431 million every year which is represented in Figure 11.

Research Visit		
Number of Tourist	85	per year
Selling Price	$ 20,000,000.00	per researcher
Profit Generated per person	$ 10,487,213.76	per researcher
Total Revenue Generated	$ 891,413,170	per year

Figure 11: Revenue from Research Visits

Asteroid Mining An abundant source of rare metals in rich quality can be found in the Asteroid belt near Mars. Asteroids such as Aspacia, Rudra, etc [14]. help us estimate revenues that could possibly be generated from asteroid mining. Since the mass ratios in the calculation of fuel expended in asteroid mining is not determined for these individual asteroids yet, we took the mass ratio from Ceres which is further away from the asteroids in study. The total payload from the asteroid in one mission is estimated to be 7500 tons which can be extracted with an efficiency of 75%. The frequency of the missions every year is 4. And once in four years an extra mission will be sent to Earth. The payload in every trip on the rocket is 150 tons which brings the total amount transported per year to 637.5 tons. Revenue generated through asteroid mining can be seen in Figure 12. Assuming the composition by weight of platinum from the asteroid mined, to be 25%, the remaining ore is rich in elements such as nickel, iron, copper, etc. which can be utilized on Brahmavarta for different purposes.

Asteriod Mining		
Number of Transits	4	per year
Platinum Extracted	637.5	tonnes
Selling Price on Earth	$ 30,000.00	per kg
Cost of Transportation	$ 752,250,000.00	per year
Estimated Cost of Payload	$ 19,125,000,000.00	per year
Total Revenue Generated	$ 18,372,750,000.00	per year

Figure 12: Revenue from Asteroid Mining

Tourism A total of six schemes has been proposed, A through F are named in order of the number of days of travel as shown in Figure 13. The cost price of the tickets is directly proportional to the number of days of stay, starting at 15 million for 30 days of stay and a maximum of 30 million for a stay up to 550 days. The aim behind having multiple schemes is to increase the frequency of travel to Mars. The available windows of travel for a sample year were found in [13]. The gradient in the color of the data demarcates the journeys under different schemes. A total of 1100 tourist can be accommodated a year at all our four different states each hosting different types of schemes in a shift basis. A revenue of 31 Billion is generated every year by tourism.

Plan	Earth Departure	Mars Arrival	Mars Departure	Earth Return	Stay time	Duration	Frequency	Cost/Person	Ticket/Person	Total Revenue/year
A	Jul-14-2020	Jan-15-2021	Feb-14-2021	Oct-22-2021	30	485	3	$ 6,358,581.45	$ 15,000,000.00	$ 3,293,142,467.73
	May-31-2018	Oct-13-2018	Nov-12-2018	Jul-15-2019	30	410				
B	Jul-18-2020	Feb-11-2021	Jun-03-2021	Sep-26-2022	112	800	3	$ 9,512,786.24	$ 18,000,000.00	$ 2,164,239,510.00
	Apr-05-2017	Mar-23-2018	Jul-13-2018	Feb-22-2019	112	688				
C	May-26-2018	Dec-04-2018	Apr-11-2019	Aug-03-2020	128	800	3	$ 6,410,684.64	$ 21,000,000.00	$ 5,514,761,206.71
	Oct-12-2022	Oct-31-2023	Aug-14-2024	Feb-06-2025	288	848				
D	Sep-26-2022	Sep-29-2023	Aug-14-2024	Mar-10-2025	320	896	3	$ 10,098,672.98	$ 24,000,000.00	$ 3,336,798,485.54
	Feb-01-2029	Sep-13-2029	Sep-16-2030	Apr-28-2031	368	816				
E	Dec-15-2028	Oct-15-2029	Nov-03-2030	Jun-15-2031	384	912	3	$ 8,905,334.55	$ 27,000,000.00	$ 7,053,227,573.10
	Jan-23-2014	Sep-04-2014	Nov-26-2015	Jun-21-2016	448	880				
F	Apr-02-2016	Oct-27-2016	Feb-03-2018	Aug-14-2018	464	864	3	$ 5,936,661.05	$ 30,000,000.00	$ 9,817,842,282.80
	May-26-2018	Dec-04-2018	May-31-2020	Nov-07-2020	544	896				
									Total Reveneue Generated =	$ 31,180,011,535.88

Figure 13: Revenue from Tourism

Deuterium Generation Deuterium, the heavy isotope of hydrogen is 166ppm by composition on Earth, comprises of 833ppm on Mars. Deuterium is the key fuel for both the first and the second-generation fusion reactors on Earth. On Earth, 1kg of deuterium exports priced between $10,000 - $16,000, depending on the purity. Therefore, the estimated profit per kg is around $9500, even if the cost drops due to less demand. For one ton the estimated profit is $9,500,000. As of now, on Earth, 452 nuclear power plants are either used or under construction which requires 400grams of Deuterium per day for 1500MW production. Upon calculation, it can be inferred from the Figure 14, 43.8 tons of deuterium is required per year. On transporting 90 tons of deuterium once in two years, the profit is $746.58 million.

Total D exported from Mars	87989,33 kg
Transportation costs	$133.32 million
Cost of D on Earth	$10,000
Total cost of D on Earth	$879.89 million
Total Revenue generated in 2 years	$746.58 million

Figure 14: Revenue from Deuterium

Lunar Dust In 2003 the federal government put a price tag on moon rocks. NASA assessed the value of the rocks at around 50,800 per gram in 1973 dollars, based on the total cost of retrieving the samples. That works to over $300,000 a gram in today's value [16]. Around 15 kgs of non-transferable Martian dust will be sold to specific research organization and universities as per their specific needs for research purpose on a rolling basis.

Lunar Dust		
Cost of Lunar Dust	$ 200,000,000.00	/kg
Total kgs transported	15.00	/year
Total Cost of Transpotation	$ 3,000.00	/year
Total Revenue Earned	$ 2,999,997,000.00	/year

Figure 15: Revenue from Research Visits

Broadcasting The Olympic Games in Vancouver in 2010 and London in 2012 lasting only three weeks each, yielded more than 3.8 billion US from broadcasting rights only [15]. In 2018, the total revenue from television broadcasting rights of Fédération Internationale de Football Association (FIFA) was 2543.97 million U.S. dollars. These are the trends in broadcasting on Earth for the events which had the magnanimity of the moon landings, the Olympics event. Once the settlement is completed, on every Friday, Saturday and Sunday evening, a sporting event will be covered extensively and will be broadcast on Earth. In the pre-initiative phase selective events of 6 such as the training, Arrival and the Departure of Brahmavartans and the tourism would be covered every year.

Broadcasting		
Revenue Generated	$ 183,333,333.00	/event
Number of Launch Events In Earth	6	/year
Number of Sporting Events on Mars	156	/year
Total Revenue Generated in the Pre-Initiative Phase Phase	$ 1,099,999,998.00	/year
Total Revenue Generated in the Self Sustaining Phase	$ 29,699,999,946.00	/year

Figure 16: Revenue from Broadcasting

As shown in Figure 16 while the full functioning of a Brahmavarta, 30 Billion can be earned through the sponsorship, ticketing and Tv right. 4.2.1 Cost plan

Cost Budget						
Years	Details	Outflow Cost Estimation	Total Outflow	Outflow Margin with Margin	Inflow Estimation	Total Inflow
2024-2034	MADE member's investments				$ 48,000,000,000.00	
	MADE operating cost	$ 19,200,000,000.00				
	MADE Infrastructure	$ 3,720,000,000.00	$ 34,740,000,000.00	$ 36,477,000,000.00		$ 55,490,230,870.00
	Training program at MADE	$ 2,820,000,000.00				
	Spacecraft manufacturing	$ 9,000,000,000.00			$ 7,490,230,870.00	
2036-2046	MADE member's investments				$ 48,000,000,000.00	
	MADE operating cost	$ 19,200,000,000.00				
	Spacecraft manufacturing	$ 9,000,000,000.00				
	Broadcasting		$ 210,714,500,000.00	$ 221,250,225,000.00	$ 6,599,999,988.00	$ 349,431,963,988.00
	Manned Mission	$ 53,970,000,000.00				
	Cargo Missions	$ 31,482,500,000.00				
	Martian Soil				$ 35,999,964,000.00	
	Training program at MADE	$ 97,062,000,000.00			$ 258,832,000,000.00	
2048-2058	MADE member's investments				$ 48,000,000,000.00	
	MADE operating cost	$ 19,200,000,000.00				
	Spacecraft manufacturing	$ 9,000,000,000.00				
	Manned Mission	$ 77,100,000,000.00				
	Broadcasting		$ 230,510,500,000.00	$ 242,036,025,000.00	$ 6,599,999,988.00	$ 358,727,415,422.37
	Cargo Missions	$ 28,342,500,000.00				
	Deuterium				$ 4,479,451,434.37	
	Martian Soil				$ 35,999,964,000.00	
	Training program at MADE	$ 98,868,000,000.00			$ 263,648,000,000.00	
2060-2070	MADE member's investments				$ 48,000,000,000.00	
	MADE operating cost	$ 19,200,000,000.00				
	Spacecraft manufacturing	$ 9,000,000,000.00				
	Manned Mission	$ 77,100,000,000.00				
	Broadcasting		$ 236,866,500,000.00	$ 247,334,325,000.00	$ 6,599,999,988.00	$ 372,163,415,422.37
	Cargo Missions	$ 26,342,500,000.00				
	Deuterium				$ 4,479,451,434.37	
	Martian Soil				$ 35,999,964,000.00	
	Training program at MADE	$ 103,914,000,000.00			$ 277,104,000,000.00	
2072-2082	MADE member's investments				$ 48,000,000,000.00	
	MADE operating cost	$ 19,200,000,000.00				
	Spacecraft manufacturing	$ 9,000,000,000.00				
	Manned Mission	$ 106,012,500,000.00				
	Broadcasting		$ 311,395,000,000.00	$ 328,964,750,000.00	$ 6,599,999,988.00	$ 497,319,415,422.37
	Cargo Missions	$ 26,342,500,000.00				
	Deuterium				$ 4,479,451,434.37	
	Martian Soil				$ 35,999,964,000.00	
	Training program at MADE	$ 150,840,000,000.00			$ 402,240,000,000.00	
2084	MADE member's investments				$ 8,000,000,000.00	
	MADE operating cost	$ 3,200,000,000.00				
	Spacecraft manufacturing	$ 1,500,000,000.00				
	Manned Mission	$ 19,275,000,000.00				
	Broadcasting		$ 49,115,000,000.00	$ 51,570,750,000.00	$ 29,699,999,948.00	$ 212,374,918,596.81
	Deuterium				$ 746,575,239.06	
	Martian Soil				$ 6,999,994,000.00	
	Training program at MADE	$ 25,140,000,000.00			$ 67,040,000,000.00	
	Asteroid Mining				$ 36,745,500,000.00	
	Research Visits				$ 1,782,826,340.00	
	Tourism				$ 62,360,023,071.73	
Total Money Borrowed =		$ 240,000,000,000.00	Money Left after Payback		$	479,894,284,521.92

Figure 17: Summary of Cost Plan for Brahmavarta

A Sample Balance Sheet for Every 10 years						
2200-2210	Broadcasting				$ 29,699,999,948.00	
	MADE operating cost	$ 19,200,000,000.00				
	Spacecraft manufacturing	$ 9,000,000,000.00				
	Manned Mission	$ 106,012,500,000.00				
	Cargo Missions	$ 26,342,500,000.00	$ 286,255,000,000.00	$ 300,567,750,000.00	$ 3,732,876,195.31	$ 903,074,593,200.07
	Deuterium				$ 29,969,870,000.00	
	Martian Soil				$ 311,800,115,356.76	
	Tourism				$ 6,914,131,700.00	
	Research Visits				$ 183,727,800,000.00	
	Asteroid Mining				$ 335,200,000,000.00	
	Training program at MADE	$ 125,700,000,000.00				
			Total Outflow =	$ 300,567,750,000.00	Total Inflow=	$ 903,074,593,200.07
			Profit in 10 Years		$	602,506,843,200.07

Figure 18: Cost Plan for Brahmavarta

Figure 17, which shows the summary of the differentiation of the ventures across a span of 60 years, is studied and analyzed to forecast the self-sustenance point of Brahmavarta. The cost flow is laid out based on the timeline of the project taking into account mandatory ventures and those introduced to incur profits. Since all three phases has some profitable operations, it is easier to attain the self-sustaining phase. The break-even analysis of the economic model of Brahmavarta is represented in Figure 19. The Figure 18 represents the profits generated in every ten years after the full functioning of Brahmavarta. It is clear that after the full functioning of all the profitable operations in the 2084, the initial investments can be paid off and the future profits can be solely used for the purpose of the development of Brahmavarta.

- 2024 - 2025 Formation of MADE and the feasibility analysis of the project
- 2026 - 2031 Various ISRU experiments, training of Brahmavartans will begin.
- 2032 - 2050 Initiation of Cargo and Manned missions, the construction phase of Hastinapur, Broadcasting and deuterium generation begins.
- 2051- 2060 Construction phase of Indraprasta.
- 2061- 2070 Construction phase of Kiskintha.
- 2070 - 2084 Construction phase of Dvarka.
- 2084 Functioning of profitable operations begin.
- 2086 Payment of the debt (initial investment).
- 2087 Sustainable phase begins, Settlement of Brahmavarta.

Future Scope and Economic Viability

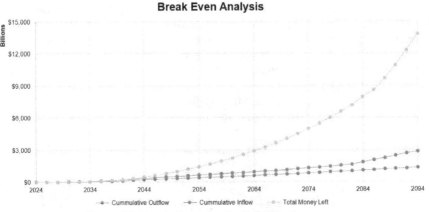

Figure 19: Break Even Analysis for Brahmavarta

In the year 2086, all the initial investments can be paid off which shows that the economic model of Brahmavarta is viable since it promises sustenance in 62 years. After 2086, MADE will become the Martian Embassy on Earth (MEE). Brahmavarta can also act as a pit stop for refueling in the future for missions like Human Outer Planet Exploration. The Martian economy shall also flourish by aiding in bio-medical research which is primarily due to the variety in atmospheric conditions that make it possible to grow and treat certain microbes that cannot be on Earth. The semi-conductive material research which requires high vacuum like conditions not available on Earth, can be smoothly regulated on Mars. There are many trips suggested between Earth and Mars in the economy plan above, which paves the way for selling water from Mars to in-orbit manned facilities. In the future there will be a triangle trade, with Earth supplying high technology manufactured goods to Mars, Mars supplying low technology manufactured goods and food staples to the Moon, and Moon sending helium-3 to Earth [19].

Social and political model

The Brahmavarta Congress (TBC)

As the population of Brahmavarta increases, a government will be formed which consists of ten ministries. Exactly when the population of a state reaches 50% the members of this congress will be chosen by the people using a democratic system. The congress is flat hierarchy model similar to the system of boards. Ten different board members will be elected to head the ministries. These ten ministries along with the six ministries in MADE will vote on the new policies to implement on Brahmavarta. The ten ministries include, Ministry of Education, Ministry of Health Care, Ministry of Science, Ministry of Food & Agriculture, Ministry of Labor & Employment, Ministry of Internal & External Affairs, Ministry of Culture & Sports, Ministry of Happiness, Ministry of Finance, Ministry of Industries. Each representative will maintain the discipline and check the functionality of Brahmavarta. The ten ministries and their members will constitute The Brahmavarta Congress as represented in the Figure 20. The responsibilities of Ministry of Education is to draw up and improvise strategies, policies and plans for educational reform and development at various levels of school; and to draft relevant rules and regulations, and supervise their implementation. The responsibilities of Ministry of Health care include contribution to socio-economic development and the development of a local health industry by promoting health and vitality through access to quality health for all Brahmavartans using motivated personnel. The responsibilities of Ministry for Agriculture and Food covers land utilization, agriculture and forestry and development of new agriculture-based industries.

Figure 20: The Brahmavarta Congress

Its objective is to safeguard the agriculture resource base, to extend the industry knowledge base and to enable nationwide economic growth and employment based on agriculture and agriculture-based products. The main responsibility of the Ministry Labor and Employment is to protect and safeguard the interests of workers with due regard to creating a healthy work environment for higher production and productivity and to develop and coordinate vocational skill training and employment services. The Ministry of Internal External Affairs is responsible for the city's security, emergency management, supervision of other ministries, conduct of elections, public administration and immigration matters. The Ministry Culture Sports is responsible to develop and sustain ways and means through which the creative and aesthetic sensibilities of all the people remain active and dynamic. It also takes care of the promotion of sports and games and is responsible to conduct various tournaments of all the popular sports. The Ministry of Happiness works to improve the levels of happiness in the country through a variety of policies measuring the effectiveness of the congress's various social welfare programs. The Ministry of Finance manages all the government financial assets, smart cards, and propose economic and financial policy, and coordinate and supervise with all the profitable operations. Ministry of Science works with the objective of promoting new areas of Science Technology and to play the role of a nodal department for organizing, coordinating and promoting Science and Technology activities in the country. The ministry of Industry overlooks and regulates the working of trade and industry and is responsible for growth and development of industry. TBC also has an Earth Ambassador, who overlooks the council but will not participate in the decision-making process. TBC will communicate with the other six ministries on Earth. Once Brahmavarta attains self-sustainability, TBC will become independent from Earth's finances.

Smart card system

The flow of money within Brahmavarta will be in the form of credits. Brahmavartans will be given a handy card with their bio-metric data imprinted in a chip. This card will be recharged everyday with credit points based on their working hours. Brahmavartans will be working 8-hour long shifts, at the end of which they will be given 240 credit points. Independent of the job, each worker will get 30 credits per hour. These credit points which is stored in a smart card will be used for buying basic utilities and other entertainment. With an average living standard, a Brahmavartan will require 126 points each day. So at the end of the day approximately 114 points will be saved. The saved points can be used during days of casualty or in sickness.

Worst case scenario

Population of 1,25,000 is quite a big to manage. If ever there is a commotion or a riot that needs to be controlled, the responsible person will be punished with a deduction of 100 credit points by Ministry of the Internal External Affairs. Still, if the person continues to cause a problem, gradually the credits will be reduced until it reaches zero and if it still continues, he would be transported back to Earth. If any of the representatives go rogue then Brahmavartans can file a petition on Human Resource Department on MADE and MADE will be given instructions for actions. These precautions will help in smooth flow of operations in Brahmavarta.

References

1]   Making Life Multi-Planetary, SpaceX. https://www.spacex.com/media/ making_life_multiplanetary-2017.pdf
2]   Paz Aaron Kleinhenz Julie E. "An ISRU Propellant Production System to Fully Fuel a Mars Ascent Vehicle". In:LUNAR AND PLANETARY SCIENCE AND EXPLORATION(2017).
3]   The ideal settlement site on Mars - hotspots if you asked a crop. https://www.wur.nl/en/newsarticle/ The-ideal-settlement-site-on-Mars-hotspots-if-you-asked-a-crop. htm
4]   Lin Wan. "A novel material for in situ construction on Mars: experiments and numericalsimulations". In:Construction and Building Materials120 (2016), pp. 222–231

[5]   Mukundan, A., Patel, A., Shastri, B., Bhatt, H., Phen, A. (2019). Mars Colonies: Plans for Settling the
        Red Planet (1st ed., p. The Dvaraka Initiative: First Self-Sustaining Human Settlement on Mars).
        Polaris Books.

[6]   Hydrosight. "Acrylic vs. Polycarbonate: A quantitative and qualitative comparison". Archived from the
        original on 2017-01-19.

[7]   WHO international news release health sheet. https://www.who.int/ news-room/fact-
        sheets/detail/healthy-diet

[8]   Astronauts urine as manure for vegetable growth on Mars. https://crowdfunding. wur.nl/project/food-
        for-mars-urine?locale=en

[9]   Abercromby, A. F., Gernhardt, M. L., Jadwick, J. (2013). Evaluation of dual multi-mission space
        exploration vehicle operations during simulated planetary surface exploration. Acta Astronautica,
        90(2), 203-214.

[10]   A 'Martian' First: Earthworms Born in Mock Mars Soil. https://www.space.com/ 38960-worms-born-
        martian-soil-experiment.html

[11]   Steven R. Oleson Mare A. Gibson. "NASA's Kilopower Reactor Development and thepath to higher
        power missions".

[12]   Boroson, D. M., Robinson, B. S. (2015). The lunar laser communication demonstration: NASA's first
        step toward very high data rate support of science and exploration missions. In The lunar atmosphere
        and dust environment explorer mission (LADEE) (pp. 115-128). Springer, Cham.

[13]   NASA Ames Research Center Trajectory Browser. https://trajbrowser.arc.
        nasa.gov/

[14]   Asterank. http://www.asterank.com/

[15]   International Olympic Committee. "Olympic Marketing Fact File". In: (2017).

[16]   How much is that vial of moon dust really worth?. https://www.tampabay.com/ news/science/How-
        much-is-that-vial-of-moon-dust-really-worth-
        169133643/#:~:text=NASA%20maintains%20that%20%22lunar%
        20material,you%20can't%20sell%20it.

[17]   Birch, Paul. "Terraforming Mars quickly." British Interplanetary Society, Journal 45.8 (1992): 331-
        340.

[18]   Transportation Options for Getting Around on Mars http://marsforthemany. com/project/living-on-
        mars/getting-around-on-mars/

[19]   Ren, Dan, et al. "Continuous production of ethylene from carbon dioxide and water using intermittent
        sunlight." ACS Sustainable Chemistry Engineering 5.10 (2017): 9191-9199.

[20]   White, C., Alvarez, F., Shafirovich, E. (2011). Combustible mixtures of Lunar regolith with aluminum
        and magnesium: thermodynamic analysis and combustion experiments. In 49th AIAA Aerospace
        Sciences Meeting including the New Horizons Forum and Aerospace Exposition (p. 613).

[21]   Howe, S. D., O'Brien, R. C., Taitano, W., Crawford, D., Jerred, N., Cooley, S., ... Werner, J. (2011).
        The Mars Hopper: a radioisotope powered, impulse driven, long-range, longlived mobile platform for
        exploration of Mars (No. INL/CON-10-20600). Idaho National Laboratory (INL).

[22]   Hashimoto, A. (1983). Evaporation metamorphism in the early solar nebula—evaporation experiments
        on the melt FeO-MgO-SiO2-CaO-Al2O3 and chemical fractionations of primitive materials.

[23]   Rangwala, S., Rangwala, K., Rangwala, P. (2003). Town planning. Anand: Charotar Publishing House.

[24]   Schleppi, J., Gibbons, J., Groetsch, A. et al. Manufacture of glass and mirrors from lunar regolith
        simulant. J Mater Sci 54, 3726–3747 (2019). https://doi.org/10.1007/ s10853-018-3101-y

[25]   Glascock, N., Huber, B., Cantrall, C., Evonosky, W., Robinson, E., Dharmadasa, Y., Baker, K. (2018).
        MAFSA: Mars Autonomous and Foldable Solar Array. New Space, 6(4), 308-319.

[26]   Kading, B., Straub, J. (2015). Utilizing in-situ resources and 3D printing structures for a manned Mars
        mission. Acta Astronautica, 107, 317-326.

6: HIVE CITY

Dina Turuk
Lviv, Ukraine
melura@ukr.net

> *"we're all mad here. I'm mad. You're mad."*
> *"How do you know I'm mad" said Alice.*
> *"You must be", said the Cat, "or you wouldn't have come here."*
> *(Alice's Adventures in Wonderland, Chapter 6)*

On the first glance the idea of Mars colonization might seem fantastic and even crazy. But in fact it is very appealing. The eternal human desire to explore new lands now is looking towards another planets, and Mars, as the most real candidate is attracting all our attention. In the nearest future Mars is unable to attract because of its mineral resources, due to high cost of logistics, but it can be very important in the scientific field, as a unique place for research and experiments. And efforts, both theoretical and practical will motivate humanity into further development. Implementing the Mars colonization project symbolize the whole new era of human civilization. This will lead not only to material expenses in order to succeed some significant result, but also to creating a lot of new and revolutionary ideas in many aspects. Colonization needs to solve a lot of problems: scientific, medical, psychological, ecological... and a whole lot of issues which can be very common for a typical Earth city, that might turn into something extremely complicated due to an unusual martial environment, and solving these problems is very important.

Who are they, the future residents of Martian city? Science fanatics, romantics, dreamers, newcomers and experienced travelers - all of them should have everything to fulfill their essential needs, to make them able work, rest and create. It is very important to make the colony quite comfortable for everyone, to make the people feel themselves at home, to love their planet, allow Mars to capture our hearts. The success of the colonization of Mars by humans directly depends on the success of the colonization of humans by Mars. This is probably the first condition for successful colonization, the so-called human factor. If we succeed in it, then we will get success in everything else. Implementation of any huge project, especially like this one - is difficult and complicated process, but nothing will be reached without faith and desire to win.

Future city - is the first real step for Mars colonization, and it is why it should help to create the new local Martial patriotism, which will help to make this colonization successful.

VITAL ACTIVITY ON MARS

Energy
In order to colonize the planet, people will need a cheap and reliable source of energy.The two most probable sources of energy are nuclear and solar.

Nuclear energy

Nuclear energy - one of the main sources of energy in the presence of sufficient local raw materials for nuclear fuel, reliable, but can be dangerous in some extreme situations, which means they will have to be some safe distance from the settlements. Nuclear reactors must be built together with industrial complexes that require a lot of energy.

Solar energy

But the main idea is, of course, solar energy. Safe and environmentally friendly solar energy can be produced both in the city and outside it on any surface or structure, although on Mars solar energy is 2.5 less powerful than on Earth, and, in addition, sandstorms are quite common for Mars (this creates additional inconvenience).

Additional sources of energy

Wind energy. It is possible to create wind generators on the surface of planet in order to use wind as a source of energy. Kinetic energy is the movement energy of an object. It can be transferred between objects and transformed into other kinds of energy. It can be created from mechanical pressure, such as walking on pavement. It could be an additional source of energy, for example used only for street light, could be used as part of the "Smart City" program [1].

MARS GRAVITY

Mars Gravity is equal to 38% of Earth Gravity. Spending a lot of time in an environment with low gravity has a negative impact on human health, decreases weight of bones and muscles, decreases level of immune system, affects lungs and brain. Without solving the problem of low gravity and overcoming its negative effects, colonization of Mars, as well as other planets, is impossible. In all parts of the colony, where people are present, gravity should be similar to the one on the Earth. And it should be some complex system for huge areas (not personal items like special beds and special individual costumes). But for fully robotic industrial enterprises, low gravity can be useful, in particular, increases the load capacity of mechanisms, vehicles and, accordingly, the load-bearing strength of building structures. In agriculture and related to its food industry - impact of low gravitation on the development of plants and bacteria might also become surprisingly positive.

RADIATION

Radiation - one of the most serious problems and barriers for colonization of the Red Planet. As there is almost no magnetic field and as a result - a very thin layer of atmosphere, a lot of sun radiation is getting to the planet. Creation of safety measures is a must for such colonization. Protection from radiation is needed for the city, for personal use, in transport, including tourist vehicles, and in any kind of industry where humans are involved. Everything that was made from local martian material should be de-radiated. There are also ideas about creating most of the colony under the surface, using martian ground as defence from radiation and low temperature. The MarsOne project proposes to construct special shelters in case of dangerous radiation flashes. All these projects are interesting and can be used for the first settlements.

Undoubtedly the most desirable for the first Martian city would be creation a reliable in terms of radiation protection, and in addition, durable, transparent, insulating material for protective domes; it may be airgel or triplex domes that protect against ultraviolet, are transparent and retain heat well.

TEMPERATURE

Temperature on Mars fluctuates according to various sources from +30 C ... +20 C (at noon on equator) to -153^0 C ...- 170^0 C (in winter at the poles). Therefore, reliable thermal protection is required, something like thermal protective sandwich panels and multi-chamber transparent structures.

ATMOSPHERE

The atmosphere is 96% Carbon Dioxide and only 0.2% Oxygen and is unsuitable for human respiration. Projects that can be used for providing colonies with Oxygen are either global, to change the planet's atmosphere and local "Oxygen plant" projects, which may become later part of the global process of creating an atmosphere suitable for breathing.

The University of Lisbon (project manager Vasco Guerra) proposes to create and split plasma Carbon Dioxide into Oxygen and Carbon Monoxide (a fuel of gas stations). Low atmospheric pressure and low temperature of Mars will create conditions that are much more favorable in comparison with terrestrial ones.

A group of physicists from the University of Porto and Lausanne also suggested getting Oxygen directly from Mars' atmosphere using low-temperature plasma. Scientists at the California Institute of Technology have created a prototype device for Oxygen production that acts as a mini-particle accelerator and converts Carbon molecules into Oxygen, accelerating them and colliding with the surface of the gold foil. Pre-ionized, electrically charged molecule of Carbon, accelerated by electric field, cuts into the gold foil and breaks into pieces, releasing Oxygen.

In MIT they are working on a device "The Mars Oxygen ISRU Experiment (MOXIE)", which converts Carbon Dioxide into Oxygen.The device accumulates Carbon Dioxide, compresses it and by electrolysis converts it into Carbon Monoxide and atomic Oxygen, which forms molecular Oxygen. It is expected that the device will significantly save money on oxygen for rocket fuel. In case of successful implementation of this project in the future it is planned to create a whole program for the providing Oxygen martial atmosphere.

American scientists have proposed to solve the problem of lack of Oxygen on Mars with help of deep-sea Cyanobacteria that survive in the most difficult conditions: in Antarctica, Valley of Death, even on the outer surface of the ISS. Due to photosynthesis, they are theoretically able to absorb Carbon Dioxide in the most extreme conditions for humans, and feed on the rocky surface of Mars, while releasing Oxygen. According to Elmars Krausz, populating the red planet with Cyanobacteria is theoretically able to start the process of Oxygen production on Mars.

Oxygen production plants are envisaged for the projected city. As additional Oxygen sources for the needs of the city can be agricultural objects - greenhouses for cultivation of different plants. Greenery within the city and plants in personal apartments will be useful for purifying the air, saturating it with aromas and for psychological comfort.

The next issue is the use of Oxygen. Breathing with pure Oxygen for more than 15 minutes. (with normal, body pressure) leads to Oxygen poisoning, and a high concentration of Oxygen is dangerous itself. By lowering the pressure to 0.21 atmospheres, poisoning can be avoided, but what can be the consequences of constant (lifelong) stay of people in such conditions - is an issue that requires further research. Another way is to make artificial air (mixture of breathing gases). Most of the known gas mixtures for respiration are a mixture of Oxygen with one or more inert gases. The problem is that there are very few inert and other gases on Mars (nitrogen - 2.7% and argon - 1.6%). Choice of safe Oxygen usage is extremely important and vital [2, 3].

WATER SUPPLY

Water on Mars is not a problem. Most water reserves on the planet are concentrated mainly in the form of ice in the so-called cryosphere with a thickness of tens and hundreds of meters. According to researchers, ice water will be a source of drinking water and production of the rocket fuel. Martian soil contains up to 60% water. Water on Mars is available even in the atmosphere, which often has one hundred percent humidity. The future city will have its own factories of water production, special equipment to prevent freezing and a system of transportation to the consumers. Sufficient reserves of water and ice must be located in the city in case of unforeseen situations [4].

MINERALS

Construction is the foundation of colonization (and civilization). Significant mineral resources are able to provide the colony with enough raw materials for their successful functioning and further development. On Mars there is Sulfur, Calcium, Aluminum, Magnesium, Sodium. Due to the lack of Oxygen in the atmosphere, the presence of metals such as Copper, Iron, Tungsten, Rhenium, Uranium, Gold is expected. Proponents of colonization believe that minerals of the planet may be suitable for the production of plastics, glass, iron, copper and steel. Martian soil (regolith) contains layers of Oxygen and Silicon compounds with Iron, Aluminum and Magnesium and can be used as a source of minerals, and waste from their extraction as a raw material for 3-D printers at construction of roads, landing sites, buildings. Construction (and other industries) for their own profitability should be as robotic as possible with widespread use of 3D printing technologies. The first Martian colonies should establish basic industrial enterprises for the functioning and development of the construction industry. This will be possible due to the significant financial costs of transportation necessary materials and equipment from the Earth. The next step should be the development of industries, whose products are focused on meeting the daily needs of people [5].

FOOD INDUSTRY

Food independence is one of the main conditions of existence of Martian colonies. Local food production should gradually catch up with the growing needs of the colony. The food industry will be based on successful results of the functioning and development of the Martian agricultural sector and the successes in its creation of artificial food in laboratories, in the chemical industry and with the help of 3-D printers.

Growing plants

Growing plants (primarily wheat, corn, potatoes, soybeans, beans) will be the basis of a colony's agriculture. Initially, growing vegetables, greens, seafood will be carried out with the help of hydroponics in underground hydroponic farms. Later - in Martian soil under the protective domes in the "vertical farms "(Martian greenhouses). Growing plants will be a key factor in the creation of a functioning waste-free system of the food industry and human life.

Jim Cleves of the Blue Marble Space Institute considers it appropriate to break the so-called "death taboo" and to feed the soil with the bodies of deceased colonists. The proposal is quite contradictory, "taboo of death" and "taboo of life" can be strangely related [6].

Artificial food

The cost of transporting animals, high energy consumption and consumption significant amounts of Oxygen, water, land, feed makes livestock unprofitable. This means lack of dairy products, eggs, meat, and the need to replace these products with artificial analogues. It is necessary to grow animal protein.

Insects

Many researchers predict that the colonists' diet will be enriched at the expense of insect breeding (based on crop production). It is offered to breed house crickets, which gives high "yield" per unit area. The idea must be carefully tested and worked out to avoid the annoying consequences of possible mutations; insects are quite agile, ubiquitous, hardy and fertile, and if they enter the city, they can quickly become its main owners, but it is possible to breed fish, frogs, snails to solve the problem. Martian cuisine should be wholesome, nutritious, tasty and exotic!

The restraining factor in food production will be the area of the territory suitable for use. According to estimates, 47 m2 of agricultural land per capita is required, but by optimizing space, due to advances in genetics and increased production of artificial products, this area can be less.

COLONY INFRASTRUCTURE. SPACEPORT. HIVE-CITY

Colony will begin with the spaceport. At first it will be a landing site, a warehouse structures and the first residential structures on Mars (protective capsules or underground bunkers). Over time, there will be spaceport buildings, warehouses, offices, customs, zones of cargo handling and quarantine, reception and passenger areas, hotel complexes, catering establishments for employees, tourists and settlers, and transport hubs. The location of the spaceport will be determined solely by functional, technological conditions, convenience of takeoff and landing of spacecraft and other specific requirements. According to sources available on the Internet, this should be a flat area somewhere in the area of the equator.

The most likely options for establishing a colony were identified during an international seminar. There are two main options: the Hellas Valley and the Mariner Valley. The valley of Hellas is a rounded plain lowland in the southern hemisphere of Mars to a depth of 9 km, with a diameter of about 2300 km. Mariner Valley - a giant system of canyons 45000 km long (up to 11 km deep, up to 200 km wide), which extends along the equator. In the valley of Hellas it is much colder, but the greatest force of gravity and rather low radiation background compared to the "warm" equatorial Mariner, where the minimum temperature is much higher. Reliable radiation protection and low gravity are still problems for humanity, and the development of thermal protection is quite successful, so - the valley of Hellas!

DELIVERY OF PEOPLE AND GOODS

Delivery of people and goods will be carried out initially by land transport (rovers) and aircraft. Eventually, aviation can become exclusive tourist transport, as well as transport for scientific exploration and in case of emergency. While departing or arriving at the spaceport it is possible to get a unique air tour over Mars (in case of good weather).

Because of the huge temperature difference, tornadoes are possible, and they can reach heights of 20 km and a speed of 300-350 km/h. Sometimes dusty storms cover almost half of the planet and can last for six months. For fast, relatively secure and a more affordable sightseeing trip from the spaceport to the city and back Hyperloop system can be used. The most reliable means of delivering goods and people, although not as attractive from a tourist point of view - in the tunnels under the surface.

Passenger and transport flows will be separated. Part of the transport will be sent directly to industrial facilities outside cities. Part of the transport and people will be delivered to the city universal transport hub, which also provides quarantine and control services (mainly for the movement of people and goods in the opposite direction). There after re-inspection, and if quarantine is not required, goods will be distributed according to the needs of the city through the same underground transport tunnels, to the consumers. People can use elevators, escalators or stairs to get into the city, directly into the pedestrian zone, or to the transport levels. The difference in height (up to 9 km) between the main surface (spaceport level) and the valley surface (universal transport node level) with a relatively small deepening of transport tunnels can be overcome by elevator systems, and in the case of external tunnels - gradual slope of the tunnel in a circle along the valley contour (additional tourist attraction) [7,8].

INDUSTRY

The main task of industry - to ensure the viability of the colony and its maximum economic self-sufficiency (supply of energy, Oxygen, water, construction materials, manufacture of machinery and equipment, agricultural products, food, chemical and light industry, waste processing, etc.). Major industrial facilities that can pose a potential threat to the city's environment and safety will be located outside the city in a safe distance. Extractive and processing industries, respectively - near the sources of useful fossils. Folding zones should be provided directly near the enterprises, and also in the city and around the spaceport (for imported terrestrial products, and subject to the flourishing of Earth-Mars trade for Martian export products).

ADDITIONAL RESOURCES

For its own safety, the city must have several autonomous sources of energy, reserves of Oxygen (air), water, food in different parts of the city.

WASTE UTILIZATION

An important element of ecology and the whole economy is the optimal industrial and household waste processing system. Processing of industrial waste is determined by their type and technology of each specific enterprise and their mutual cooperation.

Inorganic wastes of the colony according to the sorting system of different types of garbage are discharged by consumers into separate underground tanks equipped with a drying system and grinding. Through pipes (as pneumatic mail) they are sent to processing plants, where sorted garbage of each type is crushed, pressed, and packed for further transportation to factories according to the type of raw materials.

For organic waste, which is a valuable fertilizer for Martian greenhouses, it is proposed to use the sewerage system on the same principle. All organic waste is dumped in the kitchen sink or toilet, flushed into the tank, where it is automatically crushed and get into the sewer network (sewer pipe in view of the following dual use can be increased in diameter by 1.5-2 times, up to 200mm), then organic waste is dried, ground into a powder, pressed into granules and briquettes and is transported to a warehouses at greenhouses. The remaining technical water is devoid of unpleasant scent can also be used to water plants or in greenhouses within the city, or after final cleaning and disinfection is returned to the water supply system. Subject to recovery also used air and heat.

As for the disposal of large household items (furniture, household appliances, etc.), this should be done at the request of consumers, by special transport services at industrial enterprises in accordance with the type of items declared for disposal.

HIVE CITY

The first residential structures of the city will be underground, in the thickness of the slopes (not down, but sideways). There will also be the first hydroponic farms, and the necessary engineering structures (supply of air, water, energy). Gradually, the city will be selected on the surface of the valley, leaving connections with underground structures for various needs, and in case of unforeseen situations. The city is planned to be built and developed in quarters with the use of 3D printers using regolith as the main building material. The planning structure of the city is honeycomb (like hive), given the energy efficiency of the form close to the circle, offered transparent protective domes for covering complexes of buildings (such as Fuller's dome using a transparent airgel or triplex, or other durable, transparent and heat-insulating material). The apartment building will be a hexagon with a common indoor hall on each floor and with a playground for relaxation and sports on the roof; six of such different storeys of buildings will be united around a common courtyard, six of such residential structures with courtyards will be united around the public space in the primary quarter and will be covered by a single transparent protective dome. The primary quarters will be united around the primary public (administrative and cultural) center, and those around the public, educational and sports and health complex. The public center of the city and other institutions will be formed on a similar principle.

Areas of the city with a capacity of about 20 thousand inhabitants will be covered with additional external transparent domes, the same domes - for sports stadiums, parks, as well as for the needs of agriculture and farms. Additional domes mitigate the sharp temperature difference outside and inside the living space, which will prevent excessive heat loss and create the illusion of leaving the room (with a temperature of +18°C, +20°C) on the street (with a temperature of approximately +15°C, +17°C). It will be possible (when installing thermal curtains, vestibule locks) to create different climates for the different parts of the city (more energy-efficient as cool and very warm). And at the request of residents within one external dome will be possible to arrange rainy days (general watering of buildings, plants, lawns, sidewalks and people who want to walk under umbrellas), or even, symbolically, a snowfall somewhere before Christmas. If one of the domes is damaged from the outside or inside, the damaged structures are automatically isolated from the rest of the city, and residents can be evacuated using underground passages to a protected (undamaged) part of the city until the problem is resolved. External (large) domes will rest on a 2-4 storey stylobate part (open outdoor terraces), which will allow you to place several external domes on the same level, regardless of the difference in relief and tear the bases of the domes from ground level to prevent excessive soil or dust from entering the terraces and to provide convenient maintenance of the domes outside.

House. Scheme of a typical floor plan with small apartments

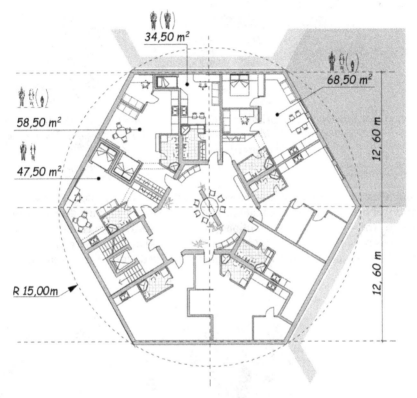

Total area - 600 m2 Approximate number of residents of the house - 105 (with an average amount of residential floors – 6).

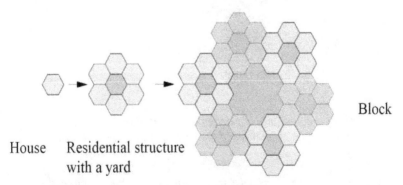

House   Residential structure
                   with a yard

Block

The scheme of formation of the courtyard. Number of inhabitants of the primary housing structure, united by a common yard – 630.

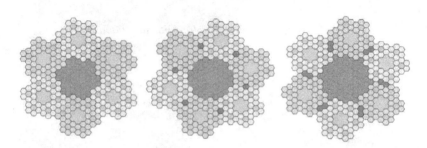

Options for forming neighborhoods in the primary community centers (the most densely built-up and with the arrangement of additional courtyards). The number of inhabitants of one structure is 22 680 (44 structures will be needed for 1 million inhabitants).

Building area of one such structure (without additional yards) - 20.58 hectares; the territory of housing and public buildings (at the maximum building density) will be equal to (for 44 structures - 1 million. population) - 905.52 hectares; the area of greenhouses (at the rate of 47 m2 per 1 person) - 4700 hectares; the area of the city without taking into account industrial enterprises and storage area - about 5610 hectares.

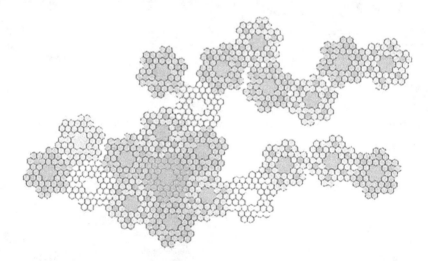

The example of the possibility of irregular layout of public housing structures (according to features of the surface and other factors).

In addition, the outdoor terraces can be used for walks in protective suits or protective personal vehicles, for temporary parking of walking rovers and small aircraft, and if desired by elevators, stairs and ramps can be descended to the surface of valley. In the lowest part, you can place auxiliary rooms, warehouses and protective shelters. Returning to the city, it should be noted that at the ground floor level there are shops, cafes, and communication nodes with the upper residential and lower non-residential floors. The ground floor is as open as possible for pedestrians, bicycle traffic (skates, gyro scooters, unicycles, etc.). From the lower (underground) floors there are exits to the level of public and individual electric transport stops in the tunnel of the loop underground transport system for electric cars and electric buses to move around the city and to facilities outside it. The lowest level is tunnels for the delivery of goods from industrial and agricultural enterprises located outside the city to consumers in the city. At the same level, unloading rooms and warehouses are provided. The lower (underground) floors, if necessary, can serve as a place of temporary stay for people in case of unusual situations with the possibility, if necessary, of their further evacuation by vehicles from different levels to safe areas of the city. For fast and comfortable movement within the city, there is also something like an above ground subway (like Hyperloop) in sound-insulating transparent pipelines at the level of 3-4 floors and above, to the level of terraces on the roofs under the primary domes. Pipelines are laid in different directions and levels, if necessary, passing directly through the buildings, stitching their architecture [9,10].

COMMUNICATION

Communication is an important sign of a healthy society. Due to maximum automation, robotics and computerization, the colonists will be deprived of episodic situational communication, which is not vital but adds color to life (a few words with sellers about the weather, fashion or their favorite cat, with a taxi driver - about politics, with waiter - about coffee, etc.) Therefore, in the planning of residential buildings in addition to a common courtyard, each floor has a spacious hall with chairs and a screen - a convenient place for closer communication between neighbors to watch news, TV series or sports programs with pizza, beer, or coffee that the robot will deliver from the nearest pizzeria or cook it.

REGARDING APARTMENT PLANNING

It is possible that the Martian standard layout of the apartment will be slightly different from the Earth version and some elements of planning and equipment from the housing bunkers of the first settlers will be borrowed with various improvements. First of all, these are sleeping capsules, following the example of sleeping modules (capsules) of the world's first Japanese capsule hotel in Osaka (Capsule Inn Osaka, architect Kisho Kurokawa). Capsule-bedroom - a bed behind the door in a capsule 1.25 m high. Such capsules can be placed in 2 tiers, if desired, they can be made double. The capsule is equipped with an individual electronic system that allows you to adjust the light level, temperature, sound background, control the TV, alarm clock, use the computer and internet. You can expand its capabilities by adding massage and solarium functions and additionally equip it with a large refrigerator with drinks and sandwiches.

Such a small private autonomous space for solitude, rest, and work. If such a capsule is made of sufficiently strong construction, with the possibility of sealing and transition to its own autonomous life support system for a certain time, then such a capsule can be used as an individual rescue module, the safest place is in the apartment in case of unforeseen situations. Thus, the plan of a small apartment for 1-2 occupants can include the following rooms: hall, wardrobe, toilet, and a common space that combines a kitchen, dining room, common room, and bedrooms with modules (capsules). This option is more energy-efficient, as the warmest rooms are the smallest (capsules and toilets). Therefore, it is possible that not all colonists will prefer traditional rooms like bedrooms. Another borrowing from the options of underground housing - the possible absence of windows in the rooms. Earth windows are traditionally used as a source of daylight, insolation, fresh air (for ventilation), for information (what is happening on the street or in the yard), and for communication (occasional communication with neighbors passing by for residents of the first 3 - 4 floors).

Over time, ventilation is supplanted by air conditioning (especially in polluted air), informativeness by video surveillance, and communication by mobile communication. All these technical means are completely available for the Martian colony, and instead of a window there can be a TV screen in the room, which in addition to TV programs shows online not only the view of the city or courtyard from the selected video camera but also breathtaking views of the planet. The psychological need for daylight can be met by luminaires that use a fiber-optic system to transmit daytime Martian light outward in the middle. The lack of insolation can be compensated by the system of automatic periodic quartzation of the premises.

It is absolutely logical that all communications and life support systems (including energy supply, water supply, sewerage, heating, air conditioning, internet, television, mobile communication, etc.) both the city as a whole and individual houses, buildings, apartments, and dwellings, maximally computerized according to the "smart city" and "smart house" systems successfully tested in earthly cities. The houses are equipped with a central air conditioner. The temperature of the premises is controlled depending on the presence of people in them and their individual needs, etc.

CONNECTION WITH THE EARTH

According to the plans of Elon Musk, Martian colonists will be connected to Starlink high-speed internet. To do this on Martian orbit will be installed several hundred satellites which will be linked to satellites in Earth orbit and will transmit information through them. The cost of the project is estimated at about $10 billion and should gain profits from the Earth's satellite Internet. It is obvious that the successful implementation of this project will bring both planets incredibly close and against the background of creating a fairly comfortable environment in Martian conditions some of the colonists will hardly notice the change in "scenery" [11].

ECONOMY. MARS EARNS

The high price of cargo delivery is the main reason of export non-profitability, especially of raw materials and products of the future Martian industry. The exception is a special export product - intellectual, one that can be created only in the unique natural environment of Mars. That's why in the future thanks to the strong scientific and technical base of the colony can be scientific, research, and experimental activities in many fields of science and technology. The tourism industry and related entertainment and service industries can make a significant contribution to the economic development of the colony over time, provided the infrastructure is well developed.   The role in tourist excitement around Mars will be played not only by a wide choice optimally planned routes to the sights of the planet but also the most exotic daily life of the colony, its entourage, cuisine, fashion, its heroes, its short history, local rules and superstitions, signs, folklore, legends, and even their ghosts. Perhaps, quite unexpectedly, the sphere of art will become a profitable sphere of activity of the colonists at a certain interest in Mars on Earth, in Martian fashion and artistic life of the colony. Mars can also be the exporter of jewelry. Such exports may be attractive, given the minimal weight and volume of such products, as well as their undeniable uniqueness and artistic value.The share of small in weight and volume can play a part in profitable exports of exclusive collectibles such as coins, postage stamps, postcards. It may be interesting to produce and electronically export products of other arts in the digital embodiment: digital art of photography, painting, and design, fashion, films and programs, scientific, popular science, educational and artistic works.

ARCHITECTURE, DESIGN, FASHION, ART TRENDS

One can predict that separated so far from Earth Martian colony can develop artistic styles in architecture, interior, design, fashion in three main areas: Nostalgic (traditional), copying perhaps even in an exaggerated-grotesque manner Earthly styles of all peoples and times. Rationalistic - pure function, and nothing beyond it, and Patriotic, Martian-futuristic concept, fundamentally unlike anything earthly, that will find his ardent supporters not only among the colonists but also among most Earthlings, in particular tourists.

ABOUT THE SAD...

Returning to the topic of death and disposal of remains. Cremation with the ritual of scattering ashes today is the most attractive from an ethical and environmental point of view. This is probably the best option of all that humanity has come up with but very energy consuming. Exotic and at the same time more utilitarian and energy saving is the proposal to use the remains for growing plants. Anyway everyone chooses what they like. The main achievement in this area should be the lack of areas occupied by cemeteries (on Earth we should also abandon traditional burials due to the constant increase of the cemetery areas). Mars must take a step forward in this area as well.

It is possible that electronic memorials will take root instead of cemeteries and it will be some kind of electronic archives (or libraries), where after each will remain except the serial number and information in the register, a small envelope (capsule) with a DNA sample and an electronic medium that duplicates the information of the general data bank: data from official sources (financial, legal, medical) and private information by will or at the request of relatives and friends (letters, posts, and comments on social networks, photos, videos). Special offer, instead of modest or pompous tombstones, funeral services bureaus will earn their money by creating modest or gorgeous memorial films, documentaries, or even feature films, for every taste and budget.

THERE WILL BE NO PRISONS

It is difficult to predict how criminogenic the situation may be in such a million-strong city far from Earth. Careful selection of candidates for the first and subsequent waves of settlers can provide relative peace and tranquility for some time, but the Martians in the second, third, and subsequent generations are not subject to selection. Law-abiding or criminal inclinations are not transmitted by heredity, only the environment, upbringing, and education can affect the development of personality and in some way adjust its behavior. But in such a delicate area, it is impossible to avoid mistakes, miscalculations and guarantee a 100% result.

On the other hand, burdening the economy with imprisonment or deporting offenders to Earth is financially burdensome and impractical. Let them sit at home in forced isolation, on state or personal maintenance (if the prisoner's financial situation allows) and work remotely. The "smart home" system is probably able to control the isolation mode. Although it is to be hoped that in the specific society of Martian colonists among scientists, creative individuals and highly qualified specialists there will be no place for serious offenses.

EDUCATION

Not only is obedience to law or criminal inclinations not inherited, but heredity also does not guarantee the transmission of genius or certain unique abilities inherent in parents, children may simply not have them. But the economy of the Martian colony cannot afford to allow retaining a large number of low - skilled workers (especially unemployed). The robots are cheaper, they do not need heat, Oxygen, food, shelter, medical care, protection from cold, radiation, etc.

Therefore, education (starting from kindergarten, then school and university) should be mandatory, effective, available to all and preferably free. Kindergartens should lay the foundations for the development of a harmonious, creative, communicative, socially adapted personality. Schools need to find, grow and develop talents. Universities need to improve and deepen skills and abilities in the chosen professional field. In order to meet the specific needs of the Martian economy exclusively with highly-skilled, intelligent individuals, the pedagogical methods of Mars must be one or two steps away ahead of Earth, Martian children must be one level above children of Earth. The main tasks of the education system will be the search for and development of abilities of all children (even those who do not seem to have special talents), education of truly creative individuals, providing them with everything they need for self-realization, happiness, as they become the pride and hope of the colony.

Such pedagogical methods have long been developed and tested and can be successfully used for the Martian colony. For example, back in the 1980s, Viktor Shatalov's system became widely known in the former Soviet Union. The Ukrainian pedagogue-innovator developed his own effective learning system using reference signals (interconnected keywords, symbols, figures and formulas with a concise conclusion). He offered the pedagogy of cooperation, game forms of classes. Thanks to this system, it is possible to teach teachers to teach all children, without exception, exact subjects (mathematics, physics, chemistry, astronomy, etc.) successfully, quickly, regardless of prior training and material level of their parents. All his pupils, without exception, showed a high level of knowledge. At that time, my mother worked as a math teacher at a specialized school with a sports bias. The school educated Future Olympic champions of the USSR. About the new method, she could read in pedagogical and popular newspapers and magazines, and what she found and understood she tried to use in her work and the result exceeded expectations! Later, colleagues and coaches even joked that now they are not sports, but sports and mathematics school, because a significant percentage of her students after graduation successfully entered technical universities.

There are the modern systems of art therapy and development of creative abilities! What amazing results of their application! Creativity inspires, gives meaning to life, stimulates the search for their place in society and improves the personality. It is clear that basic education should be free and orient students according to their inclinations and abilities to specific professions needed by the colony. Further education should also ideally be free, or with various forms of deferred payment, possibly by attracting targeted sponsorship from specific companies who are interested in their specific future employees. By and large, higher education should also be compulsory as a school, it should be a continuation of school education at a different quality level according to the specialization chosen by yesterday's school students, today's university students. The economy of the colony forecasts and shapes its demand for professions, their quality, and quantity, and distributes it according to the specialization of universities, which will form their requirements for school leavers. Everyone should have a chance (and a duty) to become a valuable, highly qualified professional, a master of his craft, to find himself, his place in a successful, unique million-strong Martian colony [12].

SECURITY. PROTECTION.

Every city, and even more so the city-state, must have police, firefighters, rescuers, and the military. In the conditions of a "smart" high-tech city-colony, equipped with automated fire extinguishing systems, notification in case of force majeure with the involvement of rescue robots, repairmen, etc., where everything is controlled by cameras, computer monitoring systems using patrol robots and drones, the number of people involved in the system of professional protection of law and order and rescue operations can be minimized. Moreover, it will be a rapid reaction force, patrols, security guards, police, military, rescuers - all in one bottle, universal defenders - a kind of professional Martian Jedi! And there will be few of them because the civil protection system must provide for regular training of citizens in case of dangerous situations and the involvement of all those capable of collective security. According to statistics for such a city, it can be about 2000 people (20 posts in different parts of the city). It is impossible to predict their exact number in a million colonies (at the beginning of the formation of the colony they will not be at all), but with the development of the colony, the number of rescuers is finally formed. There will usually be separate analytical departments for crime detection and prevention. Given the possible unique industrial and scientific developments on Mars in addition to its own security services at certain strategic sites, the city may need a separate unit to combat possible industrial and scientific espionage from some third-party companies and corporations [13].

MONEY

Of course, the colony must have their own money, maybe virtual, own currency. For small payments, for tourist attractiveness, for numismatists, as souvenirs, and as a material embodiment of an independent financial system, the minting of coins, cheap coins of everyday use and valuable collector coins can be provided. Basic payments will usually be non-cash using cards, chips, or phones. Of course, there will be banks and even your own mail as expensive tourist entertainment. It will be incredibly touching for someone to receive a letter from Mars or a package with a souvenir, regardless of the price.

CALENDAR

The Martian day (sol) is almost the same as the Earth day and is equal to 24 hours 39 minutes and 35 seconds. The number of seasons on Mars is the same as on Earth, but their duration is almost 2 times longer, because the Martian year is much longer, as much as 1.88 Earth years. At first, the Earth's calendar will work, but the less the colony depends on supplies from Earth, the more Mars feels at home, the more attractive (also for tourists) the idea of two calendars may seem (which is an exciting leisure activity for colonists - come up with and memorize names for a possible 23 months!).

RELIGION

The Internet makes it possible to attend prayer without leaving home, and more and more religions are enjoying technological advances. But it is possible that in time the Martian community will be able to afford to build a temple (temples), it would probably be appropriate for it to be a temple (temples) of all religions, following the example of the Indian temple in the city of Watunde [14].

PETS

Detachment from the Earth, a narrow circle of communication, awareness of the constant risk of living in a hostile environment, hard and responsible work, nervous tension, stress and domestic inconveniences, lack of professional psychological support in the early stages of colonization - these are the reasons that may cause first colonists of domestic animals. These creatures should be small, calm, friendly, but also unobtrusive and not too noisy, self-sufficient and warm enough, accustomed to artificial food, sedentary lifestyle in an enclosed space, and these are cats! Later, with the development of the colony, other traditional and less traditional pets may appear.

SOME ASPECTS OF THE LIFE OF THE COLONY.

In the early stages of colonization or in the comfort of a super-tech city of a million citizens, every inhabitant lives with a clear awareness of the close connection between their life and everyone around. Politeness, mutual respect, attention to one's neighbor, and mutual assistance are not only questions of pure humanism, culture, and upbringing, they are a condition for the survival of the colony. A carefully prepared terrorist attack or an element of uncontrolled rebellion can destroy the colony and undermine the very idea of colonizing the planet. Therefore, no matter what stratification of society occurs in the future, every citizen should feel protected and socially important. Probably, it will not be possible to dismiss a person without finding him a new job, preferably something better. And those who for some reason cannot or do not want to work should be guaranteed not only a roof over your head, heat, air and food, they should be engaged in social services, psychologists to help find a suitable job and their place in life. Everyone who is severely offended, deceived and disadvantaged poses a potential threat to any society, and an incomparably greater threat to a lone Martian colony far from Earth, at least until total terraforming (if any) occurs on Mars with friendly to people atmosphere, climate, radiation background. Any act of unmotivated aggression or suicide can undermine the psychological stability of the entire colony. Also an important aspect of the life of every colonist is the concern for the rational use of energy, air and water, environmental friendliness and waste-free. For this reason, using catering establishments will be a higher priority than cooking at home. In addition, it is another reason for friendly communication and informal exchange of personal information. It will probably be fashionable to grow plants at home that are decorative and edible - also not only for psychological comfort, but also to save common resources.

Modern technologies are actively displacing paper from everyday life. This process will spread even further on Mars. E-books, newspapers and magazines, notebooks, documents, letters, postcards are all we already have on Earth, then more. But a nostalgic desire to turn the pages traditionally and rustle the morning paper over a cup of coffee, or a desire to draw live with a pencil, felt-tip pen or paint, can lead to the appearance of synthetic paper, napkins or cardboard, no problems. The last pandemic drew attention to the importance of another strategic product - toilet paper, which is needed by everyone and always. We'll need artificial paper of similar quality and properties, which is soluble in water. And then we'll need diapers and dressings and napkins made of synthetic raw materials, because cotton plantations will not appear on Mars very soon.

AUTHORITIES

The social structure of the colony should be as democratic and socially oriented as possible. Management of city affairs is carried out by collective choice with the mayor. Members of the city council are elected from representatives of strategic and budget-forming enterprises, from representatives of scientific and financial circles, education, medicine, and are represented by active concerned citizens.

FULLER'S DOME

Possible long-term prospects for local terraforming.
The uniqueness of the dome design is that the more space it covers, the more reliable it is. The domes have a perfect aerodynamic shape so they can be erected in windy and hurricane areas. The disadvantages of domes in Earth conditions are that traditionally mass modern building materials have a rectangular shape (glass sheets, rolls of insulation and waterproofing, sheet metal), and to cover the faces of the dome they need to be further processed, cut to form a triangle, respectively, a lot of waste and it costs. On Mars, you can start from scratch, and with the widespread use of 3D printers, shape is not a problem. Thanks to future advances in the production of unique strength, thermal and radiation protection structures and materials, may become real the idea of local terraforming of Mars and building of a firm, transparent, huge dome over the almost round valley of Hellas, under which a securely protected, climatically comfortable area will be created with gardens and parks, with trees, flowers, birds, animals, ponds, convenient agricultural settlements, and a prosperous city.

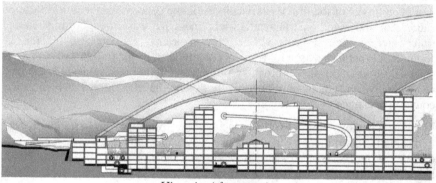

Hive city (fragment).

INSTEAD OF EPILOGUE

Looking even further.

In January 2020, at the seminar of TUBADZIN within the framework of a one-hour ecological blitz competition for ZEROwaste architects, my project of a human settlement of the distant future was among the three finalists. The participants' imagination was not limited to existing technologies and technical capabilities, so with enthusiasm and inspiration, future discoveries and achievements were borrowed from science fiction films and novels. The golden age of human civilization: unlimited energy resources (energy in the right amount can be obtained from almost anywhere in space), free gravity control, complete wasteless due to the technology of splitting into atoms and molecules, followed by the production of everything necessary, absolutely reliable protective structures and materials if necessary, both from radiation and from the aggressive environment of various types, effective heat, and sound insulation. People leave the ground, houses flying or drift above the surface (such as "flying saucers"), from time to time mooring near stationary "community centers" in their various needs (communication, entertainment, work, repairs, treatment, training, etc.), if desired, descending to the ground or the surface of the water for walking, sports or for research purposes. Earth for plants and animals! Waters for fish!

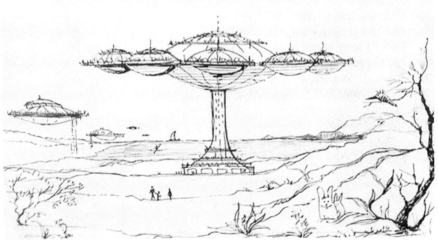

The concept of a fantastic city of the future for Earth, Mars and other planets.

Everything is environmentally friendly and safe (at the risk of natural disasters, you can fly away with your home to a safe place). This type of settlement in the distant future could be used for Mars and other planets. The capacity of settlements is unlimited in space.

Interesting, incredible journeys without interfering with the ecosystems of the planets, only study and observation. Isn't this the type of behavior inherent in mysterious UFO crews (if they really exist)? Observe, study, do not interfere, wait until humanity catches up and joins exciting journeys in search of space treasures, the most valuable of all that the universe has to offer: a unique experience, incredible impressions and boundless knowledge.

Forward to the stars!

REFERENCES

[1] Powering a Colony on Mars - Stanford University
[2] Oxygen Production on Mars - IPFN
[3] Could cyanobacteria terraform Mars? - Cosmos Magazine
[4] Water on Mars - NASA
[5] Composition of Mars - Wikipedia
[6] Henderson (Jim) Cleaves – BMSIS
[7] ESA - Frosty Martian valleys
[8] Elon Musk talks Hyperloop on Mars - Business Insider
[9] Aerogel - Wikipedia
[10] Buckminster Fuller - Wikipedia
[11] Starlink
[12] Viktor Shatalov - Wikipedia
[13] Singapore Police Force, Wikipedia
[14] Temple of All Religions - Ananda Pune

7: THE REPUBLIC CITY STATE OF THARSIS:
A Habitat of One Million People on Mars

Albert Sun*
University of Chicago
sunalber@uchicago.edu
Jessica Yuan*
University of Southern California
yuanjess@usc.edu
Lucy Shi*
University of Southern California
lucys@usc.edu

* All authors contributed equally to this work.
** Unless otherwise specified, all images are produced by the authors, and may be subject to copyright.

1 PROLOGUE

1.1 History

On May 30th, 2020, for the first time a private company accomplished a human space flight mission with a recoverable launch vehicle. Humanity saw new hopes for its grand vision of becoming a multi-planetary species. In 2035, a global treaty was signed by hundreds of countries, and humanity aimed high: to build a city-state of one million on Mars.

The region of Tharsis Montes was selected to be the destination of the pioneers and the cradle of the first human settlement on Mars. The endeavor took centurial efforts to come to fruition. The colony, later deemed "The Republic City State of Tharsis," has gone through four fundamental development stages so far: Corporatization, Sectionalization, Centralization, and Decentralization.

Corporatization was the establishment of a society structured as a corporation in the beginning stage of development. Funded by both private and public investments, the Tharsian Corporation was established in 2038, directed by the Mars Committee. Within the next 7 years, the committee selected, trained and transported 2,105 scientists, engineers, technicians, and city planners, and 42 autonomous robotic devices to planet Mars. By 2070, these people completed the environmental modification of lava tubes under Arsia Mons, built underground settlements and the prototype of the Energy System, and paved the way for businesses to enter Mars.

Sectionalization started when the basic structure of the City State had been established. The City State was divided into autonomous districts with local governments and a central government to better cope with the individual needs of every region. Prestigious universities around the world lined up to open research centers on Mars. The real estate market on Mars has bloomed. Business opportunities were at a height that was never expected. The Committee sold licenses for the entrances to Mars for personnel in private enterprises, offered loans and accelerator programs for startup companies, and guaranteed free entrance licenses for talented individuals and teams with enormous potential. The State government and district governments became well-funded through the sale of these licenses and tax collection. Many companies and individuals maintained steady growing profits renting their real estate properties to small and medium businesses. With rapid commercialization, by 2085, most of the infrastructures on Mars had been developed and 500,000 people were living in the City State.

The Centralization process, which gave the State Government the right to collect federal taxes, was initiated to resolve the remaining difficulties in constructing a self-sustainable Mars society. One such difficulty was that crucial services like the production of food, water and oxygen are difficult to privatize at low market prices. Also, the high interplanetary toll fee from Mars to Earth made it extremely difficult for preliminary businesses that exported goods to Earth to survive and forced many companies to leave Mars. Policies deployed to address those issues had been successful. By 2085, the City State had become a fully functioning colony of roughly 800,000 people. In the same year, city state "The Republic City State of Tharsis" was officially established, and all districts ratified the common constitution of the City State. The central government and district governments certified the state anthem "Life on Mars", inspired by the song under the same title by David Bowie written in August 1971.

Decentralization is the most recent stage, initiated in 2100 and remains to the present. The main objective of decentralization is to provide a free market, while ensuring the survivability of new businesses on Mars. This process is aided by industrial developments on Mars – especially the research on more efficient ways to manufacture essential products and the engineering of new propulsion methods to lower the interplanetary transportation costs. Meanwhile, the central government gradually lifted tariffs and cut taxes. Eventually, by 2120, humans have finally reached the milestone of 1,000,000 people on Mars – the goal set by far-sighted pioneers a century ago.

Up to here, humanity has learned the potential of joining hands. Now the ambitious Martians have passed the initial stages of scarcity, and capital is flowing towards research and technology that enable Martian settlers to further terraform Mars. It has been a vital steppingstone for humanity to set out further into the unknown void of the universe.

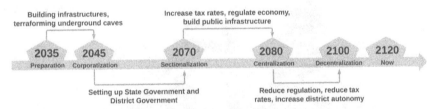

Figure 1: Development Timeline.

1.2 Foundational Concept

Signed by most countries on Earth in 2035, *Mars Advancement and Residential Settlement Treaty* (M.A.R.S Treaty) declared that the Mars City State would serve a peaceful purpose, aiming to expand the horizon for humanity. It should not spread nationalism, and it is forbidden for any country to claim land on the Red Planet [1]. Land tenure on Mars would follow the MSA standard (specified in 5.2). The rest of the land on Mars was recognized as the "*Common Heritage of Mankind.*[1]"

People who choose to permanently reside on Mars are required to relinquish their previous citizenship, and thus they would only represent themselves. They become Mars citizens and are subject to Martian laws. To ensure sustainable development on Mars, governments on Earth cannot interfere with political, cultural, and social decisions made by Mars civilization. After signing the treaty, companies from respective nations were granted the clearance to gain profits from Mars: including but not limited to the trades on Mars' resources, usage of intellectual property developed on Mars, and participation in research projects. An era of progression has thus begun.

Figure 2: The City State Flag.

2 A SUBSURFACE SETTLEMENT

2.1 Site Selection

Martian near-surface caves are ideal locations for human settlement. These underground caves 1) have sufficient lateral extent to shelter humans from surface hazards, such as micrometeorite impacts, dust storms, high fluxes of UV, alpha particles, and cosmic rays, 2) can maintain near-pristine surfaces and 3) can minimize temperature variations and create stable micro-climates [2].

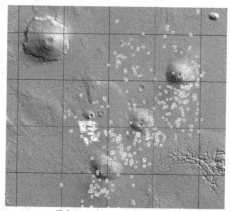

Figure 3: Cave entrance candidates detected by USGS. Image owned by USGS. City State (divided into five autonomous districts) is located to the South of Arsia Mons, the Southernmost mountain among the four.

Hence, the first million people city state is built in the region of Arsia Mons (8.35°S 120.09°W), which bears unique climate features that makes this place the best choice. First, located in the warmest temperature zone on Mars [3], it is the most energy-efficient place to maintain an optimum temperature inside the settlement. Secondly, the region consists of seven separate cave systems around the extinct volcano, along with hundreds of cave entrance candidates. Thirdly, glaciers are identified on Arsia Mons at both high and low elevations, which can be valuable resources. Fourthly, the three-mountain system on the Tharsis plain provides an accessible expansion pathway for more city states in the future.

2.2 Urban Concept

The Republic City State of Tharsis is located to the *south* of Arsia Mons, where there are denser caves and are later developed into habitable areas. Shallow and flat caves are ideal for human settlement; deeper and larger ones are usually enriched with geological evidence, making them ideal locations for scientific research [2]. Therefore, residential areas are built in underground caves while transportation systems run above the surface.

By 2120, the city state had developed five inter and intra-connected *districts*: *Flamma* (the Capital District), *Aurum*, *Lignum*, *Aqua*, and *Solum*. These regions are selected after careful assessment and consideration, including the abundance of resources and their proximity to interplanetary spaceports, to ensure the diversity of the economy.

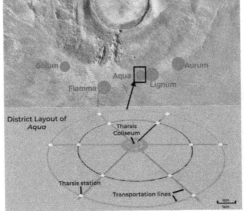

Figure 4: Layout of the 5 districts: Aurum, Lignum, Aqua, Flamma, and Solum. (Image of Arsia Mons from *Views of the Solar System* owned by Calvin J. Hamilton) and the structural layout of district Aqua.

Each district consists of two layers: *Thar-Up*, which is the surface layer, and *Thar-Down*, the subsurface layer (See Figure 6). With a fan-like layout, *Thar-Up* runs the state's fastest transportation systems; a grand coliseum is located at the center, serving as a multi-purpose entertainment center that hosts sports events and concerts. Multiple unmanned factory zones are constructed above ground, housing several 3D-printing, energy production, and resource extraction factories. *Thar-Down* hosts most human activities, including administration, agriculture, business, residential, and research.

2.3 Architecture

Thar-Down provides Martians with ample residential spaces (See Figure 6). Buildings in Thar-Down are dome-shaped to better handle the pressure difference between the Martian ambient pressure and room pressure. Domes vary in sizes, and many of them are as large as a Stadium. They are constructed by giant 3-D printers in-situ using regolith [4]. Microorganisms are used to detoxify regolith, aiming to reduce highly toxic perchlorate [5]. In addition to regolith, layers of heat insulation and air tightening materials are cast in the dome wall to enhance strength, airtightness, insulation, radiation protection, and offer pleasing aesthetics.

Serving residential, recreational, and tourism purposes, Domes are designed with compact hexagonal modules to facilitate efficient management of feed systems for oxygen, water, and energy. Lawns with a circular running track surround the periphery of each floor, providing residents with ample interaction space, which effectively soothes the *surface chauvinism* they may have. Each hexagon module consists of a small central garden and surrounding apartments. Corridors connect each module to the elevator lobby at the center of each floor. Residents can use elevators to go to different floors, the metro station in some of the sizable Domes, or go to *Thar-Up*. Domes are connected via underground tunnels through which residents can easily travel between the Domes (See Figure 6).

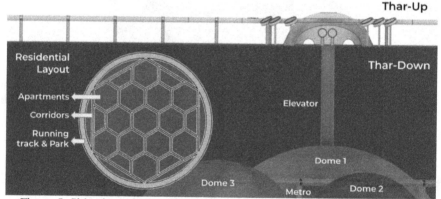

Figure 5: Side view of *Thar-Up* and *Thar-Down*. The sizes of the Station and Dome vary by the usable underground area.

Feature: Skylight
Though the lighting system could be easily implemented using LED lights (which are widely used at the beginning of the 21st century), natural sunlight is conducive to people's psychological health as well as architectural aesthetics.

A sunlight piping system conceived by James Ramsey and Dan Barasch [6] uses a system of mirrors, installed on the Martian surface and pivoting and rotating to follow the sun's path across the sky, to collect sunlight. Large parabolic dishes underground concentrate sunlight which is then funneled between a series of mirrors. Employing distributors – optical diffusers mounted to the ceiling made of anodized aluminum panels – the light is then spread over Martians' homes.

This unique system can produce a full spectrum light efficiency of about 70 percent (losing only tiny amounts of light that are absorbed instead of reflected at each mirror touchpoint). For comparison, the alternative – collecting energy from solar panels and using that electricity to power LED lights – would create efficiency closer to 7 percent.

2.4 Key Infrastructure

Transportation Tubes

To protect public transportation systems from constant dust storms, all hyperloops and railways run in closed tubes, which are mostly 3D printed with regolith. Different sections of the tube are contracted with different transportation companies, who compete to provide the best commercial services for residents and industries. Cargo tubes are shielded with aluminum to prevent charged particles from interfering with sensitive electronic devices. There are also transparent tourism tubes specially designed with radiation protection that offer tourists a wonderful panorama of the landscape.

Tharsis Station

Stations are built at the intersections of radial and peripheral transportation lines. Each station is a local center of transportation. The station has three layers. The top two layers are for commercial and tourism transportation, and the bottom layer is for private car transportation, which allows people to travel to less populous places or even the faraway Olympus Mons while enjoying driving by themselves. The skylight system is implemented around the station, collecting sunlight and directing it underground.

The Tharsian Cooperation implemented a revolutionary airlock system, consisting of *automated revolving doors* instead of two parallel gates, is widely implemented in the City State for its two major advantages:

1) The revolving door perfectly accommodates two-way traffic. Unlike the traditional airlock design of two paralleled gates, people can go in and out at the same time, making this design is significantly more time-efficient.

2) The revolving door follows a "pressurized – depressurizing – depressurized – pressurizing" cycle. Take one section of the revolving door for example. When walking out of the station, people in spacesuits enter one section of the door and it starts revolving. The controller for this section begins depressurizing the air inside, so that when this section completely faces the outside world, the pressure difference is minimal, and people can exit safely and conveniently.

Figure 6: Underground infrastructure of domes and transportation. Each dome is connected to the Tharsis Station on surface via an elevator.

3 AN INDUSTRIALIZED MARS

3.1 Automation

In 2035, a cargo ship was sent to Mars, carrying a fleet of 42 robots equipped with the most advanced AI system (at that time). They deployed themselves on the surface of the Red Planet, and first set up solar panels near the factory zone and greenhouses in the agriculture section. Then, for six months, they embarked on a journey of excavating proper materials determined by previous Mars missions. They, afterwards, are divided into different groups to work on building 3D printing factories and setting up subsurface living spaces with airlocks, nuclear plants, oxygen, and transportation systems.

Meanwhile, note that in the same way that no human is perfect, there is no error-free machine. Therefore, the robots were capable of self-repairing. Also, due to the delay in communication between Earth and Mars, each robot constantly ran a feedback loop to conduct diagnoses on itself. They bore the extreme environmental conditions on Mars, such as temperature, pressure, radiation, and dust. Importantly, the peace was carefully maintained between the first robots from Earth and their "brothers and sisters" additively manufactured on Mars.

However, with great intelligence and physical power, robots might bring security concerns. Hence, humans took extra precaution on their creations and employed the high-level concept of the *Three Laws of Robotics* suggested by Issac Asimov [7] to ensure that A.I. is designed to assist humanity and contribute to the social good.

Now in 2120, robots have a ubiquitous presence in various industries on Mars for varied reasons. In the field of resource acquisition, the toxicity of substances on the Martian surface makes human excavation, inspection, and transportation in mining extremely dangerous. Therefore, robotic dogs were designed to replace humans in these tasks. Growth monitoring robots work in the rotating hydroponics greenhouses and use computer vision to determine the growing state of each crop every six hours – something that human eyes cannot differentiate – and adjust the lighting and rotational speed according to results of numerous calculations. In energy plants, human inspection is no longer necessary thanks to the deployment of robots. Meanwhile, the extensive application of A.I. in education helps to guard against systematic bias and prejudices.

3.2 Resource Acquisition
<u>Metallic Material Extraction</u>
87 miles away from the center of Arsia Mons, Olympus Mons is in the northwest of the colony's urban area. As the tallest Mountain in the entire solar system, Olympus Mons is full of ore resources. A set of trans-mountain highways constructed at the Corporatization stage travel from Arsia Mons to where profuse resources are estimated to be at Olympus Mons. Tesla Ultra-Cybertrucks, the most popular vehicle for resource extraction, is an autopilot electric truck that feeds off its energy from the KRUSTY nuclear stations located along the road. It has four robotic arms for excavation and three automatic intelligent "robot dogs" which carry drills and explosive devices are on dock, totaling a carrying capacity of 20,000 pounds per Ultra-Cybertruck.

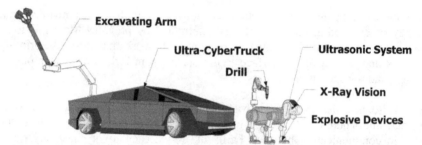

Figure 7: Ultra-CyberTruck and Robotic Dog[2].

The mining process of one site is usually carried out by an assembly of 12 Trucks. When the Trucks are driven to a site where abundant targeted resources might be found, every truck releases its "robot dogs" to the nearby locations. Incorporated with computer vision, X-Ray sensing, and metal detectors, the robots drive themselves to designated sites, drill holes through the wall, deploy explosive devices in the holes and determine the capacity of ores in that site after the explosion. If the capacity meets the designated standard, they will place beacons, which will be received by their parent Trucks and the Trucks will soon arrive and excavate the ores with robotic arms.

Raw materials are directly transported to processing facilities such as blast furnaces where raw materials go through the reduction and purification process. Refined materials are delivered to heavy manufacturing production lines through hyperloops.

Resource extraction and processing produce a substantial number of tailings. The tailings on Earth have been damaging the environment outrageously, and tailings on Mars could have done so even more due to substances on Mars that humans are not familiar with. To prevent tailings from hindering the future goal of terraforming Martian surface, the Environmental Agency on Mars has issued policies requiring tailings to be treated properly. A widespread practice is to let robotic devices decompose the tailings following the standard set by the Agency and convert them into useful materials.

Water
Water is the lifeblood of humans on Mars. The acquisition of water (and subsequently oxygen) is conducted in multiple ways. The most prominent method is to collect ice caps and produce water with the *Reverse Water Gas Shift Reaction.*

Ice caps are abundant in the Martian subsurface area, as well as the surface area of Arsia Mons and polar regions [8]. Cargo-hyperloops on a larger scale connect the City State to potential mining sites on the glacier of Arsia Mons, transporting Ultra CyberTruck and CyberDogs. The CyberDogs for ice mining are equipped with hot water drills and pump the melted ice onto the CyberTruck storage. To purify the melted water from excessive amounts of perchlorate and other salts, the CyberTrucks are equipped with desalination devices which release IX resin to capture chloride ions from perchlorate [9]. Then the water goes through a standardized purification process and be transported back to the City State.

The other way to obtain water is the Reverse Water Gas Shift (RWGS). The process is to capture the profuse amount of carbon dioxide from the atmosphere and reduce it with hydrogen to obtain water. [10]

$$CO_2 + H_2 \dashrightarrow CO + H_2O$$

The reaction is endothermic, and the equilibrium constant given by [11]

$$\log K_{eq} = -2180.6T - 0.0003855T + 2.4198$$

for such a reaction is incredibly low at low temperatures, and therefore it is ideal to combine the facility with a molten salt nuclear reactor. Instead of pushing turbines, a portion of the heat from the hot molten salt, which can get as high as 1400 degrees Celsius [12], is used to heat up the reaction chamber for RWGS, and a portion of the heat converts to electric energy to run RWGS facilities. Carbon monoxide produced is captured and become supplies for other industries, such as metal processing.

3.3 Agriculture
A Rotating Hydroponics Greenhouse System

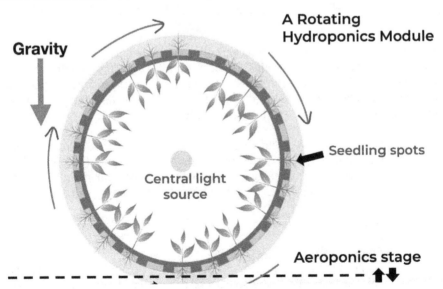

Figure 8: Rotating hydroponics System.
Upper part: a side view of the cylinder module. Plants are rotated vertically.
Lower part: a fully assembled greenhouse rendered by Omega Garden

The unstable climate, toxic soil and frequent dust storms on Mars make it difficult to grow crops in the same way as on Earth before complete terraforming. Instead of conventional space hydroponics, the Rotating Hydroponic Greenhouse System [13] is widely implemented to feed a million people. The system consists of hundreds or thousands of modules depending on greenhouse size. Each module is shaped like a cylindrical ring, with seedling spots on the inner surface. The module rotates vertically 360 degrees per hour around a central light source, which is also powered by skylight. Constant rotation activates auxins [13], the growth hormones, thus stimulate larger yields. Such rotation switches plants between hydroponics mode and aeroponics mode, saving a considerable amount of water and fertilizer and reducing the possibility of root diseases. Moreover, a cylindrical design means the plants capture most of the lumens emitted by the system's light source.

Agriculture experts have introduced "popularity" as an important metric of plant selection as population expands. The dishes served in Martian homes and restaurants are vastly diversified with the introduction of tomatoes, onions, cabbages, eggplants, carrots, garlic, spinach, pumpkin, broccoli, etc.

Food Printing and Meat Lab
3D printing and meat culture is extensively used in the settlement. Foods coming out of the harvest station of the greenhouse go straight into the processing center. A portion of the output is made into food cartridges for the 3D printing complex that produces snacks such as chocolate and biscuits, which serve as emotion-boosting refreshments.

Raising livestock on Mars is both technically and economically difficult. Instead, meat is cultured in meat labs that produce clean and healthy protein sources. Cultured meat labs take cell specimens of common livestock from Earth to generate edible meat in labs [14]. The meat produced from the meat lab tastes almost nothing different than real meat, and is widely popular for its decent prices.

3D printed pseudo-meat is another option for meat-lovers. 3D food printers can transform the texture of plant-based food into the texture of almost any meat. It is healthier, cheaper, more environmentally friendly and tastes almost no different from real meat. 3D printed meat, along with lab-grown meat, has become prevalent on Mars.

3.4 Energy
Integrated Energy System on Mars (IES-M)
IES-M is an integrated energy system that supports the life and production of one million people on Mars. The main sources of energy on Mars are solar energy, nuclear energy, and wind energy.

Every district of the colony is powered by its independent IES-M. Large nuclear power plants and solar farms make up the energy supply of most industrial productions; wind farms, mini-reactors, as well as solar balloons that are integrated into the residential grid power most of the residential area; Wind farms yield significant amount of energy during strong winds and dust storm when solar system does not function as well; transportation is powered by individual power stations supplied by mini-reactors and solar balloons, powering vehicles along the road. The electricity consumption in 2119 reached 49.2 TWh, and 48.7 KWh per capita.

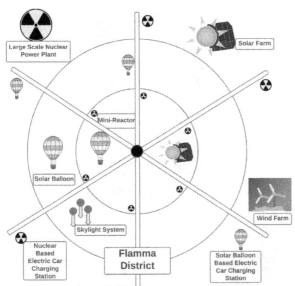

Figure 9: Energy System IES-M.

Solar Balloons

Roughly 40% of the total energy supply of the entire colony comes from solar energy. The expansive area of undeveloped land on Mars offered a tremendous spatial advantage for constructing large solar power plants. However, the periodic dust storm, which can last for days and even months, would lead to shortages of sunlight and contamination of the solar cells, defunctionalizing thousands of solar panels [15].

Therefore, the colony creatively adopted a Balloon-Based Solar Power System. The analysis and calculations demonstrated that the density for helium balloons is well below the density of atmosphere at the desired height. The system uses Solar Balloons — helium balloons covered with solar panels — raised at the height dust storms cannot reach, enabling uninterrupted sunlight reception during daytime.

A Solar Balloon consists of a section of solar panel on the balloon, a control box that automatically adjusts the air composition inside the balloon, and cables that transmit electricity to the inverter on the ground. The sphere is fixed with a rotational axis through the center. The control box feeds information about the direction of the Sun to its processor, which turns the balloon to the direction of the Sun. It also makes decisions whether to heat up or cool down the balloon to adjust the density and the pressure difference between interior and the atmosphere. The solar panels convert solar energy to electricity and transmit DC to the DC to AC inverter on the ground to power individual households or electric car charging stations.

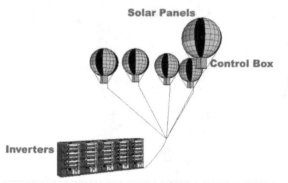

Figure 10: Solar Balloons[3].

Nuclear Energy

40% of energy in IES-M is supplied by nuclear energy. Nuclear energy is provided by both large-scale reactors built on the outskirts of urban districts and a web of mini-reactors incorporated into the residential area.

The mini-reactors are a more sophisticated version of KRUSTY (Kilopower reactor using sterling technology) [16], which is a molten salt reactor with a Stirling convertor to connect to the core, converting the heat to kinetic energy and then to storable electricity. The radiation shielding prevents any potential damage to individuals. These reactors can generate power ranging from 1 to 1000W, large enough to power an entire household. Every compartment in a structure (residential domes, factories, 3D printing stations etc.) is equipped with one such reactor, and all the reactors are in the radiation shielding room, which is connected to every individual compartment. These reactors are popular among real estate contractors to meet the MSA standard for energy provision (See 5.2).

3.5 Communication

Local Network on Mars

Constructing a local network system on Mars is more challenging than on Earth. The severe climate condition on Mars and the frequent appearance of mountains and valleys make it harder for telecommunication stations to perform optimally, plus the low curvature radius of Mars reduces the field of view of these stations [17]. While these communication methods function well in certain areas, it is difficult for them to provide a global network.

Since the City State is in the south of Arsia Mons in the low latitude zone on Mars, a group of satellites in Aero synchronous Orbits with orbital inclination 0 to 30 degrees is the most economical way to construct a local network on Mars.

Satellites in Aero synchronous Orbits are roughly 17,000 kilometers from the ground [18], forming a field of view of 23.6 degrees. The satellites, looking in the ground tracks in the "Mars Centered, Mars Fixed" coordinate, cover the whole Tharsis region enclosed in round ground tracks.

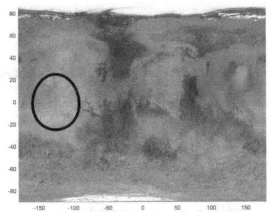

Figure 11: Tharsis region covered by satellites.

Starlink-Mars
Starlink, a project proposed by SpaceX, is a satellite constellation in low earth orbit, using laser instead of traditional radio waves to achieve higher internet speed and reduce delay of signals. Such a system is perfectly suitable for Mars in the future when the majority of land on the Red Planet is colonized. Currently, a number of engineers and technicians are dedicated to designing satellites suitable for Martian conditions to create a Starlink on Mars.

Internet on Mars
Building the Internet on Mars was a great challenge for the first group of engineers. The physical distance between Earth and Mars makes it impossible to synchronously access terrestrial internet servers on Mars. Every click comes with a waiting time of 10-20 minutes depending on the relative positions of these two planets. It is the most ideal to duplicate major internet servers onto Mars.

Beginning in the early preparation stage of the City State, for every internet server on Earth, a counterpart would be present on Mars. Every update on the internet on one planet reciprocally transmits to the servers on the other.

Traditional radio waves would not satisfy such a high volume of constant workloads due to their low data transmitting rate. Laser is a better option. Optical Payload for

LaserComm Science (OPALS) [19] has been able to achieve a multitude times of traditional speed of data transmission. When transmitting from Mars to Earth, servers on Mars uplink data to satellites on the Low Mars Orbit, which then transmits the data to the International Space Station, which downlinks the data to ground stations on Earth, and vice versa.

Solar Conjunction and Relay Stations
Oftentimes signals must travel farther to get to the other planet, especially during the Solar Conjunction when every two years Earth and Mars are on radially opposite sides of the Sun and there is a two-week long "blackout" time when most satellites become incommunicado [20]. At the early stage of the City State, it pauses most of the communication with Earth, and uses this period to repair and upgrade existing communication devices. Later, four groups of communication satellites are launched onto the La grange Points L4 and L5 of Earth and Mars, respectively. They carry signal repeaters, amplifiers and are equipped with information assurance system. These satellites serve as relay stations for signal travelling between Earth and Mars, maintaining interplanetary communication continuously.

3.6 Extraterrestrial Outpost
Phobos, one of the two Martian moons, is a destination no tourist wants to miss. It has only 0.1% of the surface gravitational pull of the Earth: a 150-pound (68 kilogram) person would weigh only two ounces (68 grams) on Phobos. In this natural microgravity environment, the customized *roller-coaster* is usually kids' favorite in amusement parks. Better yet, there are two gigantic space elevators currently under construction, which will transport people from and back to Mars while enjoying their dinner in the floating restaurant.

The Phobos Lab (nicknamed "the Crazy Lab") experiments on mind-blowing technologies, and one such undertaking is to build a small-scale model to simulate a dyson sphere which wraps up a star. Ice reservoir[4] on Phobos provides water to be made into rocket propellants [4]. Spaceships dock on Phobos Stations and refill their tanks – before starting off journeys to the moons of Jupiter or nearby solar systems.

4 AN EXO-POLITAN COMMERCIAL CENTER

4.1 Economy of the City State
The Primary Sectors
The Primary Sectors of the City State include Ore Mining, Water and Oxygen Production, Agriculture, and Manufacture. Ore Mining and other raw material extractions, processing and heavy manufacturing are mostly concentrated in Aurum, the district surrounded by abundant ore resources and connected by inter-mountain highways. Companies construct raw material processing facilities like blast furnaces. Much of the refined raw materials go into autonomous factory assembly lines. Inter-district hyperloops connect these facilities directly for distribution to every district. Many households have adopted 3D printing based customized light manufacturing in their own living space. Processed materials such

as PLA, steel, and glass are directly distributed to individual 3D printers for individual households.

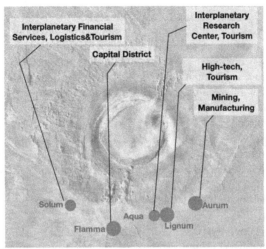

Figure 12: Districts with respective specifications in economic sectors.

The Secondary Sectors
The Secondary Sectors of the City State include advanced R&D, and cutting-edge technology industries. These mostly take place in Lignum and Aqua, which are called "twin cities" for their interconnectivity and tightly related economic activities. Lignum, utilizing its proximity to the Aurum, has been a hub of cutting-edge technology industries like Advanced Aerospace and Astronautical Industry, Advanced Agriculture and Advanced Computer and Robotic Productions. Right Beside Lignum, the "twin" Aqua is an interplanetary cosmopolitan city for higher education and research. Frequent Academic exchange programs between Mars and Earth usually take place in Aqua.

The Tertiary Sectors
The Tertiary Sectors of the City State include inter-planetary financial services, interplanetary trade logistics, tourism as well as entertainment. Every year, thousands of people and millions of tons of products depart to and arrive from Earth at the Musk Space Center near Solum. As a result, logistics companies, financial institutions and tourist agencies located in Solum have been big beneficiaries of interplanetary activities. The service industry has become a major economic sector in Solum and made up a towering percentage of Solum's GDP. Tourism is a huge economic contributor to every district. Most visitors from Earth like visiting the Capitol Building, Research Centers and Universities on Mars. The most frequently visited districts are the "Twin Cities" Lignum and Aqua: Due to the presence of advanced technology and cutting-edge research, the "Twin Cities" has drawn enormous public attention. Therefore, the entertainment industry, such as

Mollywood Martian Movie Industry (MMM), and Microtonal Martian Music (MMMc) Industry have been booming in the Twin-Cities.

Consumer Manufacture with Additive Technology
The rise in additive manufacturing changes the way people think about production on Mars. As 3D printers are being produced by 3D printers at lower prices, and convenient modular computer programs and plugins become widely available, it is immensely popular for individuals to print and produce their own everyday commodities rather than purchasing from retailers. Moreover, for complex and voluminous items, their productions are carried out in local 3D printing centers. Traditional household product businesses have not become obsolete, they are now reduced to design businesses that retail 3D CAD designs to customers. Such change also creates entrepreneurial opportunities for individual developers who may upload and sell their designs at online Design Store. On Mars, where the geological conditions make it even more expensive to construct light manufacturing factories, additive technology plays a huge role in the lives of residents.

4.2 Bilateral Currency System
The rise of decentralized cryptocurrency using the blockchain system in the 21st Century has resulted in unprecedented popularity of bitcoin on Earth. By 2050, in the United States, for every one dollar a person earns, 40 cents on average had been invested in cryptocurrency. However, due to the lack of centrality of such a currency, it is exceedingly difficult for an outside force such as the government to regulate the market on the brink of a financial crisis. The Republic City State of Tharsis adopted a Bilateral Currency System, consisting of both decentralized Blockchain Based Cryptocurrency System (BBCS) and a centralized currency system under the regulation of *The Martian Reserve*.

Galileo: A Blockchain-Based Cryptocurrency
Blockchain Based Cryptocurrency System (BBCS) is a decentralized peer-to-peer payment network. The system is created by the Tharsian Corporation, and is later ratified as one of the official currencies in the City State. Like Bitcoin, it does not have a centralized bank, and is under no control of a central government [21]. Users can process transactions to be rewarded by the system (Galileo Mining). The currency of BBCS is called Galileo (MSG). Par the M.A.R.S. Treaty, the exchange rate of Galileo to USD was 1:1 when the system was first established, effective at the beginning of Sectionalization, and is subject to fluctuation later. Galileo consists of a pair of public key and private key, which allows the coexistence of both transparency and privacy. Unlike Bitcoin, Galileo does not half, which puts no limitation on the number of Galileos in the market. Citizen Economic Key (CEK) is required for registering for Galileo wallets. Citizen Economic Key (CEK) is assigned to everyone at birth/landing to conduct economic transactions and prevent fraudulence. It ensures that Galileo is safe and portable and enables purchases through fingerprints and/or face recognition.

Kepler: A Regulated Cryptocurrency
Kepler is similar to traditional currency. Par the M.A.R.S. Treaty, the exchange rate of Kepler to USD is 1:1 when the system is first established, effective at the beginning of Sectionalization, and is subject to fluctuation later. The exchange rate to other currencies is subject to a free-floating rate system. Kepler is issued by and is under the regulation of the Martian Reserve, which is under the administration of the State government.

The mission of the Martian Reserve is to maintain monetary market order and regulate the economy. The Reserve has several primary strategic goals with interrelated and mutually reinforcing elements:

- Utilize its ability to adjust interest rates to regulate the market.
- Promoting a transparent, sustainable, and effective interplanetary trading system.
- Foster the integrity, efficiency, and effectiveness of Board programs and operations.

4.3 Dynamics Between Two Currencies

While many other cryptocurrencies are also available on Mars, Galileo and Kepler maintain their status as the dominant two currencies. The coexistence of a centralized currency system and a blockchain based decentralized currency system allows citizens to enjoy the benefits of both. Galileo and Kepler are in constant competition as well as a dynamic equilibrium with each other. When the Galileo market seems too volatile, or Galileo banks seem unreliable, people would switch to Kepler. When government regulations seem to hurt the market, or the economic opportunities in the Galileo market arise, people would switch to Galileo. The competition of two currencies allows each system to perfect themselves, and therefore fosters a more stable monetary system.

Special Features of Galileo
Product Certificate and Smart Contracts
Due to the long distance between Earth and Mars, Blockchain becomes especially vital in product transportation and logistics systems. Every product, whether shipped from Earth or Mars, bears a series of digital timestamps that indicates the information of the transactions, e.g., departure and arrival status and locations. The automatic tracking system, enhanced by the smart contract systems, which are the mutual payment agreements made by purchaser and retailer, makes Earth-Mars or intra-Mars trading reliable and efficient.

Legal documentation and Digital Voting
The tamper-proofing of BBSC protects the effectiveness of intellectual properties and legal documentation. Public office elections also adopt BBSC. While not everyone is required to have Galileos in their account, each resident may cast their votes via BBSC through which their CEKs must be verified.

4.4 Industry Profitability

The Committee determined that Mars' economy was *autonomous* in 2091, ensuring full independent sustainability of Mars even in the absence of the Earth. In achieving such autonomy, Mars' domestic economy shares a similar variety of industries to Earth, with a heavier emphasis on agriculture, manufacture, and R&D. The interplanetary economy was built upon the transportation of intellectual properties and resources, research, and tourism.

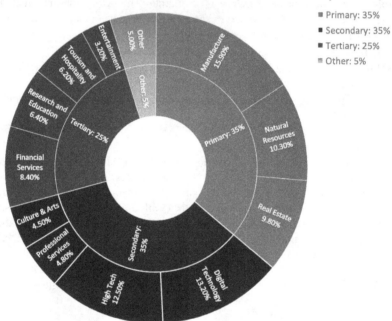

Figure 13: GDP by sector and Interplanetary Commerce.

Trade on manufactured goods

Over the 21st century, additive manufacturing technology has matured rapidly. While both Mars and Earth can produce all essential goods if needed, the cost for manufacturing certain products overwhelms the cost of producing it on the other party plus the transportation fee. Thus, interplanetary importation and exportation usually take place for light-weight products that one party has relative advantages to produce.

Trade in natural resources

The discovery of abundant natural resources on Mars triggered a "gold rush" in the 2050s. Over half a century, Mars experienced a population boom, various environmental issues, protests, and legislation of the Martian Natural Resources Law. By 2091, there were 85 interplanetary companies left that are involved in the trade of natural resources, such as deuterium, on Mars.

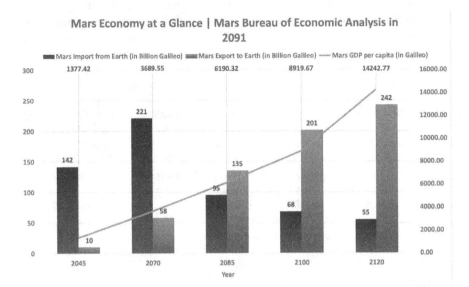

Mars Economy at a Glance | Mars Bureau of Economic Analysis in 2091

Mars Import from Earth (in Billion Galileo) ■ Mars Export to Earth (in Billion Galileo) — Mars GDP per capita (in Galileo)

Licensing to Intellectual Property

As Dr. Robert Zubrin predicted [22], Mars' economy thrived thanks to the transportation of ideas: the harshness in the new physical environment as well as a prevalent technological culture prompt Martians to continually ask themselves: *How to make it better?* Especially at the Corporatization and Sectionalization stages, science and technology were romanticized; on average every three Martians had a patent. Breakthroughs in automation, energy production, biotechnology, manufacture, quantum computing, etc. happened too often to make themselves news. These inventions brought a substantial number of profits to Mars even as they revolutionized society and advanced living standards on both Mars and Earth.

Export and Import of Artworks

An entirely new lifestyle on Mars is conducive to generating a new understanding of ourselves, our world(s), and ultimately, human condition. Martian artists have produced numerous mind-blowing literatures, music, and paintings, which sometimes go as far as to redefine the perception of *"beauty."*

These artworks have been exported to Earth in the digital form, inspiring generations of young people to be the next settlers. Likewise, the work of artists on Earth is exported to Mars, generating an artistic dialogue between planets.

Research

Due to different physical properties in the ambient environment, Mars has attracted a large number of biologists, physicists, chemists, material scientists, and geologists from Earth. Hence, the research institute to which they belong pays a fee for them to be a visiting scholar on Mars. The results of their research, such as medicines, nano-structure products, and even spices have generated a tremendous amount of revenue on Earth. Principal areas of the research on Mars has been focusing on solutions to hunger issues (agriculture), better water processing methods, sustainability and higher recycling efficiency, increased energy efficiency, better build material, more effective communication methods, etc.

Tourism

The expression of "travel around the world" was switched to "travel around the world[s]" when the number of tourism programs exploded in the 2060s. At that time, thanks to better rocket technologies, the price for tourist tickets was reduced to 42,000 USD per person, a 90% decrease from the 2030s. The Mars Committee also launched a scholarship program on Earth, which offers free tourist tickets for the scholarship's recipients, to inspire the next generation. With the increasing number of native Mar tians, visiting Earth became popular at the end of the 2070s. Its price usually varies from 10,000 Galileo to 30,000 Galileo (equivalent to 24,000 USD and 72,000 USD in 2020) depending on different tourism companies. In total, there are 87,746 tourists visiting Mars and 12,782 tourists visiting Earth in 2091.

Entertainment

Music, movies, and sports have been unprecedentedly popular after thousands of artists and athletes immigrated to Mars. Mollyhood Martian movies have become as popular as Hollywood movies during the 2000s. Movies are usually released on Mars 8 months prior to the releases on Earth. Thousands of Martian movie fans from Earth would travel to planet Mars to see the movies upon their release. Sports are huge tourist attractions and revenue generators.

Unique Martian sports draw enormous profit from interplanetary tourists. The State government constructed The Tharsis Memorial Coliseum, a sports center that has the full capability of hosting the Martian Olympics. The Center of the Coliseum is the court for Kulamu, a new sport born out of a low gravity environment (see 5.5). While VR tickets are widely available, the front row ticket price for a Kulamu match reached as high as 20,000 Galileo (equivalent to 48,000 USD in 2020).

4.5 Trade and Regulations

Interplanetary Trade

From the preliminary stages of the colony – Corporatization and Sectionalization – there was a huge trade deficit when Mars had not established its essential industries. Meanwhile, the transportation fee of $500/kg from Earth to Mars gave startup businesses on Mars considerable hardship obtaining necessary raw materials from Earth. To compensate for that, in the preliminary stages, businesses can file for subsidies for the transportation fee from the local government. The amount of compensation gradually decreased as the economy on Mars grew matured.

The transportation fee of $200/kg from Mars to Earth still poses setbacks for businesses exporting goods from Mars to Earth. To help enterprises on Mars thrive, the central government continues to provide subsidies to enterprises whose profits mostly rely on exports to compensate for the high cost of transportation.

The central government is expected to overhaul its subsidies regarding importation when Mars has developed its capacity for the research and development of interplanetary transportation industries. The high transportation fee will stimulate the market to optimize technology and find ways to reduce costs. As various interplanetary transportation methods improve, industries across Earth and Mars will eventually be in nearly perfect competition.

Price Ceilings and public owned productions

Necessities like water, oxygen, and protective garments are life-concerning for any individual. Therefore, although there is no barrier of entry for private industries to invest in these markets except that they must qualify the regulation standards, there are price ceilings for these necessities as the price surge of these products can be life-threatening on Mars. Such price ceilings would divert private businesses from participating in the production of the goods, e.g., the excavation of ice caps and the processing of perchlorate water. There the public-funded state-owned and district-owned production companies would step in. As a result, 90% of the production of water, oxygen and Martian suits remain publicly owned. There is a small percentage of private companies in the markets, endeavoring to achieve production of these necessities at lower costs and outcompeting these public owned companies. While the public owned entities ensure the provision of these necessities to the majority of people, private companies seek to provide residents with luxurious experience.

4.6 Tax System

The existence of tax revenue is dependent upon the emergence of a central governmental entity and private property. Since both conditions were met after the Corporatization stage, a tax system was developed on Mars to ensure that its government was operated based on the interest of Mars citizens, who pay for the tax.

The major components of the Martian tax system are similar to those on Earth – Enterprise Income Tax, Personal Income Tax, Land Value Tax, Consumption Tax, etc. – though the tax rate is generally more lenient compared to that on Earth. To encourage innovation and the application of innovative solutions, profits directly made by research and development centers are tax-free, and that any profits individuals/companies gain from using an intellectual property licensed in the recent ten years enjoy a tax rate cut to half.

The federal tax collected on Mars flows into the Mars Development Fund (MDF), which was set up at the beginning of the Preparation stage, receiving individual and corporate investments, crowdfund, and cash from ticket sales and bond and stock to financially support the City State's long-term development.

MDF provides residents with well-maintained city infrastructure, sources of water and oxygen and management of civil affairs, as well as loans and starter programs to startup enterprises on Mars. In the later years, MDF continues to treat tax collected as investment received, enabling the Mars Committee to write proposals on how to spend MDF every two years, and Martians' vote.

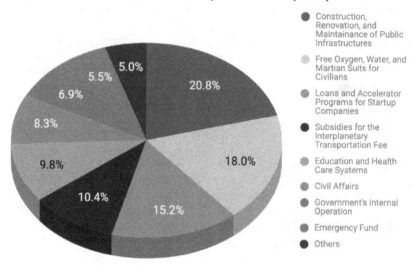

Expenditure of the Mars Development Fund (MDF) in 2120

- Construction, Renovation, and Maintainance of Public Infrastructures
- Free Oxygen, Water, and Martian Suits for Civilians
- Loans and Accelerator Programs for Startup Companies
- Subsidies for the Interplanetary Transportation Fee
- Education and Health Care Systems
- Civil Affairs
- Government's Internal Operation
- Emergency Fund
- Others

Figure 14: Expenditure of MDF in 2020. A significant amount of the MDF goes to infrastructure construction and necessity provision.

5 A VIBRANT COMMUNITY

5.1 Government Structure

The Republic City State of Tharsis is a federalist republic city state, under which the primary political entity on Mars is *District*. Each district is highly autonomous under the constitution upheld by the State Government. The State Government consists of the *State Committee*, *the Supreme Court of Justice*, and *the Commander in Chief*, as well as the *Federal Delegates* in districts. The District Government consists of *District Committee, Commission, Court* and *Citizen Legislature*.

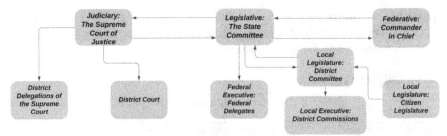

Figure 15: Government Branches.

The State Committee	Consist of members representing every district, elected every four years; Oversees federal legislation; Creates federal agencies, but executions rely on the District Delegates.
The Supreme Court of Justice	Arbiter of the law; The judges are selected by the State Committee.
Commander in Chief	Oversees the federal military with the consent of the State Committee; Deals with diplomatic relations with political entities on Earth, as well as on terrorism and civil unrest.
District Committee	In charge of district legislature in accordance with federal legislature; Each member is elected every two years.
District Court	The judges are selected by the District Committee.
District Commission	Consists of one Commissioner, cabinet members and District Agencies; Executes Federal and District Laws; The Commissioner is elected every four year.
Federal Delegates	Agencies enacted by the State Committee on district level that executes orders from the State Committee on a district level.
Delegations to the Supreme Court	Examines the consistency between district law and federal law; Arbitrates conflicts between District Commission and Federal Delegates.
Citizen Legislature	An institution on district level of which any private citizen of that district can become a member through qualification and selection; Drafts bills as a second party beside the District Committee. Bills must be approved by the District Committee to become the law, if approved by the District Committee can also be referred to the State Committee, and become federal law once approved by the State Committee and ratified by over half of all districts.

Table 1: Definitions of Government Offices.

5.2 Land Ownership

The general principle: To encourage more settlers onto Mars, individual entities or collectives can obtain the ownership of the land they make habitable. The habitability is governed by the Mars Settlement Administration Standard (MSA Standard). The tenure to land is granted to an entity or collective by assessing the portion of the land with the MSA Standard.

MSA Standard Overview

Structure stability	The structure(s) must pass the basic stability tests.
Material Durability	The material must be in good condition, and able to last for at least 50 years.
Oxygen Supply	All compartments of the structure must maintain oxygen concentration of 21%. Redundant oxygen generators must be provided and accessible all the time in every compartment of the structure in addition to the functioning one.
Energy Supply	All compartments must be connected to at least one reliable power station for the proper functioning. The minimum purchase of energy is the minimum energy consumption of all properly functioning devices in the structure.
Health and Emergency	All compartments of the structure must be connected to the emergency rooms where medicines, first-aid kits, and emergency oxygen supply must be sufficient.

Table 2: MSA Standard Overview.

Legal Protections of Properties
- The protection of land tenure is primary to governments.
- No existing sovereignty on Earth can claim land for the use of national interest.
- No land shall be claimed by any entity without being made habitable.
- An entity or collective has the right to trade the land with other entities or collectives. An entity or collective can be delegated by another entity or collective to make the portion of the land habitable, and the tenure of the land belongs to the entity or collective which pays for making the portion of the land habitable.
- The tenure of the land continues as long as the portion of the land is habitable. When the holder of the tenure is deceased, the tenure is granted to the delegate the holder appoints to. If there is no legal form of will that claims the succession of the tenure, the portion of the land will be held by the district government.
- The central government is prohibited from imposing property taxes. The District government collects no tax when no economic activity is being conducted on the portion of the land.

5.3 Individuals on Mars
One can become a citizen of the City State through birth, technical immigration, or investment immigration. Technical and investment immigrants renounce their nationalities on Earth and become full Martian citizens once they arrive. Everyone

is assigned a Citizen Economic Key, with which they can be identified in economic and political activities.

Job Opportunities on Mars are diverse and abundant. Nearly 50 percent of the citizens have expertise in the STEM field. At the same time, many of them work as managers of their neighborhoods, districts, or the entire City State. With the advancement of automation and additive manufacturing, less than 5 percent of the workforce produce all manufactured goods on Mars. Freed of heavy labor input in the manufacturing industry, a significant number of people choose to work on design and marketing. Artists, musicians, and movie producers are also coming to Mars in a quest for new inspirations.

5.4 Education
Civilization is in a race between education and catastrophe. —H.G.Wells

AI
There were two global problems in the education system at the beginning of the 21st century. First, "the rich become richer, and the poor become poorer." When education was becoming a business, the chance that a kid from a poor family receiving superior educational resources was not promising enough to confront social stratification. Second, however, while reinforcing education's role as a society's level ground, and thus placing all students on the same standard, students lose their opportunity for personalized educational development.

To solve these issues, Artificial Intelligence (AI) was introduced in the Education system on Mars. Everyone has a personal — and sometimes lifelong — AI, who possesses the entire library of educational resource that was heretofore produced. Every student is assigned access to the AI education system, which designs education plans for every learner, and adjusts according to the learner's intellectual and personal development. Education and AI bots are free for everyone to provide equal opportunities. This way, from birth, every Martian receives the resource they desire.

Principles
The initial Martian Education system was also established by absorbing research findings in neuroscience and psychology. The clear border among grade levels and disciplines was determined to be unscientific, outdated, and thus abolished [23]. Instead, students learn by exploring their curiosities and *asking questions* — What? Why? How? — and diving into them. Their personal AI acts as a mentor who guides him/her to explore questions.

The physical school was replaced by the concept of "community": students who live in the same community carry out collaborative projects. Each student spends half a day doing projects and collaborating with others. In a project, they may want to make something, improve something, shoot a film, or study a social issue, etc.

The number of people in a group ranges from 2 to 100, and is determined by the learner's own will as well as AI's observance. Each project is usually conducted over a week, but it can also be longer through group members' discussion. Sometimes like-minded people gather up and form a group, and sometimes they come from distinct backgrounds. They are also always encouraged to visit experts and scholars if they like.

Education Plan

The curriculum for students younger than 12 is composed of question-learning, general education (literacy, numeracy, and history), projects, touring around Mars, and sports. For students older than 12, they have question-learning, projects, apprenticeship — choose a favorite industry and "intern" for at least six months (if they change their minds, they can choose again afterwards) — and general education. In addition, there are also educators on Mars who actively research and improve the current system. In terms of innovation, it is also crucial to seek out faults in defaults. Therefore, the Martian kids are constantly reminded that everybody could be wrong to encourage them to actively challenge the existing framework.

5.5 Martian Sport: Kulamu

Kulamu, or Bodyball, is a new sport born out of traditional basketball in the low gravity environment. It is played between two opposing teams, each of 8-11 people, in a circular court. Three people in one team are designated as "balls." The main objective is for the "balls" of each team to jump into the opponent hoop and defend the opponent "balls" from jumping into their own hoop. Such a sport would be nowhere near possible in Earth's gravitational field, but the natural low gravity of Mars offers the ability for average people to jump 1.5 meters high and thereby making such a game viable.

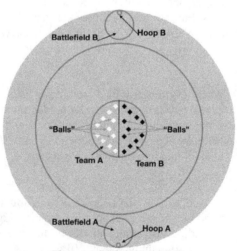

Figure 16: Kulamu Court.

A Kulamu court usually has the radius of 100 feet, surrounded by walls of 20 feet. The hoops are 20 inches in radius and are 8.5 feet above the floor. They are 4 inches thick, made of stainless steel to bear the weight of a person. The floor and wall of the circular court are made of elastic materials, making them perfect trampolines. Players are required to wear bubble suits, which contain their bodies into airborne, elastic protective suits that look like bubble balls. In such a highly elastic and low gravity environment, an average player can jump as high as 8 feet [24][25]. Players who act as a "ball" usually break through the defense of opponents, make a jump under the assistance of their teammates, grab the hoop, and then extend their bodies into the hoop to score one point for the team. Other players assist the "balls" to break through opponent defenses and defend opponents from scoring.

5.6 Emergency Management
The Mars Emergency Management Board is responsible for the overarching emergency programming, which encompasses technical development and field operations.

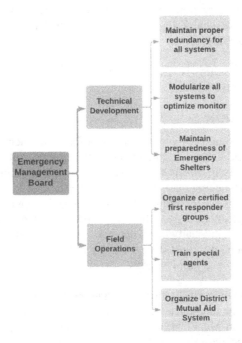

Figure 17: Responsibility of the Emergency Management Board. The Board is responsible for Technical Development and Field Operations, which branch out to many individual subsystems.

For technical development, the optimization of infrastructure is extremely important. To guarantee robust and seamless daily operations in the harsh Martian environment, *redundancy* [26] and *modularity* are two key features.

For redundancy, all critical systems such as oxygen, water, and food have independent power feeds from both solar balloons and nuclear reactors so that no interruption of electricity occurs if one power feed fails. Each district also has an Emergency Shelter which can supply oxygen, water, food, and power for all district citizens for one month.

Examples of modularity include the transportation tube system, hydroponics system, and hexagonal residential systems. When emergencies occur, locations can be pinpointed down to specific modules, reducing the time for diagnosis, and enabling a faster replacement of the malfunctioning module.

Field operations programming includes 1) certified first responder groups, such as emergency medical technicians (EMTs), who conduct small-scale operations as soon as an emergency takes place, 2) special agents who deal with problems that require advanced technical knowledge, and 3) mutual aid of supplies between districts when problems escalate.

The Management Board is responsible for the recruitment, training, and organization of all three branches.

6 FUTURE OUTLOOK

While life in this meticulously engineered subsurface settlement on the Red Planet is truly ethereal, Martians are destined to break through the soil, sprout to the surface and eventually transform this planet into the one we originated from.
The difficulties in achieving these aspirations, nevertheless, are still enormous.
Martians need to carry out a multi-centurial project to thicken the atmosphere on Mars by releasing greenhouse gases from ice caps and underground. During this process, Martians will gradually move up to the surface in settlements protected from extreme temperatures, dust storms and solar charged particles, and transform the toxic Martian soil so plants can grow.

Aside from these known difficulties, Martians must be both technologically and mentally ready for the unpredictability of the future. Should there be an unknown exoplanetary virus on Mars that humans are vulnerable to, they must be ready to take contingent measures against it. At the same time, upon the journey of searching for life, humans must ensure that Mars is protected from human contamination. No one would want to see that one day we discover life on the Red Planet only to realize they have been casualties from our carelessness.

However, challenges have only been driving humanity forward since the dawn of civilizations. We have every reason to believe that the dream of becoming a multiplanetary species will come true with our wisdom and indefatigable spirit of exploration, equipped with innovative technologies in years to come.

7 ...AND MORE

Designing a city state on Mars of one million is a challenging yet exciting endeavor. A 20-page proposal can only address the tip of the iceberg. Therefore, we decided to further our work and post updates on our website *Tharsians*: http://www.tharsians.org

ACKNOWLEDGEMENTS

Special thanks to Prof. David Barnhart at University of Southern California Space Engineering Research Center (SERC) and Prof. Michael Orosz at University of Southern California Information Science Institute (ISI) for providing professional insights.

Many thanks to our reviewers (in alphabetical order): Chenyi Zhao, Eileen Chen, Mike Ma, Tracy Yu, Victoria Yang, X Sun, Xuefei Gao, Yixuan Chen, Zijian Hu for wonderful discussions and inspirations.

SPECIAL NOTES

[1] Concept by Immanuel Kant in *Toward Perpetual Peace*.
[2] 3D models from 3D Warehouse, Contributors: JackL.,Birkholz A., Norbert S., Bart Gillespie, ///Eurasia —HHHR///
[3] Modified with CAD models from 3D Warehouse, Balloon owned by lauwtJ, Inverters owned by Mati.
[4] Phobos is similar to the C-type asteroids composed of carbonaceous surface materials. It has significant porosity, which led many scientists to think that it had a substantial reservoir of ice at the beginning of the 21st century.

REFERENCES

[1] Jacob Haqq-Misra. "The Transformative Value of Liberating Mars". In: *New Space* 4.2 (June 2016), pp. 64–67.issn: 2168-0264. doi: 10.1089 / space. 2015.0030.url:http://dx.doi.org/10.1089/space.2015.0030.
[2] Glen E Cushing. "Candidate cave entrances on Mars". In: *Journal of Cave and Karst Studies* 74.1 (2012), pp. 33–47.
[3] H. Hargitai. *Mars climate zone map based on TES data*.url: http://planetologia.elte. hu/mcdd/climatemaps.html.
[4] Laurent Sibille and Jesus A. Dominguez. "Joule-Heated Molten Regolith Electrolysis Reactor Concepts for Oxygen and Metals Production on the Moon and Mars". In: 2012.
[5] Mamie Nozawa-Inoue, Kate M Scow, and Den nis E Rolston. "Reduction of perchlorate and nitrate by microbial communities in vadose soil". In: *Applied and Environmental Microbiology* 71.7 (2005), pp. 3928–3934.

194 Mars City States: New Societies for a New World

[6] Alice Sweitzer. *New York City Is Channeling the Sun to Build the World's First Underground Park*. 2016.url:https://www.popularmechanics.com/ science/green-tech/a22642/lowline-new-york-city-underground-park/.

[7] Isaac Asimov. *I, Robot*. The Isaac Asimov Col lection ed. New York City: Doubleday, 1950.isbn: 978-0-385-42304-5.

[8] Kevin Watts Stephen Hoffman Alida Andrews. "Mining" Water Ice on Mars An Assessment of ISRU Options in Support of Future Human Missions". In: (2016).

[9] Thiruvenkatachari Viraraghavan Asha Srinivasan. "Perchlorate: Health Effects and Technologies for Its Removal from Water Resources". In: (2009).

[10] Brian Frankie Robert Zubrin and Tomoko Kito. "Mars In-Situ Resource Utilization Based on the Reverse Water Gas Shift: Experiments and Mission Applications". In: (1997).

[11] Caitlin A. Callaghan. "Kinetics and Catalysis of the Water-Gas-Shift Reaction: A Microkinetic and Graph Theoretic Approach". In: (2006).

[12] World Nuclear Association. *Molten Salt Re actors*. 2018. url: https://www. world-nuclear.org/information-library/current-and-future-generation/molten salt-reactors.aspx.

[13] Eric Bourgoin and Patrick Charron. *Orbital hydroponic or aeroponic agricultural unit*. US Patent 7,181,886. Feb. 2007.

[14] Zuhaib Bhat and Hina Bhat. "Prospectus of cultured meat - Advancing meat alternatives". In: *Journal of Food Science and Technology* (2010).

[15] M.A. Rucker. "Dust Storm Impact On Human Mars Mission Equipment And Operations". In: (2017).

[16] Gibson Et Al. "Development of NASA's Small Fission Power System for Science and Human Exploration". In: (2015).

[17] Ned Chapin. "Communications Infrastructure To Support Human Activities On Mars". In: (2000).

[18] N. Lay. "Developing Low-Power Transceiver Technologies for In Situ Communication Applications". In: *The Interplanetary Network Progress Report* (2001).

[19] Biswas1 Et Al. "Optical Payload for Laser comm Science (OPALS) Link Validation". In: (2015).

[20] *Mars in our Night Sky*. NASA.url: https://mars.nasa.gov/all-about- mars/night sky/solar-conjunction/.

[21] *Bitcoin Frequently Asked Questions*.url: https://bitcoin.org/en/faq#general

[22] Robert M. Zubrin. "The Economic Viability of Mars Colonization". In: 2018.

[23] K. Robinson and L. Aronica. *Creative Schools: The Grassroots Revolution That's Transforming Education*. Penguin Publishing Group, 2015.isbn: 9780698142848.

[24] *2019 NBA Draft Combine-All Participants*. 2019. url: https://stats.nba.com/draft/ combine/.

[25] *Mars Fact Sheet*. NASA.url: https://nssdc.gsfc.nasa.gov/planetary /factsheet/marsfact.html.

8: TOLMA CITY

Anton Aubert, Eliott Bourget, Raphaël Dehont, Pierre Fabre, Manal Gharrabou, Nathan Goy, Maxime Jalabert, Ziad Kheil, Julie Levita, Alicia Negre, Amaury Perrocheau, Marco Romero, Tanguy Sommet, François Vinet
Team ISAE-SUPAERO
Toulouse, France
fvinet07@gmail.com

Figure 1: a luxury hotel concept for Martian tourists.

I - OUR VISION OF MARS COLONIZATION

Over two millennia ago, human curiosity led us on a journey to understand the origin and organization of planets surrounding Earth. Although the conclusions were wrong, humans were trying to scientifically comprehend the starry sky. Since then, this desire to understand and explore the heavens has never left us. During 15th century Galileo invented the telescope, which in return resulted in the time of observation. Humans gradually understood Earth's place in the solar system and began to build models aimed at explaining what surrounds our planet. During the last century, the unthinkable happened: mankind landed on the Moon, this paved the way for space exploration and nothing could stop it. Following the exploration of our moon and several thousand probe launches later, Mars has imposed itself as the next step in our epic journey. And this time, it is not just to explore it but to settle there. After the time of observation, after the time of exploration, it's time for colonization. We are delighted to present our team vision for this new milestone of space exploration through this document. We sought to detail our idea of the colonization of Mars through the construction of a city of one million inhabitants. The establishment of a colony on Mars means that man will cease to be an Earthling and become an inhabitant of the Solar System. This is a powerful act in the history of mankind because for the first time, we are leaving the planet where we were born. Bravery and boldness are a must for such an accomplishment. In light of this, we have named our colony "Tolma", which means "daring" in Greek.

The word "colony" for Tolma is actually inaccurate, we would like to view colonization under a novel form. Mars will not be used as a subject of economic or political arguments. This "colony" is meant to be a collaboration between countries, companies and organisms in order to make science go forward, and let humanity extend itself without destroying and exploiting the rest of the Solar System. Tolma is the result of human evolution. It is the continuity of our civilization. It is not only Humans in space but Space-Humans, Martians, Tolmans. It is the future of humanity. Tolma is a new chapter in the fabulous history of Space and Humanity.

II - THE TEAM

We are a team of 14 master students. We all study aerospace sciences at ISAE SUPAERO University, European leader in aerospace engineering. We learn engineering, manned flight and planetology.

As soon to be post-graduate, taking part in this competition was a challenge for us. Not having the knowledge to develop a concept of a Mars colony on our own, we called upon many specialists. We would like to thank all the NASA engineers who took the time to answer our emails and guide us in our work. We also conducted a survey among our university students to determine their expectations for a Mars colony. Indeed, the opinions of potential customers must always be taken into account when designing a product. This project was an opportunity for us to learn more about Mars, but also about the technological concepts developed today for the exploration of tomorrow. We have also opened up to other fields that are not our specialty such as economics, politics… We would like to thank Paolo Guardabasso, Mehdi Scoubeau and Mohammad Iranmanesh for their sound advice which helped us a lot and Stéphanie Lizy-Destrez for her support.

III - TOLMA CITY GENERAL OVERVIEW

Before diving into the details of the design of the city, we wanted to give you a general overview of the result of our work. For this purpose, we have chosen to present a typical day for 3 inhabitants of Tolma.

As in every city, there will be students in Tolma who will need to be educated and to acquire the necessary knowledge to work in the colony. Our first inhabitant, **Elliott**, is one of them. Elliott was born on Mars. He learned the basics in a nursery, like children do on Earth. Then he spent 3 years in a school for 2 to 5 year old children where he began to discover corporal expression through sports and art. After those years, he went to primary school where he continued to receive basic education. His days were spent work out in the gymnasium, reading in the library, or studying in class. Before the age of ten, this Martian had a good overview of what life in Tolma was. He also thought of what he would like to do to help the colony. In Elliott's case, it was botany! With those ideas he went to the closest middle school from his house. There he had a vast choice of topics (mathematics, history, languages, physics, biology, computer science, music, drawings, politics, mechanics…), and had to do his first internship to have a crystal clear view of his career choices. Finally, he went to one of the 14 high schools of Tolma and chose one domain in which he would like to work: botany! This choice was possible thanks to all the internships he did and continues to do.

Today Elliott begins his last internship of the year in a greenhouse. He hops on the NEWAY, the colony's underground transportation system. In the wagon, he is surrounded by many workers and engineers on their way to work. After 2 km, he gets off and enters the 4A control center. This center, which belongs to the Ministry of Agriculture, specializes in potato production. In the control room, engineers monitor the lighting power of the plants, the temperature and the quality of the irrigation water. All these parameters are continuously optimized to improve crop productivity. In an adjacent room, the team of analysts set up and verify mathematical models to improve the efficiency of the aeroponics. On the upper floor, botanists monitor the quality of the harvested plants. They collaborate with biological engineers for the permanent improvement of the plants. Elliott is not surprised to see no gardeners. Indeed, the plantations are managed by robots. Our student enjoys the atmosphere at the center. Many different teams work together to feed the colony and he wants to be a part of it. But before, he will have to go to a college to end the learning of his job !

Our second inhabitant is **Mike**, a father of 2 children. Like every morning, Mike starts by watering his plants. The presence of plants at home is greatly encouraged by the government. Mike collects the water under his toilet. The Nano Membrane Toilet is able to process urine and feces on site into energy and clear water [1]. Then he connects to his computer to start his day of teleworking.

Mike works in the events and tourism industry. Incidentally he is one of the organizers of the annual great Martian rover race, which is broadcast on Earth. This race, which spans over several weeks on a magnificent Martian route, is a huge media success among Martians, but especially among Earthlings who are numerous to watch it every year. Thanks to the tourists it attracts, as well as its rebroadcast on television, this rover race is a great source of income for Tolma, which is why its organization is important every year. Mike's work is therefore essential. He is in charge of managing the advertising before the race, the preparation of the TV broadcast, the recruitment of the drivers, and the distribution of the stages of the race route.

After his day of work, he'll go to one of the 40 theatres in the city. He is indeed an actor with the Florento Company, which is currently performing the musical "The sound of music" every Saturday night at the Shakespeare Theatre. In Tolma, the distinction between paid jobs, which are jobs that are directly useful to the smooth running of the city, and unpaid jobs, which the inhabitants do as a leisure activity, is very important. Mike is an actor and has taken singing lessons from a very young age, rehearses every Wednesday and Thursday night out of passion, and does not earn money doing it.

Figure 2: an example of a typical living room in Tolma City.

Our last resident is **Laura**. She works as a water extraction officer. Like every morning, she travels to the outskirts of the colony to the nearest astroport. She puts on her similar BioSuit and boards one of the rovers that takes her to the extraction site. You would be surprised at quite how freely and comfortably she can move in our BioSuits. Once on site, Laura enjoys the impressive view for a few moments. Hundreds of wells connected to pipelines permanently supply the Tolma's water reservoirs. Then she reaches the control center. Although all the Rodriguez wells (in charge of water extraction) are actually robots, Laura's work is indispensable. Each time the artificial intelligence detects an anomaly, Laura goes to the well in question to repair it. She therefore works hand in hand with artificial intelligence. Following the repair on the well, Laura encounters a team of workers and scientists travelling by rover to the end of the exploitation area. They are in charge of probing the subsoil to define the future exploitation area. The lifespan of a well is indeed not unlimited !

After her working day, Laura plays Martian basketball in the Galaxy Basketball Club, one of Tolma's 20 clubs. Every day after work, she goes to the Coubertine gym, one of the 500 gyms in the city, for a 3 hour workout, and plays her games on the weekends. Martian basketball is a slightly more acrobatic version of the basketball known on Earth, due to the lower gravity on Mars: the baskets are much higher, and it is possible to use the walls to reach them. Once her training is over, she usually joins some friends at Nuage City, an arcade, which is one of Tolma's 50 game rooms. In particular, she loves virtual reality games, which she and her friends can play in large adapted rooms.

IV - TOLMA CITY AS A HUMAN COMMUNITY

Living on Mars is very difficult. That's why the life of the inhabitants of Tolma is based on solidarity, benevolence, generosity, equality and kindness. The survival of the colony depends on the stability of the political, social and economic systems. Citizenship will be essential and people in Tolma will be educated and heartened to debate and discuss about all the aspects of democratic, public and private life in order to increase the quality of life on every side. Debates allow people to meet each others, sharing their life and learning to live together. Tolma City won't be a part of a country or a company. Tolma will be its own territory with a culture (based on multiculturalism and cultures all around the world), and values. The principal values will be solidarity, respect and care about the colony first. However, people could practice the religion of their choice and follow the culture of their choice as long as it does not disturb the stability of the colony.

Education

Education is a pillar in the life of the colony. The objectives of education are multiple:

- To inculcate in young people the values of the colony and to teach them how to live in the society.

To achieve this goal, kindergarten, primary, middle and high schools are filled with hours of debate and discussion. Upon leaving the educational system, Martians will require being able to listen, to pass on their ideas, to debate in a healthy way and to form an opinion. Teamwork will be strongly emphasized and its usefulness demonstrated through the team sports that will be taught at school till the end of their learning years.

- To give children the tools to understand the world around them.

Indeed, future Martians must be able to understand their history from the birth of life on Earth to the arrival of Man on Mars. They must have the minimal scientific knowledge to be able to understand the movement of the planets as well as the way plants are produced for food... For that purpose, they will do a lot of sciences (maths, physics, chemistry, botany, computer science...) from primary school till they start to work.

- To train students for their future profession.

At the end of his training, the Martian must be able to practice his chosen profession. In order to make this choice students will carry out 5 internships from age 15 through 19 in order to discover the trades. They will also be able to choose options according to their tastes at middle and high school in order to begin to prepare them for their work. Trades are all paid in the same way, so the choice of career will be based solely on the tastes and abilities of the students.

Political System

All decisions concerning the development of Tolma City and the lives of its inhabitants are taken by the board of directors. It consists of a bureau and of 5 representatives per professional category. All the members of the bureau are elected by universal suffrage, in a vote held every 2 years. A proposal for a change in law can be proposed by any member of the board of directors, discussed at the meetings, and then voted on by all the members of the board. The aim of this council is that each socio-professional category is represented, so that all the problems and proposals put forward by inhabitants working in different environments can be brought up.

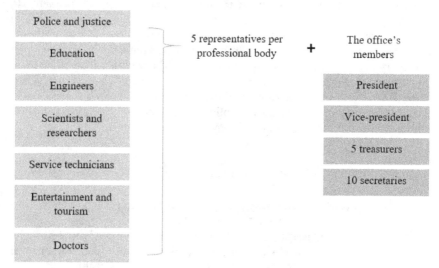

Figure 3: constitution of the Mars village board of directors.

Here is the detailed constitution of the office:

The president and the vice-president: they are in charge of chairing the meetings, giving the floor to the members requesting it to maintain a correct organization, in all neutrality. They are also responsible for maintaining good communication with Earth. In case of emergencies (epidemic...) they obtain the right to take decisions without going through the board of directors. They must always act as a duo, neither can make a decision alone and their roles are exchangeable.

The 5 treasurers (and their teams): they take care of all the money coming in and going out of the city's account. They manage the distribution of the monthly salaries of the inhabitants of Tolma, the financing of public infrastructures and special events, but also the profit made by the city thanks to tourism, and their various markets which will be exploited on Earth (see part V).

The 10 secretaries (and their teams): 2 of them are in charge of taking note of the proceedings of each council meeting, the 8 others are in charge of registering births and deaths, and managing all the city's administration.

Legal system

Tolma's judicial system is inspired by the French judicial system. When judging a misdemeanor or a crime, the court is composed of 3 professional judges and 10 jurors, randomly chosen among Tolma citizens over 20. These 13 people will make the final decision as to whether the accused is guilty or not, the sentence being determined at the end by the 3 judges in charge of the case. The accused, like the defense, will be able to have 2 lawyers. Finally, the case will be monitored and commented by a prosecutor, who will be responsible for establishing the maximum sentence in case of guilt. Life imprisonment in one of the two prisons of the city is the maximum penalty the accused can incur, the death penalty does not exist in Tolma. Lawyers, judges, and prosecutors work for the State, in particular lawyers cannot be freelance. In the colony, the law is enforced by about 5,000 police officers, who are assisted by artificial intelligence, without infringing on everyone's privacy. The crime rate is extremely low, and crime solving leaves no room for error.

Work

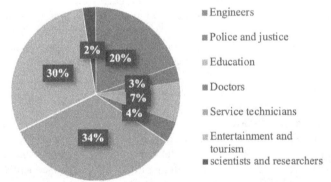

Figure 4: distribution of professional categories in Tolma City.

The vision of working on Mars will be very different from what we have on Earth now. Indeed a Martian does not work for himself. The goal is not to earn as much money as possible to live in luxury. On Mars, the inhabitants work to ensure the survival and well-being of the colony. Each job is vital to the colony, they are all equally important. Therefore, all workers will be paid the same regardless of their occupation or experience. In this way, we abolish social classes. All professions are respectable and respected by all the inhabitants because everyone knows that they are working for the colony and for the others.

V - TOLMA CITY AS A GROWING ECONOMY

The Martian Dollar

The Martian city must be self-sufficient and must have a sustainable economy. Therefore, it needs to be independent and ruled by people on Mars. The Martian city would be an autonomous country with its own money (the Martian Dollar) to limit the influence of economic crisis from earth. In order to avoid wasting materials, hard money wouldn't exist on Mars. Indeed, all the transactions would be computerized which means dematerialized. To avoid having social classes and homeless people, we would implement the universal income which means that everyone in the city would receive the same amount of money each month.

Figure 5: The Martian Dollar.

Health

The health of the people is really important in Tolma City. The danger of an epidemic shouldn't be underestimated. Moreover, each worker is useful for the city so if the worker is ill it means that he can't work and it handicaps the society. Therefore, to be sure that everyone has access to healthcare whatever the disease they have, healthcare would be free for Martian people which means that the government would pay for it. Knowing that people selected for the Martian city must be in good health, the quantity of drugs and care wouldn't be that high which would limit the government expenses in terms of health.

Public services and Education

Public services like water network, energy network, transport network and the Internet would be free. Nevertheless, the consumption of water and electricity would be monitored by the government to avoid abuses and consequently excesses would be overtaxed (for instance, a consumption 10% over the limit would result in a decrease of 5% monthly income)... Education would be free as well because it is necessary to have well educated children in Tolma City. Indeed, no one can be useless for the Martian society according to the amount of resources and money needed to live on Mars.

How to fund this?

In order just to send 1 million people to Mars, we will need $31,000,000,000. That's an example to show how expensive this project is. To fund it, there are different solutions. First of all, to have the initial investment, we have to find investors like national space agencies: NASA, ESA etc and private companies such as Space X. In fact, we would like to use Mars as a launch pad to go further in space but if countries would like to use this launch pad, they would have to pay a toll fee. Thus, to attract investors, we would propose a reduction of toll fee which would depend on the amount of money invested. Thanks to this, we would have an initial investment. To fund public services, education and health, we would use a tax on incomes and a VAT on foods, drinks and materials which would allow these fields to be free. We would also sell the knowledge and outcomes of experiments conducted on Mars or its moons which have features different from Earth.

A) Internal economy

Nationalization
The housing of the regular inhabitants as well as all common utility buildings (hospitals, schools, agricultural fields, factories, political buildings, banks, research laboratories, data centers and water supplies) will be nationalized as seen before.

The colony will have 6 hospitals with 1000 beds. Considering that one fifth of the population is treated in a hospital every year and that a treatment costs on average $2000, the health system of the colony will cost $400 million [2]. Considering that one-fifth of the population will be in school, and that a student costs an average of $9,750 per year, the total cost will be $1,950 million [3]. In addition, the leisure buildings available to tourists but also to the inhabitants will also be nationalized (swimming pool, gymnasium, library, sports field, theatre). Residents will be able to do their shopping online with a delivery service in a warehouse close to their homes. This way, a resident will be able to pick up their order without moving far from their home in order to gain time and to avoid overloading public transports.

Expenses
The state has to pay a maintenance fee:
- maintenance of the inhabitants and tourists' housings thanks to employees paid by the state
- maintenance of common-goods buildings
- maintenance of leisure buildings.

The state must pay for the development of the city: expansion and improvement (materials, architects, technicians, engineers...). It also has to pay salaries of the numerous state employees: engineers, scientists, professors, doctors, technicians, tertiary sector employees.

Revenues of the colony
Each month, inhabitants have to pay the rent and low charges up to the amount of their income (Martian housing tax), in order for everyone to finance public facilities and services. Inhabitants will be able to consume electricity, food, internet access and drinking water access free of charge up to a certain threshold beyond which they will have to pay for their consumption. The state sells access to the various leisure facilities (library, theatre, cinema, sports facilities) to tourists but also to locals. Finally, the objective of the internal economy is to circulate money between the inhabitants and the state while using the value created by the locals to reinforce the daily life of the inhabitants by improving leisure buildings, public facilities, housing, food... Moreover, all the companies belong to the government, so each

profit would directly go to the treasury of the state in order to be reused to fund public services and to keep the economy running.

B) External Economy

Expenses

Martian technology is very advanced, so there is a great need for microelectronics. Since the materials to build these systems are not available, they will be imported from Earth. These are the only goods that will be imported, the rest is manufactured locally. This expense being very important, it is important to find several external revenues.

Revenues

Firstly, tourists will play a very important role in the Martian economy. Wealthy Earthlings with a sufficiently high level of physical fitness will be allowed to travel to Mars as well:

- staying on Mars for a new launch window, 26 months on the red planet
- taking advantage of special houses designed for them (figure 1), with luxury services.

It is assumed that the Martian communication made upstream was sufficiently important to attract a maximum number of people and therefore a large flow of tourists representing one thousandth of the population. Mars does not intervene in the financing of tourist transport.

Science and research taking a very important place in Martian life, many discoveries will be made on the red planet. All these discoveries made on Martian soil will belong to the State for the good functioning of the economy. Like the scientific and economic development of the United States in the 19th century with the Europeans, patents will not be free. Thus the Earth will have to pay in order to have access to Martian knowledge.

Sports and leisure activities will be highly developed on Mars. TV rights, internet data stored in the datacenters and "Marsollywood" (a major film studio designed expressly to generate revenue for the colony) productions will be paid by earthlings.

In 2018, 613.3 million people had a subscription on a streaming platform, representing 8% of the population in that year. The benefit of these platforms was estimated at 55.7 billion dollars, which is not negligible. On a Martian scale for 1 million people this would represent a profit of 72.7 million dollars. Coupled with this we can add the profits made by its platforms on Earth accessible via the Martian internet paying for the earthlings as we mentioned, but these are more difficult to estimate not knowing the evolution of the earth's population. If the craze for the cinema remains the same as today, we can estimate a profit around a hundred billion dollars, which is very important.

Mars holds a very interesting place between the Earth and the asteroid belt. A space program made on Mars will provide insight into the exploitation of asteroids. Indeed during this program many scientific experiments (rover, probe…) will have determined which asteroids could be exploited to obtain rare metals and building materials that can be exported to Earth. These are then deposited on the Martian moons and then recovered by the Earth (Martian moons have a lower Delta-V than Mars, which saves fuel for cargo rockets). So Mars would be the ideal platform to exploit the asteroid belt. The vast majority of operations will be directed towards Earth, so the Earth would finance most of the platform's maintenance. Indeed, asteroids contain a large number of mineral resources, such as platinum, iron, nickel and cobalt. For example, NASA estimates that the asteroid 16 Psyche, located in the belt between Mars and Jupiter, could be worth 700 quintillion dollars. This represents an almost unlimited amount of resources for mankind for millennia.

VI - TOLMA CITY AS A TECHNICAL CHALLENGE

Site selection
The first constraints on the localization of our city are the latitude and the altitude. Our colony is located in the northern hemisphere, which guarantees consistent lighting throughout the year. As mentioned by the NASA Mars Human Landing Site Workshop in 2015, the altitude must be low, to guarantee good conditions of landing thanks to a sufficient density of atmosphere.

The choice of the precise localization of our colony was made according to water disponibilities. As it will be explained later, extracting water from regolith is not a viable solution to provide a 1 million inhabitants colony. Water will rather be extracted from shallow buried ice sheets. However, the localization and the characterization of such structures in the mid-latitudes is quite unknown and needs to be precise before the colony's implementation. Nowadays, only SHARAD (on Mars Reconnaissance Orbiter) and MARSIS (on Mars Express) have been used for studies of mid-latitudes ice on Mars. They found shallow ice in Deuteronilus Mensae and Utopia Planitia. Our colony will be in Deuteronilus Mensae, in a crater of 14km of diameter (Figure 6). It will benefit from a large amount of regolith and ferric oxide to build the colony, as well as iron, silicon and basalt.

Though this localization has a strong potential of scientific exploration (ancient lavas, materials from various geological periods), scientific potential wasn't a determining factor. Indeed, we assume that our colony will eventually be developed enough to enable a strong mobility on Mars and so, to move further away for scientific research.

Since right now the data on water and ore resources is incomplete, the first step of the implementation of the colony will be to check if the localization we choose satisfy the criteria of water and mineral resources.

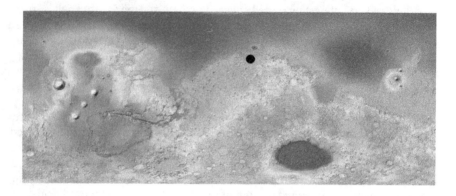

Figure 6: location of Tolma City in the north of Arabia Terra (Deuteronilus Mensae). Topography of Mars by the Mars Orbiter Laser Altimeter (MOLA).

Water extraction

Two water sources are possible to handle the needs of a mid-latitude colony: water sequestered in minerals/regolith and massive ice sheets. Both are considered viable for use by human missions but extracting water from minerals doesn't seem to be suitable for a large colony. Indeed, even if we extract water from gypsum-enriched regolith (which has a better water concentration than typical Martian regolith), it will

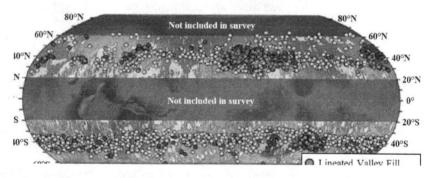

Figure 7: global distribution of glacial features on mid-latitudes (source: Dickson et al., 2012).

A more sustainable solution consists in extracting water from massive ice sheets. In the Martian mid-latitudes, terrain features called Lobate Debris Aprons (LDAs), Lineated Valley Fills (LVFs) and Concentric Crater Fills (CCFs) are present and indicate the presence of buried and discontinuous ice sheets. These structures are especially distributed in the north of Arabia Terra (Figure 7), where we plan to install our colony. A vertical profile of these ice sheets is postulated. The overlying layer (debris and ice-rich soil) seems to be just a few meters thick (<10m) and inferior to 1m in certain places. Below, there is quasi-pure water ice (mass concentration superior to 90%) hundreds of meters thick [5].

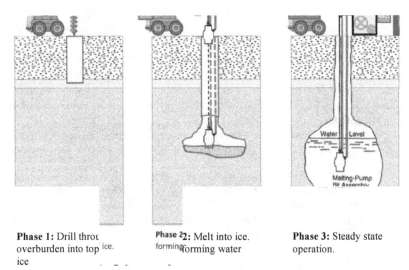

Phase 1: Drill throu
overburden into top ice.
ice

Phase 22: Melt into ice.
formingforming water

Phase 3: Steady state
operation.

Figure 8: steps to establish a Rodriguez well (source: Hottman et al., "Simulated Water Well Performance on Mars", 2018).

This almost inexhaustible source of water is easy to access, thanks to its weak overlying layer. The best solution to access it is the Rodriguez well [6]. The concept is simple: drilling through the overlying layer and creating an underground reservoir by heating the ice in place. The first step is to access the ice layer where the Rodriguez well can be used. For that purpose, a hot water drill seems to be the best solution to drill the ice-rich soil. Then, a heat source (recirculated hot water) is introduced in the ice to melt it, thus creating a natural impermeable tank. The melted water will remain liquid so long as sufficient heat is added. When the quantity of water is sufficient, pumping begins to supply the tanks of the colony in liquid water. Figure 8 illustrates the steps to establish a functional Rodwell (Rodriguez well). To control the size and the performance of the pool, the rates of melting and removal must be adjusted. The well cannot be used indefinitely in the same place. Indeed, simulations [7] have proven that the well will stop producing water. The moment when it will occur depends on the power used to melt water and on the withdrawal rate. According to this study, extracting one liter of liquid water (with a withdrawal rate of 1900 liters per day) needs $1.3.10^6$ J (which is almost 5 times less energy than for regolith). Withdrawal rate of the simulation needs to be adapted to the colony and its demands but it gives an idea of how economical this solution is compared to regolith.

The advantage of using Rodwell is that the technology is already used and mastered in the Arctic and Antarctic regions. The well just needs to be adapted to atmospheric and temperature conditions on Mars. The access hole will need to be lined by an impervious casing to avoid ice and water sublimation...

Domestic water consumption

The domestic use of water is evaluated between 25 to 30L per day. Thanks to new and economical domestic appliances (flushless toilets, water-saving showers, etc.), such a little water consumption will not represent an uncomfortable deprivation for inhabitants. For example, a shower every 2 days is in line with this consumption. The construction of the colony will be an opportunity to rethink the water networks. These are not optimized on Earth at all. By optimizing them, we expect to achieve a recycling rate of 90% for domestic use water. Toilets will treat human waste on-site without external energy or water. The urine is transformed into water for plants and the feces into energy (ex: the Nano Membrane Toilet of Cranfield University [1]).

Each house will have 2 supplies, one of drinking water (shower...) and one of non-drinking water (washing water for clothes for example). In this way, we reduce the treatments to make the water drinkable. Waste water will also be separated into 2 types: grey water and swill water. Grey water will be treated by biological membranes and then sent to the greenhouses. This process will be done locally: a treatment center will be present for each group of 1000 inhabitants and the treated water will be sent to the greenhouses in the surrounding area. Only swill water will be collected colony-wide for treatment. However, this represents only 15% of wastewater.

Sources of energy

We need power on Mars for several reasons: power for habitats, creating oxygen, purifying water, making liquid oxygen and propellant for returning to Earth (even though the colony does not aim to leave the red planet, spacecrafts must be sent back home). There are three ways of generating power on Mars: solar panels, Radioisotope thermoelectric generators (RGT) and nuclear fission reactors.

What about solar panels ? Mars' surface presents major challenges for solar. Energy intensity from the sun on mars is about 589.2 W/m^2 (whereas it is about 1.3 kW/m^2 on earth orbit). Besides, there are significant variations in solar energy by latitude: it is extremely difficult to provide year round power at 40 degrees latitude (which is where most of the easily accessible water appears to be). Last but not least, potential for long-term dust storms (months or even years in length) requires substantial over-capacity and infrastructure. These points make solar panels a bad candidate... In fact, there is only one viable solution to produce energy for one million people on Mars: nuclear fission reactors.

Working in partnership with NASA, Los Alamos scientists recently unveiled a nuclear reactor concept that is very simple and uses a minimum number of parts in order to produce power on Mars. This nuclear fission reactor produces 10 MWe. Thus, it is called Megapower. The reactor consists of an enriched uranium core undergoing fission decay. It generates heat, and that heat is delivered to a heat pipe made of tubes full of potassium that are boiled. Then, that heat is delivered to a gas turbine which converts it into mechanical energy. Finally, a generator converts mechanical energy into electricity. The whole system is self-regulated and really simple compared to our classic nuclear plants.

Can such nuclear reactors be sent on Mars? Assuming a spaceship like SpaceX Starship could bring a hundred tons on Mars, neither the weight nor the dimensions of Megapower would seem to be a problem. Given that one nuclear reactor is about 35 to 40 tons, that means we could put at least two nuclear reactors in a spaceship [8].

Then it is settled ! The next question is: how many reactors have to be sent on Mars ? A million inhabitants colony would require 4 000 to 5 000 reactors. Given that life expectancy of a reactor is about a decade, it also means that every ten years, stocks of uranium will have to be sent from earth. That sounds reasonable for such a big colony and widely feasible over a century given Starship capacity.

To sum up, Mars explorers could benefit from Megapower as it is: small, fully-automated and designed to operate continuously for decades.

Food production
One of the most important needs of any form of life is food and humans do not escape that rule. A colony of 1 million people must produce their own food because they can't rely on cargos that have a launch period of about two years. This colony must aim towards a more sustainable lifestyle than on earth with limited waste and consumption of resources such as air and water.

A vegetarian diet has been proved to be the less consuming one: the water footprint of a vegetarian diet is 40% less than a meat-eaters diet. So the columns will be vegetarian and must grow their food in this particularly tough environment that is Mars.

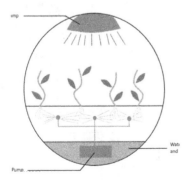

Figure 10: diagram of a tunnel for aeroponics.

The less water consuming mean of production is aeroponics, which is a soil-less form of agriculture. It consists in spraying water and nutriment directly on the roots of the plant. This technique allows the roots to capture oxygen more efficiently and it reduces the amount of nutriment (up to 60%) and water (up to 95%) needed for the plant to grow compared to traditional agriculture. Infected seeds can easily be separated from the rest of the production. Aeroponics could take place in huge lava tubes or in tubes dug by the settlers, this will allow a greater control of the environment (light, temperature, pressure...). If all food was produced in 3.5m diameter tunnels with a flat floor at 1/4 of the tunnel height (Figure 10), then 14,500 km of tunnel segments (which can be stacked vertically) would be needed.

Due to its position in the solar system, Mars receives about 2.32 times less sunlight than Earth, which is why the plants would probably need another source of light: UV lamps, especially during sand storms. These UV lamps will allow Martians to control the frequency and the intensity of the light which will result in better yields. Martians will have to grow food as nutritious and as tiny as possible. We can think of beans and potatoes for instance. Mushrooms could be grown on the waste coming from non-edible food. However, greens such as lettuce are not nutritious enough but will help the Martians to diversify their consumption, which is something important if someone wants to settle on the red planet.

To conclude agriculture on Mars requires absolute control over the production of huge quantities of water and energy, the building of completely new types of structures and the waste management.

Materials

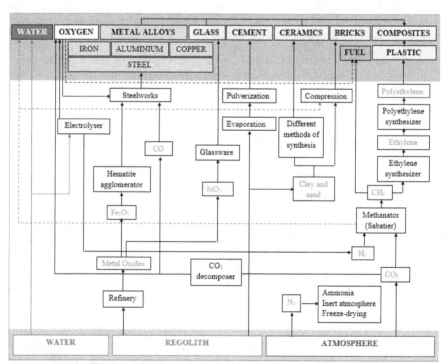

Figure 11: network explaining the ISRU strategy.

Regarding materials, our plan is based on two main techniques: In Situ Resource Utilization (ISRU) and 3D Printing. ISRU is crucial to extract the resources on Mars and to build the colony on the planet. The only things we will bring from Earth will be advanced electronics components. 3D Printing will be used extensively to build all the pieces and structures we need. Additive manufacture will be performed by robots.

Figure 11 explains how to obtain all we need from regolith and atmosphere. We will build plants, factories and other infrastructures and connect the places where we extract the raw materials to the factories (pipelines, transport vehicles...). We will manage the resources, we will store them and we will recycle everything we can.

Colony Structure and Habitability

It takes a lot more than an architect to overcome the challenges imposed by the extreme environment on Mars and ensure the habitability and comfort of the entire population in Tolma City. Since the beginning, engineering has been the main basis for the vector of human and robotic exploration of the solar system. Hence the projection of the city was made using communion between the knowledge of architecture and Systems Engineering.

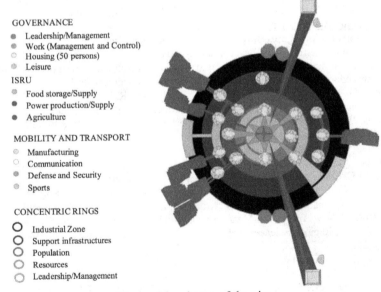

GOVERNANCE
- Leadership/Management
- Work (Management and Control)
- Housing (50 persons)
- Leisure

ISRU
- Food storage/Supply
- Power production/Supply
- Agriculture

MOBILITY AND TRANSPORT
- Manufacturing
- Communication
- Defense and Security
- Sports

CONCENTRIC RINGS
- Industrial Zone
- Support infrastructures
- Population
- Resources
- Leadership/Management

Figure 12: schema of the city.

Beyond mimicry in nature some references and inspirations were used to dimension Tolma city. We took the best of each one and created our own space model to come up with the ideal balance between resources, logistics, accessibility and sustainability in harmony with the entire population of Tolma. More specifically our man-centered permanent settlement on Mars is the result of the hybridization of several Rings Theories and Sierpinski Gasket Fractal Algorithms in a Model based System Engineering of each infrastructure [9].

The city's architecture needs to be articulated between socio-spatial organization, human ecology and easy access to resource production systems. TOLMA's 5-ring system (Industrial Zone/Expansion, Infrastructures, Population, Resources and Leadership/Management) allowed the city to be divided into the main centers of gravity distributed on the Martian soil isotropically using von Thünen's Mathematical Equation Model. This distribution in rings takes into account the shortening of the tunnels of underground displacement, the distribution channels of the ring of resources, having the leadership in the center of the city.

The first central ring will be occupied by the leadership elements of the colony as they are able to converge all the information and have easy access to the other rings and thus be able to better manage the colony. From this central ring, we can find the corridors of transport of resources and access to the launch zones strategically placed in the zones more to the east of the colony.

The storage of all consumables will be made in "green belts" that following a hybrid urban plan will occupy the 2, 3 and 4 rings forming a triangle.

REFERENCES

[1] The Nano Membrane Toilet, 2020, Bill & Melinda Gates Foundation and Cranfield University, http://www.nanomembranetoilet.org/
[2] Key hospitalization figures, July 2019, regional health agency
https://www.ars.sante.fr/les-chiffres-cles-de-lhospitalisation
[3] Key figures of the education system, 2020, Ministry of National Education and Youth,
https://www.education.gouv.fr/les-chiffres-cles-du-systeme-educatif-6515
[4] Abbud-Madrid, A., D.W. Beaty, D. Boucher, B. Bussey, R. Davis, L. Gertsch, L.E. Hays, J. Kleinhenz, M.A. Meyer, M. Moats, R.P. Mueller, A. Paz, N. Suzuki, P. van Susante, C. Whetsel, E.A. Zbinden, 2016, « Report of the Mars Water In-Situ Resource Utilization (ISRU) Planning (M-WIP) Study", 90 p, posted April, 2016 at
http://mepag.nasa.gov/reports/Mars_Water_ISRU_Study.pptx
[5] Rummel, J.D., D.W. Beaty, M.A. Jones, C. Bakermans, N.G. Barlow, P.J. Boston, V.F. Chevrier, B.C. Clark, J.P.P. de Vera, R.V. Gough, J.E. Hallsworth, J.W. Head, V. J. Hipkin, T.L. Kieft, A.S. McEwen, M.T. Mellon, J.A. Mikucki, W.L. Nicholson, C.R. Omelon, R. Peterson, E.E. Roden, B. Sherwood Lollar, K.L. Tanaka, D. Viola, and J.J. Wray, 2014, "A New Analysis of Mars ''Special Regions'': Findings of the Second MEPAG Special Regions Science Analysis Group (SR-SAG2)", Astrobiology, 14 (11): 887-968, 2014
[6] S.J. Hoffman, A.D. Andrews, and K.D. Watts, 2018, "Simulated Water Well Performance on Mars"
[7] S.J. Hoffman, A.D. Andrews, B.K. Joosten, and K.D. Watts, 2017, "A Water Rich Mars Surface Mission Scenario", IEEE Paper 2422, IEEE Aerospace Conference 2017, Big Sky, MT, March 4-11, 2017.
[8] C. Tyler, 1663, Megapower, February 2019,
https://www.lanl.gov/discover/publications/1663/2019-february/megapower.php
[9] Theoretical-methodological issues between economic geography and regional development, July 2015, p.5-21, www.revista.fct.unesp.br

9: PAX ARES SETTLEMENT

David Spencer, Erin M. Edwards, and Ben Drexler
University of North Dakota
david.c.spencer@nasa.gov, erin.edwards@und.edu, ben.drexler@gmail.com

1.0 INTRODUCTION

The idea of exploring and inhabiting Mars is an exciting prospect that has captivated humanity for many years. The feasibility of such a venture remained the realm of science fiction for much of that history, that is until recently. Humanity stands on the verge of making our collective visions of exploring another planet a reality in a few short years. Getting to that planet is one thing; creating a permanent and sustainable home on Mars is another matter entirely and the complexity of that challenge cannot be possibly addressed in a twenty-page document, but we hope that the broad ideas presented here can be used as part of the important conversation about humanity's next giant leap – Mars!

Given that a Martian settlement of one million inhabitants is likely several hundred years off in the future, it is assumed that the settlement will incorporate as yet unheard-of technologies. Therefore, it is not the intent of this paper to design the equipment or technologies utilized on Mars. It is assumed that future technologies will be at least as capable as modern ones; where necessary, current technologies will be used to estimate the capabilities and efficiencies of Martian equipment. Due to both the futuristic nature and scale of the settlement, this paper focuses on large-scale design aspects based on current understandings of physics and psychology as opposed to the in-depth design of individual components. Unless stated otherwise, all references to timekeeping units refer to the passage of time on Earth, with a "year" defined as one astronomical Julian year, or exactly 86,400 SI seconds. All currency is assumed to be in 2020 United States dollars. This allows ease of comparison for valuing economic growth and debt measured over centuries. Prices taken from data sources prior to 2020 are still assumed to be in 2020 US dollars, as inflation was not taken into account over these relatively small timeframes. It is assumed that any physiological concerns with humans living long-term in a low gravity environment will be minimal or easily offset by medical technology. Additionally, this paper only provides a snapshot of how the settlement will appear at one stage of its development (with a population of one million). The authors recognize that a plan for a settlement of one million must consider the prior development that leads to that point and also that this plan is not an end goal, but merely a step to a more fully populated Mars. Lastly, the authors recognize the difficult history surrounding the terms colony, colonist, colonization, settler and settlement in Western colonial exploration; however, for simplicity we may utilize those words solely in the context of inhabiting Mars.

2.0 SETTLEMENT TIMELINE

Before colonizing Mars proper, crews will arrive at its larger moon Phobos to begin operations. These adventurers will not be permanent inhabitants, but will complete various tasks to prepare for the first wave of colonization. There are several reasons to establish a base on Phobos before colonizing Mars. A mission to and from Phobos requires even less $\Box v$ than a mission to and from Earth's moon, meaning that travel there is relatively inexpensive. Phobos has extremely little gravitational pull and no atmosphere, so a spacecraft on the surface is essentially docked to a large asteroid in low Mars orbit. No landers or atmospheric entry vehicles are required to dock there. The first crewed mission to Phobos will take a solar-powered excavation robot to dig holes large enough for habitation modules to fit inside so that they can be covered in regolith to provide radiation shielding and insulation. Solar panels will provide power for these habitats.

These first teams of specialists will consist of mechanics/engineers, scientists and medical staff. The mechanics/engineers will erect the base and keep its systems running. These personnel can also fill in communications and piloting roles as needed. It is assumed that aside from the colonization efforts, government crewed and robotic expeditions will be ongoing at this time. The mechanics/engineers will also support other missions by piloting rovers on the Martian surface with minimal time delay. The base itself can be used as a safe haven or resupply point for other missions. This will further support the exploration of Mars as well as begin generating revenue for the settlement. Scientists will experiment with growing plants in near micro-gravity and attempt to provide food for the expedition crew, although provisions will also be on hand. The scientists will study Phobos, as it is a worthy scientific objective in its own right. However, their main objective will be to study Mars from above and choose sites for further exploration. They will be supported by experts on Earth who will analyze the data they provide. As cargo modules are emptied, they will be utilized for other purposes, such as additional hydroponics space, equipment storage, or work space.

The second crew will depart in the next launch window with additional habitats and cargo modules, as well as two reusable atmospheric entry vehicles (RAEV). The first crew will return to Earth in one RAEV and an attached habitat module. With landing sites chosen based on data provided by the first crew, the second crew will use the remaining RAEV to explore the Martian surface and return to Phobos. The following first wave of inhabitants will focus on establishing a permanent base and quickly set up the first industries to permit growth and reduce reliance on Earth resupply. Agriculture and construction facilities will need to begin production quickly, not only to provide for the needs of the first inhabitants but to allow for rapid expansion of production in preparation for further colonization efforts. Initially, all fuel, food, equipment, and construction materials will be supplied by Earth. As industry is established, many of these supplies will be produced locally.

Paralleling the history of human civilization on Earth, agriculture will be among the most critical industries to establish early on. Early construction materials will consist of inflatable plastic structures brought from Earth. Regolith will serve as a simple construction material as it will be piled on top of these structures for radiation protection. Resources and goods that can be produced with minimal processing steps include water, carbon dioxide, and regolith bricks. Equipment brought from Earth can begin producing oxygen, hydrogen, and methane with minimal supervision. More advanced products such as steel, plastic, glass, paper, and textiles will begin production as more inhabitants arrive. Eventually, it could be possible to manufacture even such advanced goods as electronic components either from local resources or from material extracted from asteroids, but with a population of one million it is likely that such products will still come from Earth.

The first inhabitants will likely need to be technical specialists: engineers, scientists, mechanics, and doctors. They will primarily be concerned with ensuring the initial settlement's survival and maintaining basic functional operations. As the population increases into the hundreds, there will be a demand for all manner of occupations including: trades-people, utility workers, water/sanitation workers, merchants, dentists, psychologists, service workers, educators, information specialists, construction workers, and manufacturers to name a few. Eventually, many occupations that exist on Earth will also exist on Mars, though a significant use of automation is anticipated. Police, firefighters, transportation workers, athletes, lawyers, entertainers, artists, architects, logisticians, nurses, daycare providers, and writers will all find their way to Mars.

Mission year	Major Milestone
1	Establish base on Phobos
3	First human footprints on Mars
5	First settlement established on Mars
6	First birth
9	First death
19	50% chance of fatal launch accident occurring by this time
27	Population surpasses one thousand
59	Birth rate surpasses immigration rate
71	Martian net worth surpasses cost of colonization
94	Population surpasses ten thousand
259	Population surpasses one hundred thousand
459	Population surpasses one million

Fig. 2.0 Projecting major events in Martian history based on population growth and economic models.

3.0 EARTH TO MARS TRANSPORTATION, TRANSIT, AND CARGO STAGING

3.1 Interplanetary Transportation System

The first wave of inhabitants will travel to Mars in something similar to NASA's proposed Deep Space Transport (DST). This consists of an Orion capsule, which can accommodate up to six persons, attached to a yet-to-be-designed habitation module. The habitation module provides additional cargo capacity and living space for the multi-month journey to Mars. The DST will leave Earth with six inhabitants onboard. Upon arriving in orbit over Mars, the Inhabitants will depart in their capsule for the surface (via Phobos) while the habitation module either remains in Mars orbit waiting for the return window to Earth or slingshots around Mars on its return journey. Capsules can be utilized as housing until semi-permanent inflatable structures are erected. Methane/oxygen fuel produced in-situ (discussed in more detail in Section 6.3) will return the capsules to orbit. There, they will rendezvous with the habitation modules and return to Earth to pick up additional Inhabitants. Once local industry and agriculture is established, these missions can be resupplied on Mars, reducing the launch demand for Earth-based rockets.

The number of Inhabitants and supplies bound for Mars aboard these DST-like missions is limited by the number of launches that the world's spaceports can support during the launch window. Even sending a hundred inhabitants per launch window (roughly fifty inhabitants per year, as launch windows occur about every two years) would require multiple launches from each of the world's major spaceports over the course of a few weeks. While this is within the realm of possibility, a different approach that would be less taxing on launch infrastructure would be to use cyclers. The use of cyclers will eliminate the need to wait for launch windows and a fleet of them can support a constant stream of settlers and supplies to Mars. The launch window between Earth and Mars only opens every 2.2 years. Using a fleet of eight Aldrin cyclers (which have a period of just over two years) would allow a transport ship to arrive at Mars roughly once every three months. Multiple capsules filled with Inhabitants would leave Earth and rendezvous with the cycler bound for Mars. The cycler would provide additional living space for the Inhabitants during the voyage. Upon arrival, the capsules would detach from the cycler and transport the inhabitants to the surface (after stopping at Phobos). When the cycler returns on its way to Earth, the refueled and restocked capsules will take off from Mars and dock with the cycler for the return journey. Crew or robots will then transfer supplies from the capsules to the main cycler body. At Earth, the capsules will return to Earth's surface and be refitted for another wave of inhabitants. Obviously, reusable spacecraft are critical for this plan to work.

For mining operations beyond Mars, the DST-like architecture could also be used for personnel traveling between Mars and the mining sites in the asteroid belt. Craft equipped to haul ore would also be required for these missions and could either take ore to Mars or directly to Earth.

3.2 Martian Surface Transit

An electric rail network will be the primary transportation system on the surface. These trains will transport both passengers and materials within the main base. Section 5.0 discusses the settlement layout in greater detail, but for brevity, communities of ~900 inhabitants will connect to central village hubs via rail. These trains will constantly shuttle between the hub and the outlying communities. Assuming that half of all inhabitants are working adults, all of these workers are rail commuters, and the morning commute "rush hour" lasts half an hour, two train cars similar to the R160 design used by the New York City subway system would easily provide enough capacity for each community.[i] Each village hub will contain a metro station that services 6,300 inhabitants. With 999 community shuttles in the settlement, two cars per shuttle, and an assumed 10% additional cars to account for maintenance and down-time, the settlement will need 2,198 tram cars. Comparing the settlement's 166 metro stations to the Washington Metropolitan Area Transit Authority's 91 stations and 1,144 cars gives an additional 2,093 cars needed to maintain the same car/station ratio.[ii] This brings the total number of passenger rail cars needed to 4,291.

Fig 3.1 A typical neighborhood street with hydroponic plants along the walls and ample space for both pedestrians and electric utility vehicles. (Artwork by Ben Drexler.)

In addition to the rail network, the settlement will be connected by roads. With the prevalence of public transit throughout, personal vehicles will not be necessary, but may exist to a certain extent in the form of taxis. Most vehicular traffic will be for emergency services, utilities, or goods deliveries. Vehicles will either be electric or run on biofuel. Due to the limited use of wheeled vehicles, it will not be worthwhile to build separate roadways for vehicular and pedestrian traffic. Public pathways should be at least eight meters wide in order to allow sufficient space for pedestrians and two-way vehicular traffic, should the situation arise. Lights could warn pedestrians of approaching vehicles. Passageways will be large enough to allow delivery vehicles or ambulances to drive all the way to someone's front door. In addition to the pedestrian walkways open to vehicles within communities, train tunnels will also have roadways to allow rapid vehicle travel over longer distances. Akin to highways, these roads are necessary to allow emergency vehicles to arrive on scene quickly without having to drive through winding neighbourhood pathways.

In addition to the use of public transit, walking and biking will be encouraged within the settlement. All communities within the settlement will be accessible via pedestrian tunnels. Moving walkways in longer tunnels will also encourage more pedestrian traffic. Additionally, ramps, moving walkways, and easy access to rail connections will aid mobility restricted or injured inhabitants in their transportation needs.

4.0 SITE SELECTION

Three criteria were used to determine a location for the main settlement site. First, the main base should be located in an area where a sufficient amount of either exploration or remote study has occurred. Second, the site should be located near resources essential for survival, construction, and industry. Third, locations closer to the equator are preferred for solar power availability, moderate seasonal weather changes, and lower fuel costs for spacecraft launches.

Fig. 4.0 False-color photograph taken by Opportunity *of Cape Tribulation (left), Solander Point (center background), and Botany Bay (center foreground), features of Meridiani Planum within Arabia Terra. Photograph courtesy of NASA.*

Based on these criteria, Meridiani Planum was selected as the location for the main settlement. The rover Opportunity operated in the region for over fourteen years, collecting large amounts of data on the local environment. Opportunity located hematite (a source of iron), basaltic sand (including feldspar and glass), and rocks rich in silica and sulfates.[iii] Other important elements discovered in various forms include aluminum, calcium, and magnesium. Additionally, the region lies along the equator. Thus, Meridiani Planum satisfies all three criteria for a successful settlement location. Additionally, the site is not only located along the equator but is also within a few degrees of the prime meridian. Thus, the current latitude/longitude system devised for Mars would essentially give coordinates in relation to the settlement. An observatory located at the prime meridian could serve the same function for Mars that the Royal Observatory in Greenwich, England serves for Earth in terms of official global timekeeping.

In addition to the main settlement, Phobos will serve as a gateway to the settlement for both inhabitants and cargo. The gateway will serve as a quarantine facility to screen for and treat any contagious diseases and check cargo for invasive species. Phobos will not have a permanent population, but will be operated by Martian astronauts who are regularly rotated back to the surface.

5.0 STRUCTURES AND ARCHITECTURAL CONSIDERATIONS

There are several unique considerations for Martian architecture that will differentiate it from Earth architecture, three of which we will discuss here as they are of significant consequence. Those main areas of concern are: atmospheric composition and decreased pressure, radiation exposure, and the psycho-social effects of isolation.

The thin atmosphere on Mars has a significantly lower atmospheric pressure (0.088 psi) and a high carbon dioxide concentration (about 95%),[iv] making the Martian atmosphere immediately toxic to humans. Space-ready airtight structures are already commonplace in habitat construction, so designing an airtight Martian building should be simple utilizing both inflatable technologies pioneered by companies like Bigelow Aerospace[v] and the folding technologies in their conceptual stage such as those by Saga Space.[vi] The compactness and modularity designs of both these companies are preferred as transporting objects to Mars is both expensive and mass limited.

Radiation exposure is a significant risk in any plan that involves humans on the surface for any amount of time. As such, Martian buildings will require significant radiation shielding to protect the inhabitants. Recent data suggests that a human on the surface of Mars for 500 days would be exposed to 320 mSv of radiation, and an additional 662 mSv during the transit;[vii] for comparison, the average yearly dose on Earth is approximately 2.4 mSv.[viii] Radiation is known to cause a host of human health issues, cancers being one, that would make long term stays extremely dangerous to inhabitants. For example, NASA recommends a lifetime radiation limit of a 55-year-old female astronaut of 3000 mSv.[ix] Martian regolith is a prime candidate for radiation shielding as it is readily available on the surface. A 1 m layer could reduce exposure to approximately 81 mSv per year and a 3 m layer down to 2.9 mSv. Compressing the regolith into bricks is also a possibility, as increasing the density theoretically reduces the exposure.[x]

Lastly, a unique requirement of the Mars settlement is accounting for psycho-social effects of living somewhere so novel - meaning as best as can be replicated, the settlement will need to feel like a home such that inhabitants can adapt more readily, not simply attempt to tolerate their surroundings. Designers and engineers will need to account for the basic needs and safety of inhabitants but it must be underscored that designed space and architectural cues have a significant impact on the expected human behaviour within the space.[xi] Some have even argued that societal stability or instability is closely linked to how the spaces around us are organized.[xii] As such, they should be designed in such a manner that makes the daily lives of inhabitants as efficient and pleasing as possible, with some options for customization within private dwelling spaces to account for personal and cultural variations. The previously mentioned company Saga Space and AI Space Factory[xiii] both list the human perspective of mental well-being and social sustainability as core tenets of their design processes. Leveraging these companies and upscaling their proposals would be a good start forward to ensure the designs are indeed inhabitable for the long term with fewer negative psychological effects.

5.1 Construction Concepts
While using inflatable structures initially is efficient, it will be unfeasible and expensive to scale up due to prohibitive launch costs. The Martian city will require permanent long-term building techniques that utilize native materials as much as possible.

Building underground (or mostly underground) by trenching is the most cost-effective way of shielding and insulating Martian structures. Additive construction (3D printing) is already used in construction and would be useful to quickly build structures on Mars from local resources (producing building materials locally is discussed in Section 7.0). Currently, AI Space factory has been testing full scale 3D printed structures using Martian simulant regolith with promising results, and has also received funding from NASA to continue their investigation in ISRU building techniques.[xiv] In structural components that may be difficult to print, regolith bricks could be manufactured in an autonomous facility and could be employed as required for additional strength. These bricks need only to be compacted at specific pressure to perform as structural components making their autonomous mass production relatively easy.[xv]

The majority of structures in which humans will spend significant amounts of time (living areas, gathering places, research facilities, offices, pathways, etc.) will be partially buried or trenched to take advantage of the properties of the regolith mentioned previously without the added complexity and expense of underground tunneling work which requires significant reinforcement with expensive finished materials.

Initially, the soil excavated during construction will be used to cover or print cap pieces, or create structural bricks for the upper level portions constructed using either the 3D printing method or a compaction method. Much of this excavation and printing of the individual units, hubs and tunnel structures could feasibly be done autonomously prior to the arrival of the first crews. Inside these printed shells, initial crews then need only select the correct inflatable section and make the appropriate connections and a viable settlement could come together relatively quickly, without the added time and complexity of sealing and testing each autonomously printed section. Using pre-fab techniques, inflatable sections and connector pieces could be built and shipped to parking orbits or to Phobos prior to arrival and then these building supplies could be deorbited when and where needed. As the settlement grows and can maintain full time construction crews and industry, the building techniques could shift to completely sealed, total ISRU methods, thus ending the reliance on Earth to send habitat modules and pieces.

5.2 Community Design Considerations

As mentioned in the outline, a significant consideration in Martian architecture will be the psychological health of the inhabitants. This idea extends to all aspects of the Martian design, not just to private dwellings.

Natural light is an important factor in this design consideration for many reasons; one being that the diurnal regulation of human biological rhythms is based on light and dark cycles.[xvi] Where appropriate, windows or light tubes should be installed to bring natural light into as many areas as possible. Light tubes can reflect light down below the surface, reducing the demand on electrical lighting during the day in those areas and bring vital light to plants. When windows are used, they should be coated in radiation-blocking film, or potentially a sandwich of film and transparent peroskovite solar cell technology adding power back into the grid. Peroskovite is an extremely versatile, flexible and relatively cheap photovoltaic in development that can be sprayed onto surfaces.[xvii]

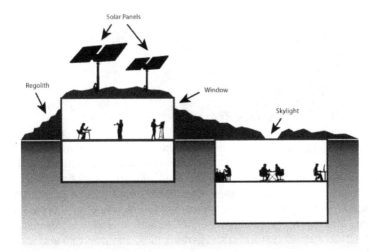

Fig 5.1 Cutaway diagram showing offset upper and lower housing levels. (Artwork by Ben Drexler.)

Community will be central to the design philosophy of the Pax Ares community. In a spoke and wheel fashion, individual units will be built around neighbourhood hubs, which will be centered around larger community hubs, all connected via tunnels, thus removing the need to don pressure suits to simply go about daily activities. This design also allows for differing depths of social interaction. While any community will develop organically based on local needs and topography, for the sake of using a repeatable layout to estimate interior volume, habitable area, and transportation needs a hexagonal expanding "snowflake" design was used for the settlement layout. The smallest "snowflake" is the neighbourhood hub with ten tunnels or "arms" branching off. The ten arms are arranged in two stories with the upper and lower levels offset from each other so that the lower levels still have access to natural light from above. Each arm has five residential units and with an assumed average family size of 3 people per unit, this gives a population of 150 inhabitants per neighbourhood. Six neighbourhoods will in turn form a community, also laid out in a hexagonal pattern. Each community will house 900 inhabitants. For purposes of the transportation network, seven communities will be organized into a village of 6,300 Inhabitants, with the central village serving as the transportation hub connecting villages to each other.

Building/Space	Area
Residential	49.0 sq m
Roadways	45.0 sq m
Hydroponic agriculture	28.0 sq m
Mechanical (mostly under roadways)	45.0 sq m
Commercial	2.3 sq m
Office space	1.0 sq m
Educational	0.3 sq m
Healthcare	0.3 sq m
Assembly	1.0 sq m
Open space	10.0 sq m
Industry	8.0 sq m
Storage	7.0 sq m
Traditional agriculture	7.1 sq m

Table 5.1 Utilization of available area on a per-inhabitant basis.

This specific arrangement of the hub and spoke pattern was arrived at as a tradeoff between balancing an accurate ratio between residential area and area utilized for other purposes, transportation efficiency, and inhabitant psychology. Necessary space per person for various uses was based on a NASA study on space settlements.[xviii]   Forty-nine square meters of living space is allocated to each inhabitant, leading to an average of 147 square meters per residential unit. Subtracting the area needed for roads, hydroponics, and mechanical infrastructure leaves 37.1 square meters per inhabitant to be utilized in the hub areas. Table 5.1 shows how this area is utilized. An additional 9.6 square meters of agricultural area per inhabitant is necessary to meet the demand for plant products (food, fiber, and fuel). This can be accomplished either through increasing available area by building multi-story shops and offices or by finding additional hydroponic space on the outside of commercial buildings or inside public transportation tunnels. This plan assumes a worst-case scenario where additional areas are built solely for agriculture. Using this plan, one million inhabitants would live in 159 villages with an additional 8 village-sized areas dedicated to agriculture.

Roughly ten minutes was the maximum amount of time considered as acceptable for residents to have to walk anywhere before reaching public transportation. Increasing the number of residential units per neighbourhood would potentially make the transportation network more efficient by increasing the number of daily commuters per metro station, but would affect the ratio of residential to non-residential area usage, meaning that more hubs would have to be built dedicated solely to non-residential usage. In terms of total available space, this tradeoff largely cancels itself out. Two levels is viewed as the maximum acceptable depth for residential areas, as building any deeper would make it difficult for the lower levels to receive any natural light.

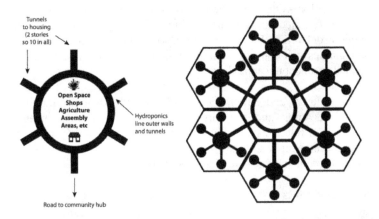

Fig 5.2 Layout diagram of community organization. The neighbourhood hub on the left is the core of one hexagon on the right, that shows a larger community hub. (Artwork by Ben Drexler.)

Greenspace and community gardens will be added to essentially any area that is suitable; this will ensure a level of community involvement in crops and plant care which will be vital to the settlement, and make efficient use of all indoor space. Spaces such as the connecting tunnels, hubs, and pathways for pedestrian or electric vehicular traffic will have some measure of green space. As it is quite an ordeal to head outside on Mars, these dispersed green spaces will act as a sort of "outdoor" space. Lining the passageways and thoroughfares of the settlement, these crops will ensure all space has a useful purpose, distribute an essential component of the oxygen generation system around the settlement, expose the inhabitants to broad spectrum light needed to grow the plants, and make the "outdoors" truly feel like the outside world. Martians will need to feel like the planet is their home and not just an alien world that they have to survive within a sterile environment. Exposing the inhabitants to abundant plant life throughout the settlement will reinforce the idea that the settlement is a welcoming, livable environment and help to reduce conditions such as claustrophobia, depression, or seasonal affective disorder that could be exacerbated by living underground.

The central area in the middle of a neighbourhood will host public space that can be utilized for small neighborhood markets, parks, open areas for gatherings, restaurants, athletics, clinics, or anything else that inhabitants may need in the immediate vicinity of their homes. Each of these neighborhood public spaces will connect (via tunnel) to a larger central community hub utilized by multiple neighborhoods. Even the most distant housing unit will be within a ten-minute walk from the community center, and ramps and moving walkways will allow access for the disabled or injured inhabitants as well. This larger centralized area in the middle of a community will house office space, various shops and retailers, local services, greenspace, theaters, athletic fields, schools, and access to public transportation. The main transportation system will be electric trains; however, tunnels will be designed to allow for electric vehicular traffic for emergency services, public utilities, and goods deliveries. Inhabitants will also be encouraged to ride bicycles or walk. Tram shuttles will connect community hubs to village hubs, where larger metro stations will connect to each other.

The efficient modularity and simple design of the individual-sized units, the neighbourhood-sized units and the larger community-sized hubs mean that expansion can happen relatively quickly without the need for timely redesign of each unit. Planners can simply add the appropriate units where and when they need them. Various settlement statistics are displayed in Table 5.2.

Population Stats	
Population	1,000,000
Population density (people per sq km)	5,483
Surface land area (sq km)	191
Habitable volume (cubic meters)	309,300,000

Table 5.2 Various settlement statistics. The population is similar to that of San Jose. Population density is similar to Boston. Land area (on the surface) is less than that of Las Cruces.

6.0 POWER, WATER, ROCKET FUEL, AND ATMOSPHERE GENERATION

6.1 Power
Electricity production options are limited, but not lacking in capability. Keeping the goal of maximizing self-sustainment in mind, importing fossil fuels from Earth is unviable and counter to the self-sufficiency goals of Pax Ares. While nuclear power is not sustainable, it is incredibly energy-dense. It also does not rely on favorable weather or daylight. However, until uranium can be mined and processed on Mars, nuclear fuel will require importation. Solar power is clean, sustainable, and available in abundance on Mars. Although Mars is farther from the Sun than Earth,

its atmosphere is much thinner, meaning that a higher percentage of sunlight that reaches the planet makes its way to the surface. Available solar power at the Martian surface is roughly two-thirds of that available on Earth. The entire surface of Mars is prone to dust storms, meaning that solar power could be reduced for long periods of time. Also, providing enough power for a large settlement relying only on solar power would require vast areas dedicated to solar farms. Geothermal power may be a potential candidate in some regions, but at this point in time the existence of subterranean hot springs has not been proven. Biofuel is the final option for energy production. Like solar power, biofuel is sustainable. It does require significant space to produce, and providing enough energy for an entire settlement strictly using biofuel is not likely feasible. For this plan, nuclear and solar power will be the two primary methods of power production. The combination of solar sustainability and low cost with nuclear reliability and vast energy potential means that they complement each other well. Biofuel will play a minor role for use in combustion engines or emergency backup generators should they be required, but the vast majority of vehicles and equipment in the settlement will be powered either directly by the electric grid or with rechargeable batteries.

Energy sector	Power demand (MW)	Percent of total
Industrial	348	39%
Transportation	119	13%
Residential	228	26%
Commercial	196	22%
Total	891	100%

Table 6.1 Estimated load for various sectors of the Martian power grid.

Estimating that the settlement demands a constant power production rate of 891 MW, this can be achieved using 132 square km of solar farms. The total surface footprint of the settlement is 191 square km, so the area devoted to power generation will be roughly 70% of the land area (as construction will be largely underground, solar panels will simply be built on the "roofs"). This assumes that average modern-day solar panels are used,[xix] and not higher-end or more advanced options that will likely be available in the future. The figures from United States data were adjusted to account for both the decreased solar capability on Mars and the increased performance due to being located on the equator. Two small 720 MW nuclear reactors can provide 100% redundancy.[xx] An average power consumption per person of 2 MWh/year (Earth year) was chosen based on the average residential consumption of a UK citizen.[xxi] This rate will likely be lower in reality, as subterranean construction and insulation will reduce the need for heating dwellings. In addition, waste heat from the industry sector can be transferred to areas that require heating via hot water pipes. This conservative estimate accounts for residential power consumption. Power demands for the industrial and commercial sectors were assumed to be in the same proportion to residential power as in the 2018 data provided by the U.S. Energy Information Administration.[xxii] The transportation sector was adjusted to account for the higher efficiency of electric vehicles, which will make up the vast majority of vehicles in the settlement.

Initially, solar panels brought from Earth will provide power for the small settlement. As the population and industry rapidly grow, importing solar panels will become impractical and a more energy-dense power source will be needed. The small nuclear plants shipped from Earth will power the growth of Martian industry. As the settlement's industrial capacity grows, it will eventually begin producing its own solar panels. Martian-produced solar panels will make the settlement truly energy independent and offset the need to import nuclear fuel from Earth. The two reactors will remain on standby in case of severe disruption to the solar power grid. Batteries will be essential to maintain power capability through each night.

Power source	Power produced (MW)	Percent of total capability	Percent of total demand
Nuclear	1,440	59%	162%
Solar	927	38%	104%
Biofuel (ethanol)	66	3%	7%
Total	2,432	100%	100%

Table 6.2 Components of electric power production and percentage of total capability and total demand for each.

6.2 Water

Excavated regolith will be heated to produce liquid water from the permafrost. In order to reduce the demand for water production, the settlement will recycle what water it uses. Assuming that the average inhabitant uses 300 liters of water per day,[xxiii] the settlement will need to reclaim 53 million liters of water every day. This is possible with a modern water reclamation plant, such as the one in Rochester, MN.[xxiv] Industrial processes and energy production will also require water. The industrial water supply can be separated from the reclaimed water supply for use in residences and agriculture, thereby negating the need to reclaim industrial wastewater. If water is completely recycled throughout the settlement once per 24-hour period, then the settlement will require approximately 772 million kilograms of water, some of which will be in the form of atmospheric water vapor. Industrial dehydrators will return water vapor from the settlement atmosphere back to the liquid water supply.

In addition to residential and industrial processes that utilize water and produce wastewater, some processes will actually change the total amount of water in the system. Photosynthesis turns water (with carbon dioxide) into sugars, and electrolysis turns water into hydrogen and oxygen. Most of this water loss will be compensated for by processes that produce water. Hydrogen produced by electrolysis will combine with carbon dioxide in the Sabatier process[xxv] to produce methane and water. Plant sugars will eventually produce water through metabolic processes, decomposition, or combustion as biofuels. It is assumed that some degree of water loss through chemical reactions or leakage is unavoidable; melted permafrost will replace losses.

6.3 Rocket fuel

The methane produced by the aforementioned Sabatier process will be utilized as rocket fuel for powering trade with Earth. The oxygen produced by electrolysis will provide the oxidizer for the rockets. As just discussed, one of the ingredients for the Sabatier reaction, hydrogen, will come from electrolyzed water. The other ingredient, carbon dioxide, is the primary component of the native atmosphere. Thus, rockets will be powered entirely by Martian air and ice.

6.4 Atmosphere

Oxygen is essential for a breathable habitat atmosphere. Settlement plants will produce oxygen through photosynthesis, but the main source of oxygen will be through mechanical electrolysis. Oxygen production must compensate for losses to aerobic respiration and combustion. Like on Earth, the atmosphere in the settlement will be mostly nitrogen and oxygen, although at reduced pressure. Carbon dioxide will also be a component as crops need it for photosynthesis. Carbon dioxide already exists in abundance on Mars and will be produced by respiration of the inhabitants, fermentation of agricultural products, and combustion of ethanol. Any need for excess carbon dioxide will be readily supplied by the native atmosphere. However, it is likely that carbon dioxide will be produced in excess inside the settlement. It will be scrubbed from the interior atmosphere to an acceptable level and then released into the Martian atmosphere. The largest component of the atmosphere, nitrogen, is not strictly needed in gaseous form as it is largely inert. However, the nitrogen cycle is critical to producing fertilizer for plants. Nitrogen-fixing bacteria will turn atmospheric nitrogen into ammonium for use in generating fertilizer. These bacteria will utilize the small amount of nitrogen present in the Martian atmosphere to provide a source of nitrogen production as well as recovering nitrogen from the settlement atmosphere. Nitrogen (and other nutrients) will also be recovered from wastewater during the treatment process and reused for fertilizer production.

Component	Partial pressure (kPa)	Total mass (kg)	Percentage by mass
Oxygen	20	87,214,205	27.1%
Carbon dioxide	0.4	2,398,381	0.7%
Nitrogen	60	228,937,288	71.1%
Water vapor	1.4	3,340,440	1.0%

Table 6.3 Components of interior settlement atmosphere.

The resulting settlement atmosphere is shown in Table 6.3. Oxygen is maintained at 20 kPa to allow for sufficient respiration. Nitrogen is maintained at roughly the same proportion to oxygen as in Earth's native atmosphere to allow sufficient levels for nitrogen fixation to occur. Carbon dioxide will be maintained at 0.4 kPa or below to prevent carbon dioxide poisoning. Humidity will be maintained at 60%. Higher levels could lead to issues with condensation and electrical problems. Lower levels would cause dehydrators to operate inefficiently.

7.0 SUSTAINMENT

7.1 Food

In their paper *Feeding One Million People on Mars*,[xxvi] Cannon and Britt determined the figures for providing enough calories to sustain one million Martians entirely from local resources. With a diet composed of 50% plant-based nutrients, 25% insect proteins, and 25% lab-grown meat, one million Martians could be fed with forty-six square kilometers of agricultural area. Ranching would not be economical on Mars, as farm animals would also require their own water and food crops, which in turn require additional water. Raising insects for food has the advantage that they would not require additional food crops, as they can be entirely sustained with waste and the portions of crops unused by humans. In addition to providing food, plants can also be used to produce textiles for clothing and furniture, biofuel, and bioplastic. Plants would also be critical components in the oxygen generation and carbon dioxide scrubbing processes.

7.2 Essential Goods Production

Building materials can all be locally produced on Mars. Bricks and concrete can be produced from sand and sulfates. The regolith of Meridiani Planum is rich in iron and the atmosphere contains plenty of carbon. Together these are the ingredients for producing steel. A variety of plastics can be produced from biomass in fermentation tanks or from combining hydrogen and carbon dioxide. Local silicates will be used to produce glass. One notable construction component missing on Mars will be wood. Some trees will be grown inside the settlement, but they will largely be for agriculture and aesthetics. While it will not be feasible to grow a large enough forest to sustain an industrial-level lumber industry, it may be possible to sustain a limited paper industry, especially if bamboo is used instead of tree pulp. As older trees within the settlement die they could be utilized for woodworking, providing a limited supply of wood products that would likely be luxury items.

Construction materials such as plastic and glass can also be used to craft furniture. Textiles and fabrics will come from agriculture. Undoubtedly, Martians will desire personal electronics just like modern Earthlings do. While many, if not all, components of electronics and household appliances can be manufactured on Mars, so many different elements are needed to build contemporary consumer electronics that it may not be economical to develop a local production chain to support one million people. This reliance on Earth imports will be reduced by recycling materials instead of importing replacements.

8.0 SETTLEMENT GOVERNANCE

While the major space treaties cite global cooperation as a core tenet, perhaps one of the most pressing issues to address in the near term is that of the legality of a Martian settlement. Currently, Article II of the 1967 Outer Space Treaty specifically states, "...outer space, including the moon and other celestial bodies, is not subject to national appropriation by claim of sovereignty, by means of use or occupation, or by any other means."[xxvii] One possibility would be for the countries interested in pursuing a settlement to request amendments to the treaty, which has been seen as outdated given the most recent advancements of space technology. But this would leave sovereignty as an issue that remains, as several countries would be pursing this venture together. One possibility would be that the settlement operates under similar principles to the International Space Station, wherein each individual nation involved is participating under the cooperative International Space Station Intergovernmental Agreement.[xxviii] Of note, however, countries remain legally responsible for their citizenry while in space, and most countries it seems are somewhat subordinate to US procedures in space. However, much past initial exploration stages, it is hard to see how this would be applicable and administrable so far from Earth.

It is not unreasonable to then assume that a local government, perhaps operating under a Mars Charter of Rights agreed upon by OST signatory nations with a mechanism to trigger a process for the settlement to be recognized as a sovereign state at some pre-determined level of self-sufficiency (such as the population reaching a certain size) would be a start. This would mean that with amendments to the OST the Martian inhabitants would at some point become sovereign citizens of Mars. Collectively creating a Mars Charter that borrows much of the cooperative, equitable and human-rights-centric language would help enshrine a community and sustainably focused government from the outset. It is envisioned that similar to the governance models in use by UN member states that some form of a representative democratic practice would be utilized post initial stages, (though the authors recognize that there is much to be discussed on this front which is beyond the scope of this project).

In Pax Ares, this could break down to local small-scale governance at the neighbourhood hub level, to community concerns at the community/village hub level, to large scale settlement management (health care, disaster response, economics, Earth relations, etc.) with 159 village representatives forming a sort of legislative body. Civic engagement and responsibility would be greatly emphasized in early years education, and it is hoped that this ownership over personal and collective fate would deter serious criminal activity, but some sort of means to arbitrate crimes or disputes would be indicated. Depending on their severity, this could be done directly by involving the community and utilizing restorative and transformative justice practices, which are significantly less punitive as they more constructive and healing in their expected outcome for both perpetrator and victim.[xxix]

Selection of personnel and inhabitants could be accomplished at a UN Council for Martian Activities, via call outs through appropriate means in participating countries. The selected participants would then attend UN selection and training until such a time that the colony is self-sustaining or has formed its own government and has the means to manage its population.

9.0 ECONOMY

9.1 Funding
Colonizing Mars will require a huge upfront cost. This does not mean that colonization is impossible or will not generate income, but the costs are beyond the capability of most private organizations to absorb. The Mars settlement will have a strong economy, but it should not be viewed as a get-rich-quick scheme. It should be viewed as an essential extension of human civilization. To avoid past mistakes from Earth's colonial history, the Mars settlement will not be owned by Earth and it is important to develop an economic model that reflects that. One possibility considered was the idea of having the settlement itself assume the cost. This debt would be sold in the form of bonds which would then be paid off over time. The problem with this model is that, due to the enormous upfront cost of the project and the time needed for the Martian economy to grow, it would take hundreds (or potentially thousands) of years for the settlement to pay off its debt, even with low interest rates, high taxes, and a significant portion of revenue dedicated to bonds payments. While the model would generate a great amount of wealth for those purchasing bonds, it would mean Martian indebtedness to Earth long after achieving a GDP much greater than the initial colonization cost.

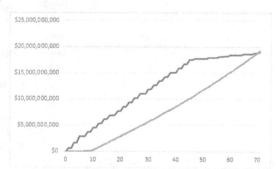

Fig. 9.1 Colonial net worth (bottom line) and running cost (top line) for the first 70 years of colonization. After approximately fifty years, the estimated value of the settlement exceeds the total cost of colonization. Note that spacecraft purchases significantly drop after mission year 42.

An alternative option is to use a combination of public and private funding sources to pay the upfront costs. National governments have deep pockets (particularly the wealthier nations that already fund space agencies) and could easily absorb the upfront cost for spacecraft, equipment, and supplies. Aside from the obvious benefits to humanity as a whole from developing a second home for human civilization, nations would see long-term benefits in developing a new planetary trading partner. As will be discussed, establishing a settlement on Mars will also open the door to accessing rare minerals in the asteroid belt. Additionally, funding can be obtained from private organizations in exchange for exclusive trading rights. For example, a corporation may agree to offset the cost of a spacecraft launch in exchange for the right to sell their products on Mars.

Figure 9.1 shows that the Martian settlement should pay for itself in roughly 72 years. Costs do not include the internal costs of running the settlement, as a dollar spent on Martian infrastructure or goods is a dollar contributing towards the Pax Ares GDP. Costs include fuel (until mission year 20), operations ($6 million/year), spacecraft purchases, and advanced goods imports. The spacecraft cost was reduced from $600 million per vehicle (based on the current cost of an Orion capsule),[xxx] to $200 million each. This assumes a significantly reduced unit price based on mass production. After 42 years, the rate of spacecraft purchase drops and stops altogether after 58 years. In order to both reduce the total cost spent on hardware as well as to limit the stress on Earth's launch infrastructure, the immigration fleet reaches max capacity at 60 vehicles (30 re-entry capsules and 30 deep space habitats). Imports are conservatively valued at $1,200 per immigrant (based on the $1,344.55 per American consumer per year spent on media),[xxxi] and at $120 per inhabitant per year assuming a 90% recycling rate.

9.2 Internal Economy
Most space colonization plans assume that wealth will have to come from an external source in exchange for exports. However, simply by existing and producing the resources and goods needed to survive and be comfortable on Mars, inhabitants will be creating wealth. Mars does not have to produce goods for export in order to become wealthy.

The economic growth in Figure 9.1 is based on a simple model. Martian income includes rocket fuel (after mission year 20), agriculture, construction, goods, and services. Crops are valued at $150/ton based on the current market price of hay.[xxxii] Construction work is valued at $160,000 per "building" (defined as one residential unit or another construction of equal size). Goods production is valued at $14,700 per worker per year, and services are valued at $25,700 per worker per year. These figures are based on the goods and services components of the first quarter 2020 United States GDP.[xxxiii] While the United States has a powerful economy, these estimates are still conservative as the figures used were determined on a GDP per person basis while the Mars data is calculated on a per worker basis (assuming that half of the total population is in the workforce). Additionally, the Mars economy will by necessity be high-tech and highly automated, and thus advanced. Data and information produced by the settlement is valued at a flat rate of $250 million per year. This is a crude economic model and is meant only to give a ballpark estimate of how long it will take for the settlement to generate positive net wealth.

It should be remembered that once the settlement pays for itself, that does not mean that those who paid the price will have seen a return on their investments. National governments will largely foot the bill, but the wealth accumulated will belong to the Martian inhabitants. For the previously mentioned reasons, nations should view this as a long-term investment for the betterment of the world as a whole and for their national self-interest in developing a future trading partner and establishing ease of access to the asteroid belt.

The three main components of the economy will be the people, government, and businesses. Money will flow from the government to the people via a living wage. This living wage will be enough to cover essentials (food, water, air, healthcare, clothing, electricity, heating, information access, etc.). Additionally, housing, education, police, and fire services will be provided. As each inhabitant will receive a stipend, it is important that everyone contribute to the workforce. The job and housing markets will be tightly regulated to ensure that there is enough work and living space for everyone. In addition to the living wage, inhabitants will receive wages for their work, whether it is in the public or private sector. In order to work toward retirement, inhabitants will earn social credit for working in the public services or in key settlement-sustaining industries, or for otherwise contributing towards the success of the settlement. This "credit" will not be transferable to cash, but can be "spent" on benefits such as additional education or retirement. Optionally, inhabitants can work for the private sector, choosing to earn higher wages at the expense of having to pay for their retirement. Recycling will be another source of income for inhabitants. It will be critical to recycle everything possible to reduce the need to import goods from Earth. In exchange for providing materials that will be used in production, inhabitants will receive payment for their recycled goods. Finally, Martians will earn money from stocks, savings, and bonds if they choose to invest. The safety net of a living wage and guaranteed home will encourage Martians to make investments that they otherwise would not be willing to risk, leading to bigger payoffs and more rapid economic growth. The government will generate revenue through luxury, corporate, and capital gains taxes. Tax revenue will pay for wages and services as well as subsidies to critical industries (life support, sanitation, transportation, etc.). In addition to subsidies, businesses will earn revenue by providing goods and services to inhabitants.

One potential for Martian economic development that is left out of this plan is land development. Currently, it is internationally illegal for anyone to claim territory on Mars.[xxxiv] While the value of construction work was included in the economic model, that value only included the value of the living space itself and not the value of the land it occupies. Once a permanent human presence is established on Mars, the treaties could change to allow treatment of Mars as a second Earth as opposed to an unclaimable celestial body.

9.3 External Trade
Eventually Mars will become a hub for trade between Earth and the asteroid belt. This could very well happen by the time there are a million inhabitants on Mars.

Due to its low gravity and proximity to the asteroid belt, Mars would make an ideal location for supplying mining expeditions and possibly even refining ores before shipping them to Earth. While Mars has all the essential elements to support the local population, it is unlikely to have many unique resources that would be economical to produce for shipment to Earth. Its true value lies in its location next to the asteroid belt and in its low gravity making space missions more affordable. An early objective of the settlement should be to utilize Mars' position, with its thin skies and unprecedented view of the asteroid belt, to catalogue asteroids and determine which ones make the best targets for mining operations. Another possible Martian export would be satellites. Fuel can be produced directly from the Martian atmosphere (as discussed in Section 6.0), and it is actually more fuel-efficient to launch satellites into Earth orbit from Mars than it is to launch them from Earth (due to a significantly smaller gravity well). Information will also be a valuable export. Like money, ideas can cheaply be transmitted between planets without need for fuel. Martians will likely rely heavily on technologies such as robotics and nuclear power, so advances in these fields would be valuable to Earth industry. In addition, Mars provides fertile ground for providing unique insight into geology and astronomy.

10.0 SETTLEMENT HEALTH AND CULTURE

10.1 Societal Characteristics
Determining the social-cultural characteristics of a society that does not yet exist, would be as challenging as it would be misinformed. However, given the early focus on scientific exploration, the necessity of self-sufficiency, and the need to work in community in order for the Martian society to survive, it is likely that those initial characteristics will inform a proto-Martian society. Inhabitants will have to be thrifty and resourceful. Martians will need to adopt a "fix it" versus a "replace it" consumerist attitude. Due to the fact that everything (food, air, water, hard and soft goods) will come at such a direct cost in terms of labour and time to produce, Martians will need a deep understanding of their relationship to their environment; as such a cornerstone of Pax Ares society will be social responsibility and sustainability in order not to waste precious resources. Thus, it will be likely that Martians are heavily dependent on artificial intelligence, automation, and other technologies. Inhabitants will need to be highly educated and technically skilled. A strong respect for education and personal growth will also be essential to developing a civically and critically engaged citizenry for a democratic government to function. Another aspect of Martian society will be its diversity; this is due to the fact that it is unlikely that only one nation will be sending citizens initially, and it is envisioned that Mars will not be an outpost of American or even Western civilization, but rather a new form of global cooperation. Martians will come from a wide variety of cultures, religious traditions, and backgrounds. The culture that will naturally develop will be representative of that blending and the adversity that the inhabitants will face collectively to overcome.

10.2 Cultural Practice and Recreation

Science, economics, and civics may all pose fascinating questions for a Mars settlement to solve but there is an even bigger question looming behind them that impacts the lives of the inhabitants on a daily and frequent basis: recreation. Inhabitants cannot spend all their time working in remote conditions; as such significant investments must be made into leisure and recreational activities for their mental and physical health. Most sports would be quite unpredictable in lower-G environments with perhaps one unexpected exception: swimming. As it mostly depends on buoyancy (a function of density) and viscosity, swimming represents a wonderful opportunity for Earth-like exercise that can be managed in smaller and controlled environments.[xxxv] Be it swimming or other forms of recreation, participating in community wellness is likely to be of high importance for social stability in the difficult conditions that Mars presents.

It seems plausible that Pax Ares inhabitants will enjoy some of the same forms of entertainment that modern Earth inhabitants currently experience. Inhabitants will take their favourite examples of art and culture with them, but there will most definitely be uniquely Martian music, literature, architecture, paintings, sculpture, and theater that develop over time. However, architecture and sculpture specifically represent two areas in which there could be civic opportunities to invest in projects that are imaginative and colorful to contrast the stark environment outside the settlement. Musical instruments will likely have to be created using resources on-site with iron or silicate-based brass and woodwind instruments being the easiest to fabricate.

10.3 Health

Selecting healthy individuals initially and a culture of preventative health practice will be vital for the Martian city-state dwellers; this includes optimized nutrition and large volumes of exercise – most especially for the early waves of explorers that will be exposed to the withering effects of micro-gravity for long periods of time – and there is an anticipated significant focus on mental health support requirements. The crews that initially arrive on Phobos to begin surveying the surface remotely will likely need to rotate out to avoid serious complications from extended zero gravity and radiation exposure. To accommodate the immediate and emergency health requirements of early waves of inhabitants, doctors and medical staff (and cross-trained crew members of other disciplines) will have to manage in extremely remote circumstances. Telemedicine and advancements in AI health diagnostics will likely play a vital role in the early stages. Chronic conditions and severe traumas resulting in massively reduced abilities to perform primary functions in early explorers would likely result in that member making the journey back to Earth given the lack of resources and personnel available to care for them. As the settlement grows, it is likely that Martian-centric medical research would develop as a result and all manner of conditions could be supported by the settlement, and alternate work/accommodations found for ill and injured Martians. To support this, it is proposed that medical school and research facilities be housed in some sort of post-secondary institution on the settlement. Given the significant importance of each member's contribution within Martian society to its continued day-to-day functioning, and the presumed focus on health and well-being from the outset, health care would be provided by the local government to residents to avoid inequities in health access or Martians delaying seeking medical attention which would impact their effectiveness and potentially harm the settlement stability.

11.0 CONCLUSION

In this paper, the authors have attempted to both realistically and optimistically lay out our broad vision of what one million humans living on Mars will look like. It is no secret that colonizing Mars will be an expensive endeavor, but it also has the potential for incredible benefits for society as a whole. Mars should not just be colonized for the economic benefits, but for the benefit of all humanity. By colonizing Mars, humans will no longer be restricted to one planet and will truly become a space-faring species. It is important to avoid the mistakes of the past, and the settlement should not be thought of as a "colony" in the traditional sense, where its sole purpose is to benefit the colonizing power (in this case, Earth). The Mars settlement must be an independent branch of civilization, and the economic, government, and social aspects of its planning should reflect that. The people who choose to make the journey and make Mars their home will be opening new perspectives for all of humanity to ponder.

ACKNOWLEDGEMENTS

The authors would like to thank the following people for their contributions and/or lengthy discussions about how they would do their jobs on Mars: Mark Bohme, Andrew Carrier, Erin Plate, and Amy Rossig. Ad Ares!

REFERENCES

[i] https://www.nycsubway.org/perl/caption.pl?/img/cars/sheet-r160.jpg
[ii] https://www.nycsubway.org/perl/caption.pl?/img/cars/sheet-r160.jpg
[iii] Ruff, S.W., Christensen, P.R., Glotch, T.D., Blaney, D.L., Moersch, J.E., & Wyatt, M.B. (2008). The mineralogy of Gusev crater and Meridiani Planum derived from the Miniature Thermal Emission Spectrometers on the *Spirit* and *Opportunity* rovers. In J. Bell (Ed.), *The Martian Surface: Composition, Mineralogy, and Physical Properties* (pp. 315-338). Cambridge, UK: Cambridge University Press
[iv] Franz, H.B., Trainer, M.G., Malespin, C.A., Mahaffy, P.R., Atreya, S.K., Becker, R.H., Benna, M., Conrad, P.G., Eigenbrode, J.L. (1 Apr 2017). "Initial SAM calibration gas experiments on Mars: Quadrupole mass spectrometer results and implications." Planetary and Space Science. 138: 44-54. DOI: 10.1016/j.pss.2017.01.014.
[v] https://bigelowaerospace.com/pages/firstbase/
[vi] https://asaga.space/projects/mars-lab
[vii] https://www.businessinsider.com/mars-radiation-levels-spacex-nasa-exposure-2018-4
[viii] https://www.iaea.org/Publications/Factsheets/English/radlife#dose
[ix] https://www.nasa.gov/sites/default/files/atoms/files/space_radiation_ebook.pdf

[x] https://www.scienceintheclassroom.org/research-papers/curiosity-tells-all-about-mars-radiation-environment

[xi] https://www.archdaily.com/936027/psychology-of-space-how-interiors-impact-our-behavior

[xii] Jacobs, Jane. *The Death and Life Of Great American Cities*. (1992). Vintage Books, NY, NYC

[xiii] https://www.aispacefactory.com/

[xiv] Ibid

[xv] https://www.nature.com/articles/s41598-017-01157-w

[xvi] https://www.ncbi.nlm.nih.gov/pmc/articles/PMC6751071/

[xvii] https://www.sciencedaily.com/releases/2016/12/161201114543.htm

[xviii] NASA. (1977). *Space Settlements: A Design Study*. R.D. Johnson & C. Holbrow (Eds.). NASA SP-413.

[xix] Ong, S., Campbell, C., Denholm, P., Margolis, R., & Heath, G. (2013). "Land-Use Requirements for Solar Power Plants in the United States". Technical Report NREL/TP-6A20-56290.

[xx] https://www.nuscalepower.com/environment/carbon-free-energy

[xxi] http://shrinkthatfootprint.com/average-household-electricity-consumption

[xxii] U.S. Energy Information Administration. (2020, May 7). "U.S. energy facts explained". https://www.eia.gov/energyexplained/us-energy-facts/ (accessed Jun 3, 2020).

[xxiii] United States Geological Survery. "Water Q&A: How much water do I use at home each day?". https://www.usgs.gov/special-topic/water-science-school/science/water-qa-how-much-water-do-i-use-home-each-day?qt-science_center_objects=0#qt-science_center_objects (accessed Jun 3, 2020).

[xxiv] https://www.rochestermn.gov/departments/public-works/wastewater-management/water-reclamation-plant

[xxv] https://marspedia.org/Sabatier/Water_Electrolysis_Process

[xxvi] Cannon, K.M., & Britt, D.T. (2019). Feeding One Million People on Mars. DOI: 10.1089/space.2019.0018

[xxvii] Treaty on Principles Governing the Activities of States in the Exploration and Use of Outer Space, including the Moon and Other Celestial Bodies, Jan 27 1967, 610 U.N.T.S. 205

[xxviii] http://www.esa.int/Science_Exploration/Human_and_Robotic_Exploration/International_Space_Station/International_Space_Station_legal_framework

[xxix] https://www.justiceeducation.ca/about-us/research/aboriginal-sentencing/restorative-justice

[xxx] https://www.nasa.gov/press-release/nasa-commits-to-long-term-artemis-missions-with-orion-production-contract

[xxxi] https://www.forbes.com/sites/tonifitzgerald/2018/11/29/how-much-does-the-average-person-spend-each-year-on-media/#7803f3e3319d

[xxxii] https://www.nass.usda.gov/Publications/Todays_Reports/reports/agpr1118.pdf

[xxxiii] https://apps.bea.gov/iTable/iTable.cfm?reqid=19&step=3&isuri=1&nipa_table_list=6&categories=survey

[xxxiv] Treaty on Principles Governing the Activities of States in the Exploration and Use of Outer Space, including the Moon and Other Celestial Bodies, Jan 27 1967, 610 U.N.T.S. 205

[xxxv] https://what-if.xkcd.com/124/

10: THE SUSTAINABLE OFFWORLD NETWORK:
THE NÜWA CONCEPT

Authors: Guillem Anglada-Escudé [1,2], Miquel Sureda[2,3], Gisela Detrell[4],
Alfredo Muñoz[5], Owen Hughes Pearce[6], Gonzalo Rojas[5], Engeland Apostol[7],
Sebastián Rodríguez[5], Verónica Florido[5], Ignasi Casanova[3], David Cullen[8],
Miquel Banchs-Piqué[9], Philipp Hartlieb[10], Laia Ribas[11], David de la Torre[3]
Contributors: Jordi Miralda Escudé[12,13,2], Rafael Harillo Gomez-Pastrana[14],
Lluis Soler[3], Paula Betriu[3], Uygar Atalay[3], Pau Cardona[3], Oscar Macia[3], Eric
Fimbinger[10], Stephanie Hensley[15], Carlos Sierra[1], Elena Montero[16], Robert
Myhill[17], Rory Beard[18]

Affiliations (1) Institut of Space Sciences - CSIC, Spain; (2) Institut d'Estudis Espacials de Catalunya,
Spain; (3) Universitat Politècnica de Catalunya, Spain; (4) Universität Stuttgart, Germany; (5) ABIBOO
Studio, USA/Spain & intl.; (6) OHP+, UK; (7) WestonWilliamson+Partners, UK; (8) Cranfield
University, UK; (9) University of Portsmouth, UK (10); Montanuniversität Leoben,Austria; (11) Institut
de Ciències del Mar – CSIC; (12) ICREA, Spain; (13) Universitat de Barcelona, Spain (14); Bufete Mas
y Calvet, Spain; (15) Stephanie Hensley, LLC, USA; (16) Sentido Psicología, Spain; (17)
University of Bristol, UK; (18) Babylon Health, UK.

e-mail: contact@sonet-hub.com, website : www.sonet-hub.com

PROLOGUE

*The spacecraft in which we have traveled for eight months has just landed gently
on the surface. Passengers stared at each other in disbelief, still feeling numb from
the brutality of the supersonic retro propulsion during the braked descent. It seems
unbelievable... We are on Mars! Many of us have arrived in recent years.
Thousands upon thousands of people traveling from our comfortable blue home to
the uncertain red dream. The promise of a new world. An opportunity to rethink
everything, in order not to repeat the same mistakes. Beyond the polarized glass
windows, the light of the Martian dawn spills over Tempe Mensa. The Sun is a small
golden circle, surrounded by two bright points: Venus and the Earth. A few
kilometers from the spaceport, the gigantic extensions of the greenhouse's domes
are outlined. Beyond, the land descends steeply down the south-facing cliff. On its
vertical walls awaits Nüwa, the wonderful city that leads the accelerated settlement
of Mars. Its name recalls the goddess who protects humans. And, certainly, the city
protects and provides us with everything we need. Now, I'll become part of Mars;*

I will dissolve into Nüwa and join all those that give life to it. And we will become Nüwa, forever...

The Nüwa concept is not only an urban solution for a city on Mars; it is an attempt to sketch a long-term plan for a new human society. It is a society that uses data to make decisions and applies science to understand, tame, and cherish the world. This is a society whose infrastructure emerges from air, water, rock, and light, but that is shaped by our ingenuity and our collective desire to work towards a better, more sustainable future...because if we can restart society on Mars, Earth problems can be solved too.

Sustainability, but especially self-sustainable development is at the core of the Nüwa concept. As an example of a key concept of sustainability that must be incorporated at design level is how energy in consumed and incorporated into the infrastructure. For being self-sustainable, a settlement on Mars needs to be able to obtain all resources locally. Energy is the underlying resource that ALL processes need. A fundamental characteristic of real-world energy sources is their *Energy Return on Investment* (or EROI); the ratio between the amount of usable energy and the amount of energy used to that end. This concept can be applied to whole societies too, which allows investigating minimum EROI required to thrive [1]. For Earth, this minimum EROI is estimated at around 12-14:1 (5:1 in severely impoverished countries) [2]. The minimum EROI of a Martian society must be higher than that on Earth (Mars is a hostile place), and the EROI values of the more mature Martian energy options (nuclear & solar) are around the so-called *energy cliff zone*, where small changes in EROI carry large changes in energy available to society. This has two significant consequences: 1) it is critical to keep EROI high; 2) a large percentage of the economic activity must be on energy production. As discussed later, establishing the energy and power requirements generates quite some interdependencies that, at design level, must be solved iteratively.

In addition, settlements need the ability to cope with potential future uncertainties; at the risk of creating assets that become prematurely obsolete [3]. Resilience can be improved by redundancy, meaning that no function is covered by a single solution. This is our effort into designing one of these futures, including considerations on sustainability as much as possible. The principles of self-sustainability can be expanded into these top-level functional requirements:
- Collect and process its own resources (primary sector)
- Sustain its own growth by manufacturing all its parts (secondary sector)
- Maintain a safe and good living standard of the population (services)
- Provide mechanisms for individual and collective improvement (governance)

Primary sector functions include all functions related to resource gathering, material processing using physical and chemical processes, and energy production including generation, storage and distribution. Compared to Earth, secondary sector functions must follow 1) use of local inputs only, 2) enable reusability and recyclability by design, and 3) self-replicability of the whole infrastructure. In terms of capabilities, they shall be able to use metals, polymers, ceramic and glassy materials, perform product and manufacturing transformations (physical & chemical), produce machines, make advanced electronics, construction materials and textiles, among others. Finally, end use functions shall include the manufacturing of components for both *infrastructure* and operations of other systems.

Some service functions consist of essential human support functions such as housing in various forms, and life support including atmosphere, water, food and waste management. Services shall also cover other *human factors* related to the wellbeing of the citizens at individual and community level (noise, privacy, etc.) via communal space and mechanism to identify possible needs, dedicated professionals & infrastructure, and the promotion and development of social welfare. In terms of safety legislation, the city should enforce construction and manufacture standards to protect its citizens against the reality of life on Mars (fire, pressure loss, structural, chemical, mechanical hazards), which shall also follow stricter rules of resilience and redundancy. Additional services include city operation functions such as distribution (water, air, energy...), distribution of goods, human transportation, social welfare and education facilities, support to surface operations, access to network and computing services, communications, and logistics management. In terms of governance, the city shall provide institutions with executive, legislative & judiciary functions, law enforcement mechanisms, digital administration, transparency & anti-corruption, law codes, and diplomatic representation.

THE NÜWA CONCEPT
1 - Development Concept and Timeline
After a short initial phase relying on capital investments and supplies from Earth, a large-scale urban development to accommodate one million people shall be able to sustain its growth with local resources only. Our underlying economic model is built around the concept of the *City Unit* (or CU), which corresponds to one human and all the material budget associated with it. That is, for the city to grow one inhabitant, its infrastructure must be able to *ingest and incorporate all the resources* of one CU. Assuming that the CU production rate r is proportional to N (number of existing CU), and that there is a constant flow n_i of CU imported from Earth; the rate of change of CUs can be written as $dN/dt = N/\tau + n_i/\tau$, $\tau = 1/r$ is the characteristic growth timescale and it is a function of the boundary conditions, and n_i/τ is the rate of imports from Earth. Imposing an initial population of $N_0 = 10^3$, that N must be 10^6 after 50 years (or yr), and that n_i becomes small compared to N quickly, we obtain a minimum growth rate of 0.12 CU/yr as illustrated in Figure 1. Assuming that each CU needs to be renovated every 20 years, we add 0.120 CU/yr · 1/20 = 0.006 CU/yr to the required rate reaching a development threshold value of $r_t = 0.126$ CU/yr. This means that every eight citizens should be able to contribute about one CU per year. Qualified immigration is necessary to keep the growth rate, so providing good live standards is essential. The focus of the nascent Martian economic activity is likely to be around **Maintenance** (life support, energy, consumables and operations, ~20%), **Development** (resource collection, transformation, manufacture and construction, 50%), **Innovation & research** (to

develop solutions for increasing r, 10%) and **Services** (20%, which is much lower than in a modern western country). Surpluses are important because they can accelerate the growth rate, improve the life quality of its inhabitants (incentive to entrepreneurship and personal progress); and as mitigation against contingencies (dust storms, turmoil on Earth, and disasters).

The city shall also engage in minor economic activities especially oriented to services for Earth entities (private & institutions), which shall be spent in supporting Earthbound activities, developing the Earth part of the Earth-Mars transportation system, and importing goods of high-density value, such as electronics and rare catalysts. For the model to work, all the material cycles and costs must be designed following standards of economic circularity and Energy Return on Energy Investment (EROI) considerations. That is, once a material is integrated into the system, it shall not leave it.

2 - Development of Phases, Institutions and System of Value
Figure 1 also shows that there are three natural development phases associated with our model. These phases set the pace on how the governance of the cities and the Martian economy evolves over time.
• **Phase 1 – Corporate (years 0 to ~10)**. The growth supported by imports dedicated to high-value machines and components. This phase can afford subpar production, and it ends when the 0.126 CU/yr growth rate threshold is exceeded (pop ~10k, after about 10 years). All colonists must be experienced and motivated individuals (age ~'30s). Individual freedom and privacy are guaranteed, but citizens are formally workers for the Mars Enterprise and remain citizens of their countries. The Mars Enterprise is managed like a large public/private consortium with an approximate power balance of Earth governments (40%), private interests (30%), and the United Nations (UN, 30%) via the UNOOSA (United Nations Office for Outer Space Affairs.). The UN participation formally acknowledges that the resources of *space* belong to the humankind, and effectively implements a similar arrangement given to countries hosting scientific infrastructures. Operations are regulated under the **Mars Colony Act (MCA)**. MCA follows the spirit of both the *Agreement for the development of the International Space Station*, and the (updated) *Planetary Protection Policy* issued by the COSPAR (COmmittee of Space Research) or equivalent. The act must include essential rights and obligations, controversial disputes system, criminal regulation and the first approach to licensing for natural resources, goods, trade and import-export regulations, and the framework to regulate private property. In all practical aspects, daily governance resembles that of an Antarctic base. The Mars Enterprise selects colonists, covers Earth-Mars transport expenses, provides housing and all the basic needs, but not all welfare services are fully developed yet. As an incentive, these settlers are given one **Mars City Share (MCS)** each, which will be later used to distribute surpluses. No return trips are possible yet.

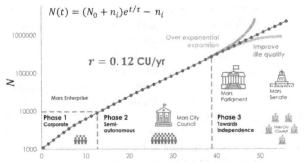

Figure 1. *The Nüwa development concept.*

• **Phase 2 – Semi-autonomous phase (years 10 to ~45)** This is the central and most crucial phase of the development. Earth financed imports become unnecessary (end of Earth investment), and the growth becomes exponential. In addition to the development of the Mars cities, growth is also bolstered by the development of the Earth-Mars Transportation system (EMT), and the corresponding reduction of immigration costs in Earth currency. Launches from Earth are still needed, which shall be self-financed by motivated applicants attracted to the wonders of Martian life, which will then be selected via lottery. The Mars Enterprise will set quotas to ensure a balanced workforce, but that is also representative of the Earth's population and cultures. One Mars ticket will have an approx. price tag of 300k USD (200k USD + 100k for Earth operations and financing imports), and it includes: a one-way trip, one residential unit (~25.5 m^2/person), full access to common facilities, all life support services & food, and a binding work contract to devote between 60% and 80% of their work time to tasks assigned by the *Mars City Council* (MCC). By default, all newcomers also receive ½ MCS, and up to one additional MSC can be purchased with Earth currency before departure. Even if immigration slows down, the city is self-sufficient already. The MCC is a new institution formed by democratically elected representatives. It manages and legislates daily life on Mars, and it is promoted and legally bound to the higher authority of the Mars Enterprise. Most of the population growth is in the initial center (Nüwa), but the construction of a second urban center begins nearby (Fuxi). The construction of a polar water mining settlement (Abalos) starts once Mars can locally manufacture nuclear reactors and fuel. Plans for two additional urban centers near the equator (Marineris and Ascraeus) also begin, but they shall not start until the initial city reaches a critical population size of 200k. A system of currency is created in this phase as well. For each surplus CU, one million crypto tokens informally called **micros** (or **Micro City Unit Credit**) are issued and proportionally distributed among MSC holders. *Micros* can be used to purchase part of the surplus infrastructure from the Mars City Council, providing intrinsic asset value to the currency. MSC can be traded with *micros*, but no citizen can hold less than ½ MCS and no more than 5 MCS to avoid the rise of severe inequality. Similarly, the maximum amount of total assets held by one individual should also be capped at the equivalent of 5 CU (or *5 million micros*). Only Martian companies with participation of the Mars City Council or the Mars Enterprise (which should hold at least 40% of the shares &

decision power) can hold values above this threshold. While basic services are still provided by the city, additional services can now be traded in the open market. This trade is the seed to entrepreneurship, which shall now freely develop. Since all the transactions are digital, they can be automatically taxed, with a practical implementation depending on the politics of the moment. The issue of the first *micros* marks the beginning of a true Martian economy. The *Mars Technical University* is created with the goals of training the workforce based upon need, fostering innovation to increase the development rate r, and support basic research activities which shall be mainly driven by Earth's vaster knowledge and intellectual base. Return trips can now be purchased to the MCC at a cost in *micros*. Remaining MCS can be sold back to the Mars Enterprise at a non-negotiable rate in Earth's currency. An independent judiciary branch of the government, and a law enforcement body is also established in this phase.

• **Phase 3 – Towards independence (year 45+).** After three decades of exponential growth (and a total Mars population reaching 500k), experience and innovation shall lead to a rate of unit production well over the minimum threshold, and the core of the free-market economy based on the *micro* shifts naturally to other activities. Mars is still a frontier territory, so all citizens should still contribute to its sustainability (20% mandatory community work to cover essential tasks). A complete *Body of Law* for questions including a *Constitution*, shall be created. It establishes the creation of a *Mars Parliament* which shall absorb the legislative powers of the Mars City Council and the Mars Enterprise, and act as the institutional representation of the people of Mars. The parliament will then elect a chancellor and a ministerial cabinet to deal with executive matters and implement diplomatic relations with Earth. At this point, the relation with Earth requires redefinition. The opportunity is used to set up a legal framework for other similar initiatives in *The Solar Federation Treaty (SFT)*. A possible implementation would be the creation of a second chamber composed of locally elected representatives (60%), and a fixed number of representatives from Earth designated by UNOOSA (40%). The chamber shall have veto powers on new legislation, and it shall arbitrate in case of conflict between institutions (executive, parliament, City councils). The debts and commitments with the Mars Enterprise shall be settled, and then the Mars Enterprise shall be dissolved.

3 - A Place to Live: Nüwa and the Cliffs of Mars
Establishing a permanent settlement of one million people on Mars requires access to a diversity of resources. Therefore, our proposal divides human settlement into five cities. Nüwa is the capital, with a population between 200,000 and 250,000 people. Its name has its roots in the mythological goddess that is the protector of Humans, who melted five stones to give robust societal pillars. Nüwa provides the environment for a prosperous Martian settlement by providing 1) Atmosphere; 2) Food & Water; 3) Shelter; 4) Energy; and 5) A higher purpose of communal existence. Although we are proposing five cities, our proposal strives to offer a highly-scalable and flexible solution that can be implemented in many locations across Mars. The urban and architectural solution achieves:

1. Total protection from ionizing radiation.

2. Access to indirect sunlight.
3. Efficient use of resources.
4. A low-cost solution for the skin of buildings that solves the difference in pressure between the inside and the outside air.
5. A sustainable settlement that integrates local conditions and that is dense enough to minimize its environmental and economic impact.

Mars has hundreds of cliffs, many with inclinations higher than 45 degrees. A cliff provides a broad, structurally stable "vertical" surface, which is a unique opportunity to create a "vertical city" inside the cliff, providing 24/7 protection from radiation, while still having many perforations in the cliff wall's face to bring indirect sunlight inside. All five locations proposed have a direct orientation to the Sun (south-facing cliffs if located in the northern hemisphere), so each urban sector is maximizing access to sunlight. As these spaces are inside of the rock, the pressure from the inside of the buildings is structurally absorbed by the surrounding rock, reducing both cost and risk of failure of the buildings' skin. Building into the cliff provides proximity between the city and all of its supporting infrastructure. Cliffs have large horizontal areas at the Mesa (top of the cliff) and the Valley (bottom of the cliff), which in this proposal houses the city's supporting buildings. This solution creates an efficient and concentrated urban development that reduces the amount of infrastructure required to provide a fully functional city. Although craters and other geological features may also offer some of the advantages of cliffs, they usually do not provide the rock's structural stability, unidirectional orientation to the Sun, and the opportunity for a dense urban development that cliffs provide.

Figure 2. Conceptual section of a Nüwa cylinder and a Green-Dome with natural light acting as buffer to the hostile environment outside (left). Interior view of the inside of a Green-Dome (right).

3.1 The City Configuration

Nüwa and all of its sister urban developments integrate their presence in the *Mesa*, the *Wall*, and the *Valley* of cliffs. Additionally, each city *comprises five main urban elements* that are highly interconnected, and that can be summarized as follows;

A. Macro-Buildings are mixed-use excavated structures inside the Wall of the cliff. Each accommodates about 4,440 people. The elevation of the *Macro-Building* can be inscribed in a rectangle of 750 meters in length by 200 meters in height. Each one is a self-contained construction that comprises six Residential-Modules and six Work-Modules. Each of these Modules is an intricate net of three-dimensional ten-meter diameter tunnels that go as deep as 150 meters into the cliff. Most of the inhabitants do not need to leave their *Macro-Building* to perform their daily activities unless desired. Each Residential-Module includes *Housing-Quarters*, *Residential-Facilities*, *Urban-Orchards*, and *Art-Domes*. Each Work-Module includes *Work-Quarters, Office-Facilities*, and *Urban-Orchards*. Additionally, all modules include *Green-Domes* that are excavated at the cliff wall's face and that connect to common areas of the modules, providing natural light and lush gardens. Each module also includes a **Snow-Dome**, which dissipates heat and cleans the air (described in Life Support System). Each module also includes horizontal *Corridors* and *Vertical-Cores* to connect the modules vertically. There are three different types of Residential-Modules and three kinds of Work-Modules (Fig.6). These six "building-modules" provide standardization, reducing complexity, cost, and construction time. Thanks to the modular nature of the *Macro-Building*, the solution is flexible, so no two *Macro-Buildings* are the same. This flexibility is also critical when adapting to the local geological and geometrical conditions with essentially the same machinery.

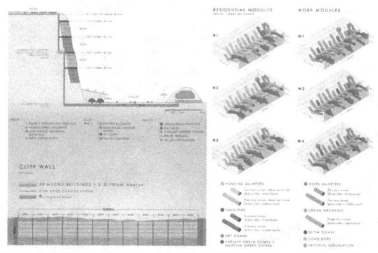

Figure 3. Conceptual section (left top) and front elevation of the cliff (left bottom). Type of Modules that comprise each Macro-Building (right).

B. Transportation Infrastructure within Macro-Buildings: The *Macro-Buildings* are linked with *High-Speed Elevator Systems*, like those of skyscrapers. This infrastructure connects the top Mesa with the bottom Valley and stops in the middle at *Sky-Lobbies*. These intermediary lobbies provide access to the adjacent *Macro-Buildings* and their communication systems. The transportation infrastructure on both sides of each *Macro-Building* creates a city vertical-grid of *800-meters* that can be replicated as needed. The transportation infrastructure acts as the bones of the new "urban body," giving structure and flexibility for growth.

C. Harvesting. Agriculture & Energy: Our proposal locates Farming, Energy Generation, and their related Industrial Processing on the Mesa, the substantial flat area on the top of the cliff. Agricultural and energy production facilities require direct access to sun radiation but do not need a shield from radiation, as only maintenance personnel and robotics will be operating there. Moreover, the inner air pressure can be substantially less than in the human premises. As a result, it is more economical and efficient to build on the Martian surface than to provide these activities below ground. As the Mesa of a cliff is often flat right from the edge of the cliff, it is beneficial to locate these facilities on the Mesa. These facilities connect with *Low-Impact Industrial Buildings* that process the food and energy and distribute the outcome to the rest of the city. These Processing and Manufacturing buildings connect with the vertical city's infrastructure. The *Agricultural Buildings* spread longitudinally in the direction of the cliff. Next to these agroproduction facilities, Nüwa includes the solar power generation infrastructure, which is of three types: *Concentration Towers, Parabolic through fields,* and *Photovoltaics.* Further from the cliff, we propose a ***Nuclear Plant*** and the ***High-Impact Industry***.

Figure 4. *Excavated components of a Macro-Building (left), Marco-building openings on the Cliff (top right), a tunnel-Garden connecting to the High-Speed Elevator System (center right), and conceptual elevation of a Macro-Building of twelve modules (bottom right).*

D. City Communal Areas: Humans are social beings, so providing ample spaces for social gatherings is essential for the wellbeing of the citizens. Additionally, in terms of the quality of the space, it is vital in a harsh environment, like the one on Mars, to provide large green areas where humans, plants, and animals interact similarly to how we do in city parks on Earth. It is also critical to offer long-distance views of the outside landscapes, especially when many daily activities will happen underground. To achieve this objective, Nüwa includes ***Pavilions***, which are large domes in the Valley of the Cliff. The proposed solution ensures protection from radiation by having large, low-cost canopies at a distance over these Pavilions. The material obtained from excavating the tunnels is reused on top of these canopies. The domes' main structure absorbs the pressure from the inside of these Pavilions. Thanks to the mentioned canopies, the Pavilions' building skin can be transparent or translucent in an economical manner. Some of these Pavilions include nature and are called **Green-Pavilions.** In contrast, other types of these dome-like structures are the **Urban-Pavilions**, which provide plazas for public activities and act as the Agoras of Nüwa. These Pavilions connect with other spaces called **Galleries**, which are vast underground spaces, such as sports arenas, music halls, and light train stations to communicate with the **Space Shuttle Hub.** The large volume of air required inside of these communal areas is solved with 30-meter-diameter underground tunnels. The *Galleries* and *Pavilions* connect underground to the *High-Speed Elevator Systems* through additional buildings that are part of these 30-meter-diameter tunneling systems, which comprises of **Arcades**, that include shopping and entertainment areas, and **Tunnel-Gardens**, which are underground gardens at the base of the *High-Speed Elevator Systems.* On the surface of the Valley, and connected to the underground *Galleries*, the city included support buildings, such as **Infrastructure-Pavilion**, and **Advanced Components**. Finally, an artificial mountain on the Valley created with the material extracted from the excavation, which includes **Auxiliary Energy Systems**, such as fuel cells, **Low-Pressure Tanks**, including ice and solid CO_2, and **High-Pressure Tanks** for CH_4 and Liquid O_2 (LOX), among others. The artificial mountain also includes **Storage of Raw Material, Storage of Processed Material**, and **Parking Lots**, for Rovers and Trucks. This mountain reduces the impact of temperature variation and provides a visual framework for the arrival by land to the city.

E. Landscape: The landscape is a fundamental element in Nüwa and its sister cities. The location itself, being able to live inside a cliff, is a powerful emotional experience. The integration of the buildings with the landscape transforms the city into a land-art, creating a unique identity for its citizens. Nüwa also includes artificially created landscapes at the *Green-Domes*, located on the cliff wall's face. These spaces are unique to Mars, but at the same time, bring memories of nature on Earth, creating an emotional and historical link for the citizens in these Martian communities. There are two very distinct types of *Green-Domes*. The first type is called **Earthly-Green-Domes** which act as neighborhood parks. Here citizens come to socialize and enjoy the vistas from the cliff. The second type is called **Martian-Green-Domes**, which cannot be accessed due to their lower atmospheric pressure to recreate conditions for experimental vegetation more suited to an Earth-Mars intermediate environment. The considerable diversity of environments at the *Green-Domes* ensures that each module has its own personality, identity, and sense of community, even though the building modules' architecture is standardized for scalability and efficiency. Another different type of landscape is provided at the **Urban-Orchards** inside the Residential and Work Modules. These spaces are underground and integrated with the residential and office quarters. Like small community gardens, they include animals, small bodies of water, and limited plantation to provide a physiological benefit to the citizens. The **Green-Pavilions**, located on the Valley, act as artificially lighted greenhouses that include lush greenery and water, creating diverse ecosystems. These provide space for large gatherings and also create the ideal environment for the farmed animals to flourish. Finally, the **Tunnel-Gardens** create a cave-like landscape that would be very unique to Mars.

In summary, the fourth dimension (time) plays a critical role in the landscape solution of Nüwa, because depending on the movement of its citizens, the architectural experience and its surroundings will be completely different and unique.

3.2 Building Standards
Key safety issues result in unique building types and building standards within the city. Horizontal tunnels include shafts at intervals to regulate atmospheric pressure and provide an escape to refuge areas if needed, so in the case of a structural failure or fire, these elements act as the first barrier, preventing further compromise to the rest of the tunnels. Similar to shafts, the *sky lobbies* are at elevation intervals at the *High-Speed Vertical Elevator* stack, they provide public shared spaces and gardens, but also act as a compartmented space between the macro-buildings. In the unlikely event of a structural failure or fire, the Sky Lobbies lead to refuge areas designed to protect inhabitants for a prolonged duration before rescue services arrive. Air showers will be included at the entrance to each Macro-building to sterilize and clean garments. Artificial Intelligence (AI) algorithms combined with personal monitoring devices will also play a critical role in Nüwa's building standards to help manage the optimal conditions and minimize risk.

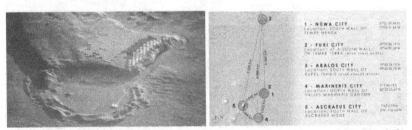

Figure 5 (left). *Aerial view of Nüwa, showing nuclear energy plant (top left), solar energy and food production areas (top center) and space shuttle hub (bottom center), and proposed location of the five centers on Mars (right).*

4 - Location (primary) and Nearby Resources
4.1 Location
Mars receives less irradiation from the Sun than the Earth. Additionally, the slightly higher orbital eccentricity results in broader differences between the irradiation received at the aphelion (furthest distance from the Sun) and the perihelion (closest distance to the Sun). Nüwa, Fuxi, Marineris and Ascraeus are all located between the equator and 30° north latitude, avoiding extreme environmental conditions (see Figure 5). Five locations instead of one are chosen to improve resilience, long term easy access to resources and to add mobility options to the citizens of Mars. The chosen location for Nüwa - the first city - is on the southern rim of Temple Mensa (28° N, 288°, see Table 1 for an overview of the environmental conditions), and the second urban center (Fuxi) is in the same general region but 170 km to the North-East of Nüwa. The region was chosen for its easy access to sources of water (clays), insolation levels, and abundance of near vertical walls of strong rock.

Physical quantity	Summer / Autumn	Winter / Spring
Atmospheric temperature	200-240 K	170-190 K
Atmospheric pressure	500-600 Pa	600-700 Pa
Atmospheric density	11-18 g/m³	15-21 g/m³
Wind speed	0-20 m/s	0-20 m/s
Soil temperature	195-265 K	170-210 K
Sky temperature	130-175 K	120-135 K

Table 1. *Environmental conditions at Tempe Mensa*

The other cities in the Arcadia Quadrangle exhibit a climatology quite similar to Nüwa, with some minor differences due to the different latitudes. The polar location of Abalos, however, is quite different. Located at 79° north latitude. There, the daily variations of atmospheric temperature, pressure, and density are almost nonexistent, but these quantities vary significantly throughout the year. This is especially the case with the atmospheric temperature, which remains at an almost constant 150 K for most of the year except for summer, when there is Sunlight on the poles, and it raises up to 200 K. Thermal designs shall account for these differences.

4.2 Nearby Resources

	Nüwa/Fuxi	Abalos	Marineris	Ascraeus
Water	In clay minerals [1]. Estimated 1-5% by mass [3]. Probably in permafrost (>10%), as evidenced by lobate structures around craters made by impact melting	Ice layer of ~420 m in thickness	In sulfates (gypsum), up to 21%	Unknown
Sulfur		Gypsum dunes in Chryse Planitia, containing 23.5% sulfur	Widespread sulfate deposits [1]	Unknown
Iron ore	Iron oxides present everywhere in regolith at moderate concentrations (~20% in mass)	Iron oxides present everywhere in regolith at moderate concentrations (~20% in mass)	Crystalline gray hematite (Fe2O3, with up to 70% iron) has also been discovered in numerous small (5–20 km) areas in Valles Marineris [4]	Unknown
Nitrogen	Found as nitrates in aeolian deposits at de ~0.1% level [5], similar to Earth's Atacama Desert			
Phosphorous	Widespread in regolith at ~1% (P2O5) in Ca-phosphates resulting from alteration of volcanic rocks			
Chlorine	Widespread in regolith at ~29% (Cl) in perchlorates (toxic)			
Regolith	Readily available throughout the planet's surface, in sizes ranging between a few tens of microns to sand and gravel			
Comments	Southernmost end of Tempe Terra. By North Kasei Channel, one of the largest outflow channels on Mars, emptying into Chryse Planitia		Near Tithonium Chasma. West end of Valles Marineris.	Rounded terrace-like structures arranged concentrically around the summit of the volcano, with heights of about 3 km

Table 2 (left). *Key resources availability in each region.* **Figure 6 (right).** *Aerial view of possible sister city locations.*

Key resources (especially water) should be located near the cities, to ensure accessibility and ease of extraction. Table 2 summarizes the availability of water, Sulfur, Iron, Nitrogen, Phosphorus, Chlorine, and regolith, for each of the five cities. As documented, all the cities have access to Nitrogen, Phosphorus, Chlorine and regolith, which are crucial resources to develop fertilizers and produce aggregates for construction. Water is available in Tempe Mensa (as clays) and Marineris region's subsoil, although the most convenient source for mass extraction would be the one coming directly from the polar ice around Abalos. Sulfur can be collected in Marineris and Abalos. Minerals rich in Iron oxides can also be found everywhere, but the highest quality ores are likely to be found around Marineris. We assume that most of the main resources necessary to maintain population growth on Mars can be initially found around all the city locations, but that the latest development phases shall require the use of the richer deposits of each city.

5 - Keeping People Alive: Life Support Systems & Biosystems

A Mars civilization should be independent of Earth resources and sustainable in the long-term. Thus, a closed Life Support Systems (LSS), where all produced bio-waste is collected and transformed in fresh consumables for the humans is needed. To reproduce Earth biodiversity is a complex task and cannot be easily done in a restricted volume (in this case 187,500,000 m^3 of breathable air for a 200,000 inhabitants city). However, several living organisms will be required to fulfill the tasks needed for full recycling and to ensure a rich and balanced diet for the city inhabitants. Figure 7 shows the scheme of the Nüwa LSS.

Food management. Crop cultivation will be the main food production source, providing 50% of the human diet, while processing CO_2 into O_2 and taking part in the water processing system. Although crops can provide a tastier and more varied diet than microalgae, those are more efficient in terms of space and resource utilization, while also contributing to atmosphere revitalization and water management. Thus, microalgae are selected for 20% of the human diet.

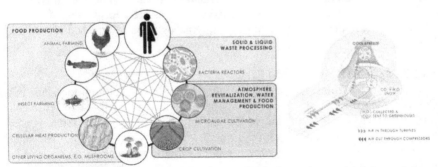

Figure 7. *Nüwa Life Support System concept(left). Snow domes (right) precipitate Water and CO_2 and regulate the temperature in the human dwelled volumes.*

Typical farm animals such as pigs, chicken, or fish will also be included, but in very small amounts. Such animals are very inefficient, but provide a high psychological value, and can also serve as buffers in the system. They will represent 4% of the human diet. More efficient systems such as insects or cellular meat production will be included, representing 10% and 16% of the human diet, respectively. Other living organisms, such as mushrooms, should also be included.

Waste management. Bacteria will be in charge of processing both the solid and liquid waste products from the humans but also from the other living organisms in the system. A combination of several bacteria species will be required in order to process the bio-waste into useful resources. They are a key element since they will provide the required nutrients for the system.

Water Management. Similarly, as we have on Earth, there will be a water-cycle in Nüwa. Humans, animals, and plants will add water to the atmosphere through respiration/perspiration/transpiration. This water will be collected by condensation through the ventilation system, thus providing pure water. With proper usage of water, requiring the usage of bio-friendly products for washing and cleaning, the wastewater produced can be used for gardening and farming. Urine from humans and animals will be treated by bacteria. It is crucial to control and eventually filter the substances that go into the water system, for example, antibiotics, since those could disrupt microbial ecosystems. The reduction of water usage at home, compared to current values in western countries on Earth, is a key aspect to sustainability in Nüwa. Mist showering or alternative cloth washing systems will ensure that the required amount of hygiene water per person is decreased to about 7 liters per day.

Air Management. Both the microalgae and crop photosynthesis will produce the oxygen required to replace the oxygen consumed by humans and animals, but also the one required for other LSS subsystems, for example, some bacteria. Besides those food-producing-plants, green areas on the human spaces will also contribute to the reduction of the Carbon dioxide levels and production of oxygen. Carbon dioxide, excess air humidity, and heat will be extracted from the human spaces via *Snow Domes* and delivered to the microalgae and crop cultivation areas. Additional temperature, humidity control and ventilation will also be available for redundancy and local fine-tuning. An inert gas (a combination of Nitrogen or Argon) will also be present in the atmosphere to ensure oxygen levels are kept within safe limits. These gasses are not consumed and will only needs to be refurbished if there are leakages.

5.1 Life Support System (LSS) Facilities

Agricultural Modules: Crops will be cultivated in the agricultural modules, in a CO_2-rich atmosphere, with a total pressure of 25 kPa. This will not be breathable for humans, requiring a fully automated cultivation. A hydroponic system is used to increase cultivation efficiency, requiring, for example, less water and space than soil-based systems and avoiding the use of pesticides. A cultivation surface of 88 m^2 per person is required, which will provide food both directly for human consumption and for the other animals in the city. Figure 8 shows the internal distribution of the modules. Some crops will receive direct sunlight complemented with artificial light. Some crops will be stacked in vertically distributed trays, thus requiring a dedicated LED panel. Algae, cellular meat, and bacteria reactors are also located in these modules, either in dark or with 24/7 artificial lighting, depending on the type of reactor. Those will occupy a surface of 4.5 m^2 per person.

Figure 8 Nüwa non-human dwelled agricultural Modules (left). Human dwelled farming area in the habitation areas for recreational purposes (right).

CO_2 extraction in habitation modules & Snow Domes: The atmosphere in the human habitats will have a total pressure of 75 kPa, providing a proper Oxygen partial pressure for breathing, but keeping the oxygen volume under 30% to reduce fire and explosion risk. The CO_2 produced by the inhabitants will be collected through cold precipitation of the CO_2. In addition to CO_2 removal, all spaces also require air circulation and heat removal, which is achieved by expanding the pressurized air into large, but closed volumes exposed to the surface of the planet called *Snow Domes*. The expanding gas goes through a turbine producing mechanical energy, losing heat, and its temperature further drops (Joule-Thomson effect). As a result, CO_2 and air humidity freeze into *snow* that falls to the bottom, and it is transported elsewhere. The other gases (O_2, N_2, and Ar) are cooled but do not become liquid, and they are reinjected with a compressor to the pressurized environment (the compressor uses energy from the turbine, see Fig.7).

Farming areas: Ground and water lakes for animal farming, as well as for insect farming, will be placed in the *Green-Pavilions* since those require a similar atmosphere as humans to breathe. Some animals will also be in the *Earthly-Green-Domes* and the *Urban-Orchards*. Less than 1.5 m^2 of animal/aquaculture/insect farming area is required per person, due to the low amount of food consumed directly from animals.

5.2 Required Life Support System Resources
In order to initiate the LSS and keep it running, besides the required seeds and cultures, 8tn of water, 100 kg of nutrients (mainly nitrate and phosphate) and 120 kg of CO_2 per person are required. The initiation of the system is a critical aspect, since plants and animals require some time to grow before they can be consumed, thus time planning will be crucial. To fill the atmosphere of the habitat and the agricultural modules 240 kg of oxygen and 490 kg of Nitrogen per person will be required. We estimate that the LSS requires 37 kW of continuous power per person, which is mostly necessary for the crop cultivation lighting system. The human requirements, as well as data for the design of the different elements in the LSS are based on available literature [9,10,11,12,13,14,15,16,17,18].

Volume of spaces suitable for humans in Nüwa (for 200,000 people)	Floor area (m²)	Volume (m²)
Macro Buildings (inside the walls of the cliff)	30,000,000	125,000,000
High-Speed Elevator Systems + Sky lobbies (inside the walls of the cliff)	125,000	25,000,000
Tunnel-Gardens (lower part of the cliff)	200,000	2,000,000
Galleries & Arcades (underground, lower part of the cliff)	2,470,000	23,000,000
Pavilions (lower part of the cliff)	250,000	4,500,000
Low-Impact Industrial Buildings (top part of the cliff)	750,000	7,500,000
Space Shuttle Hub and its related buildings (at center)	30,000	500,000
Volume of spaces not suitable for humans in Nüwa	**Floor area (m²)**	**Volume (m²)**
MEP Domes & Snow Domes (inside the cliff), MEP Pavilions	4,500,000	45,000,000
Energy Generation Buildings (Solar Generation + Nuclear Plant)	300,000	3,000,000
Agricultural Buildings	15,000,000	90,000,000
High-Impact Industrial Buildings	1,000,000	10,000,000
Storage and Support Building (Closed area at artificial mountain)	1,000,000	2,400,000
TOTAL:	**55,675,000**	**337,900,000**

MESA - 20 million m³ · excavation
WALL - 190 million m³ · excavation
VALLEY - 23 million m³ · excavation

MESA

WALL

VALLEY

Table 3(left). *Built-up areas and Buildings volume per 200k inhabitants.*
Figure 9(right) *shows a conceptual section with the required excavated volumes.*

6 - Resource Collection

Table 4 (next page) documents a list of the most common materials needed for manufacturing human items, and it identifies their Martian counterparts. The production energy costs for some key materials dominating the CU budget are shown in the last column.

Extracting minerals and digging tunnels is a central activity in creating city units at Nüwa. With a road header like the Sandvik MH621, digging relatively hard rock at a rate of 20 m³ per hour, and consuming 150 kW of continuous power (50% of max capacity), only one machine every 400 inhabitants is needed to maintain the production rate. Digging rock consumes about 0.03 GJ/m³. So, if we need excavating 800 m³ per CU, we obtain that the excavation contribution of the road headers to the CU energy budget is 21.6 GJ/CU. The same logic is applied to all the other systems when deriving required materials and energy. Surface miners, which are similar machines optimized for surface activities, will be used to extract clays and mine soft rock at the surface (riverbed around Tempe Mensa). Smaller units, in a swarm-like mode of operation (such as Robominers - https://robominers.eu/) shall be imported during phase 1. The amount of mineral needed for metal smelting (about 300 tn/CU) is much lower than the excavation requirements for the city, so the same road header machines used to excavate habitats can be temporarily assigned to work on nearby mining sites (~100 km distance). A fleet of transportation trucks will be needed. Assuming a truck transporting 20 tons per day translates to one truck for every 48 citizens. Assuming a 20-ton truck, these contribute 0.4 tons of steel to each CU, and about 40 GJ/CU. Some additional materials for the electric power and batteries might be needed but they remain as minor contributions.

Figure 10. *Exterior (left) & interior (right) views of an Earthly-Green-Domes*

EARTH RESOURCE	MARS RESOURCE	USES	COMMENT	Energy in GJ/TN
O_2	CO_2	Life support	Most O_2 produced as by-product of CO_2 reduction	-
N_2	N2	Life support, ammonia production (industrial)	Inert/carrier gas, by-product of atm. capture	-
CARBON	Atmospheric CO_2	Electrical devices and infrastructure, Structural material and reinforcement, Graphite	Electrodes for speciality metals , Carbon fibers, Carbon nanotubes	
OIL	Atm. CO_2 and Water	Production of syngas ($CO+H_2$) and Methane (CH_4)	Reverse Water Gas Shift ($CO_2 + H_2O$ to syngas + O_2), Water Electrolysis + Sabatier (CH_4 and O_2).	
POLYMERS AND ORGANICS	Atm. CO_2 and Water	Use syngas ($CO+H_2$) and Methane (CH_4) & greenhouses	Very energy intensive	99
PERFORMANCE POLYMERS	UHMWP	Tensile strength material and reinforcement (80% of Kevlar's).	Ultra-High Molecular Weight Polyethylene – Polyethylene with very long chains. Metallocene Fe catalyst. Domes and pressure vessels.	200
WATER	Solid water (mixed with solid CO_2)	Various including life support, rocket fuel, energy storage (fuel cells). Main source of H, used in all organic chemical chains as input.	Polar caps. About 30% by mass. Assumed resource in Dev. phase 3.	
	Permafrost (regolith + ice)		Water 10-25% by mass. Near the surface (few cm) at latitudes >60°. evidence possible presence at Tempe Mensa latitudes (>30 m depth)	
	Mineral water		In clay minerals and poly-hydrated sulfates. ~5% by mass, evaporates upon heating above 150°C. Main water source assumed for Tempe Mensa sites. One tn water leaves ~20tn gravels	
AGGREGATES	Regolith, by-products		Same use as in Earth construction materials	
CEMENT	Native sulfur	Construction material	Strength comparable to concrete. Extracted from sulfates. Avoid in areas with heat/flame as it melts at 130°C	
CONCRETE	Sintered regolith or aggregates, or sulphur concrete	Construction material	Produce blocks of arbitrary shapes. Requires heating at 1200°C, but little processing, and much less energetic than actual concrete when adding all the implicit material costs.	2
GLASS & CERAMICS	Glass & ceramics from regolith	Construction material	Glass (mostly SiO_2) is a very inert material. A bit brittle, it shall be used sparingly. Main uses are internal transparent windows, glass covers for photovoltaics and Solar Concentrated Power technologies.	26
IRON & STEEL	Iron, Steel	Hardware manufacturing, construction material, electrical conductor at initial phases, vehicle	Iron Smelted using hydrogen reduction. Workhorse metal.	56
ALUMINIUM, TITANIUM & LIGHT METALS	Same	Used very reduced (always try using iron/steel if possible) as smelting is very energy intensive	Smelting using thermal energy and electrolysis using graphite captured from atmosphere, CO separated from syngas, or processed Carbon materials from greenhouse waste.	200
FERTILIZER	Nitrates, Atm. N_2		Limited availability. Directly exploitable from soil.	
	Sulfates		Widespread deposits. Directly exploitable from soil.	
	Phosphates		Locally available as soil alteration products (locations with ~5% P_2O_5 identified)	
	Chlorine	Chlorination, some industrial uses, sanitation	Widespread chloride salt surface deposits.	
ADV. ORGANICS AND DRUGS	Bioproduced, greenhouses, synthetic organics	Medicine for human inhabitants and the small animal stock of the city.	Low diversity of medicines initially. Bulk drugs (painkillers), produced in dedicated factories. Others in flexible chemical labs, and genetic engineering techniques (use of bacteria and gene editing techniques)	500
DISINFECTANT	Methanol and similar	Used in sanitary facilities as the bulk disinfectant		
TEXTILES	Natural fibers & synthetic fabrics	Clothing, construction materials, decoration & light management.	From synthetic organics (syngas and CH_4)	99
SEMICONDUC-TORS (SILICON)	Organic semi-conductors, pure silicon	Solar photovoltaics, Electronics, LEDs for greenhouses	SiO_2 abundant but hard to refine at required purity. Some imported. Alternatives not ready yet, but likely (Perovskites, quantum dot)	1600
NUCLEAR FUEL	Thorium and/or Uranium ores	Fuel for Nuclear power plants	Thorium & Uranium enrichment & reactors on Mars will require extensive R & D	?

Table 4. *List of the common materials needed for manufacturing human items.*

The Martian atmosphere is mainly composed of CO_2 (95%), N_2, and argon (at about 2% each) and trace gases. The Carbon is essential for the biological systems, it is needed to manufacture polymers and advanced organic materials (e.g. disinfectants, and drugs), and can be used in electrodes to smelt specialty metals such as Titanium, Tungsten, or even Al, Mg and Si. We estimate that 114 tons of CO_2 need to be captured per CU for materials usage. To estimate the scale of the machines needed to achieve this, we extrapolate from a commercial compressors (0.3 tn of steel, and a few more materials) with a compression capability of 0.88 tn/hr/bar. One such compressor would be needed for every 40 citizens. The CO_2 needs to be then stored and sent to the transformation industry. We assume 27 m^3 of tanks prepared to contain gases and liquids per CU. These tanks can be built with high-performance materials (workhorse material is UHMWP, a very high tensile strength polymer) shielded against the environment by solid sintered regolith. This solution should also be used for pipes transporting gases and liquids under pressure. Water is the main (only) source of hydrogen on Mars. We estimate that around 80 tons of water are needed per CU, where 8 tons are for life support and the rest comes from creating organic materials and polymers. In Tempe Mensa, the preferred extraction method exploits clays, which are hydrated minerals with about 5-10% of mass in water that is released upon heating to about 400°C. This results of about 1000 tn of to be mined per CU. Water is then extracted using residual heat from other processes and concentrated solar light. A polar Martian city should be able to access frozen water from the surface, thus requiring no mineral processing. Because of the high latitude, such a city would need nuclear power to be viable by mid-phase 2. There are reports of solid water reservoirs in the Tempe Mensa region. Determining the most optimal way to obtain this water will a core task of the initial colony. Given current information, we conservatively assume the clay method as the baseline one.

7 - Chemical Transformation

Gases & Carbon. Many chemical processes must be enabled to implement the self-sustainability principle. Here we mention the most important ones that initiate the chemical pathway providing feedstocks to the manufacturing industry (syngas, methane, H_2, and pure Carbon). *Reverse Water Gas Shift* (or RWGS) produces syngas ($CO + H_2$), which can then be used as the precursor for more advanced materials such as polymers and organic chemistry. Oxygen is separated at formation, greatly simplifying the management of the output gases [19]. *Water electrolysis* is the best-known reaction to produce molecular Hydrogen from water. A suitable implementation for Mars consists of water in alkaline solution at 80°C using a Ni anode that also acts as a catalyst. The containers in the reactor must be made with corrosion-resistant materials (e.g. glass), and on Earth, the stacks need to be replaced only every 10-12 years of continuous operation. Water electrolysis is also used to store energy in fuel cells [20]. Sabatier reaction is used to obtain Methane from CO_2 and H_2 using catalytic methods. In the most used form, the reaction requires moderately high pressures (about 20 bars) and is exothermic so the excess heat needs to be removed and can be used to process clays for water extraction. In all the reactions involving Carbon oxides, there is a risk of accumulation of pure Carbon onto the catalysts (*Boudouard reaction*). Despite being problematic in some cases, this process executed under controlled conditions forms pure graphite, which shall be used in electrolytic smelting of specialty metals. In both *Sabatier* reaction and *RGWS*, the goal is converting H in a readily usable form, and they both follow similar energetic efficiencies (200 GJ are needed to produce one ton of H_2). While all the C and H remain fixed to organics, a large surplus of O_2 is left as a by-product (130 tons of O_2 per CU), while less than 0.3 tons are needed for life support. While it can be kept in tanks, venting most of it is more economical. CH_4 and liquid O_2 (or LOX) are the preferred choices for rocket propellants that shall be used by the Earth-Mars transportation system. Implementation of these reactions requires a set of specialized chemical factories whose mass and energy budgets are included in under *manufacturing infrastructure*.

Metals. The workhorse metal for Mars shall be Iron and its alloys. Iron oxides are very abundant on the Martian surface, and the smelting of the ores into the metal can be done using *thermal reduction with Hydrogen*. In this process, Hydrogen always remains in the cycle, so no water is consumed in the process. Iron produced this way shall consume about 20 GJ/tn, and steel (which requires more treatment) consumes about 56 GJ/tn. Other metals such as Ti, Al, or Mg can be smelted using processes based on closed cycle electrolytic Carbon reduction, using graphite cathodes. Deposits of Ti ore (Ilmenite) should not be very rare, and Al and Mg oxides are present almost everywhere, but beneficiating them is difficult and energy intensive. Small amounts of Copper and other metals are likely to be produced as a by-product. As an estimate, we assume that 200 GJ are needed to produce a ton of a specialty metal, although this number could be considerably higher in some cases. Regarding nuclear material ores, there is not much information on deposits of uranium on Mar. However Thorium is relatively abundant on its surface. Thorium reactors also need salts to operate, but the amount needed would be small compared to other ores. For resilience issues, it is highly desirable to develop the processes to enable nuclear power options on Mars as soon as possible. As for the gases, the CU budget for smelting metals are included in the *manufacturing infrastructure*.

8 – Manufacture

Given the range of products needed to run a complex system like a City, a broad range of manufacturing capabilities are required with equivalence to Earth heavy, medium, and light industries. A key context for appropriate-scale manufacturing capabilities on Mars is that a capability does not have to be developed within an Earth-legacy context. Many of the present-day Earth manufacturing trends towards a sustainable and renewable manufacturing implementation are highly relevant to Mars, i.e. sustainability, circular economies, life cycle analysis and embodied energy consideration.

A manufacturing capability on Mars is expected to exhibit the following generic features. 1) choices must be driven by minimizing energy usage, 2) embrace the circular economy concept and have a high degree of recycling of technical materials at product end-of-life. 3) embrace many of the concepts of Industry 4.0 and thus implement flexible, re-configurable, robotic manufacturing to maximize the range of product outputs for a minimized infrastructural cost, 4) The product design must incorporating a wide and unique set of Martian context requirements including; the use of relevant Martian raw and technical materials, different environmental requirements (e.g. reduced load requirements, different external atmospheric chemistry – very low O_2 and H_2O concentrations, lower but variable temperatures, etc.).

Some examples and consequences of the preceding manufacturing context follows. Wherever possible, products should try to exploit minimally processed regolith as a manufacturing material due to low embodied energy, e.g. sintering it at high temperatures (~1200°C) or used with a suitable and available low embodied energy binders such as $CaSO_4 \cdot \frac{1}{2}H_2O$ (Plaster of Paris) [21], or Sulphur. As discussed earlier, Iron-based alloys should be more widely used compared to other metals. Aluminum and other specialty metals should only be used where a full product life cycle analysis indicates this is the most energy-efficient solution. A wide range of products will require access to organic Carbon-based materials. Physical/chemical processing as well as biotechnological (photosynthesis) approaches, are expected to be used, although the provision of suitable photosynthetic biological growth environments on Mars imparts a significantly higher embodied energy cost than the equivalent use of biology on the Earth. After the primary capture of organic Carbon, further processing steps will produce a range of organic-based materials, including bulk polymers and functional materials. Given the future timescale for a Mars manufacturing capability implementation, it is expected that current research trends in molecular electronics, efficient Carbon capture and use, advanced biotechnology, advanced nanomaterials, etc., will have been matured and that they will offer optimized solutions for Martian use, such as organic semiconductors, Carbon nanotube-based electrical conductors, etc. Overall, the manufacturing capability shall be driven by simplicity in used materials, and a "low-energy economy". In this first iteration of the design, we applied some of these principles in obvious situations (replacing use of metals), but many other situations (use of polymers in constructive elements) would require a much more in-depth analysis.

For the current study, to estimate the required mass of expected manufacturing infrastructure, a case study of an Earth automotive manufacturing facility/factory was chosen [22] and scaled to a factory lifetime of 50 years to give an estimated value of 0.05 tonnes of infrastructure required per tonne of "medium" industrial manufactured product. For heavy industries, it was assumed the manufactured product mass throughput per unit mass of infrastructure would be higher (estimate as a factor of 5 higher) giving a value of 0.01 tonnes of infrastructure required per ton of "heavy" industrial product. For the embodied energy resulting from the manufacturing process only (i.e. not including the embodied energy resulting from the basic chemical processing of the material) and using the same case study, a value of 27 GJ/tonne of the manufactured product resulted.

9 – Energy

Energy is an essential "nutrient" to both keep the city alive and to enable the city's growth. As any other resource, it needs to be collected, transformed, and stored. Given enough power, one can reproduce any of the functions we do on Earth. However, scaling the energy production facilities creates a recurrence in the design. If we need more power, more materials need to be extracted and manufactured to provide the power, which leads to even higher power needs. To address this aspect in the design, we started with requirements of the Life Support System (37kW per citizen) and added plus an allowance of 5kW per citizen (personal and services). We then added power in 10 kW, recomputed the resource budget and CU production rate, and repeat the process until enough energy was produced per year to achieve the citizen production goal of 0.126 CU/yr. Fortunately, the process converged for the three energy systems considered to a value of around 110 kW per capita. This number is quite high (Earth power per capita is between 5 and 10 kW per capita in developed countries), but unavoidable. The chosen distribution of power in a hybrid system of Photovoltaics (PV), Concentrated Solar Power (CSP), and Nuclear is discussed later. For resilience and efficiency considerations, the three technologies should be implemented, so there is always spare power to run the LSS. Any excess of energy should be stored chemically in (for example) H_2 and O_2, solid C, and liquid fuels (e.g. Methanol), or even metals.

Concentrated Solar Power: Although the EROI values of CSP (between 4 and 15) for electric energy generation might not be as high as some initial (naïve) estimates, relatively low tech is needed to make its components (mirrors, boilers, pipes, tanks, and pumps). Therefore, CSP is the only technology that can be built for sure with in-situ resources only and current technologies. A heavy industry that is designed to operate with CSP (e.g. processes that mostly require heat) could save large amounts of electric energy too. In this sense, our CSP solution would consist of *parabolic trough fields* for electricity production, and Concentrated Solar Power furnaces in towers for industrial processing. We initially considered coupling solar concentrators to greenhouses directly. Although it initially seemed a good idea, not enough energy could be provided to make efficient use of the surface around the city, requiring vaster extensions of crops and mirrors. Although the trade-off needs to be studied in more detail, for simplicity, we now assume that the energy needed to 'feed' the plants comes from electricity.

Photovoltaics: We allocate a smaller fraction of power to PV because - although this is a Mars tested technology - the in-situ construction of solar panels may prove challenging due to the difficulties in obtaining high-quality semiconductors. Even on Earth, silicon-based photovoltaics don't seem to have EROIs higher than 5, so it is not really that competitive against Concentrated Solar Power options.

Nuclear power plants are an obvious choice, especially if the high-tech components of the reactor cores can be shipped from Earth together with the fissile fuel (e.g. solution based on the design by the company Terrestrial Energy, adopted by the Mars Colony studies *Star City* and *Menegroth*, [32]), at least during phase 1. Except for the reactor core, a Nuclear power plant is essentially a big mass of steel and concrete, which can be locally sourced. It is unclear however, whether current reactor designs shall be economically operable on Mars (aka, they have EROI larger than 1). While there are functioning reactors in military vessels, experimental devices in extreme environments such as Antarctica have not been very successful. We speculate that the most promising option in the long term would consist of developing Thorium-based reactors. These reactors have been a promising technology for some decades, and several countries with abundant reserves of Thorium (e.g. India) are developing promising prototypes. Thorium reactors use molten salts in the reactor instead of water, requiring lower water pressures, and the waste has shorter radioactive lifetimes, possibly leading to higher overall EROI.

For all our calculations and to avoid projecting personal biases, we conservatively assume an EROI of 5 for the three methods, keeping in mind that the actual EROI of each technology needs to be measured in operational conditions. The energy transformation costs of materials that we use throughout the study are based on those provided in De Castro et al. [23], with some overheads added to account for the extra energy needed to produce some of the input products. The materials needed for the energy distribution grid were also derived from the same reference. Due to space constraints in this document, we have not explained in detail solutions for the storage of energy. In general, batteries are needed to store energy and react to quick changes in demand. Instead of Li ion (which may be hard to find and process), these could be made using Zinc-air batteries, flow batteries, or fuel cells (water electrolysis plants operated in reverse). In terms of strategy, during phase-1 both high-performance PVs (thin semiconductor substrate), and nuclear power solutions (only high-tech elements of the core) shall be imported from Earth while starting mass production of Concentrated Solar Power solutions. In phase 2, the PV infrastructure can start growing if a solution is found for easy manufacturing of semiconductor (organic semiconductors, or simpler processes to produce Solar panel grade Si), while reserves for nuclear fuel ores on Mars are surveyed.

Figure 11. Aerial view of the Mesa, Cliff & Valley of Nüwa; and general views of Abalos Polar city, and Marineris City.

10 - The Development of a Martian Culture and Education
Martian society must be fully supported by the concept of community. Thus, education (formal and social) is designed to overcome the traditional duality of me (self) and them (community). Mars is a hostile planet, and when you are risking your life every single day, you learn to rely on the others around you. Here, the African proverb *"if you want to go quickly, go alone; if you want to go far, go together"* dramatically changes to *"if you want to stay alone, stay on Earth; if you want to come to Mars, join us"*. Ultimately, no matter how advanced the technology is, it is not enough if it fails to foster a healthy community. This is also embedded at the administration level where all citizens are shareholders of the city, which in turn provides for all the basic needs. All the people are part of the city, and the city is everyone. Additionally, living on Mars will force a radical change in the way of understanding the relationships between humans and their natural environment. On the Earth, one can forget that the subjectivity of each individual is intimately connected with the environment and the social relations. On Mars this relationship is evident in everything that is done. People there do not adapt the environment to their comfort, but are forced to adapt themselves. And this must lead to greater respect for what Mars offers, ensuring the sustainability of the trio me-community-environment. In this way, Martian society will evolve and differentiate from ours, until it becomes a mirror in which to look at itself from Earth. Earth's environmental problems are results of the evolution of society in its economic, political, social, and educational aspects. Therefore, the circle could be closed by inheriting a social model based on full integration in the community and on material and environmental sustainability. The educational model shall fully embed Mars born children in this cultural mindset. In its practical implementation, the educational model is organized in three stages: Early years, Intermediate years, and University.
Early years. Mars immigrants are typically in their 30s, so many of them may want to start families. Human procreation is also an essential function required for the long-term survival of Nüwa. Consequently, children's care is addressed by the

community. Early education consists of social services for newborn to preschool age (i.e., up to 5 years old). These programs will be founded by specialized methods such as Montessori, Waldorf or Kirchner, which are committed to developing children physically, socially, emotionally, and cognitively. Universal access to childcare satisfies a dual function: it allows the parent to keep up with their tasks, and it nurtures the spirit of shared communal responsibility (collective education). In all the educational stages, there will be a consistent and persistent commitment to co-education programs to abolish gender & diversity biases[24,25].

Figure 12. View of Residential-Quarters(left) & view of an Art Dome (right)

Intermediate years. Adult personalities are forged during this period of life. The educational program for children between 5 and 16 years old, will follow the directives of the *Incheon Declaration*, which has been designed to satisfy the UN 2030 Agenda for Sustainable Development (or the future version of it). In this sense, education must be accessible for all, must have a global scope, and must train the future adults into operating in collaborative and inclusive environments. The curricula should be generalist in the sense of learning skills, and training in the usage & triage of information rather than mere acquisition of knowledge. All students need to learn how to both play leadership roles and how to become productive team members, as they will have to play both roles during their adult lives. Strong emphasis shall also be given in teaching how to learn new skills from available resources (books, digital world, mentors), and how to integrate different skill sets in teams to complete tasks. This kind of education is an important asset to both the city and the pupils, given that the tasks and work distribution may be changing rapidly over time. As in the early ages, the teaching materials & practices must be designed to abolish implicit gender/sexual orientation/cultural biases and toxic cultural heritages that go against the free development of the self. Embracing teamwork practices shall also be oriented towards promoting the sense of shared communal responsibility towards the others, and the common project that is Nüwa.

University (Mars Technical Universities, MTU). Due to the economic focus of the city towards development, the MTU (which would also include studies typically associated to professional and technical schools) shall be focused on preparing specialists to the most critical technical and management tasks (e.g. it does not train miners, it trains mining engineers, mechanical engineers and software developers to make and operate machines). An important aspect that will also be nurtured will be the Arts and a self-expression education. Arts will be a critical part of Society on Mars and the urban planning accounts for having *Art-Domes* in each Macro-building to inspire the citizens of Nüwa, its sister cities and help in the creation of a variety of local identities. The academic staff shall be composed by a full time teaching academics that shall manage and organize the courses (30%), a small fraction of pure academics tasked with the development of innovation programs (20%), and the rest shall be associated members with a high professional profile that are offered to spend 50% of their usual workload to teaching duties. Basic research is not a core activity at the MTU. In this aspect, the role of the MTU is to provide access to Martian unique environments to scientists from abroad, and formulate engineering and scientific challenges so they can be addressed by the much larger intellectual resources of Earth. A program of visiting scientists (for minimal time periods of 6 years) from other locations of the solar system shall be implemented with co-funding initiatives. All Mars citizens shall be able to apply for studies at the MTU at any time (no age restriction). Apart from a multi-year degree, the MTU shall prepare short term intensive programs (similar to summer universities on Earth, but all year round) to foster interaction among citizens of different sectors and encourage innovation. MTU will also provide administrative support to entrepreneurs willing to start business cases on Mars, which require considering technical aspects more complex than those on Earth.

11 - Services & Recreation
11.1 Social Welfare Infrastructure
Access to medical care is a fundamental right on Mars. As such, the health facilities are maintained by the city, and all citizens have free, equal access to them. A Personal Monitoring Device will also send periodic data on basic vital information of each citizen to a centralized computer, so urgent help can be sent when certain conditions are met (pre-specified by the user, and/or automatically by the system as a function of the location).

Hospitals and residences. Based on the Organization for Economic Co-operation and Development (OECD) data from a total of 42 countries worldwide, the number of beds per 1000 inhabitants ranges from 0.6 (India) to 12.5 (Japan). The mean of the number of beds for those 42 countries is 4.4 per 1000 inhabitants. It is estimated that for a 200K inhabitant city a ~880 number of beds in hospitals are required. The number of beds available for people requiring long-term care in institutions (other than hospitals), are 752.69 per 100K inhabitants in EU member states based on data collected in 2014 (European Health Organization Gateway). Thus, for a 200k city, the total number of beds required in residences is ~1500. Hospitals and residences are situated at the *Galleries* and *Urban-Pavilions* at the Valley.

Drugs and Biochemicals. Production of medicine and drugs is needed to preserve the citizens in good health. To account for them in the CU budgets, we estimate the amount required per person using Earth data. Concerning analgesics, these can be classified into three types of groups: non-opioids, and mild and strong opioids (World Health Organization's, WHO) and their consumption is highly variable on Earth. In Mars, it is estimated that about ~40 DDD (Defined Daily Doses per 1000 inhabitants per day) such as paracetamol, ibuprofen or aspirin will be consumed [26]. Opioids will also be produced, and it is estimated that the consumption will range between 80-250 morphine milligram equivalents (MME) per capita [27].

The number of antibiotics consumed will range between 4 to 64 DDD (defined daily doses) per 1000 inhabitants per day, and the absolute overall weight will vary from 1 tonne to 2225 tons (tn) per year [28]. For CU accountancy reasons, we estimated the amount of medicines consumed over 20 years per average person. Assuming this, we added 0.5 tn of advanced medicine at an energy cost of 500 GJ/tn, and 1.5 tons of basic medicine at 100 GJ/tn for basic medicines to each CU. Methanol, Ethanol, and Hydrogen peroxide in solution with water shall be used as the workhorse disinfectants in homes, public spaces and health care facilities. A material budget account for all the disinfectants required by a hospital bed is added to the CU budget at 0.8 tons of simple polymers and organics. Both methanol and ethanol can be derived from bio-products (e.g. sugar cane) or from chemical processes. Hydrogen peroxide can be chemically produced easily from H_2 and O_2 using relatively simple catalytic processes.

Early years schools. The number of enrolled children in early childhood education programs in Europe is about half of those enrolled in primary schools (Early Childhood Education and Care in Europe, ECECE). Each urban center of 200k inhabitants shall offer around 3000 nursery spots spread over ~40 nursery centers. Nurseries will be situated in well *Facilities* inside the *Macro-Buildings* that are directly connected to *Earthly-Green-Domes*.

Schools and Universities. According to 2017-2018 data, in the US, with a population of 328.2 million people, there were a total of 130,930 schools, 87.498 were elementary schools, and 26,727 secondary schools (National Center for Education Statistics, NCES). In a city of 200K, ~80 schools for both elementary and secondary will be built. Currently, there are a total of 3,228, 2,596, and 2,725 universities in the USA, China, and Europe, respectively (NCES) in a total population between the three regions of 2,462.6 million people. Thus, for a Mars city of 200K inhabitants, the equivalent of 0.35 large universities (10 000 x 0.35 = 3500 student placements) are needed. Schools are integrated inside the *Work-Quarters*, inside the *Macro-Buildings*, and share the *Facilities* and *Urban-Orchards* with the offices and research hubs located in the modules. Each Urban center includes three University campuses that are located at the Galleries and Urban-Pavilions in the Valley.

Figure 13. *Interior view of a Green-Pavilion (left), and interior view of the Galleries connecting with the Space Shuttle Hub (right).*

Recreation facilities engage citizens in physical activities such as sports, games and fitness, social activities, camping, and arts & crafts activities. As described in the urban design, Nüwa has many volumes dedicated to leisure. In New York City there are a total of 1,700 parks for 18 million citizens, corresponding to 18 large recreational parks for a population of 200k. These recreational facilities are distributed at the Mesa inside the *Galleries, Arcades, Tunnel-Parks, Green-Pavilions,* and *Urban-Pavilions*. Additional smaller social areas for the local communities are provided at the *Earthly-Green-Domes*, in the *Macro-Buildings*, inside the cliff (Wall).

11.2 Ground Transport

Intracity. The transportation in the vertical direction of the Wall (inside the cliff) is carried out with elevators in the sky-lobbies and shafts. At the base and top of the Wall, a system of light trains and buses is used to move in the longitudinal direction of the cliff. The underground spaces at the base of the Wall, adjacent to the *Tunnel-Parks* and the *Arcades*, include the train stations that connect with the *Space Shuttle Hub* that, for the case of Nüwa, is located inside the big crater in contact to the west side of the Mesa. All transportation inside the city happens within pressurized spaces and consists of electrically powered vehicles. To estimate the energy and mass budget of the fleet of vehicles, we assume one electric bus/wagon for every 400 inhabitants (a bit higher than Earth average). Out of dome operations shall be executed via pressurized rovers, which shall all be equipped with enough pressure suits and backup air to sustain their inhabitants for at least 10 hours, even in case of full vehicle failure. Each city also includes two large helicoidal ramps that connect the Mesa and Valley vertically and that are mainly used for vehicular logistics, during the construction/excavation of the "vertical city" inside the cliff and for emergency situations.

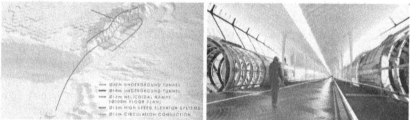

Figure 14. Transportation network (left), and view of a station (right).

Intercity transport is done with trains of buses/wagons on paved roads. The roads shall be built using excess gravel and sintered regolith (pavement layer) enabling travel speeds of the order of 100 km/h. These buses are electric, they are charged at the origin stations before departure, but they also have photovoltaic panels for emergency backup. All human transportation shall be done in daytime. Most harmful radiation outside the city comes from the zenith. Therefore, all wagons have an extra cover of polymer materials on the ceiling but shall have transparent windows to enjoy the views. Each urban center will have a central receiving station for all intercity transport, located at the *Galleries* that connect with the artificial mountain. As for the intercity transport, we assume one of such wagons for every 400 inhabitants should be enough to cover transportation demands. More advanced methods of transportation (*Maglev* trains) may eventually develop between cities, but not during the initial development phases of the Martian infrastructure.

11.3 Personal Monitoring and Digital Administration

All citizens must always wear a Personal Monitoring Device. This has the function of 1) Activating the correct safety protocols in case of emergency, 2) Monitor the health of the citizens and detect patterns of possible city wide-spread issues, 3) Localizing key citizen at any time (cases of extreme emergency). The device may be physically embedded in the body or can be a wearable piece (watch, bracelet, necklace, earring), and relies on the radio Wi-Fi network. Very clear legislation shall exist in terms of how the personal information gathered by the Personal Monitoring Device can be used. Anonymization of the information and legislation shall be strictly followed. In addition to this, all citizens will be given a Portable City Terminal (PCT), which shall feature all the common functions of a modern cell phone device. Citizens keener on advanced technology can spend their *micros* on more sophisticated hardware. The Portable City Terminal shall be carried along to all public places, as it doubles as an ID system and as a communication tool. All city administrative matters will be managed through the PCT or a computer. To account for all these computing needs, 5 kg of electronics is assumed per CU.

11.4 Communications
Communication services shall ensure a constant link between different parts of the city, between different cities on Mars and between Mars and the Earth. **Intracity communications.** Broadband data will be disseminated via a careful design of the artificial lighting inside the city following the Li-Fi concept. For safety and need of redundancy, LiFi shall be complemented by lower bandwidth radio-link Wi-Fi in all human spaces. **Intercity communications.** Initially, radio links shall be used between cities, with repeaters installed every few tens of kilometers in high-terrain locations. Eventually, fiber optic lines would complement the system, providing high bandwidth communications, especially between nearby cities. Fiber optics are not immediately implemented because, while they have the advantage of being low power and low latency, their manufacture on Mars might require significant amounts of energy-intensive materials (polymers, specialty glasses and semiconductors). For longer distances (connections with Abalos city near the pole), direct satellite links are considered more efficient. This system is mainly supported by the Areostationary network of satellites. **Mars-Earth communications.** Consists of permanent links with Earth's Deep Space Network (DSN). Two systems are used. *1) Direct-to-Earth comms:* Mars cities communicate directly with the DSN through Laser Band, ensuring a very reliable system, but only when the city and the Earth are in view. Uplink and downlink velocities can reach 680 kbps and 44 Mbps, respectively; *2) Relay comms:* The Areostationary network satellites are used as a signal relay, ensuring constant link with the DSN. In this case, a combination of UHF and X-band antennas are used to obtain uplink and downlink velocities of around 44 Mbps per satellite.

11.5 Farewell Facilities
Death is a necessary part of life on Mars. For efficiency and to avoid biocontamination of Mars, corpses will be processed (composted, or incinerated in a dedicated Solar Concentrator Tower), and their biomass reincorporated into the system (small samples can be kept by close-ones). Some of the *Urban-Orchards* and *Green-Domes* will be dedicated as a remembrance space for the departed ones.

In terms of inheritance, part of the *micro* and *Mars City Shares* will be passed to children, spouse or designated heir, and the rest shall be returned to the city.

12 - The Earth-Mars Transportation (EMT) System
A robust and scalable Earth-Mars transportation system is vital to ensure the success and sustainability of Nüwa and the other cities on the red planet. Once the first city of 1,000 inhabitants has settled, the landing of new people from Earth will grow exponentially. In addition, an economy based on Mars-Earth relations and the exploitation of resources throughout the Solar System will begin to emerge. Taking all this into account, the Earth-Mars transportation system main objectives are: 1) Allow the flow of passengers from Earth to Mars at the right pace to meet the exponential demand of inhabitants (1,000 to 800,000 people from 2050 to 2100); 2) Allow the flow of passengers from Mars to Earth; 3) Allow the flow of goods between Mars and Earth (and vice versa), key to economic relations between the two planets; 4) Operates as a transport hub between the inner and outer Solar System. All these objectives shall be accomplished taking into account the following top-level requirements for the Earth-Mars transportation system: 1) Be manufactured mostly on Mars with ISRU technologies; 2) Be managed and operated from Mars; 3) Work efficiently, requiring the least amount of fuel and materials possible; 4) Be scalable and adaptable to the needs imposed by the future space economy. The optimized solution is based on two different systems working together: a regular shuttle and a station periodically orbiting between the Earth and Mars. This combination is able to adapt the overall system to the demanding amount of people that need to be transported to Mars between 2040 and 2050, taking advantage of shared components and synergies between them.

12.1 Shuttle
A shuttle system is developed to transport people and cargo aboard a reusable spacecraft (S/C) able to be launched and land both on Mars and the Earth (see Shuttle architecture in Figure 15). The S/C is launched and placed into Low Earth Orbit (LEO) using a reusable booster. Once in LEO, it is refilled with Liquid O_2 (LOX) and Methane (CH_4). Then, a Trans Mars Injection makes the S/C escape the Earth's gravitational influence, entering into an elliptical orbit around the Sun, directed towards Mars. After an average time of 250 days of coasting, the S/C lands safely on the red planet, using a heat shield and supersonic retro propulsion. In Mars, ISRU technologies allow the extraction of O_2 and CH_4 from atmospheric CO_2 and water, using the Sabatier Process (see Section 7 - Chemical transformation). Now, departing from the red planet is easier, thanks to its lower gravity. Therefore, the shuttle S/C does not need a booster and can go all the way from the Mars surface to the Earth surface by itself (direct ascent architecture). The shuttles will depart from the Earth and Mars when the relative position of both planets allow for minimum fuel requirements. These launch windows open approximately every 26 months and last between one and three months. For example, a window to travel from the Earth to Mars opens the 15th of September 2054 and closes after 28 days, the 13th of October 2054. The next window opens the 6th of September 2056 and closes after 65 days, the 10th of November 2056, and so on. Earth and Mars launch

windows are crucial to optimize the amount of S/C and boosters to assure the right flow of people and cargo between the two planets. The study of the interplanetary trajectories allows obtaining a set of analytical results to analyze the mission feasibility in terms of propulsive requirements. The Earth-Mars transportation system optimization finally results in a total of 151 units of Shuttle_1 (200 PAX) and 200 units of Shuttle_2 (500 PAX) (see Figure 15), over the 50-year period.

12.2 Earth-Mars Station

To complement the Shuttle service, an Earth-Mars Station (EMS) will start operating from the year 2075, to be used as a large-capacity transport system. The EMS is placed in a Cycler Orbit that periodically moves between the Earth and Mars [29]. By the year 2100, the EMS must be able to carry up to 135,000 people at its higher occupancy rate. The station can be a big S/C or an asteroid placed at the Cycler Orbit. Schematics of the EMS architecture can be seen in Figure 15. The station orbits between the Earth and Mars, performing a fly-by around each planet every time it reaches them. This is very convenient since almost no fuel is needed to keep such a station on its path. However, passengers and cargo need to be moved from the surface of the planets to the station and back. For this, dedicated S/C and boosters are used. These S/C and boosters are new versions of Shuttle_2, adapted with fewer thrusters, smaller fuel tanks, and a less demanding Life Support System. Also, more than one S/C shall be able to dock with the station simultaneously, to facilitate the boarding during the 70 – 100 minutes fly-by. All in all, a total of 467 S/C and boosters are needed to feed the station between 2075 and 2100.

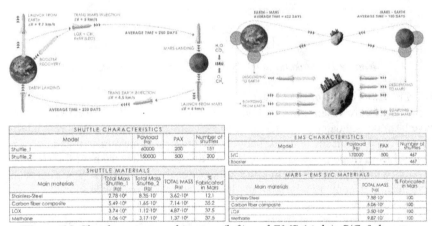

SHUTTLE CHARACTERISTICS

Model	Payload (kg)	PAX	Number of shuttles
Shuttle_1	60000	200	151
Shuttle_2	150000	500	200

EMS CHARACTERISTICS

Model	Payload (kg)	PAX	Number of shuttles
S/C	170000	500	467
Booster			467

SHUTTLE MATERIALS

Main materials	Total Mass Shuttle_1 (kg)	Total Mass Shuttle_2 (kg)	TOTAL MASS (kg)	% Fabricated in Mars
Stainless-Steel	$2.78 \cdot 10^8$	$8.35 \cdot 10^7$	$3.62 \cdot 10^8$	12.1
Carbon fiber composite	$5.49 \cdot 10^6$	$1.65 \cdot 10^6$	$7.14 \cdot 10^6$	35.2
LOX	$3.74 \cdot 10^9$	$1.12 \cdot 10^9$	$4.87 \cdot 10^9$	37.5
Methane	$1.06 \cdot 10^9$	$3.17 \cdot 10^9$	$1.37 \cdot 10^9$	37.5

MARS – EMS S/C MATERIALS

Main materials	TOTAL MASS (kg)	% Fabricated in Mars
Stainless-Steel	$7.38 \cdot 10^7$	100
Carbon fiber composite	$6.06 \cdot 10^7$	100
LOX	$3.50 \cdot 10^9$	100
Methane	$9.87 \cdot 10^9$	100

Figure 15. Shuttle system architecture(left) and EMS (right). S/C & booster images based on Starship and Superheavy from SpaceX. In both cases material masses are estimated using [30]

12.3 Earth Mars Transportation system: Material budget

Figure15 includes the total mass of the main materials needed for the Shuttle system and EMS, respectively. Taking into account that the S/C and boosters operating all their lifetime between the Earth surface and LEO are built on Earth, the materials needed to fabricate all the elements of the transportation system on Mars are Stainless-Steel $1.2 \cdot 10^8$ kg, Carbon fiber composite $3.1 \cdot 10^8$ kg, LOX $2.2 \cdot 10^9$ kg and CH_4 $6.1 \cdot 10^8$ kg.

13 - Distribution of Services

Since the city's tunnel system provides access and reaches to all parts of the city, it makes it sensible to also integrate its services within them. As a default, and as an overarching framework within the entire city, a Level 1 channel for distribution of services is proposed. Here it is assumed that the distributor plays a significant role as it manages the supply and delivery of building services, in the same way it manages the logistical activities related to the distribution of goods and passengers within the transportation network. While this creates a scenario where the entire infrastructure is wholly managed by a single stakeholder, it reduces system-wide operational risk. This provides an opportunity for further layering of Level 2 and 3 channels within the overall ecosystem.

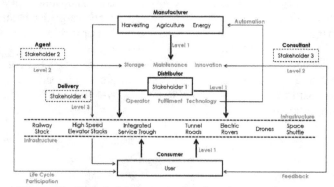

Figure 16. *Stakeholder scheme for distribution of resources*

14 - The Emergence of the City

A pictorial representation of the budget per CU is shown in Figure 17. Materials have been consolidated into categories to aid in the visualization. As a reminder, a CU (or City Unit) is the resource budget required to add one human to the city. The budget has been prepared consolidating all the information from all the solutions developed in the previous sections in a rather extensive spreadsheet. Some iterations were needed to converge the solution due to recurrent dependencies mostly caused by the energy production infrastructure. As seen in Fig 17, the major contributors to the budgets are the construction of human spaces, greenhouses, and the energy and manufacturing infrastructures. The bottom left of the figure also shows the bulk amount of minerals and gases that need to be ingested to the system. The manufacturing energy block corresponds to the energy required to manipulate products and transform them in user-ready goods (see Section 8- Manufacture). Important observations to be made: **Gravels are needed in large amounts**, but they are cheap in terms of energy. **Digging activities such as tunneling, and surface mining are not as energy intensive** as they may initially seem. Deploying machines with digging capabilities shall be a priority in the early development phases. **We based our material estimates on Earth's constructive style.** This results in the use of significant quantities of polymers (aka plastics). Most of the polymers do not go to domes, but into classic construction elements. Reduction in usage of polymers (& also steel) would propagate strongly and reduce the costs. For example, reducing the use of polymers would also remove most of the water budget, as all the Hydrogen atoms incorporated in organic molecules originate from it. One may think that the CU budget is too large, but this shall be compared to the equivalent one on Earth. In developed countries, the number of minerals mined per person/yr is about 10 tn. If we multiply this by 70 years of life, this leads to 700 tons of mineral, which is very consistent with our CU budget estimate, especially if we consider that most of the digging comes from clays to produce. In summary, despite the average power requirement is high, the result indicates that with clever implementation of current technology and refinements in the design, the implementation of self-sustained growth concept is be physically viable on Mars.

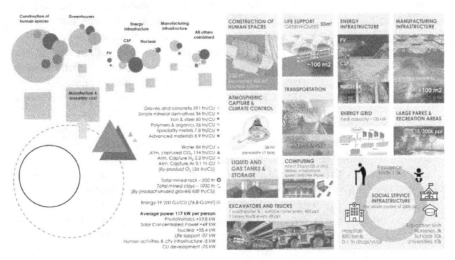

Figure 17. Mass, energy, power and raw material summary budgets per CU(left) and summary of physical assets(right).

15 - Operating the City, Artificial Intelligence (AI), and the Role of Innovation

Considering that Nüwa needs 117 kW per inhabitant compared to the 11kW used by humans on Earth developed countries, a Mars citizen would have about ten times higher workloads than Earth ones. This shall be mitigated by standardization, automation, and the use Artificial intelligence (AI). We are still very Earth bound to the methods that we can think about. Practical experience and locally-driven innovation shall both improve productivity and produce surpluses fast. Currently AI designers have the challenging task of getting AI agents to work effectively in environments not originally built with them in mind. However, if this is taken into account at design level, then the sophistication of their reasoning can be greatly reduced, e.g. limiting locomotion to tracks, having to distinguish only a few types of clearly marked object types, not requiring ability to circumnavigate unexpected objects (possibly humans). Consider the ensemble of robotic arms and conveyor belts used in fabrication factories - these don't require AI. AI shall also be used in services and digital administration tasks if possible. However, any automated decision by AI that affects the citizens will be auditable and the proper policy making will be in place to ensure that AI design is unbiased with regards to the values of Nüwa, such as equality among the members of the Settlement.

CONCLUDING REMARKS

This design proposal was *initiated and promoted by SONet (the Sustainable Off-world Network), which is a community of mainly European professionals interested in multidisciplinary approaches to sustainable exploration of space.* For a while, we have been asking ourselves the question "what would it take for an off-world infrastructure to be able to sustain its own growth"? The Mars Society competition gave us the chance to focus on a well-defined design exercise. After several brainstorming sessions, we established a simple development model, that pointed to a solution based on exponential growth and three phases. We then summarized the functions of the city, and proposed technical and urban design solutions that should address all of them. More than **twenty-five** people participated in the discussion sessions and contributed to a first 100+ pages document, which has been distilled in this report. Besides infrastructure, the city also needs citizens to grow. The socio-cultural aspects of the design naturally came into play when **building a compelling case to attract the future Martians**. We made no especial effort in assessing the cost of the development in Earth currency, except for the cost to transport people. This was intentional. Of course, this leaves open the *small issue* of funding for the initial colony. After making various assessments, we concluded that - as of today - **there is no clear Earthbound business case** for a colony on Mars that does not rely on wildly uncertain assumptions and correspondingly high risk. However, anyone would agree that economic returns are bound to happen when an economic activity enters exponential growth. If our self-sustainable development concept could be demonstrated (even on Earth), this would provide incentive for governments and the private sector alike. After analysis, design changes, and interactions, **we conclude that, in principle, the Nüwa concept could become a reality**; but additional design work is still needed.

Building a sustainable and socially cohesive society has strong synergies with solutions to Earth's problems, so solutions to both challenges can be worked out together. It is clear to all of us that a highly multidisciplinary effort would be needed to move this initial sketch into something more tangible. In future iterations, we hope to engage a more diverse community and this design concept as a starting point. In the meantime, we hope to meet again soon *"...I'll become part of Mars; I will dissolve into Nüwa and join all those that give life to it. And we will become Nüwa, forever..."*

Figure 18. Aerial view of the access to Nüwa via the valley floor (top), interior view of an office in a Macro-Building (bottom).

Team contributions. Project Coordination, Economic model & High-level concepts (**Guillem Anglada-Escudé**); Co-coordination, Space, Earth-Mars transportation & Socio-economics (**Miquel Sureda**); Life Support, Biosystems & Human factors (**Gisela Detrell**); design Strategy & Coordination (**ABIBOO Studio**); Preliminary Architectural Analysis & Urban Configuration (**Alfredo Muñoz, Owen Hughes**); Detailed Architecture & Urban Design (**Alfredo Muñoz, Gonzalo Rojas, Engeland Apostol, Sebastián Rodríguez, Verónica Florido**); Identity & Graphic Design (**Engeland Apostol, Verónica Florido**); Mars Materials and Location (**Ignasi Casanova**); Manufacturing, Adv. Biosystems & Materials (**David Cullen**); Energy and Sustainability (**Miquel Banchs-Piqué**); Mining & Excavation systems (**Philipp Hartlieb**); Social services, Life support, Biosystems (**Laia Ribas**); Mars Climate and Environment (**David de la Torre**); Ground Transport, economy (**Jordi Miralda Escudé**), Politics & Space law (**Rafael Harillo Gomez-Pastrana**); Chemical processes (**Lluis Soler**); Topographical analysis (**Paula Betriu**); Location, temperature & Radiation analysis (**Uygar Atalay**), Earth-Mars Transportation (**Oscar Macià**); Resource Extraction & Conveyance (**Eric Fimbinger**); Art Strategy in Mars (**Stephanie Hensley**); Electronic Eng. (**Carlos Sierra**); Psychology (**Elena Montero**); Mars science (**Robert Myhill**); Artificial Intelligence (**Rory Beard**).

Acknowledgements. This work has been possible thanks to resources, salaries and grants from Institut de Ciències de l'Espai (ICE) & Institut de Ciències del Mar (ICM) both part of CSIC (Spain), Ramón y Cajal & Retos/JIN from MICINN (Spain), Univ. Politècnica de Catalunya (UPC, Spain), ABIBOO Studio (USA/Spain & intl.), Cranfield University (UK), Universität Stuttgart (Germany), Institut d'Estudis Espacials de Catalunya (IEEC, Spain), Montanuniversität Leoben (Austria), and University of Portsmouth (UK). All participants contributed with their own time and effort to the discussions that led to this manuscript.

REFERENCES [1] Hall, C. A. S. et al., Energies, 2: 25-47, 2009, [2] Lambert, J. G. et al., Energy Policy, 64: 153-167, 2014, [3] Boyko, C. T. et al., Global Environmental Change, 22: 245-254, 2012; [4] Ehlmann, B. L. et al., Ann. Rev. of Earth & Planetary Sci., 42: 291–315, 2014; [5] Brothers,T. C. et al. " 43rd Lunar and Planetary Science Conf., 1452, 2012; [6] Milliken, R. E. et al., J. of Geo. Research Planets, 112, 2007; [7] Christensen, P. R. et al., J. of Geophys. Research, 106: 873–23, 2001; [8] Stern, J. C. et al., Proc. Natl. Acad. Sci. USA, 112:4245–4250, 2015; [9] Anderson, M. S. et al. "Life supp. baseline values and assumptions doc.," 2018; [10] Berkovich, Y. A., et al. Adv. in Space research, 44.2: 170-176, 2009; [11] Boscheri, G., et al. 46th Intl. Conf. on Env. Systems, 2016; [12] Cannon, K. M., et al. New Space, 7.4: 245-254, 2019; [13] Detrell, G., et al. in, from *Biofiltration to Promising Options...*, Elsevier; [14] Dobermann, D., et al. Nutrition Bulletin, 42.4: 293-308, 2017; [15] Duatis, J., et al., 40th Intl. Conf. on Env. Systems, 2010; [16] Hanboonsong, Y., et al. RAP publication, 3, 2013; [17] Wheeler, R. M., et al. NASA/TM-2003-211184, 2003; [18] Zeidler, C., et al., Open Agriculture 2.1: 116-132, 2017; [19] Fu, Q., et al. Energy & Env. Science, 2010; [20] Gotz, M., et al. Renewable Energy, 85: 1371-1390, 2016; [21] Reches, Y., et al. Cement and Concrete Composites, 14, 2019; [22] Gebler, M., et al. J. of Cleaner Prod., 2020; [23] De Castro, C., et al. Energies, 13, 2020; [24] Tyack, D., et al. *Learning together: ... in K-12*, Russell Sage Foundation, 1992; [25] Goodman, J., et al., Palgrave Macmillan, 2010; [26] Hider-Mlynarz, K., et al. British J. of Clin. Pharma., 84: 1324-1334, 2018; [27] Scholten, et al., 2019. AJPH PAIN MANAGEMENT, 109, 2019; [28] Guven, G. S. et al., Journal of Hospital Infection, 53(2): 91-96, 2003; [29] Byrnes, D. V. et al., Journal of Spacecraft and Rockets, 30(3), 1993; [30] Wertz, J. R. et al., Space Mission An. and Des., Space Tech. Library, 1992. Many ideas were inspired by solutions presented in *Mars Colonies, Plans for Settling the Red Planet*, by *Dr. Frank Crossman/The Mars Society 2020*; especially from *Aeneas Complex* (by J. Little), *"Team spaceship" Engineering Requirements* (by R. Mahoney, A. Bryant, M. Hayward, T. Mew, T. Green & J. Simich), *Star City, Mars* (by G. Lordos & A. Lordos); and *Project Menegroth* (by W. Stine).

11: PHRONESIUM: A SELF-ORGANIZED MARTIAN CITY STATE

Michael Bouchey, PhD
Taylor Dotson, PhD
The New Mexico Institute of Mining and Technology
michael.bouchey@nmt.edu
taylor.dotson@nmt.edu

INTRODUCTION

Our Mars city-state takes its name from *Phronesis*, a Greek word meaning "practical wisdom." The starting point for our design is the recognition that predictions of technological futures are often very wrong. What matters more for achieving a workable Martian colony are overarching social, economic, political, and cultural systems that maximize the generativity of new ideas, empower residents to experiment and tinker, maximize the intelligence of the innovation process, help colonists productively navigate technical and political disagreements, and meaningfully engage residents in cooperatively building their own society. If a million person Martian colony is to be not only feasible but flourish, it must be built upon democratic principles.

TECHNICAL DESIGN

Our technical designs will focus on overcoming the main barriers to life on Mars. Notably, Mars lacks oxygen and a sufficient atmospheric pressure, liquid water, and food sources. Citizens will require habitats that shelter them from the harsh Martian environment. Finally, all of these systems will require power generation. Each of these barriers will require creative and functional engineering designs for denizens to overcome them.

It is impossible to know which of the various technical alternatives are feasible for a 1M person colony now. Rather than detail a specific final design, we instead propose a system for assuring that the best technical possibilities are realized by the colonists as quickly and reliably as possible. Consider NASA's brief foray in wind turbine construction in the 1980s. NASA engineers thought they understood wind power far more than they really did, quickly scaling up their designs to large megawatt prototypes that failed within a mere dozens of hours. They were outdone by small-scale Danish agricultural manufacturers, who instead learned gradually from experience, scaled-up their designs incrementally, and cooperatively shared what they learned through a widely circulated journal [1]. We propose that a successful Mars colony can be nothing less than the product of highly organized hands-on learning by diverse groups of colonists.

The Mars colony will be divided into individual and isolatable "pods" of roughly 30-50 people, each utilizing different technologies for necessities like oxygen, water, food production, and the development of economic goods for export. Any particular technology would at first be used by at least a dozen or more pods, to ensure a diversity of experience, and will be supported for as long as feasibly possible. This reflects the strategy used during the Manhattan Project. It was unclear which uranium enrichment technology would succeed, so project managers explored all the options simultaneously. Every pod will include emergency and rescue equipment and sufficient space to host some of their neighbors in the case of technological failure. Pod groups will share their findings and meet periodically for scientific and engineering conferences, where breakthroughs can be shared and, eventually, the best technical designs determined.

Oxygen

The production of oxygen is the first barrier to overcome. NASA accounts for 0.84kg (590 L) of oxygen per day for astronauts [2]. Thus our city will be consuming 840 thousand kg (590 million liters) of oxygen per day. This estimate only includes oxygen for human consumption. Oxygen for industrial purposes, like LOX rocket fuel, are not included. We also assume oxygen recovered from the air in the habitat will supplement oxygen from atmospheric CO_2. Current oxygen recovery systems utilize the Sabatier system [3] and recover around 50% of oxygen used on the ISS [4]. We will assume 50% recovery. 420 thousand kg of oxygen will come from recovery, 420 thousand kg will come from processing atmospheric CO_2.

There are several possible methods which might be able to extract oxygen from atmospheric CO_2. Water can also be separated into hydrogen and oxygen and Martian regolith can be refined into oxygen and various other metals.

Atmospheric carbon dioxide

One method of attaining oxygen might be separating atmospheric CO_2 into oxygen and carbon monoxide. There are three main ways of transforming CO_2 into O_2: the Sabatier process, the Bosch process, or electrolysis.

Sabatier process and electrolysis

The Sabatier process [5] follows the chemical reaction

(1) $CO_2 + 4H_2 <-> CH_4 + 2H_2O$ (2) $2H_2O <-> O_2 + 2H_2$

Assuming an 80% efficient reaction, we estimate 95.2 kcal/gm mol of O_2 for both reactions. Assuming 50% oxygen recovery, 60.6 MW is required for oxygen recovery and the same for harvesting oxygen from atmospheric CO_2, totaling 122.2 MW. A disadvantage of this process is that it is not a closed loop system. Hydrogen will either have to be imported from Earth, or hydrogen will have to be reclaimed from the CH_4 via pyrolysis [6]. This increases the power by 14.2 MW for reclamation, and the same for atmospheric CO_2, totaling 28.4 additional MW. Thus we estimate that this closed loop system will require 150.6 MW. Citizens will have to weigh this process against others, as well as determine how much oxygen will be reclaimed versus harvested from atmospheric CO_2, and how much, if any, hydrogen to reclaim via pyrolysis. We do not presume to be able to predict the answers to any of these decisions, nor the factors that will be most important to citizens in making these decisions. As will be discussed in more detail, we will rely on innovations to decision-making processes to aid in this endeavor.

The Bosch process
The Bosch process [3] utilizes the following series of chemical reactions:

(1) $CO_2 + H_2 <-> CO + H_2O$ (2) $CO + H_2 <-> C + H_2O$
To give the overall reaction
(3) $CO_2 + 2H_2 <-> C + 2H_2O$ (4) $2H_2O <-> O_2 + 2H_2$

Assuming 80% efficiency for the reaction, the overall process requires 117.5 kcal/gm mole of O2 produced. Assuming this process only produces 50% required O2, the other 50% coming from reclamation, that totals 74.9 MW for 420 thousand kg of O2/day. Although this is more energy intensive than the Sabatier process, it produces all the hydrogen needed for its input as an output. It has nearly identical power needs to Sabatier plus pyrolysis. The catalyst is also iron, which is abundant on Mars. However, the process of reaction (2) leaves a solid film of carbon precipitate which would foul the catalyst. It would be very energy intensive to remove the carbon and, although some might be useful, the 158 metric tons/day it would produce will likely pose a serious disposal problem. Martian citizens will have to be able to decide between the relative trade-offs of increased energy consumption and carbon disposal vs hydrogen imports.

Carbon dioxide electrolysis
Carbon dioxide electrolysis directly converts CO2 to O2. By sending CO2 over a catalyzed cathode surface under an applied electric potential, Martians can split CO2 into carbon and oxygen by the reaction

(1) $2CO_2 <-> 2CO + O_2$

We estimate that this process operating at 80% efficiency will require 9310 GJ/day, or 107.8 MW to produce 50% of the city's oxygen, with the other 50% coming from recovery. It is the most energy intensive, having the highest temperature requirement [7]. Such a system will be tested using the Mars Oxygen In Situ Resource Utilization Experiment (MOXIE) [8] on the Mars 2020 mission. The MOXIE is a 1% scale model of an oxygen producing plant that is designed at a rate of 20 grams per hour. A full scale plant based on the MOXIE, would produce 48 kg/day. It would take 8750 such plants to convert enough atmospheric CO2. Development of industrial scaling of this process may be possible by the time the city is fully functional, however.

Oxygen from water
We do not yet know whether citizens will struggle more with water acquisition or oxygen production. If water acquisition proves to be easy and there is a surplus in water, Martian citizens can use water electrolysis as a way of extracting oxygen. This would be identical to the last reaction of the Sabatier or Bosch process, simply allowing colonists to skip the first steps in those reactions. See the analysis for both of those reactions for more details. We mention the possibility here simply because of the unknowns involved in estimated resource surpluses and deficits. Rather than suggest a definitive solution about which methods Martian citizens should prefer, we will later in the paper discuss systems for decision-

making that will maintain the flexibility and intelligence necessary to navigate these complexities.

Regolith refining

A final method may be to refine oxygen from the Martian regolith itself. The Red Planet is so colored precisely because of the abundant iron oxide on the surface. How can those oxides be turned into breathable oxygen?

Regolith can be refined into high grade metal oxides and oxygen [9]. The metal oxides, such as iron, alumina, magnesia, and calcia, can be used in construction materials, structural components, and insulation. Oxygen is a byproduct of metal oxide reduction [9]. Once the regolith is extracted, it is dissolved in a sulfuric acid solution. The undissolved residue can be sintered to create structural materials and glass. The dissolved material can be precipitated or crystallized, initiated by pH adjustment with a magnesium hydroxide base. The precipitates from this process are calcium oxide, iron oxide, aluminum oxide, and magnesium oxide. These materials can be used as is, or reduced into their pure metallic form with oxygen as a byproduct. This method may be necessary for producing useful material products regardless of oxygen needs. We do not know what the requirements for material products will be, so we also do not know how much oxygen will be a byproduct. But this could be a useful process, splitting the energy required between oxygen and material production and reducing the energy needed for the other processes. However it may be unfeasible to move enough material for this process and the process of extracting water from regolith.

Water

Liquid water on the surface of Mars is negligible, but other sources of water may be able to supply Martian citizens. It may be the case that absorbed water in the Martian regolith, hydrated minerals, or ice located in the subsurface, mid-latitude glaciers, or poles [10] can provide all the water necessary to support a city. How much water would a city of 1 million require every day?

The average adult male requires 3.7L of water per day, while the average adult female requires 2.7L per day [11], averaging to 3.2L per day with equal numbers of male and female adults. For 1 million people, that amounts to 3.2M liters per day. Food production also requires water. We will assume relatively water wise crops, needing no more than 1000mm of water per growing season (approximately 90 days) along with $46*10^7$ m^2 of total crop area [12]. These numbers factor in the benefits of hydroponic technology as well as of vertical farming [13]. Food production thus requires approximately 5.2M liters per day with a total requirement of 8.4M liters per day. To account for industrial uses and other household uses will we assume double this value, 16.8 M liters per day. Much of this water will be reused. The ISS reuses 93% of its water [14], but due to the large scale required for a Martian city, a better assumption for reuse is about 80% [15]. Thus, the total daily production requirement of water, we estimate, is 3.36M L/day.

Mining ice

On initial inspection mining and melting subsurface ice seems to be the most energy efficient source of water. There are two main challenges to mining ice. First, subsurface ice is only found in certain locations on Mars. Second, deposits of water ice are typically covered in 1-10m of debris and dust known as overburden which must be removed.

Location

Subsurface ice is abundant in high latitudes, with decreasing abundance and increasing depth from the surface nearing the equator. See Figure 1 for details.

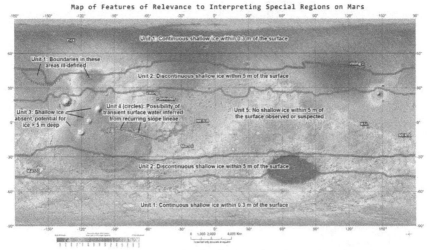

Figure 1: Abundance and depth of subsurface ice by location [10]

Unfortunately, landing and takeoff from Mars is easier at the equator and more difficult at higher latitudes. NASA has set preliminary boundaries for crewed missions at no more than 50 degrees [10]. Therefore, there is an inverse relationship between ease of Terrestrial trade and ease of access to ice. Furthermore high latitude locations have reduced prospects for solar power. Martian citizens will likely have to compromise and live between 30-50 degrees latitude.

Extraction

The major barrier to use of ice for water extraction is the layer of overburden covering the surface. If removing the overburden only allows access to less water than processing an equivalent volume of regolith, it may not be worth pursuing. How deep is too deep? The volume of overburden that must be removed depends on how the pit is dug. Figure 2 shows a NASA estimate from the Mars Water In Situ Resource Utilization Planning Study (M-WIP) [16].

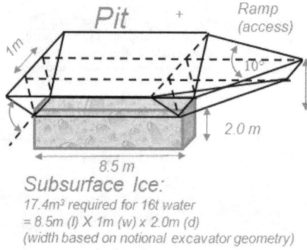

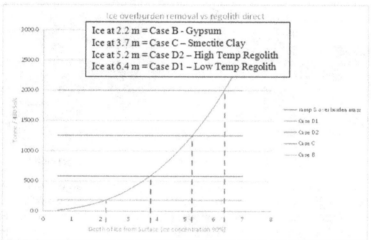

Figure 2: Size of pit for ice mining and relative amount of material moved by depth of pit [16]

Ice at 2.2m or greater requires excavating enough material that an equivalent amount of water could be processed from Gypsum rich regolith. Ice at 3.7m or greater yields an equivalent amount of water to Smectite rich regolith, and ice at 5.2m or greater yields an equivalent amount of water to typical regolith. Thus, from the perspective of the amount of material, only near surface ice yields benefits compared to regolith refining.

However, from the perspective of power capacity, melting ice is superior to refining water from regolith. Although moving large amounts of material has its own challenges, accessing ice at depths beyond those listed above may still be more energy efficient than water from regolith. The primary power draw for regolith processing comes from heating material. At 0C and 1atm, the latent heat of fusion for water is 333 J/gram. Thus, to gain the 3.36 million liters per day to supply the city, the total energy cost to melt the ice would be at least 1118 GJ. Spread out over the course of the day this is 13 MW. Based on this initial estimate, ice mining presents substantial energy savings over any kind of regolith refining.

Water in the regolith

Although mining ice is likely to be the primary source of water for a Martian city, access to ice may be initially difficult. The ability to refine water from regolith may be an important step prior to water from ice. NASA currently considers centralized processing and collection of regolith from water refining to be the baseline setup for such a water collection system [17].

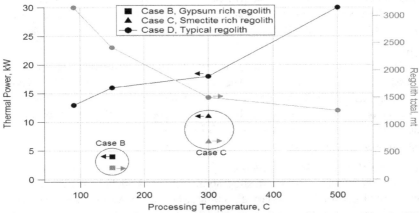

Figure 3: Graph of thermal power and mass of regolith required to produce 1.36kg of water based on three regolith compositions [17]

Martian regolith is an amalgamation of different minerals and is not homogeneous. Each material has different thermal power requirements to produce water. So how much energy and regolith will it take to satisfy the city's needs assuming various temperatures and compositions? Figure 3 provides a measurement assuming the production of 1.36 kg of water.

Assuming average regolith composition, heating the regolith to 500C will take about 3.08 billion metric tons of regolith expending 74 GW of power. Heating the regolith to under 100C will take about 7.7 billion metric tons of regolith expending 32 GW of power. However, we will examine Gypsum and Smectite rich regolith deposits as well.

Smectite and Gypsum rich regolith are substantially more water rich and therefore require less energy and less regolith for the same amount of water, as can be seen in Case B and C in Figure 3 above. Smectite rich regolith need only be heated to 300C to extract water, which is about 2.74% water by mass. Utilizing the data presented in figure 3 above, using smectite rich regolith reduces the regolith quantity needed per day to 1.73 billion metric tons, expending 27.2 GW of power. Gypsum rich regolith needs only to be heated to 150C and has a total water content of about 9.08% by mass. Utilizing the same data, the total mass of gypsum rich regolith needed to supply water for one day is approximately 618 million metric tons, expending 9.88 GW of power. While it is feasible to produce this much power, it is substantially more than melting ice. Additionally, one of the largest mines on Earth, the oil sand mine in Alberta, Canada, moves on the order of 200 million metric tons of material per year. Regolith operations on Mars will have to move several orders of magnitude more material. This pattern seems fairly typical for most of the technical systems necessary to support a Martian city.

Food

Unlike oxygen and water, which can be refined from existing Martian material, there is no easy way to create food from what already exists on the red planet. But an independent Martian city state will require an in situ food supply chain. The main challenges are minimizing water and power usage and selecting food sources suitable for the challenges of Martian life.

Crops

We recommend that underground growing chambers are likely to be the most successful setup for Martian crops. The surface of Mars is very hazardous to crop growth, and natural sunlight under even transparent protective coverings will not be sufficient for most crops [18]. Instead we recommend that crops be grown underground in tunnels with high strength LEDs as the primary light source, and supplemental sunlight piped in through fiber optic cables.

Cannon and Britt's model predicts 46×10^7 m^2 of total growing area to supply food for 1 million people [18]. Their suggestion is that 14,500km of 12m diameter tunnels could sustain the necessary agriculture. The largest vertical farm facility in the world, Mirai in Japan, has about 25000m^2 of growing area [19] and uses 369 MWhr/month [20]. The Mirai farm grows only lettuce and so growing a larger variety of crops is likely to have different energy requirements per growing area. However, if we assume similar overall energy use per growing area, the Martian city state will require 930MW of power for crop growth.

A more thorny technical problem is the growth solution for Martian crops. We recommend a soilless system for several reasons. Such systems require substantially less water per yield. Moreover soil is a highly dynamic living system. Attempts to create soil from inert Martian regolith simulants have not yet been able to support even basic life such as earthworms, much less grow crops. While, at least initially, hydroponic equipment will have to be shipped from Earth, biological material for the creation of soil would likely also have to be shipped. It is possible that by the time a Mars city reaches 1 million people it could produce some of its own farming equipment, and focusing on soilless systems does not preclude conducting a trial and error approach to creating productive Martian soil in the meantime.

Protein

Animal based proteins are the most common sources on Earth, but taking animals to Mars would be both difficult, and waste valuable resources and space for relatively low caloric return. The main potential alternatives are plant protein, lab grown animal protein, and insect protein.

Some varieties of plant based foods are both low water use and high in protein. Quinoa, black-eyed pea, chia, and tarwi are examples. The use of gene editing techniques could also increase the variety of plant based proteins available on Mars. The protein rich, water wise legume, Vetch, normally produces the toxin GBCA that limits potential human consumption. Gene editing techniques could eliminate GBCA production. New studies into plant protein BAG4 show how manipulating this protein can reduce water requirements for existing crops. Applying this change in other high protein crops that would otherwise consume too much water could also increase Martian citizens' access to protein. Lab grown animal protein has been lowering considerably in price such that it is now a feasible alternative. In 2013 the price of the first lab grown meat was $1.2M per pound. A year ago the price for 1kg of lab grown meat was $800. Today it is $100/kg. The primary barrier to further price reductions is limited scale due to easy access to traditional meat, which would not be present on Mars. Finally, insects are a far more efficient source of protein than other animals. Many cultures are not put off by eating insects, and have a variety of ways to prepare them. But processing the proteins could also allow insect proteins to supplement more widely accepted dishes. Products such as some protein bars use insects as the source protein. Flour substitute and other products could also utilize insects to increase Martian protein options without forcing citizens to eat anything they find unpalatable.

Additional growing scenarios
Greenery and plant growth provide demonstrable psychological benefits, making it desirable throughout the city, not just in dedicated farms. Combining these green spaces with some food production is likely to offset the inefficiencies of conducting some food production outside of dedicated farms. We suggest Martian green spaces be organized like community gardens.

Highly successful community gardens like Art in the Garden of Pittsburg, PA, BUGS Indoor Vertical Garden in Brooklyn, NY, and The Bronx is Blooming in Bronx, NY offer good examples of the benefits of the community garden model. Community gardens would reduce barriers to organic waste reuse/composting. They increase the physical activity of participants, as well as improving diets. They often incorporate education into gardening practices which provides better understanding of a variety of biological and environmental subjects by participating citizenry, improves employment prospects, and provides other practical skills that can apply to a variety of employment. This, in turn, decreases unemployment. Unemployed or less fortunate citizens can also use community gardens to supplement income or as a source of food which reduces expenses. Thus a tertiary consequence is crime reduction. Community gardens also tend to reduce crime by reduce vacant space and providing alternative activities for idle residents. Moreover, working together on community gardens can build close relationships and bonds between members of a community, engendering mutual support outside of the garden as well.

Habitats

Mars is cold, has an extremely thin atmosphere, and lacks a magnetic field. Wilderness survival on Earth prioritizes shelter as the first concern, this is even more necessary on Mars.

Structural challenges

Habitat structures on Mars have several forces that they must contend with. Internal pressure, low gravity, thermo-elastic loads, and potential micro-meteoroid impacts all pose threats to Martian habitats. On Mars gravity is ⅓ the force of Earth, and the atmospheric pressure is very low, so the primary force there will be internal pressure. The pressure differential is 2090 psf on Mars if the inside of habitats is pressurized to Earth's atmospheric pressure [21]. For comparison, airplanes on Earth must contend with only 1400psf. In addition, due to Mars's thin atmosphere, the thermal gradient due to the sun is 148F [21]. Not only will any structure have to support expanding and contraction due to this gradient, it will also have to provide sufficient insulation to maintain a constant temperature on the inside. Finally, a habitat punctured by a micrometeorite will release air into the Martian atmosphere. This could deny air to citizens and collapse structures supported by internal pressure. Lacking any magnetic field, citizens living on the surface of Mars would receive 1 Sv of radiation exposure every 500 days [22]. Combining all of these challenges severely limits the types of structures that can compose the city.

Structural solutions

Simply locating habitats inside lava tubes solves many problems. First, they would protect citizens from the harsh radiation on Mars's surface. Radiation exposure underground would be negligible. Once the habitats are heated, basalt has a thermal conductivity of $2x10^{-3}$, an order of magnitude better than most terrestrial insulations, and are 1-15 meters thick. Plenty of protection against the Martian cold, and fairly easy to maintain a constant temperature, especially if combined with passive geothermal systems. This would also provide a more than sufficient barrier against micrometeorites.

To make the lava tube air tight, Martian citizens will still need to install inflatable barriers. The tube itself provides most of the containment strength, so the material can be light and low bulk, acting merely as a liner. It would not have to resist any of the harsh forces on the surface. The tubes will also need airlocks. The openings may be irregular, so any airlock device would have to fit in any shape of opening. First, an inflatable airlock may be possible. It would conform to the shape of the opening, but would have to be more similar in material to that of inflatable habitats on the surface than to the lava tube seal. Another alternative is to use standard rigid airlocks with a hardening foam to connect the opening to the lava tube, although such a foam may degrade fairly quickly on Mars. Due to flammability issues, the atmosphere cannot be 100% O2. The Martian atmosphere contains 2.7% Nitrogen and 1.6% Argon. Some of that nitrogen may be necessary for making crop nutrients, and since nitrogen in the air may induce microbial growth via nitrogen fixation processes [23], we recommend utilizing argon in at least equal measure to nitrogen to achieve atmospheric pressure.

Power

Each preceding section has discussed the power requirements of these necessary systems on Mars. This section will discuss how much of which sources are necessary to meet the minimum energy requirements presented here and will also discuss some of the technical and political barriers to the use of those sources.

First, it should be noted that the Martian environment limits the available sources of energy production without relying on continuous imports from Earth. Mars cannot independently support hydroelectric power, geothermal power, wind power, coal, natural gas, or oil. This leaves solar energy and nuclear energy as the two primary sources of power production to consider.

Solar energy

Solar has several advantages. It is a well established technology in space applications. Because the city is underground, panels could be placed above the city. The main drawback is the distance from the sun dramatically reduces the potential power from solar energy. For our assessment of solar energy, we are assuming photovoltaic technology rather than solar thermal.

Designs and installation

First, because the solar intensity is so much lower on Mars than on Earth, the panels will likely require a tracking mechanism to continually optimize their angle relative to the sun. Second, dust accumulation will require a system to regularly clear dust from the surface of the panels, either permanently affixed or moved between panels. Third, solar panels will require substantial diagnostic equipment in order to indicate loss of efficiency both from dust or age related degradation. Fourth, each panel would also have to be periodically replaced as they age and efficiency gets too low. Finally, Each of these systems will themselves need a maintenance system, as well as cleared land around each panel to provide maintenance access.

Power estimates

We will assume the solar intensity averaged over the entire surface, each season, and the day night cycle. For mars this amounts to approximately 100 W/m^2. Solar panels currently used by NASA are 30-40% efficient [24], so we will assume 30% efficiency, meaning the available power from solar panels will be 30 W/m^2. Furthermore, each panel will not always be operating at full capacity, thus we will also assume a 90% servicing factor. Our final estimate for the available power from solar panels is 27 W/m^2 or about 37000 m^2/MW. This represents a square area of 192m on each side. The total surface area, including spaces between panels, will be a square 250m on each side: 62500 m^2/MW of Martian surface area.

In addition to our existing estimates for other technological systems, citizens will still also need domestic power. We will assume a per capita energy estimate of 13

MWhr/yr, similar to the United States. Thus, the city of 1 million people will require about 1484 MW for domestic use.

Table 1 shows the solar panel area and ground area required to power each subsystem we have so far discussed.

Subsystem	Power (MW)	Solar Panel Area (km²)	Ground Area (km²)
Domestic	1484	55	94
O2			
Sabatier	136	5	8.5
Bosch	143	6.5	8.9
Electrolysis	176	6.5	11
Ice Mining	13	0.48	0.81
Food	930	34.4	58.1

Table 1 Subsystems Area

Nuclear energy

As Table 1 shows, the total surface area necessary to power the Martian city is exceedingly large. It would take up a land area the size of Las Cruces, NM. Nuclear power of some variety is likely necessary to supplement solar. Mars has no tectonic activity and most of the planet likely lacks a water table. Thus we predict that nuclear waste storage will be substantially easier on Mars. As long as the power stations are kept outside of the city, the radiation protection offered by locating within a lava tube should alleviate the consequences of any potential meltdown.

Several barriers to nuclear energy still exist, however. First, from a political and legal standpoint, it seems unlikely that nuclear material will launch from Earth. Mars does have viable deposits of nuclear material, but this does mean that all of the necessary infrastructure for refining raw nuclear material into usable fuel will have to be developed in situ. In addition, the only well practiced nuclear technology on Earth, light water reactors, are extremely water intensive. In the United States, nuclear energy production uses approximately 400 gallons of water (1516 L) per megawatt hour [25]. Liquid salt and gas cooled reactors would alleviate this problem, but such technologies have not yet been commercially implemented, so we recommend caution, flexibility, and an eye towards learning if these technologies are to be implemented on Mars without being able to learn lessons from Terrestrial implementation.

POLITICAL SYSTEMS

As may be becoming clear, following the discussion of the technical challenges of maintaining a 1 million person city state on Mars, the existence of such a city is, in many ways, tenuous. The operation of the technical systems of such a city will have to be very precise, and there is little room for error. Moreover, the lifestyles of its citizens will have to be very different from Terrestrial lifestyles due to the difficulties of maintaining life on Mars. The social, cultural, and political institutions of Martian society will also have to be organized to reflect these challenges.

Citizens will have to organize together to make a variety of different sectors of Martian life function well. The governance structure of the Martian city will have to organize technological development, the economy, community life, and systems for developing rules and resolving conflicts. We propose that, while each of these tasks need not, and likely should not, be done by the same organization, that they should all be conducted democratically.

Technological Development
As previously discussed, we envision the development of the Martian city to proceed primarily through several small pods which each run semi-autonomously and experiment in cautious and incremental ways to discover what the best blend of technological systems will be to make the Martian city as successful as possible. But once the city reaches a population of 1 million and the selections of technological systems have been made, how will the city govern these systems and ensure a continual stream of path-breaking but safe innovations?

One existing system that is targeted at creative policy innovation used throughout the European Union is an Enquete (Inquiry) Commission (ex. [26]). These commissions consist of half subject-matter experts, half citizens and are tasked with considering alternative solutions to thorny technological problems, such as nuclear power policy. Similarly, in the Martian city, technological problems will be broken down into a number of subcategories. These may align with our above technical categories or include others depending on the needs of the city. Each category will have its own inquiry commission. These commissions will determine policies such as which technological developments to pursue, how much funding to provide for each potential pathway, when to increase the scale of a particular technology, and when a particular technology is not worth pursuing. The membership on these commissions will rotate periodically and members will be fairly compensated for their time to ensure that anyone can serve while preventing career politicians.

Such a system will improve the intelligence of technological decision-making and therefore reduce both the frequency and intensity of technological mistakes. By incorporating citizens, it includes a wider variety of interests and ideas that may not be included with only experts deliberating. It organizes development by starting at small scale and scaling up slowly to maximize flexibility, as well as including ongoing institutional evaluation to learn from trial and error. It serves much the same purpose as the Institute of Nuclear Power Organization does in the United States. The INPO incentivizes firms to share information, set their own high safety standards, and sanctions firms that don't conform to them. On their own, firms are incentivized to do the opposite, but the consequences of a catastrophic accident are such that it is better for all firms to coordinate in this way. Likewise, this level of technological coordination will be necessary on Mars, where mistakes are also likely to be catastrophic.

Economic Development

As we will discuss in the economy section, much like on Earth, the economy will mostly consist of autonomous firms. We will answer the question of how firms will be run below, but how will the Martian government coordinate those firms? All firms will require infrastructure, a financial sector must regulate the currency, some system for investment needs to exist, and wholesalers will have to be organized to ensure efficient and timely distribution of goods. And each of these tasks has the extra challenge that, unlike in Terrestrial markets, supply chain, infrastructural, and financial failures can result in the loss of function of life critical systems.

Infrastructure

In the same way that Terrestrial manufacturers rely on publicly funded roads and rail to get their products to market, Martian industries are likely to require some form of support from the broader Martian public. While individual firms will best know their own needs in this regard, support from other Martian citizens depends on the latter having a substantive say.

We propose a confederation system in which representative bodies from firms and the general citizenry form a system of checks and balances for public infrastructure investment decisions. Martian firms will self organize into councils based on common infrastructural needs, i.e. oxygen companies together, mining companies together, research groups together, etc. They will have proportional representation within those councils, which in turn will provide infrastructural plans for the city as well as cost estimates and associated taxation estimates. Economic councils then evaluate these plans. These councils are formed out of a random selection of Martian citizens and advised by a diverse group of experts on the industry in question and on finances. They will approve taxation levels to fund infrastructure programs. This system provides an easy avenue for firms to coordinate the construction of infrastructure in a way that targets their specific needs, but also checks against the potential for offloading costs to all citizens without their representation. This system utilizes the checks and balances of systems of confederation used to some success both in the United States and in earlier societies, like the Haudenosaunee.

Financing and investment

Financial institutions have often created negative outcomes in Terrestrial economies. One need only look at the 2008 financial collapse to see how the decisions of a few powerful financial executives were able to sink many otherwise productive and healthy industries. Given the need for self-governing firms, we propose that financial institutions follow a similar model. Take the Mondragon cooperative group in Spain as an example. This cooperative group also runs their own bank, the Labor Credit Union (Caja Laboral). Jointly run by the workers in a variety of different interrelated cooperatives which are part of the group, their credit union can engage in independent investments, but its charter requires a certain percentage of funds be invested in the Mondragon group or in the community i.e. home loans, and minimum percentage of value kept in liquid capital. These practices helped see growth during the 2008 recession. Establishing similar joint financial ventures on Mars is likely to help maintain financial stability and protect against the collapse of essential industries without having to expend public money on bailouts.

Market investments, along with recruiting new members are two ways that self-governing firms could raise capital, but occasionally firms require public funds to conduct business with broad benefits that might not otherwise be profitable. In such cases we recommend that these public funds be given to firms under the same conditions of other investments. Public entities will simply buy memberships in accordance with market value and be endowed with equivalent voting rights for a predetermined period to ensure that public funds serve public interests.

Wholesaling

In traditional economies, producers often sell to wholesalers rather than directly to retailers. On Mars, wholesale organizations will be made quasi-public [27]. Such an organization will not be concerned only with economic considerations, but will want to take into account public considerations, for example overuse of necessary resources, impact on quality of life, or risk to the city's viability. They would then use their economic position to alter production patterns via market mechanisms rather than state mandates. For example, if a product contained a potentially harmful material that was cheaper than the safer alternative, a democratic wholesaler could selectively order only products with the safer alternative whereas a private wholesaler has little incentive to do so. Such an entity could be run by elected representatives or randomly selected citizens or could, itself, be a self-governing firm reviewed by public councils.

Community Development

Many Martian citizens are likely to be involved in areas of life that are outside of technological and economic development. Care workers, community garden organizers, and others who serve more social roles within communities. Moreover things like public spaces, schools, and recreation spaces are important but fall outside of the purview of economic activity or technical development. How are these spaces and people to govern themselves?

We recommend that community oriented development be organized in similar ways to the other categories. While local communities can govern themselves through deliberative mechanisms such as New England town halls [28], such deliberations would need ways to operate on larger scales in order to foster coordination. We suggest for these purposes a series of nested assemblies that each deliberate over community development. Groups of neighbors will collectively select one from among them to go to a local assembly, which selects a single member to represent them at the district level. Each assembly will focus on the relevant problems of their level. Perhaps the local assembly simply chooses who to send to the district assembly, but the district assembly might determine resource allocation between member neighborhoods, decide policy regarding green space, or organize social events between member neighborhoods. If each assembly consisted of a mere 10 people, the entire city would be represented by only five levels of assembly, and each representative would have only 9 constituents. This cuts out many of the potential pitfalls of intractability due to loyalty to one's political "team." It ensures that community development is also supported at a variety of different scales. Mars is a particularly hostile environment. If 1 million people are going to be convinced to live underground, isolated with one another for their entire lives, they will need democratic mechanisms to build levels of community support that are absent in most places on Earth.

Rules and Conflict

Living underground, with limited green spaces, in a hostile, if not outright lethal environment, is bound to be stressful. Much of what will make life on Mars pleasant comes from the company of fellow citizens. But the high stakes are going to result in serious conflicts that citizens will have to be able to resolve without killing one another or polarized gridlock. The various democratic systems that we have already suggested for other areas of governance are necessary for this. Democracy in general is a uniquely suitable form of governance for dealing with conflict. In general, Martian citizens should establish governing institutions that conform to the rules of democracy [29]:

1. Equality: when making decisions, each citizen should have an equal say in determining the outcome.
2. Effective Participation: Throughout the entire process, each citizen should have adequate and equal opportunity for expressing their preferences.
3. Enlightened Understanding: Given the limitations of information and time, each citizen should have adequate and equal opportunity to consider and make their own judgment about the most desirable outcome.
4. Control Over Agenda: All citizens should have equal and exclusive authority to determine what matters are or are not decided by democratic processes.
5. Inclusion: Citizens should include all people subject to the laws/outcomes of its democratic processes except those interests opposed to its existence.

But it is not enough that only governing institutions be more democratic. Citizen participants must, themselves, be more democratic [30]. Given that any Martian colony will teeter ever so close to the precipice of existential destruction, Martian citizens must be able to tolerate and productively resolve disagreements far better than people currently do. As we note in the Community section, the overall culture of the Martian colony needs to support high reliability organizations and high levels of political community. What can help Martian citizens to act democratically? What supporting institutions promote these outcomes?

Democratic Schooling

Martian citizens will have to be trained to participate in governance to ensure that no talent goes to waste, and disagreements do not result in catastrophic consequences. This training must start early in each citizen's life. Education needs to be a means of apprenticing children on the processes of negotiation, compromise, and unencumbered explorative thinking that will be necessary for the functioning of the city's political institutions. Citizens must be educated so as to avoid the polarization death spiral that plagues many terrestrial countries today.

The curricular structure of Martian schools will be substantially more open than traditional schools. Students will be expected to take responsibility for their own learning. Attendance will be optional. Formal courses will not be the basis for learning and will be optional as well. Teachers will act as fellow learners using their experience and expertise to guide students along in their inquiries. But primarily students will be expected to learn from each other through play, exploration, socialization, and working together. Students will also create their own curriculum.

The administration of Martian schools will reflect these practices. They will be run by members - the staff, students, and parents - who will meet periodically, perhaps once a week, to make important decisions about the school, such as policy and staffing decisions. Each member, including students will have an equal vote at these meetings. Of course the members will delegate particular tasks, for example budgeting, to smaller groups with relevant expertise.

These schools have often been highly successful on Earth, in the limited capacity to which they have been tried. Their benefits are not just theoretical. Democratic schools have been shown to foster traits such as personal responsibility, initiative, curiosity, communication skills, especially the ability to communicate well with a large diversity of people from many backgrounds and situations, and generally a greater appreciation and practice of democratic values compared to traditional schools. A longitudinal study showed students in the most well known of these schools, Sudbury Valley School, had no difficulty in adjusting to the demands of higher education or employment in a wide variety of fields [31]. Moreover, those students found that their democratic education advantaged them compared to their traditionally educated peers. These are precisely the skills each student will need as they enter adulthood and contribute to the survival of the Martian city.

ECONOMIC SYSTEMS

The economic system for the colony will be as decentralized and democratic as the technical and political design. It will employ a market system [32], but far differently than is typical on Earth. In the pursuit of profits and/or efficiencies, market activity on our planet often sacrifices resilience, leaving little to no slack for necessary goods and services or allowing critical infrastructures or environmental services to decay until they are almost unworkable. For example, shortages during the recent COVID-19 pandemic demonstrated that most contemporary market systems emphasized the resource efficiency of supply chains rather than their resilience. Climate change in general illustrates how vital life support systems on Earth end up being neglected because they have no explicit "price." We recommend that the economic design of a Martian colony should prioritize resilience, work satisfaction and buy-in, decentralization, and self-sufficiency.

Imports, Exports, and Industries

A Mars city state will need to be self-governing and agile. The first priority is balancing trade with Earth. If the cost of importing the necessary goods from Earth is greater than the price received from exports, whatever organization makes up that deficit, be it some terrestrial state or company, will have an inordinate amount of control over the economy of the city state. Martian residents, in that case, would begin to feel less like colonists settling a planetary frontier and more like the indentured subjects of the outer space equivalent of the East India Company.

Martian citizens will have to consider two main factors. First, minimizing imports. Martian citizens will need to carefully consider the pros and cons associated with local production and Earth imports. Second, what products and/or services will be most valuable to provide to Earth? Which of these will be easiest to produce on Mars? Again, abstract calculations made in advance of settlement are unlikely to prove true. Needed most is a flexible system for quickly learning via experience. Moreover, providing every citizen with an ownership stake in the colony will make thinking about how to improve the trade balance the job of every resident.

Importing from Earth

Much of the cost of importing goods to Mars from Earth will be transportation related. We are assuming $500/kg from Earth to Mars. At the outset, a great deal will have to be imported from Earth. It is likely that resources such as food, water, and raw materials like building materials and metals could be produced in situ. All other supplies, including the machinery required for producing these initial resources, will need to come from Earth, much of it will need to arrive before settlers. One estimate of this initial mass is 500MT [33]. Thus the initial investment is $250M for transportation plus the procurement costs of the supplies. These costs will vary considerably, being relatively low for items like food, but very high for technically advanced supplies like construction and refining equipment. We will estimate a total initial cost of $1B.

Ensuing shipments from Earth will include manufacturing equipment, like 3D printers, in order to help achieve some self-sufficiency. These will have to be designed with large enough tolerances in order to function in the harsh and low-gravity Martian environment. As energy is key for setting up the rest of the colony, early importations will include specialized parts to begin work on constructing a nuclear reactor and setting up the mining and refining of nuclear fuel.

Exporting from Mars

It is difficult to anticipate what Mars colonists could produce that will be valuable enough to ship back to Earth. We only have very rough data on the mineral composition of the Martian surface and subsurface, with some potential for extracting metals that are relatively rare on Earth. And some possibilities, such as whether certain manufactured goods or biotech products might be best or only feasible to make in Mars' low-gravity conditions, are almost unknowable in advance. The question is equivalent to asking an early American colonist what goods the United States would best excel in making after the industrial revolution.

As we have proposed for the technical design of the colony, we recommend provisioning pods with equipment for prototyping and experimenting with various means of economic production, including 3D printers and other low-weight equipment. Most of all, colonists would need to be afforded the time to experiment and debate with one another. Discerning the most lucrative economic activities on Mars can only be the result of dedicated hands-on-learning within an environment that fosters scientific learning and small-scale technical innovation.

Democratized Economy
As discussed above, a far more important consideration than attempting to divine the right combination of imports and exports is how to devise an economic system so that it arrives at the optimal mixture. We believe that an economic system with high-levels of democracy are essential for this task, which have the additional benefit of providing great worker buy-in and equal or greater efficiency.

First, it is already well-known on Earth that stark economic inequalities often lead to dangerous levels of political instability. Since we want a Martian colony to be resilient by design, rather than reliant on the use of force, the economic system must self-correct to a state of relative equality. Any Martian citizen falling through the economic cracks not only has their own life put into jeopardy--given that there are no "free" life support systems on Mars--but could be a risk to others. As such, our Martian market is based on small cooperatively owned firms and providing all residents an ownership stake in the colony as a whole. These firms would not only exchange goods and services through a market but also send representatives to economic counsels and co-governed credit unions to coordinate collective infrastructure projects, decide prices for necessary goods and disposal of waste, and set workplace-related regulations and laws.

Cooperative ownership is similar to individual ownership, except ownership rights are vested in workers as a collective rather than individually. Rights are not connected to ownership of shares, but directly to membership in the cooperative. Members have equal say in decisions, and rather than shares which represent the value of the workers portion of the enterprise's assets, each employee has an account to which surplus or loss is allocated periodically. The members of the enterprise can set the initial membership fee to whatever they wish, and that fee sets the initial balance of the account. The accounts are non-transferable, thus protecting from hostile takeover, but can be drawn on to certain limits set by the enterprise members to protect liquidity. Cooperative ownership protects against many of the trade-offs of individual ownership and can expect a much greater degree of equality than any traditional corporation.

A further justification is that a Martian colony cannot buffer a volatile labor market. Everyone capable of working must work. Even the relatively low "natural" unemployment rate that pervades economies based on non-cooperative firms will be too high for Mars. What is a person to do if they cannot work towards the success of the Martian city? We want to avoid a class of residents being supported by all the others, unless they are elderly, sick, disabled, pregnant, or caring for those groups. Everyone must be afforded the ability to labor under good conditions. Cooperative organization ensures a mutual commitment between firms and employees.

Second, a decentralized economy of small-scale firms, while sacrificing economies of scale, will help ensure the resilience of the entire colony. If the colony's oxygen production were located in one area, which suffered an accident or disaster, the entire population would be put at risk. And the physical design of the colony will make sure that needed infrastructure cannot be hoarded by any one group of citizens but is equally available.

Third, by keeping firms small and providing all workers with an ownership stake, we avoid the risk of residents becoming alienated or demoralized. They cannot see themselves as "wage slaves," because they are also proprietors. While there is tolerance on Earth for employees who absent-mindedly run through their work tasks and shirk as much as possible or labor away at "bullshit jobs," [34] there is no room for such waste of human potential on Mars. The economic system needs to maximize the degree to which workers find their labor to be meaningful. And affording all workers the ability to contribute consequentially to how their workplaces are run and to be incentivized to innovate new products and techniques does just that.

Data from plywood cooperatives in the pacific northwest indicates that employees are more likely to sacrifice earnings or wages to save a firm than investors or private owners [35]. The Mondragon group in Spain grew four times faster than comparable corporations during Spain's boom economy. And during the recession, they actually increased investment, growing the company while other corporations cut labor. Cooperatives in former Yugoslavia also tended to favor investment over increasing wages. The evidence clearly demonstrates that worker owned and operated firms are optimal for situations where total "buy-in" is required for success. And it is clear that workers self-educate to a greater extent, rise to meet the challenges presented by self-management, and achieve equivalent efficiencies [36].

Furthermore, a decentralized network of small producers will ensure a rapid pace of generating new innovations and ideas. One of the major challenges for the Martian settlement is to recreate the intelligence and ingenuity of several billion Earthlings with only one million colonists. Our colony will simply be too small and precarious to tolerate monopoly or oligopolies. The barriers of entry cannot be too large, so overly scaled and centralized infrastructure is to be avoided. Just as networks of small-scale Danish wind turbine-builders out innovated NASA [1], so too would a web of artisan manufacturers on Mars working together be expected to figure out what will invariably elude even the smartest Earth-based analyst.

Finally, we want to continually reinforce the skills necessary for residents to productively work through conflicts. As discussed above, organizations that work constantly on the edge of disaster cannot afford members who are unable to work cooperatively. They must embrace the inevitability of conflict and be capable of working through disagreements productively and without irreparably harming social relationships. Acting as a citizen of an enterprise as well as a citizen of a state will improve each citizen's capacity for doing the activities of governing. Organizing the economy through small-scale cooperative firms will ensure that citizens' apprenticeship in democracy will continue on from their schooling years into their adult lives.

COMMUNITY, LEISURE, AND AESTHETICS
A million-person Mars colony poses particular challenges in terms of social organization. The environment is similarly insular and isolated as a commune or Antarctic research station, while facing the hazards and existential threats seen in submarines in wartime, nuclear reactors, and in other extreme environments. Acts of sabotage, or even mere negligence, can fundamentally undermine the survival of the entire group. Paramount in these circumstances is ensuring a thick communitarian society with high levels of social trust [37].

Stable Social Networks
Martian society will need to be arranged very differently than contemporary Western nations, which now face critically low levels of public trust [38] and fanatical polarization [39]. The first feature is ensuring that social ties are dense, multiplex, and systematic. That is, Mars citizens will know most of the other residents in their area, interact with them in multiple facets of daily life (work, leisure, errands, church, governance, etc.), and localities are meaningfully nested into larger forms of organization. Individuals and families from each pod would meet to discuss issues they face, which would inform discussions at the level of the pod-district and eventually the colony as a whole. This classical topology of community ties ensures that citizens do not feel isolated but rather meaningfully engaged into social life at multiple levels.

As most people know, social ties do not maintain themselves. They are reinforced by the interactions and exchanges of everyday life. Even if economic provisioning could be theoretically more efficient if centralized. Far better for sustaining community life is to ensure the existence of small-scale shops and even mutual self-provisioning. If residents grow some of what they eat in collective hydroponic gardens or have their tools repaired by local mechanics, they are put in more frequent contact with each other--cultivating a sense of belonging.

Pleasant "Third Places"
Also vital is affording and encouraging collective leisure and pleasant conversation. Human speech is not a mere vehicle for transmitting information but an embodied mechanism for social bonding [40]. Pods and pod-districts should

contain the equivalent of pubs, cafes, and parks, small social spaces that entice relaxed chatting and gossip [41]. And such spaces must offer something of beauty, whether by its internal architecture and presence of artwork, provision of a window to observe the stars, or by the presence of cultivated plants and animals. In a colony of 1 million residents, there should be a small botanical garden for every 500 residents and countless systems of windows and mirrors for stargazing while also avoiding dangerous cosmic rays.

Group leisure activities would also be plentiful. No doubt some individual weight training may be necessary to ward off the negative effects of low gravity, but the design would include spaces for playing cards, video games, and highly social sporting activities (e.g., bowling, curling).

Moreover, despite the economic pressure to economize on space, the colony should contain a broad mixture of private, semi-private and public space. Packing people in like sardines can work for a six-month shift on a nuclear submarine, but not indefinitely. Colonists will need access to private sleep quarters, no matter how small, and pods should contain the equivalent of a "front porch," a space where colonists could linger to converse with passing neighbors.

Economic and Political Togetherness
The economic arrangement that we have proposed will also aid in binding Mars citizens together. The key is to support not only mutual aid but also market exchanges that bolster social connection. Most people have experienced this when frequenting certain farmers market stalls, cafes and diners, bike shops, or mechanics. By getting to know and bond with the proprietors, tips become larger and small services come without charge. Owners and customers satisfice rather than maximize economic utility in order to maintain a trusting socioeconomic relationship.

Robert Nisbet defined a community as the result of members working alongside one another to solve problems, fulfill common objectives, and jointly write the rules that governed their lives [42]. In order to fully support thick community life on Mars, colonists should not be ruled by a distant Earth-based bureaucracy, but afforded the maximum ability to develop the laws, processes, and norms that will stabilize their society. We have already discussed above how democratic processes will govern the colony. For the relatively mundane conflicts that residents are likely to face, governance should be designed around the principle of subsidiarity. That is, decision making should devolve to the lowest level possible. Coordination across pods and pod-regions would be accomplished by electing representatives for higher level governing bodies.

Managing the larger risks associated with maintaining resilient life support systems and ensuring the integrity of the colony's physical structure will require designing processes according to the principles of a High Reliability Organization, which include rewarding rather than punishing the reporting of mistakes, prioritizing ongoing learning, empowering lower-level workers to halt processes in case of danger, and cultivate a broader culture of excellence [43].

Moral Community

One often overlooked requirement is the need to bind Mars colonists together via some shared understanding of values, identity, and morals. The risk to the colony posed by the potential for polarization and division is too great, for the same environmental buffers and slack that make survival on Earth relatively straightforward are not present on Mars.

This is key for the correct political functioning of the colony, as High Reliability Organizations are "total institutions." While underlying democratic processes ensure success in managing risk, the organization is held together by an unerring commitment to the mission. Aircraft carriers are safe because sailors are socialized into a brother-/sisterhood and an identity that shapes their lives on the ship, a collectivity that is held together via ongoing rituals and mythmaking.

Additionally, studies of communes find that what sustains them are rituals that sacralize the collective commitments made by members, the creation of a sense of "we" that justified their roles within the community and in reference to something bigger than themselves [44], [45], if not other-worldly. A shared notion of a "we" helped these groups weather economic hardship and disputes. Given the harshness of the Martian environment and the daily risk of existential destruction, the early stages of Mars colonization may well require participants brought up with a belief system that designates voyagers as a chosen people on a sacred mission for humanity as a whole. This facet of Martian life may be deemphasized as the technical foundations for the colony become more assured and resilient, but it is unwise to begin without it.

Setting a foundation for a strong shared moral sense of "we", however, *does not* mean the Mars colony resembling an authoritarian cult. Democratic processes are what assure safety within a High Reliability Organization and the legitimacy of a political community. Plus, commune studies find that coercive and illegitimate forms of authority tend to be destabilizing.

In any case, we believe that a strong emphasis on building thick communal life on Mars will help compensate for deficiencies in the pleasantness of the natural environment, limitations that are unlikely to be overcome. More and more research finds that citizens of most developed nations are increasingly isolated, anxious, depressed, and lonely. The proposed Mars colony would be designed to provide a much more fulfilling experience of togetherness than most people currently enjoy, making Martian life, in some ways, better than that on Earth.

CONCLUSION

Our design emphasizes the importance of process in realizing intelligent outcomes. The limits of prediction need not limit future Martian settlement as long as we develop processes that promote colonists to develop their own solutions incrementally, flexibility, and intelligently.

Mars is inhospitable, seemingly bent on ending any human life there. Yet despite this formidable challenge, the sociopolitical conditions on Mars can be set up to enable humans to flourish in aspects of being that are relatively underutilized or inequitably enjoyed on Earth: constructive democratic dialogue, empathic community building, deeply engaging and meaningful work, the ability to substantively co-shape one's own built environment, economy, and society, and the opportunity to develop practical wisdom or *phronesis* in all facets of life. The right socio-political institutions can make life on Mars perhaps even more fulfilling than on Earth.

REFERENCES

[1]   M. Heymann, "Signs of Hubris: The Shaping of Wind Technology Styles in Germany, Denmark, and the United States, 1940-1990," *Technol. Cult.*, vol. 39, no. 4, pp. 641–670, 1998, doi: 10.2307/1215843.

[2]   "Survival! Exploration: Then and Now." NASA and Jamestown Education Module, Sep. 25, 2006, Accessed: Jun. 30, 2020. [Online]. Available: https://www.nasa.gov/pdf/166504main_Survival.pdf.

[3]   Z. W. Greenwood, M. B. Abney, B. R. Brown, E. T. Fox, and C. Stanley, "State of NASA Oxygen Recovery," presented at the International Conference on Environmental Systems, Albuquerque, NM, Jul. 01, 2018, Accessed: Nov. 22, 2020. [Online]. Available: https://ntrs.nasa.gov/citations/20180006356.

[4]   G. Anderson and J. Atkinson, "NASA Announces Awards to Develop Oxygen Recovery Technologies," NASA, Text 17–014, Feb. 2017. Accessed: Nov. 22, 2020. [Online]. Available: http://www.nasa.gov/press-release/nasa-announces-awards-to-develop-oxygen-recovery-technologies-for-future-deep-space.

[5]   C. Junaedi, K. Hawley, D. Walsh, S. Roychoudhury, M. B. Abney, and J. L. Perry, "Compact and Lightweight Sabatier Reactor for Caron Dioxide Reduction," presented at the International Conference on Environmental Systems, Portland, OR, Jul. 17, 2011, Accessed: Nov. 22, 2020. [Online]. Available: https://ntrs.nasa.gov/citations/20120016419.

[6]   P. K. Sharma, D. Rapp, and N. K. Rahotgi, "Methane Pyrolysis and Disposing Off Resulting Carbon," presented at the In Situ Resource Utilization (ISRU III) Technical Interchange Meeting, Denver, CO, Feb. 1999, Accessed: Jun. 30, 2020. [Online]. Available: https://www.lpi.usra.edu/meetings/ISRU-III-99/pdf/8008.pdf.

[7]   NASA, "MOXIE for Scientists," *NASA Science Mars 2020 Mission Perseverance Rover.* https://mars.nasa.gov/mars2020/spacecraft/instruments/moxie/for-scientists/ (accessed Jun. 30, 2020).

[8]   NASA, "Space Technology Game Changing Development The Mars Oxygen ISRU Experiment (MOXIE)." NASA Jet Propulsion Laboratory, Accessed: Jun. 30, 2020. [Online]. Available: https://www.nasa.gov/sites/default/files/atoms/files/fs_moxie_150908.pdf.

[9]   M. Berggren, R. Zubrin, C. Wilson, H. Rose, and S. Carrera, "Mars Aqueous Processing System," in *Mars: Prospective Energy and Material Resources*, V. Badescu, Ed. Berlin, Heidelberg: Springer, 2009, pp. 563–586.

[10]   S. Hoffman, A. Andrews, and K. Watts, ""Mining Water Ice on Mars: An Assessment of ISRU Options in Support of Future Human Missions," Jul. 2016, Accessed: Jun. 30, 2020. [Online]. Available: https://www.nasa.gov/sites/default/files/atoms/files/mars_ice_drilling_assessment_v6_for_public_release.pdf.

[11]   Mayo Clinic Staff, "How much water do you need to stay healthy?," *Mayo Clinic*, Oct. 14, 2020. https://www.mayoclinic.org/healthy-lifestyle/nutrition-and-healthy-eating/in-depth/water/art-20044256 (accessed Jun. 30, 2020).

[12]   C. Brouwer and M. Heibloem, "Crop Water Needs," in *Irrigation Water Management: Irrigation Water Needs*, vol. 3, 3 vols., Rome, Italy: United Nations Food and Agriculture Organization, 1986, p. Part I, Chapter 2.

[13]   "Hydroponics: A Better Way to Grow Food," *U.S. National Parks Service*, May 21, 2018. https://www.nps.gov/articles/hydroponics.htm (accessed Jun. 30, 2020).

[14]   N. Atkinson, "An Inside Look at the Water/Urine Recycling System on the Space Station," *Universe Today*, Apr. 26, 2013. https://www.universetoday.com/101775/an-inside-look-at-the-waterurine-recycling-system-on-the-space-station/ (accessed Jun. 30, 2020).

[15]   "Improving Water Efficiency: Residential Water Recycling," *American Society of Landscape Architects*, 2020. https://www.asla.org/waterrecycling.aspx (accessed Jun. 30, 2020).

[16]   A. Abbud-Madrid *et al.*, "Mars Water In-Situ Resource Utilization (ISRU) Planning (M-WIP) Study," California Institute of Technology, Apr. 22, 2016, Accessed: Jun. 30, 2020. [Online]. Available: https://www.researchgate.net/publication/301614744_Mars_Water_In-Situ_Resource_Utilization_ISRU_Planning_M-WIP_Study.

[17]   J. Kleinhenz, "ISRU Soil Water Extraction: Thermal Challenges," presented at the Thermal and Fluids Analysis Workshop, Galveston, TX, Mar. 29, 2019, Accessed: Nov. 24, 2020. [Online]. Available: https://ntrs.nasa.gov/citations/20190002020.

[18]   K. M. Cannon and D. T. Britt, "Feeding One Million People on Mars," *New Space*, vol. 7, no. 4, pp. 245–254, Aug. 2019, doi: 10.1089/space.2019.0018.

[19]   K. Benke and B. Tomkins, "Future food-production systems: vertical farming and controlled-environment agriculture," *Sustain. Sci. Pract. Policy*, vol. 13, no. 1, pp. 13–26, Jan. 2017, doi: 10.1080/15487733.2017.1394054.

[20]   V. Mendez Perez, "Study of the Sustainibility Issues of Food Production Using Vertical Farm Methods in an Urban Environment within the State of Indiana," Purdue University, 2014.

[21]   V. Sumini and C. T. Mueller, "Structural Challenges for Space Architecture," *Structure Magazine*, pp. 42–45, Dec. 2017.

[22]   D. Schmid, "SwRI Scientists Publish First Radiation Measurements From the Surface of Mars," *Southwest Research Institute*, Dec. 09, 2013. https://www.swri.org/press-release/swri-scientists-publish-first-radiation-measurements-surface-mars (accessed Jun. 30, 2020).

[23]   P. Boston *et al.*, "Human Utilization of Subsurface Extraterrestrial Environments," Complex Systems Research, Inc., Boulder, CO, 2004. Accessed: Jun. 30, 2020. [Online]. Available: http://www.niac.usra.edu/files/studies/final_report/710Boston.pdf.

[24]   "High-Efficiency Solar Cell," *NASA Technology Transfer Program*, 2020. https://technology.nasa.gov/patent/LEW-TOPS-50 (accessed Jun. 30, 2020).

[25]   J. H. Styles, "Nuclear Power Plant Water Usage and Consumption," Mar. 19, 2107. http://large.stanford.edu/courses/2017/ph241/styles2/ (accessed Mar. 30, 2020).

[26]   J. Conrad, "Future nuclear energy policy—the West German enquete commission," *Energy Policy*, vol. 10, no. 3, pp. 244–249, Sep. 1982, doi: 10.1016/0301-4215(82)90103-3.

[27]   E. J. Woodhouse, *The Future of Technological Civilization*. University Readers, 2014.

[28]   A. Crawford, "What I Saw When I Participated in One of the Truest Forms of Democracy," *Slate Magazine*, May 22, 2013. https://slate.com/news-and-politics/2013/05/new-england-town-halls-these-experiments-in-direct-democracy-do-a-far-better-job-than-congress.html (accessed Jun. 30, 2020).

[29]   R. A. Dahl, *Dilemmas of Pluralist Democracy: Autonomy vs. Control*. Yale University Press, 1983.

[30]   J. Drydyk, "When is Development More Democratic?," *J. Hum. Dev.*, vol. 6, no. 2, pp. 247–267, Jul. 2005, doi: 10.1080/14649880500120566.

[31]   P. Gray and D. Chanoff, "Democratic Schooling: What Happens to Young People Who Have Charge of Their Own Education?," *Am. J. Educ.*, vol. 94, no. 2, pp. 182–213, Feb. 1986, doi: 10.1086/443842.

[32]   C. E. Lindblom, *The Market System: What it Is, how it Works, and what to Make of it*. Yale University Press, 2002.

[33]   R. Heidmann and P. Brisson, "An economic model for a martian colony of a thousand people," *Planete Mars*, Sep. 26, 2016. https://planete-mars.com/an-economic-model-for-a-martian-colony-of-a-thousand-people/ (accessed Jun. 30, 2020).

[34]   D. Graeber, *Bullshit Jobs: A Theory*. Simon and Schuster, 2019.

[35]   D. Zwerdling, *Workplace Democracy: A Guide to Workplace Ownership, Participation & Self-management Experiments in the United States & Europe*. Harper & Row, 1980.

[36]   T. Malleson, *After Occupy: Economic Democracy for the 21st Century*. Oxford University Press, 2014.

[37]   T. Dotson, *Technically Together: Reconstructing Community in a Networked World*. MIT Press, 2017.

[38]   "Public Trust in Government: 1958-2019," *Pew Research Center - U.S. Politics & Policy*, Apr. 11, 2019. https://www.pewresearch.org/politics/2019/04/11/public-trust-in-government-1958-2019/ (accessed Jun. 30, 2020).

[39]   "Partisan Antipathy: More Intense, More Personal," *Pew Research Center - U.S. Politics & Policy*, Oct. 10, 2019. https://www.pewresearch.org/politics/2019/10/10/partisan-antipathy-more-intense-more-personal/ (accessed Jun. 30, 2020).

[40]   J. L. Locke, *The De-voicing of Society: Why We Don't Talk to Each Other Anymore*. Simon & Schuster, 1998.

[41]   R. Oldenburg, *The Great Good Place: Cafes, Coffee Shops, Bookstores, Bars, Hair Salons, and Other Hangouts at the Heart of a Community*. Hachette Books, 1999.

[42]   R. Nisbet, *The Quest for Community*. Open Road Media, 2014.

[43]   S. D. Sagan, *The Limits of Safety: Organizations, Accidents, and Nuclear Weapons*. Princeton University Press, 1993.

[44]   R. Sosis and E. R. Bressler, "Cooperation and Commune Longevity: A Test of the Costly Signaling Theory of Religion," *Cross-Cult. Res.*, vol. 37, no. 2, pp. 211–239, May 2003, doi: 10.1177/1069397103037002003.

[45]   S. Vaisey, "Structure, Culture, and Community: The Search for Belonging in 50 Urban Communes:," *Am. Sociol. Rev.*, Jun. 2016, doi: 10.1177/000312240707200601.

12: MARTIAN COSMOCITY

Firmansyah Bachtiar
Astrid Hapsari Rahardjo
Dedi Burhanudin
Egi Kurniawan
Reza Rezaie
Architecta-Tanri Abeng University, Indonesia
firmansyah.bachtiar@gmail.com

DREAMING MARS

The success of SpaceX as the first private venture to send a manned-mission the International Space Station (ISS) is seen as an important milestone in realizing what has been said as the unfulfilled dream to send human's mission to Mars and build a colony that will eventually reach one million population of human existence on the Red Planet.[1] Many ideas on how a one thousand population of Mars colony have been proposed through a variety of competition in the past years, including the Mars Colony idea competition held by the Mars society in 2019. Thus, the Star City concept– a city that is conceptually built in a crater was born.[2]

The idea to develop a city as a dwelling place for a population of 1,000 times greater than the early colony will take a long time. This vision can be seen as a dream to make Mars into the New Earth, which has been imagined through the work of Kim Stanley Robinson in his award-winning science-fiction novel, The Trilogy of Red Mars, Green Mars, Blue Mars. It is through this idea, that in the Red Mars timeline that the city of one million population is being projected. Most inhabitants in this City are originated from Earth. They come from multiple country and carry along a considerably diverse cultural background. Thus, the city on Mars will be called the Martian Cosmocity as it will be a place of representation of people with a variety of backgrounds, living in one universal metropolis.

History has shown the ability of humans in the ancient time to build many cities and inhabitations under extremely hostile conditions. The record of their success in adapting and making their places can be found in the patterns found in many ancient cities on Earth. Most of the ancient cities were established in a circular form, such as Baghdad, Mari, and (presumably) Atlantis. The circular city form is beneficial in term of cutting travel times, which in turn increasing the effectiveness in terms of city movement, distribution, and arrangement, and most significantly providing a 360-degree view angle to identify possible external threats. The Martian Cosmocity will be developed based on several circular subcities/ colonies with different themes or key functions to make a reasonably easy setup of the Cosmocity's facilities and infrastructures. The early development strategy of the Martian Cosmocity is built by taking advantage of the planitia, vallis, crater and lava tube as the base canvas to start the colonies so that the environment will automatically be protected from the risk of radiation.

Humans will start a new life on Mars, but in some ways, will still be carrying on the same routines and quality habits from their lives on Earth. Through time, new routines and habits will take place as a part of their adaptation process. Though this will never mean that some of the way human live on Earth will be long forgotten. In terms of population, not only the Mars colonization will potentially help reduce the rising number the human population on Earth, but it will also open up the possibility of the many technology inventions that helps the livelihood of the people on Earth. In the beginning, some of the first commodities on Mars will be produced on Earth by way of reusing and manufacturing wastes into usable materials to be used on Mars. Every non-organic refuse and waste on Earth along with the space debris from previous rocket launches on the Earth orbit are to be recycled and manufactured into construction materials that will be used to build the colonies structure, shelters, and infrastructures of the Cosmocity.

CONNECTING MARS

International Alliances
Since one of the objectives of the Cosmocity is to having a diverse cultural life under one city, this mission needs an organizational body to oversee and manage the collaboration amongst many nations that make contribution to this mission. The United Nation will bear the role in developing a separate organization called the United Nations Office for the Outer Space Affairs (UNOOSA). This organization will act as a liaison between the government space agencies, the private sectors and the civil society. In its development, there should be a special board council, the United Nation Space Exploration Council – UNSEC, whose members will include representatives of the pioneering countries in the space missions. These members will hold a special right of order. When Cosmocity is formally inhabited by one million people there will be an elected governmental body. Then, there will be the Earth and Mars Embassies to act as a bridge in facilitating the administrative relationships between the government of Mars and the government of Earth.

Citizen Recruitment
Ahead of the launch of Citizen Zero, it is crucial to have a careful selection process in place to determine the first sixty people to be sent off to Mars. These individuals will be considered as the first colonist and will open the way for the next inhabitants. The recruitment process of Citizen Zero will take about five years prior to the departure, which will include vigorous training and preparations. The main criteria of Citizen Zero as the Cosmocity pioneer will be based on many aspects including physical strength, gender, country of origins, and expertise/ competencies.

The diversity of country of origins will help to determine the seeds of diversity of the early colony. Culture blending and assimilation is very much expected to provide a way for the new cultural embryo to emerge on Mars. At least a representation of every major cultural group on Earth will be a part of the first colony. Competency in various fields will also be the main requirement with certain disciplines being highly preferred in order to help develop the base colony. After the send-off of the first colony, periodical send-off will be carried out for the next colonies (Citizen X) to further broaden the diversity of age, culture and expertise. Thus, there will be more representations of other cultures included and more Professionals needed such as traders, teachers, athletes, artists, bankers, and so forth.

Launch Scenario

It is estimated that 80% (800,000) of the first generation of Mars population will originate from Earth whereas the rest (200,000) will be born in Mars. Therefore, it will take around 20 years to reach the number of 800,000 inhabitants, and another 30 years to reach the total population of one million in Cosmocity. One of the way to achieve this is to expand the existing space ports on Earth. These space ports will also be supported with several facilities in the surrounding cities such as space tourism, space research center, astronaut training grounds, and so forth) and will take place in twelve locations in ten different countries across five continents.

The launch to Mars will be limited to every 26 months using the Mars Direct Scenario as the distance between Earth and Mars is at their closest.[3] Each departure period will take thirty days from twelve ports. This will see the maximum number of three launches per day in each port. Hence, the maximum number of launches will be around 1.000 spaceship launches during each launch period.[4] That, in total, will add the average amount of 108,000 colonists per launch period starting from the eighth year.

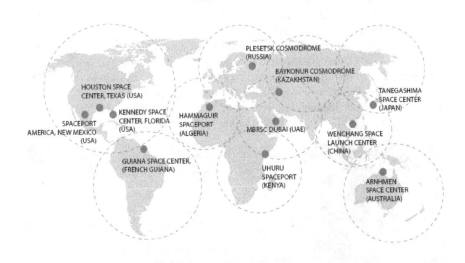

Launch Port		Launch Windows (Colonist)										
	year	1	2	4	6	8	20	12	14	16	18	20
	fleet/port	1	2	30	60	90	90	90	90	90	90	90
Kennedy Space Center (USA)		60	200	3.000	6.000	9.000	9.000	9.000	9.000	9.000	9.000	9.000
Space Port America (USA)		-	-	3.000	6.000	9.000	9.000	9.000	9.000	9.000	9.000	9.000
Houston Space Center (USA)		-	-	3.000	6.000	9.000	9.000	9.000	9.000	9.000	9.000	9.000
Guiana Space Center (French Guiana)		-	-	3.000	6.000	9.000	9.000	9.000	9.000	9.000	9.000	9.000
Tanegashima Space Center (Japan)		-	-	-	6.000	9.000	9.000	9.000	9.000	9.000	9.000	9.000
Wenchang Space Launch Center (PR of China)		-	-	3.000	6.000	9.000	9.000	9.000	9.000	9.000	9.000	9.000
Arnhem Space Center (Australia)		-	-	-	6.000	9.000	9.000	9.000	9.000	9.000	9.000	9.000
Baykonur Cosmodrome (Kazakhstan)		-	-	3.000	6.000	9.000	9.000	9.000	9.000	9.000	9.000	9.000
Plesetsk Cosmodrome (Russia)		-	-	-	6.000	9.000	9.000	9.000	9.000	9.000	9.000	9.000
Hannamuir Space Port (Algeria)		-	-	-	6.000	9.000	9.000	9.000	9.000	9.000	9.000	9.000
MSBRC Dubai (UEA)		-	-	-	6.000	9.000	9.000	9.000	9.000	9.000	9.000	9.000
Uhuru Space Center (Kenya)		-	-	-	6.000	9.000	9.000	9.000	9.000	9.000	9.000	9.000
		60	200	18.000	72.000	108.000	108.000	108.000	108.000	108.000	108.000	108.000

Total Earth Colonist: 846.260
Mars Born Population (after 50 years): 200.000
Total Population: 1.046.260

STARTING MARS

Site Selection

The site selection process for the development for Cosmocity takes several aspects into consideration, such as (1) the availability of ice deposit, which contains extractable elements like water and oxygen to support and maintain the survival of living organism, (2) the existing gas and minerals to be further mined and processed so they will be ready to use, (3) the geological terrains that will provide protections from strong wind gusts and sand storms, especially for the early inhabitants, (4) landing safety issues, and (5) easy access for further explorations. Based on these points, added with the topographical and geological conditions and the ice deposit mapping analysis, the Gusev Crater area (14.5°S, 175.4°E, 166 km in diameter) in Aeolis Quadrant has been selected as the location for the first Martian Cosmocity. The starting point of the development is the Mars Base Alpha, which is located in the Thira Crater (14.47°S, 175.98°W, 22 km in diameter).

Pre Colony Mission

Approximately three years prior to the Citizen Zero send-off, a pre-colony mission will be dispatched to prepare a living place for the Citizen Zero members. This mission will incorporate several unmanned crew launches to carry the necessary gears and equipment, which will include advanced robotic technologies, rovers, 3D printers, drones, ISRU and habitat modules. The construction process will therefore be controlled remotely from Earth. This is done in such method so that there will be basic infrastructures for energy resources, shelter bases, and water and oxygen reserves readily available for them. The first landing will be done in the inner part of Thira Crater. Afterwards, site clearing will be performed and followed by the construction of the Mars Base Alpha and its initial infrastructure. Along with that, the rovers and the drones will be dispatched remotely to explore other areas for finding the locations of minerals, ice deposit, the regolith, and also for installing solar panels to generate energy from the sun.

Mars Base Alpha

The Mars Base Alpha is the location for the first colony in Mars, which will be the embryo of the first development of Gusev Cosmocity. The design concept will adopt the Foster + Partner concept of the Mars Habitat[5], which uses a concept of manufactured regolith shelter by means of 3D-printing to protect the inflatable modules for human living vessels, greenhouses, and research stations. One habitat module will be suitable for six colonists. So, there will be a total number of 10 habitat modules formed in the early phase of the Mars Base Alpha.

Colony

The expansion of Mars Base Alpha will be started by establishing new landing pods for further Citizen X arrivals. The next colonies will begin around the landing pods in the form of landed and underground housings. These housings will accommodate approximately 260 people. Further colony expansion and development will be allocated at the Thira Rim Crater. The Thira Crater Rim will be drilled to accommodate the next expansion for colonies. It is planned that the Rim Colony area would be ready to accommodate nearly 6.500 inhabitants in the sixth year. The first colony would then join the succeeding colony in the Rim Colony. Ten habitat modules in the Mars Base Alpha will be transformed into a public facility module to support the Rim Colony. This type of colony development is actually adopted partially from the Star-City design concept, which utilizes the rim crater wall as a means of protection or a shield in the early colony phase until there is a more advanced technology to build a giant structural dome.

Subcity

The arrival of the next colonies will eventually add the population number to 100,000 within the next years, which will form a wider hierarchy to the city structures and functions. The City area will ultimately reach the Thira Subcity, which will be under the Thira Crater. The Subcity will become the Municipal Subcity, where the Government and City Administration for the whole Cosmocity takes place.

The former area of the Mars Base Alpha will be developed into The Old City. This area will be regarded as the historic area to preserve the record of the first colony footprint on Mars. The expansion of the Subcity can be done in a form of an independent Colony Cluster, which is completely separated from the Subcity. This cluster will become a suburban area to be used as residential expansion and other functions necessary for the Subcity

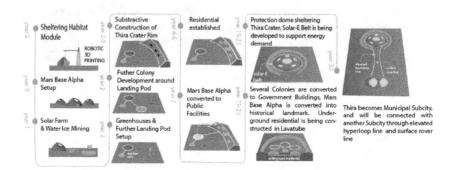

PLANNING MARS

Mars Millenium

The macro development period on Mars will be conceptualized as the Mars Millenium as this is estimated to take place within the next 1,000 years. The concept is divided into three phases, as follows:

1. The Red Mars phase. This period will be carried out in the first 200 years. The emphasis on this phase is the development of Cosmocities on the whole planet.
2. The Green Mars phase. This period follows the previous phase into the 600th year. The phase emphasizes on the terraforming the surface of the planet so that the cycle of new ecosystem can start to live.
3. The Blue Mars phase. This last period, which runs since the last year of the previous phase to the 1.000th year, will see a formation of a more livable climate on Mars for humans.

This plan is made such to prevent over-exploitation of the planet (unlike that of Earth). In order to do this, a Planetary Zoning Plan is developed. This Plan will map out not only the areas that are suitable for living but also the areas that will be preserved for terraforming in the Green Mars and Blue Mars periods. This plan will help determine which areas to be developed further into rivers, lakes, sea and hills in the Blue Mars period.

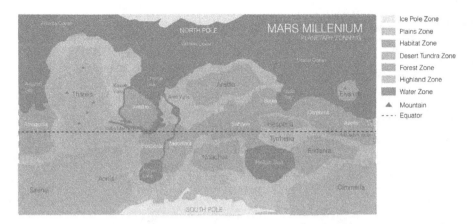

Cosmocity

The Cosmocity is formed through interconnectivity amongst several Subcities. Each Subcity will hold different themes or key functions, such as government, agriculture, energy, commercial, and so forth. The connection will therefore allow the Cosmocity to function as a city in its entirety. The diversity from one Cosmocity to another does not only derive through its functions, but it is also shown through the population background profiles and the special commodity carried by the place. The Cosmocity will have a total population of 1,000.000. The ratio of the development expansion will reach maximum about 20% of the whole City area.

The Gusev Cosmocity will be generated through the expansion of several Subcities in the Mars Base Alpha in Thira Crater. These early Subcities will have the Municipal Subcity as the government center in the early colonization period of Mars. The distribution of the Colony Clusters in the Gusev Crater will be organized for the need of exploration and the mining the minerals found on Mars. This Colony will become the embryo of the next Subcities, which will hold different roles and functions. Within 50 years, there will be several Subcities developed to entirely form the Gusev Cosmocity in the Gusev Crater.

The development proses of the Gusev Cosmocity is based on the infrastructure grid divisions that divide the Crater surface into several urban quadrants, each with the area of 4 km^2. Each quadrant line will be used to connect energy resources and the Cosmocity utility system toward each Colony Cluster and Subcity. The infrastructure of the quadrant will also be used as a way to control the development expansion. Should there be further expansions from the quadrants, they will be treated as satellite developments and therefore, it will only be allowed to be done in the next "empty" quadrants. This satellite Colony Cluster can only be inhabited by 1,000 to 2,000 people per colony and it will be completed with public and basic commercial facilities. If an expansion is initiated, then it will only be allowed to expand to a maximum three different quadrants within the closest range. Aside from the satellite Colony Cluster, the development at the Crater Wall will be allocated to house up to 300,000 inhabitants. The Rim City will be built by taking advantage of the crater cliffs as to protect the inhabitants from any possible radiations. The housing module will be developed underground while completed with basic public facilities.

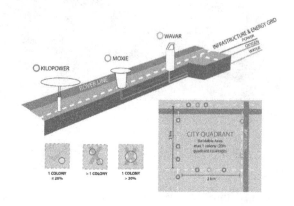

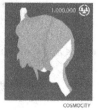

MARS BASE ALPHA   SUBCITIES   COSMOCITY

City Structure	Supporting Facilities	Population/ Inhabitant (estimated)	Area in km² (estimated)
Municipal Subcity	residential (20%), commercial (10%), public facilities (10%), government (50%), Social Eden (5%), farm (5%)	100.000	433,52
Agriculture Subcity	residential (20%), commercial (5%), public facilities (5%), farming (60%), Social Eden (5%), farm (5%)	50.000	107,51

City Structure	Supporting Facilities	Population/ Inhabitant (estimated)	Area in km² (estimated)
R & D Subcity	residential (15%), commercial (5%), public facilities (5%), research (30%), higher education (25%), training academy (10%), Social Eden (5%), farm (5%)	50.000	104,11
Central Hub Subcity	residential (15%), public facilities (10%), trade (30%), office (20%), entertainment (5%), Social Eden (15%), farm (5%)	200.000	865,14
Leisure Subcity	residential (15%), commercial (15%), public facilities (5%), recreation (40%), sport & culture (15%), Social Eden (5%), farm (5%)	50.000	134,09
Industrial Subcity	residential (10%), commercial (5%), public facilities (5%), industry (40%), warehouse & manufacture (35%), Social Eden (5%), farm (5%)	100.000	63,45
Trans Subcity	transit compound & residential (15%), commercial (10%), public facilities (10%), launch port (30%), space elevator (10%), hyperloop hub (20%), Social Eden (5%)	50.000	157,06
Energy Subcity	residential (10%), commercial (5%), public facilities (5%), energy reactor (40%), utility (30%), Social Eden (5%), farm (5%)	50.000	85,45
Colony Cluster (satelit Subcity)	residential/industrial 75%), commercial (5%), public facilities (5%), Social Eden (10%), farm (5%)	50.000	350
Rim City	residential (75%), commercial (5%), public	300.000	300

City Structure	Supporting Facilities	Population/ Inhabitant (estimated)	Area in km² (estimated)
	facilities (5%), Social Eden (10%), farm (5%)		
City Quadrant	infrastructure & energy line (5%), open terrain (90%)	-	25.955
	Total	1.000.000	28.556
	Built Environment Coverage	2.600,33 km2 (9,1%)	
	Future Development Land	49.110,87 km2 (10,9%)	

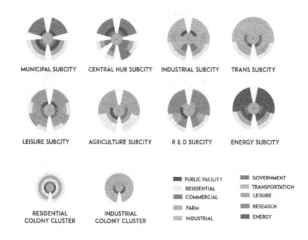

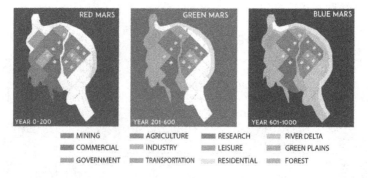

Cosmocity will transform into forest and water areas. The quarry region will be converted into conservation woodlands and so will the Husband Hills, which will be planted with trees that have been genetically modified to adapt to the altered climate on Mars. A sea will be formed in Zephyria Mensae and Apollinaris Sucii in the Blue Mars era, with the outflow channels directed pass the Thira Crater to liven the river path. This will form the Gusev Crater into a waterfront city that stretches along the Ma'adim Vallis.

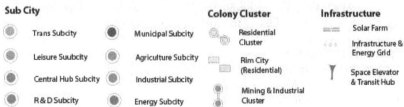

Sub City

- Trans Subcity
- Leisure Suubcity
- Central Hub Subcity
- R & D Subcity
- Municipal Subcity
- Agriculture Subcity
- Industrial Subcity
- Energy Subcity

Colony Cluster

- Residential Cluster
- Rim City (Residential)
- Mining & Industrial Cluster

Infrastructure

- Solar Farm
- Infrastructure & Energy Grid
- Space Elevator & Transit Hub

SURVIVING MARS

Oxygen

In the beginning, oxygen for the early phase of Mars Base Alpha will be produced by way of ISRU (In Situ Resources Utilization) through the activation of solid oxide electrolysis reaction to extract oxygen from carbon dioxide in the atmosphere through MOXIE (Mars Oxygen In-Situ Resources Utilization) technology. At first the MOXIE technology is sent along with the pre-colony mission. This mission will also contain the extracted oxygen for the first Colony in the Thira Crater. Furthermore, the MOXIE technology will get expanded into a larger scale oxygen extraction facility and connected to the energy grid and nuclear reactor to cover the City's needs. The outcome will be distributed to every Colony and Subcity in Gusev Cosmocity. Afterwards, this technology will slowly be alternated naturally through biological means by way of using microalgae that is being cultivated in every Farm zone in Subcities. In the Green Mars phase, the microalgae will become an integrated part of the Subcity and Colony dome structure. Also, with the increase of atmospheric temperature triggered by terraforming activities, microalgae and moss will grow on the Mars surface, which will make them as the base stimulants for local (genetically-modified) plants and vegetation to grow on their own. Next, the empty quadrants that are formed through the previous energy grid will be used as green parks and plains. In the Blue Mars phase, the parks and plains will grow into forests, which will produce more oxygen at a higher rate to match the oxygen level on Earth. This circumstance will make Mars become more habitable.

Protection

The Colony protection will mainly be against the atmospheric condition on Mars such as the solar flare, the cosmic rays, and the dust storm. In order to provide protection, the habitat module structure and materials will be covered with anti-radiation shield. In Mars Base Alpha, the inflatable module will be fended with the Martian concrete and brick, which will be formed from the regolith using both the 3D printing technology and the shield aerogel technology brought from Earth. In the next years, the effort of creating building materials will be by using a more advanced technology. The raw materials such as sulfur, iron-oxide, and silica glass will be mined and extracted on Mars to create construction materials such as glass and metals. Those materials will be used to make the shelter structures for the Subcities. A one-meter water or ice sheath will be combined with the dome surface for means of protection from radiation. The water can also be used for the aquaponic culture within the dome shelter.

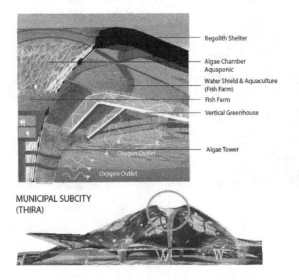

MUNICIPAL SUBCITY
(THIRA)

Food Supply

The food supply for the first Colony will depend 80% on dry food that has been brought from Earth. The rest will be produced through the space vegetable technology during the journey from Earth to Mars. When there is sufficient oxygen and water, farming will get started using the aquaponic technology. This technology is employed in the Agriculture Subcity as the center of food production, and locally in each Subcity. The Agriculture Subcity will also become the research center for Agriculture affairs as well as the place to nurture nutrition needed for the regolith soil using the land terracing toward the underground. The Agriculture Subcity will also distribute the seeding materials to every smaller farming zone in other Subcity. In terms of protein needs, the early phase will use plant-based proteins as well as use an aquaculture technology for fish farming. The fish used for this will be the ones that need less oxygen, such as cat fish and snake-head fish. Not only that they are more efficient because of the less oxygen required but also, they need less space to live in (one fish require only about 1 liter of water in volume). As the transportation technology gets more advanced, transportation and distributions of small farming animals for protein consumptions will be conducted between Earth and Mars. Other animals such as insects will be transported to create a balanced cycle of life in the Subcities. The space cargo mechanism used for this transportation will simulate the 1g environment to suit the adaptation process of the animals. Once they arrive, they will be genetically procreated in the Agriculture Subcity. During the Green Mars phase, the availability of plants and farming animals will be more wholesome. But the breeding process will still be done under a separated and protected dome as a safety measurement. It is in the Blue Mars period, where the climate has become more adaptable, the animal farming can be done outdoors in several quadrants outside the Subcities.

Water

Mars already has water available in many forms such as gas in the atmosphere, regolith, permafrost, and as ice deposits under the planet surface. The presence of ice deposit can also be seen on the Mars pole (JD Williams et all, 1999). The water in the frozen form can be found under the Mars surface, especially around the crater areas (Grecius Tony, 2019). It is based on these findings that the Gusev Crater was chosen as the location to start a life on Mars. It is easier to thaw and extract the oxygen and use the water and channel it to be used by the Colony when it is closer to the place of the source. The water extraction in the early stage of the Colony will be processed through the Water Vapor Absorption Reactor (WAVAR) technology. This technology will take advantage of the CO2 already existing on Mars as a part of the elements that creates water. Water will also be produced by melting the ice deposit on the Mars pole by using the satellite expansions, combined with solar reflectors, aimed directly to the pole. This is done in a safe distance from the Cosmocities as to eschew any potential of radiation from potentially contaminating the inhabited areas. In the long run, water will be provided through terraforming since terraforming also increases the atmospheric temperature. This will encourage the forming of clouds and therefore creates rain in the Green Mars period. The rain water will then be collected in the crater to create lakes, which will become the water supply for the next lives during the Blue Mars period.

MANAGING MARS

Energy & Information Technology

Energy will be generated to meet the requirement of the colonist by taking advantage of the sun to generate solar energy. Solar panels will be installed in the Subcity scale (around the energy belt location with the Subcity) and in Cosmocity scale (as a solar farm). As the sun does have its limitation in providing energy on Mars, nuclear reactors will be built as a means of alternative energy production. The reactors will be built as Kilopower, which will be installed along the energy & infrastructure grid in Gusev Cosmocity. A larger scale reactor will be developed in the Energy Subcity. This reactor will be used as the main electrical energy supply in Gusev Cosmocity. Further in the Green and Blue Mars phases, it is expected that the climatic condition will be more stable and adaptable. Therefore, wind turbines will be used as an alternative backup for energy resources. Information technology will be applied to provide communication relay and internet connection on Mars as well as for communication between Mars and Earth. With the latter, the operation will be connected to relay satellites on the orbital paths between Mars and Earth.

Mobility

The mobility in the Mars Colony in the early phase will be depending entirely on rovers. Afterwards, all transportation mode will be based on the hyperloop and maglev train. This will then see Rover used only as a vehicle designated for cargo and logistic distribution between Subcities and Cosmocities. The Mars-Line connection will be elevated from the ground and the path will also be in integration with the Cosmocity's utility ducts. The path used for the Rover will be built right under the elevated hyperloop line. This path will be developed into Highway in the future. There will be personal/ private vehicles allowed (as electrical rover) in the future, but this will only be allowed when the population number of the Gusev Cosmocity reaches about 500,000, which by then the use of public transportation will have become saturated. The hyperloop (Mars-line) will be interconnected with Mars X Maglev Train in Transit Subcity to serve connection with other Cosmocities. Airship transportation will also be provided and it will be centered in the Transit Subcity. The facility will include a launch pod and runway for planetary ship (between Cosmocities). A Space Elevator to replace the surface landing pod will connect Gusev Cosmocity with Phobos Space Terminal in Mars orbit. The Phobos Space Terminal will circle around the Phobos and its development is not only to make the transportation of people and cargo to Earth/Mars rather easier but also to reduce the load and friction caused by landing onto the surface of Mars.

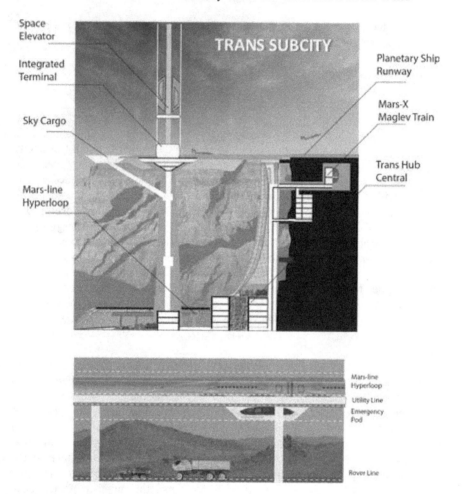

Waste Management

While waste is inevitable, it is the management that needs to be properly addressed as the living condition on Mars is completely different from those of Earth's. The waste processing and management for Mars Colonies should be made effortless by way of carrying recyclable and reusable materials from Earth. This should be done since the first day of launch. All recyclable and reusable materials should be researched and determined prior the first departure as to ensure that the material characteristics can be remanufactured or reprocessed to meet the needs of the Colony for the foreseeable future. With that in mind, any non-organic waste from the Earth must be made into a type of light waste block/sheet module for easy transportation to Mars. Aside from that, the space debris in the Earth orbit can also be collected to be reused and remanufactured on the Moon Recycling Station before it will be transported to Mars.

The dry food packages and containers and other imported products that are brought from Earth into the ship by the Colony members will be repurposed through thermal repolymerization process. Plastic and metal wastes will be reprocessed into furniture materials on Mars. Other types of waste will be collected and incinerated to generate CO_2 to help raise the Mars' atmospheric temperature. Any organic waste such as human waste will be processed and utilized as fertilizers for the Farming Zone in the Subcity, as fuels (methane-based fuel), and as bioreactors for algae (derived from urine).

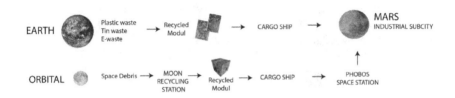

DEVELOPING MARS

Education
Education system on Mars will adopt that of on Earth. The curriculum however, will emphasize more heavily on the fields of engineering, applied sciences, medical, law, architecture, biology and agriculture, and social sciences.

The elementary to high school education facilities will be located in the public zone in every Subcity. However, there will be a special curriculum designated to each Subcity according to each of the Subcity role within the Cosmocity, for example the education in the Agriculture Subcity will stress on biology, agriculture, and farming sciences. Kindergartens will be treated differently due of the needs of the children as well as its different running hours. Therefore, kindergartens will be located in each Colony and integrated with the residential zones.

Higher education facilities (colleges and universities) will be centered in the R & D Subcity. This means that every educational activity on Mars will be paired with research to develop further and better life on Mars. The higher education program will prioritize on the fields that are directly beneficial to the reformation and adaptation toward life on Mars such as aeronautics, robotic technology, biomedical, construction, artificial intelligence, economics, geology and mining, and agriculture and farming. Other programs will be related to law, pedagogy and kinesiology. There will also be Skill Academy that will be located in the R & D Subcity.

Housing
Residential facility in Mars will have several typologies. In the early colony phase, the inhabitants will use inflatable module which will be protected by the regolith structure. As there will be more colonists arrive on Mars, the residential provision

will be developed further underground by using the existing lava tube space or digging underground. Construction will be done by latching to the cliff structure of the crater rim. The housing module will have a single living unit typology and family living unit typology, which can easily be expanded when required. The residential zone in every Subcity will cover about 10-20% of the whole Subcity area. There will be a priority list of who can live in the Subcity with the highest priority given to the workers and their families whose work directly related to the which role the Subcity plays in the whole Cosmocity scheme. Housing facilities will also be expanded through the independent Colony Clusters (the satellite clusters). This will be supported with the underground housing within the shelter dome and connected with one of the Subcity. A greater housing area will be allocated in the Rim City (Thira Crater), which will have a variety of residential typology such as owned housings, rental housings, and hotels.

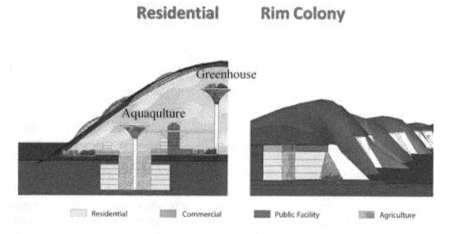

Tourism

Tourism business will be developed in Gusev Cosmocity after the population reaches a number above 100,000 colonists and will be centered in the Leisure Subcity. The facilities will be focused on the sports and recreational elements such as aquatic pools, sports field and stadium as well as conservatory halls, amusement parts, interactive museum, and virtual reality studio. The latest will show a simulation on life on Earth as well as a visualization of the Future of Mars. During the Red Mars phase, all tourism activities will be emphasized in a smaller scale (within the Cosmocity) and will be carried out under a protected shelter. In a micro scale, every Subcity will be completed with recreational neighborhood around the residential zone. There will also be a Subcity park in the form of Social Eden (a garden combined with a social interactive and entertainment facility) located in every Subcity center. Interplanetary tourism will be developed in the Green and Blue Mars phases. This is when the space ship technology can take tourists and visitors from Earth to Mars. The planned prominent tourism spots will be Vallis Marineris, Olympus Mons, Ma'adim Vallis, and Schiaparelli's Crater.

Commercial

The commercial facility developed in Mars will include offices, shops and entertainment. These developments will be allocated started from the Subcity level to the Cosmocity. At the Subcity level, the commercial facilities comprise of convenient stores, restaurants, bar and services. Those will be located in its own commercial zone. In the urban scale, the business and commercial center will be developed in the Central Hub Subcity and it will be built as rental offices, large retails and wholesales, and services such as banks, hotels, and commercial housings.

Industry & Manufacture

Industry in Mars will be based on the process and production of minerals found on Mars such as deuterium, silica, iron, and so forth, and the development of robotic technologies. The industrial Subcity will be located in near proximity from the mining and excavation areas for a relatively unproblematic material transportation to the processing facilities. Aside from the industrial facility, the Industrial Subcity will also be completed with manufacturing facility and warehouses. This area will also be in close proximity to the living space of the industrial workers. The industrial functions will also be developed in the smaller Industrial Colony Cluster to accelerate productions in the early phases of the colony arrivals. It is expected that the industrial activities will be well distributed in over Cosmocities during the Green Mars phase so that the early industrial area in the Colony Cluster can then be altered into a green plain

Healthcare

In the Colony level, the healthcare facilities will be developed as a medical bay wih several medical staff to treat minor illnesses, injuries as well as the first-responder to the local and immediate communities. A public and special-care hospital (such as maternity treatment) will be placed in the public zone in every Subcity. As for the special-case hospital and rehabilitation center will be located in the Commercial Subcity. The need for therapy and biological rehabilitation due to gravitational adaptation process will be accommodated with several Earth-gravity treatment centers, which are located in the underground structure in the Central & R & D Subcity. Every Colonist will have a free medical access to any healthcare facilities up to the Subcity levels. All medical records will be integrated with the Mars ID system, which will be injected under the skin. This will help monitor the health condition of every colonist as well as a way to prevent viral and bacterial pandemic. However, all shelters in the Subcity and Colony will be integrated with a disinfecting piping line to be used when required.

HUMANIZING MARS

Social Life

The social life on Mars is still founded around how it is run on Earth, where background diversity of the colonists will be managed so as to blend and be well-distributed in several Subcity. The objective is to prevent a sociocultural domination blockings. With the diversity of backgrounds then communication will use English as the already universally recognized language on Earth. The ethnic and cultural communities that grew on Earth will flourish from the colony level. This way, the culture and language brought from Earth will assimilate and eventually shape a new type of culture on Mars. Every colonist will gain a digital ID in the form of an injected mini card that is integrated with the City central data system. As an incentive, the pioneering colonist will also receive a life-time benefit and a given living property ownership.

Inter-colony interaction will be conducted in open social and ecological areas. In the colony scale, such open space will be provided in the neighborhood parks completed with garden and social-interacting areas. Whereas in the Subcity scale, the open space will be called Social Eden, a multifunction park and garden that is integrated with public and commercial facilities. This Social Eden will also be encrusted with oxygen-producing plants. The Social Eden Area will also function as a public transportation stop such as the hyperloop Mars-Line that will pass by every Subcity.

Government
The establishment of government will be developed in within the Cosmocity scale in form of Municipal Council. The counselors will be voted directly by the colonist. During the early phase, the establishment of the Municipal Council will still coordinate their policy making with the United Nations Space Exploration Council (UNSEC) on Earth. When the colonist population reaches one million in number, the government in Cosmocity will be stated as independent. The Council will have representations from each Subcity. The Gusev Cosmocity government will be located in the Municipal Subcity (Thira). In the time when there are more Cosmocities on Mars, the government body will adopt the early Municipal Council in a larger scale. Here, there will be political parties that represents every

Cosmocity. Each party will have their representing councilors. An election will be held every 5 years and all colonist of the age of above 18 years old will hold a voting right. Voting will be conducted online through their Mars ID.

In terms of law and justice along with the enforcements, there will be a Municipal Justice Court that will oversee violations against the law on the Cosmocity level. Every law offender will be faced with various sentences according to the type of violation, starting from city imprisonment located in the underground area around Husband Hills to extradition to Earth. For any high crime or violations starting in the Green Mars phase, there will be an Orbital Prison in Phobos where prisoners will be transported from Cosmocity using the Space Elevator.

FINANCING MARS

The Mars Living Mission will see the development of the first city on Mars before there will be more cities being built. With respect to that, such massive development will indeed be a high investment to make. The Colony launch about 18 years after the first departure will absorb the larger capital. During the process, private sector investments will be required to cover the cost of the massive space ships and cargos launches to further develop and expand the Colonies and Cities. The investment will meet the break-even point approximately 18 years after the initial launch, during which commerce and trading of commodities on Mars will have run for several years.

Trading and all transactional activities on Mars will use one digital currency. The result of mineral exploration will also find a highly-valued mineral to be used as an investment medium, similar to gold on Earth. In order to create a micro economic climate in Gusev Cosmocity, the colonists will be supported through university education and research in the R& D Subcity to further develop their entrepreneurship skills. They then will be able to market their creative commodities locally in the Subcities, regionally in the Cosmocity, and inter-Cosmocity, and furthermore, it can also lead into exports to Earth.

OUTFLOW	year 0	year 2-6	year 8-12	year 14-18	year 18-20	year 21-50
1 Transport Cost	800	1.600	18.000	36.000	36.000	80.000
2 Salary & Incentives (Colonists)	60	120	1.800	4.000	8.000	65.000
3 R & D	-	-	350	350	350	10.000
4 Habitat Module & Equipment	250	1.000	-	-	-	-
5 Recycled Waste		100	200	450	500	12.000
6 Dry Food Supply	20	40	100	-	-	-
7 Tourism Marketing	-	-	15	30	50	15.000
8 Power Plant & Reactor	1.000	2.000	4.500	4.500	4.500	45.000
9 Manufacture Equipment	-	350	3.200	3.200	3.200	55.000
10 Mining Machine	300	500	1.500	1.500	1.500	-
11 Colonist Training on Trainer	60	100	250	250	250	-
12 Subcities Development	-	-	15.000	15.000	35.000	550.000
13 Infrastructure	300	2.000	8.000	12.000	12.000	350.000
14 Transportation Infrastructure	-	2.000	8.000	8.000	15.000	120.000
Total Outcome	2.790	9.810	60.915	85.280	116.350	1.302.000

INFLOW	year 0	year 2-6	year 8-12	year 14-18	year 18-20	year 21-50
1 Broadcasting	150	150	300	300	300	7.500
2 Colonist Application Fee	-	800	1.200	2.400	2.400	-
3 Mineral Export	-	500	2.000	6.000	6.000	70.000
4 Fuel & Energy Production	-	700	3.500	4.000	5.000	20.000
5 Advanced Technology	-	-	-	5.000	5.000	25.000
6 Licensing	-	-	-	100	100	3.500
7 Merchandise	-	-	200	500	1.200	4.000
8 Tourism	-	-	5.000	10.000	25.000	100.000
9 Citizen Tax	-	1.000	12.000	30.000	45.000	200.000
10 Property Asset	-	-	-	-	-	300.000
11 Domestic Trading	-	500	1.000	3.000	4.500	300.000
12 Industrial Partnerhsip	-	-	-	10.000	15.000	100.000
13 Foundation	-	500	1.000	1.000	-	-
14 Space Agencies Contribution	500	2.000	2.000	4.000	2.000	-
15 Transportation Tax	-	-	-	10.000	10.000	50.000
16 Industrial Tax	-	-	500	2.000	4.000	200.000
Total Income	650	6.150	28.700	88.300	125.500	1.380.000
Net Inflow-Outflow	- 2.140	- 3.660	- 32.215	3.020	9.150	78.000

*in millions USD

ENVISIONING MARS

Red Mars (0-200 years)

The Cosmocity of Gusev will reach a population of one million in the fiftieth year. During this time, the climatic condition of the planet will not be livable for the inhabitants and therefore there is the need to build shelters to protect the Cosmocity structures. The space ship technology will be more advanced so there will be a Phobos Space Elevator functioning for cross-planetary transportation. Skyscraper buildings used for business and office functions will exist in the Commercial Subcity as the material technology will have been more advanced and protection against radiation will be blended in the building materials. Residential buildings, however, will still be dominated with vertical buildings in the underground, above ground and in the Rim of Thira Crater.

Green Mars (201-600 years)
After 200 years, the Gusev Cosmocity will have been expanded so that more
development can be executed in the previously empty area of the City Quadrants.
The areas that have grown previously as the mining and industrial areas will be
vacated and developed further into green open space. Some microorganism will
start to grow on the planet surface as the overall temperature rises. This will make
the soil and the atmosphere more suitable for some types of plants to grow. The
Colony cluster will be developed further around the Green Plains area. Within the
next 200 years another Cosmocity will be developed. The expansion of the
residential areas will be directed toward the Ma'adim Vallis as the cliff edge
residential will be used as a transition zone toward the other Cosmocities. This will
make the total population of Gusev Cosmocity up to five million people.

Blue Mars (601-1000 years)
During the Blue Mars period, the atmosphere condition will reach a livable
condition so that some activities including farming can be done outdoor. The
Cosmocity will be appear more developed and more integrated with the woodlands
in the undeveloped urban areas. Rain will start to fall and this will lead into the
forming of water bodies in the crater including lakes, sea and river. Therefore, new
Colony Cluster will occur around those water bodies. During this phase, the shelter
protection can be released and opened gradually so that all structures and lives will
get direct exposure with the air and atmosphere of Mars. Gusev Cosmocity will
expand toward Ma'admi Vallis, which will increase the population to ten million
colonists. Then, the Gusev Cosmocity will become a beautiful tourism designation
with a number of potential natural features such as delta, canyon and the Ma'adim
river that separates Gusev Cosmocity from Old Town Thira.

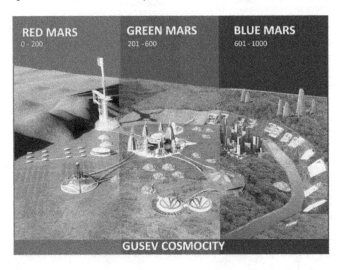

REFERENCES

[1] Wall, Mike. (2016). SpaceX's Mars Colony Plan: By the Numbers. Space.Com.
    https://www.space.com/34234-spacex-mars-colony-plan-by-the-
    numbers.html. (Accessed : June 11, 2020)
[2] Lordos, George & Alexandros Lordos (2019). Star City : Designing a Settlement
    on Mars. Conference: 22nd Annual Mars Society Convention, Oct 2019.
[3] Zubrin, Robert & Robert Wagner (2011). The Case for Mars: The Plan to Settle
    the Red Planet and Why We Must. Free Press.
[4] Kooser, Amanda (2020). Elon Musk Drops Details for Spacex Mars Mega-
    Colony. C.Net.Com https://www.cnet.com/news/elon-musk-drops-details-
    for-spacexs-million-person-Mars-mega-colony/. (Accessed : June 25, 2020)
[5] Weller, Chris. (2015). The Architects Behind Apple's New Campus Have
    Released Designs for A Mars Settlement. Business Insider.
    https://www.businessinsider.com/nasa-mars-settlement-designed-by-foster-
    and-partners-2015-9?r=US&IR=T (Accessed : June 25, 2020)

13: NERIO, UNDERGROUND CAPITAL OF MARS

Paul Meillon MSc.
James Bay, northern Canada
Ouroboros555@gmail.com – 2020

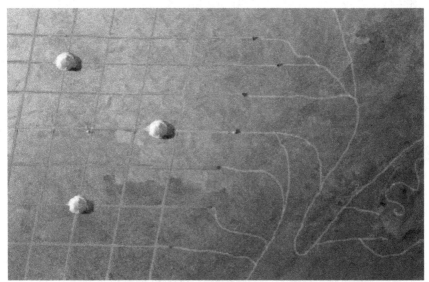

"If you want to build a ship, don't drum up people to collect wood and don't assign them tasks and work, but rather teach them to long for the endless immensity of the sea." - Antoine de Saint-Exupery

Mars is the prize of our age, an undertaking that is a testimony to humanity's instincts for daring in the 21st century. The historical moment we will establish a beachhead there will mark a break in our species history, with life becoming multi-planetary. This enterprise is not only as a hedge against existential risks, but the noblest incarnation of the spirit of exploration. The adversity encountered on the red planet will constitute a crucible within which the most awe-inspiring technological and scientific advances will spring forth.

Nerio, consort of the god Mars and name of the proposed city, personifies the warm, the nurturing and the feminine, qualities that we should aspire to integrate into colony designs, so that the human race can flourish on the forbidding, masculine, formidable conditions present.

Technical Design
Martian design opens up the possibility to start a technological civilization anew, and it is paramount to be untrammeled by convention and earth-bias. All efforts must be made to maximize use indigenous resources, as the very long supply line from earth should not be counted on. Life on Mars will be terribly expensive, so

every minor efficiency must be exploited and amplified, with the intent of attracting the most people there. The lower the cost of living, the more people will be able to flock to the adventure of a lifetime.

Below you will find the outlines of how to purvey the five major consumables needed for settlement: habitat, water, air, energy, and food. This will be followed by some comments on industry and urban planning. In designing the following city plans, all attempts have been made to curtail resource use and be frugal.

Habitat-At its essence, a Martian home is simply and enclosed space that can withstand 1 atmosphere of pressure at room temperature, and that can provide radiation shielding. This can all be achieved *without* fabricating walls, floors or ceilings, by directly carving it out of bedrock. Living space will have to be robotically extracted from outcrops, and as the homes will be filled with warm, humid air; water would freeze and seal any fine cracks to the exterior (see Illustration 1). Michelangelo famously said "Every block of stone has a statue inside it, it is the role of the sculptor to discover it". It falls upon our shoulders to discover the hidden city lurking within the Martian bedrock.

Illustration 1: View of an alley dug in bedrock; a glass skylight covers the alleys, enclosing the pressurized environment and providing sunlight. A portcullis with guide rails is seen behind the first arch, which can close vertically, and seal a section of the alley if a depressurization incident occurs. An emergency pressure suit is stored in the black enclosure to the right, for individuals caught in the alley during such an incident. Whitewash (slacked lime or chalk) paint will likely be fashionable.

Habitat will necessarily have to be cheaper than housing on earth, *a lot* cheaper. With North American housing at $ 100-1000 per ft^2, the objective would be to contract the price by 2 or 3 orders of magnitude, to $ 1 per ft^2 on earth (equivalent to 1 m^3 of tunnelled rock habitat). Since the majority of building cost are for paying wages, and since the proposed methods use no labour and barely any materials, this objective is considered to be reasonable.

In order to reach these prices, it is necessary to mine soft rock which does not require an extensive mining "toolkit" or wear down equipment. Depending on the mineral, large pillars would have to be left in place for support, and epoxy anchors might be needed (see below for epoxy and plastics production). Gypsum, the main mineral in drywall, and epsomite salts (calcium and magnesium sulfates respectively) are highly enriched in some sedimentary deposits on Mars, and should be mineable, using less resources than the basalts that dominate most of the surface. These salts are visible in near-infrared spectroscopy, so they have been identified and mapped in many places around Mars[1]. The spectroscopy data clearly shows where the salts are present, but the information does not permit to determine now much of the rock is comprised of salt, this would have to be confirmed by surface exploration. Gypsum dunes are present near the north pole, close to water resources; this area would make an ideal spot for a scientific outpost, if gypsum bedrock could be found. There are significant rock layers of sulfate in the Valles Marineris canyon walls, where another outpost could be constructed, but water would have to be imported from other locations. Clay minerals, or phyllosilicates, if present in large quantities may also be dig-able, and in some locations they constitute 20 to 30% of some rocks, as the Mars Science Laboratory has seen.

Illustration 2): Artist's conception of an alleyway cross-section, with a colonist taking a night stroll by the moonlight of Phobos and Deimos. Alluring European cities with delightful vegetated alleys, can guide our design for high density living

space, with skylights covering the top of the streets. Paris and Barcelona, which millions of tourists visit every year, are prime examples. The illustration on the first page of this document displays an aerial view of the skylights topping the alleys. An occasional above-ground domed public parc, an expensive luxury, dots the city grid. Wheel tracks at the end of the alleys mark where the mining vehicles exit to the exterior, to deposit the waste rock in piles situated on the outskirts of the city.

Living space and work space could be extracted by using the BADGER drilling technology currently undergoing field trials by the EU[2]. The segmented robot has the ability to clamp itself to the borehole, map and navigate underground, and tunnel its way forward using a combination of impact-drilling and ultrasonic tool to pulverize the rock (Illustration 3). For more massive enclosures, tunnel-boring machines or roadheader robots could be used. Drilling "moles" would have to be invented, using a drill motor attached a flexible shaft, in order to drill the plumbing, including siphons, and connect the lines to the mains. A snaking robot would have to direct the drill bit and push out the rock dust, say by Archimedes' screw or a peristaltic movement. Pneumatic snaking robots already exist, such as the BioSoftArm by Festo.

Illustration 3: Habitat extraction robot, based on the BADGER technology being developed by the EU. It uses georadar to navigate underground. Several different types of robots could carve out habitat autonomously, using artificial intelligence techniques such as deep learning.

Dwellings need to be provided with fresh air, clean water and power, and maybe a little sunlight. The lodging's "circulatory system" comprised of vents, plumbing, and sunlight tubes, drilled *in situ*, need additional treatment to avoid water infiltration through fractures. Several companies use an elegant strategy for revamping deteriorated pipes and conduits, by using a method called forced air epoxy lining (ie. Nu Flow Technologies). Lining the drilled conduits would be necessary to seal rock fractures in the plumbing and create a smooth, cleanable surface in the ductwork. The liner could be applied directly onto the rock, thus the material savings of not having to fabricate plumbing and ducts are prodigious. Epoxy, or a compound fulfilling the same function, would be the focus of the nascent chemical industry (along with soap).

Water-Life on earth was created in the oceans, and we will desperately thirst for water on mars, using it for fuel, industry and personal use. The following describes water usage, from mining to its integration into a closed-loop system.

Most of the planet's water is stored in solid form, and water extraction from ice deposits will be necessary. Obtaining water from atmospheric water vapour, groundwater brines, adsorbed water or from hydrated minerals is not considered practical[3]. Landing and ascent technology options place boundaries on surface locations for the city, leading to a preference for mid-to low-latitudes and mid-to low-elevations. Within these latitude and elevation limits, the presence of soft unfractured rock close to water resources will determine the location of the colony. Areas of interest will have to be explored by exogeologists by drilling and using ground penetrating radar for ice detection for example. In the mid latitudes discontinuous shallow ice is encountered within 5 m (16 ft.) of the surface; this is where extraction efforts would be concentrated. Evidence of frozen water has been observed in these areas, where new small impact craters expose ice in crater interiors and ejecta. Ice depth progressively increases at lower latitudes.

The debris and sublimation till layer overlaying the ice, is likely to resemble terrestrial glacial till, and would need to be drilled through. Direct push technology such as the Geoprobe Systems drill rig could be used to install a Rodriguez well (Rodwell) using a hammering action. Since the probe rods are hollow, no well casings would be necessary in the overburden, and this technology could attain depths of up to 18 m (60 ft.).

The robotic drilling system would inject steam into the borehole and melt the surrounding ice, at a sufficient depth so the cavity thus created does not collapse. The probe rods and piping assembly would be constructed in such a way as to provide a seal to the underlying chamber, to avoid atmospheric losses. The ice would be melted/sublimed, then cold-trapped in a hopper on the drill rig. Automated transport trucks would the pickup the ice from several drill rigs and ship it back to the city for further processing, such as perchlorate removal. Ideally, a radioisotope thermoelectric generator would provide the power requirements for the drill rigs. The trucks could be powered by fuel cells, since they would be going back to a fuel source in the city.

Illustration 4: Duckweed raceway (artificial wetland) used for a portion of sewage treatment and O_2 production. The other portion of sewage would be diverted to aerobic treatment and CO_2 production. Changing the ratio between the two treatment regimes could help with atmospheric rebalancing. Duckweed biomass could be used for biomethane or aquaculture production. Once the water is consumed, the sewage would go through a typical treatment chain, except that it would be diverted to both aerobic treatment and duckweed raceways (Illustration 4). This would enable sewage treatment to also be used for atmospheric rebalancing (see below) and the final effluent could be used in agriculture. Clean water could be acquired from the atmosphere in the plant chambers, with plant respiration controllable to a certain extent by raising temperature[4].

Air-Initially, the air would be made by physicochemical processes, and would subsequently be bioregenerated, with plants acting as CO_2 scrubbers and O_2 producers. Approximately half the food grown for a person can regenerate all their air needs, with additional crops leading to atmospheric imbalances[5]. If all the crops were used to bioregenerate the air, there would be an excess of O_2 and catastrophic crop failure would ensue due to a lack of CO_2. In order to avoid this calamity, two air compartments would be necessary. The first one, growing crops below the housing, would simulate earth's atmosphere, and the plant and animal kingdoms would balance each other out. The second, deeper compartment, would have an atmosphere that would maximize plant growth, but would not be appropriate for humans (see Illustration 7).

In a biologically closed system, the numbers should all add up, however in practice this is difficult to achieve, since biological systems are complex and can be unpredictable. The CO_2 to O_2 usage ratio of plants and humans are not exactly complimentary (1 and 0.7 respectively), when people are eating an omnivorous diet. Changes to the system, such as an arrival of a new wave of immigrants, will quickly create imbalances if homeostatic mechanisms are not in place, consequently a flexible system of atmospheric control using biological elements is necessary. These elements could include throttling up and down the composting of waste agricultural products, supercritical wet oxidation of organic garbage, or diverting sewage to aerobic treatment versus wetland treatment by duckweed plants. To minimize resource use and dependence on complex and onerous processes, physicochemical atmospheric processing should be curtailed, and reserved for fine-tuning, topping up leakage to the exterior, and as a backup to the biological systems.

The physicochemical processing would initially start by filtering the dust from the atmosphere by electrostatic or cyclonic action. The atmosphere would be separated into two streams: CO_2 and buffer gases, by using next-generation graphene membranes developed for carbon capture[6]. The new membrane has a six-fold higher CO_2 permeance over incumbent technology, and graphene prices have been dropping rapidly.

The classical method of CH_4 and O_2 production is through the Sabatier process and water hydrolysis. It is complex and cumbersome, so energetic R&D efforts should be brought to bear on other options. Some ionic liquids have the capability to adsorb carbon dioxide at low partial pressures, and provide a conductive medium for electrolysis of the captured CO_2 to oxygen and other products[7]. These organic liquid salts can also dissolve water, allowing for co-electrolysis of water. Using this method has several advantages: high temperatures are not needed, it has one less pump and no cryocoolers, it has 4 fewer major steps, and has half the mass and uses one quarter the power. Though the technology is not yet mature, an intensive research effort involving combinatorial techniques could quickly bring it to fruition. The air created through these processes would have a composition of approximately 20% O_2, 40% N_2 and 40% Ar.

CH_4 would need to be cryocooled and stored in liquid form, inside vacuum insulated tanks. Waste gases created from atmospheric processing could be used to actuate the airlocks, to conserve breathable air. Pressure would be sea level, since recent psychiatric data has shown increased risk of suicide at higher elevations. VOCs emitted by humans and other processes would be burned at high temperature as was done in the BIOS 3 experiment.

Energy-Frugality is a virtue when using material resources, yet the colony should be energy rich. Electricity is the lifeblood of the city, and death due to blood loss would be a calamity. Apart from a few hints of contemporary volcanism and marsquakes at Cerberus Fossae, most of the crust appears to be cold, which excludes the use of geothermal power. If future prospecting shows geothermal potential, then those resources should be decisively exploited. With the current state of knowledge, the simplest power solution would be to use nuclear power, the king of energy density.

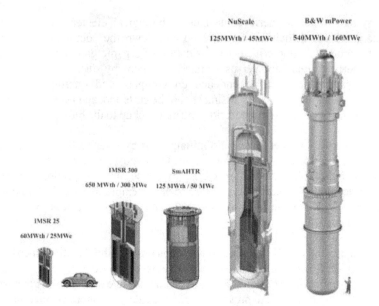

Illustration 5: In the world of small modular reactors, molten salt reactors have the highest energy density. The IMSR 25 and 300 to the left, can be imported to the colony as power demand and the population grows. Waste heat could be used in blast furnaces or district heating.

The Canadian company Terrestrial Energy is working on a molten salt reactor with sizable energy density; there have been claims that their technology is 50 times as energy dense as conventional submarine reactors (see Illustration 5). The design is inherently safe, they are self-governing and have passive cooling systems that eliminate pump failure risk. Fusion reactors can have even higher energy densities, and could replace fission reactors once the technology is ready. Supercritical CO_2 turbines could be combined with the salt fission reactors, saving an additional 10-15% in efficiency; these turbines have 100 times less volume conventional ones. These machines are still under development, with cavitation problems still needing to be overcome. All critical systems in the city should be duplicated, the redundancy providing a safety factor needed in case of breakdown, not only for the power systems but also all other life-support systems.

Peaking demand on the power grid would be satisfied using compressed atmosphere in underground tunnels and water reservoirs reserved for that purpose (see Illustration 6). Ontario company Hydrostor designs systems that uses surplus electricity to run a compressor that generates heated compressed air. Heat is extracted from the air stream and stored inside a proprietary thermal store, preserving the energy for use later in the cycle, without the use of fossil fuels. Hydrostatic pressure forces air to the surface where it is recombined with the stored heat and expanded through a turbine to generate electricity on demand.

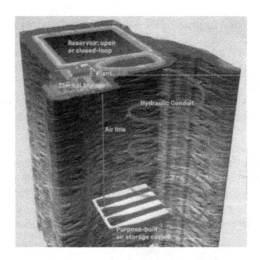

Illustration 6: A compressed air "peaker plant" can provide power at breakfast or supper time, when demand is highest. The robots that extract habitat would also extract the tunnels and reservoirs holding the compressed air and water. The reservoir would have a dual role for generating electricity and for storage municipal drinking water.

The city would at first need to use nuclear power to heat itself up from the -50 °C that prevails in the subsurface[8]. Once the core of the city is warmed, this temperature would remain constant and use less inputs, with the rock acting as thermal mass. Special reinforcement will be given to the airlocks where thermal cycling might structurally degrade the minerals. Temperature control of housing units could be accomplished with heaters using a carbon fiber heating element, capable of producing infrared heat; these could be fabricated from atmospheric carbon. Refrigeration needs for foods or industrial processes could have coolant loops circulating in the atmosphere or cold bedrock for shedding heat. This system would help with the refrigeration of produce, and with the capture of water from plant transpiration via dehumidifiers.

Food-In order to transform the first temporary outposts into permanent inhabitancy of mars, local agriculture needs to be established. The western diet contains foods such as meat, dairy, eggs that use inordinate amount of resources; our agriculture system cannot be directly transplanted off-world. Most people would not want to forgo meat, so less resource-intensive alternatives must be found. Cultured meat and insect-based products would fill the demand.

Solar lighting at the Martian equator is of equivalent strength to that of Alaska, and with the decreased transmittance through greenhouse glass (which would have to be even thicker on Mars), staple crops would not be able to grow[9]. Using subterranean farms with LED lighting, would be the ideal solution to the planet's constraints.

Efforts have been made to grow plants on high-fidelity regolith simulant, which resulted in plant and earthworm death [9], suggesting the raw dirt would have to be processed. The key would be to use soil-less systems (hydroponics, and especially anthroponics). These could work as soon as the colony is established, with a gradual phasing-out of plant nutrient imports from earth, as systems are developed for producing them on-site. Soil-less systems do require more equipment such as trays, pumps and reservoirs, these could be made of ceramics, or dug directly into the bedrock.

Martian agriculture can benefit from decades of work that have been done by NASA and others on closed systems and plants. Several authors have compiled lists of appropriate species, these include carrot, lettuce, wheat, soybean, rice, cabbage, radish and green beans. A continuous growing scheme, with small batches staggered in time, would provide food every single day, and reduce storage space for crops.

Insect farming is ideally suited for a Martian diet, rearing requires very little space and the feed conversion ratio is prodigiously high. Already several companies, such as Entocycle, are developing automated systems that produce protein for animal feed, using black soldier flies, crickets or mealworms. For those of us who are more squeamish, insects can be ground into a flour and hidden in other foods.

Recent developments in cellular agriculture make this option particularly attractive. Several startups have benefitted from heavy investments and cowless milk, egg, meat and fish products are on their way to rapid commercialization. The meat substitute quorn (*Fusarium venenatum* mushroom) has been around for decades, and carbohydrates from duckweed grown on sewage could feed the bioreactor. These systems, like insect rearing, have colossal productivity per unit area and water usage, as compared to traditional agriculture. Additionally, synthetic agriculture has appeared, with the notable case of laboratory created coffee, which will surely agree with many colonists who are espresso aficionados.

Human urine is the waste that has the greatest content of nitrogen, experimental results suggested that human urine could satisfy the demand of at least 3 to 4 out of 6 nutrients with an offset in pH and salinity tolerable by plants [10]. The term anthroponics or "peeponics" has been coined to describe nitrogen recycling by using human urine; and it could eliminate the need for mining fertilizers, which would constitute a significant saving in time and materials.

Urban and Industrial considerations- Grid city patterns have been used in urban planning since the classic antiquity and have withstood the test of time. This arrangement is proposed here, but other more haphazard design philosophies would be equally acceptable, and would give more character to living space, as can found in medieval neighbourhoods. High density Eastern European cities have population densities approximating 20,000 persons per km^2, and this was used in the following assessment. Since the room-and-pillar mining technique employed will use 25% of space for column support structures[12], a city of 1 million people is estimated to cover a square of 8 x 8 km, with 5 % of that reserved for industrial activities, as is the case for Paris or Barcelona. Additional vertical distance would be necessary between floors for support, making a city of 15 levels approximately 100 m deep (see Illustration 7). An agricultural space of 80m² per person was assumed, which is double the living space of 40 m² per inhabitant. Agriculture areas would include linear parcs, public transport tunnels, and room for plant growth, cellular vats and aquaculture. The spaces with industrial zoning would be maintained on the edge of the city, to facilitate ingress of mined materials, and minimize the interface with housing in case an industrial accident occurs.

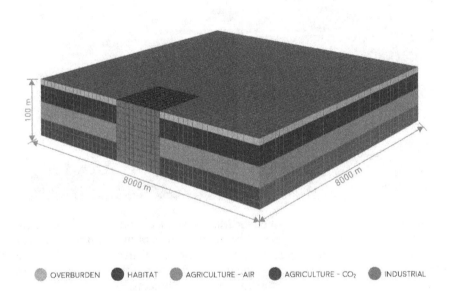

⬤ OVERBURDEN   ⬤ HABITAT   ⬤ AGRICULTURE - AIR   ⬤ AGRICULTURE - CO₂   ⬤ INDUSTRIAL

Illustration 7: 3d view of the underground city. A million people could live in a European-style metropolis, protected from radiation by the overlaying rock and overburden, and nourished by food growing on the deeper levels. Industrial areas would interface with the edge of the city, much like terrestrial cities.

Glass production will become one of the first industries, supplying glass for enclosing habitat as well as household objects such as desktops, tabletops, doors, windows and cookware. Harvesting the silica in sand is straightforward, and the iron content could be reduced using CO. The other necessary ingredients, soda and lime would need to be mined from carbonates or calcite. The metallic oxides necessary in smaller quantities would also need to be mined or harvested chemically from mobile processing units. Molten glass would be heated using CH_4 torches, and subsequently floated on a tin bath for production.

The steelworks would be implemented using hematite, common on Mars, and reduced by CO, which is easily produced from the atmosphere, and the right agent for iron reduction. The reaction is slightly exothermic, so the blast furnace could be operated at only 700 °C, using waste heat from the fission reactor(s). On earth, impurities are burned off with O_2 in a refinement unit; on Mars both procedures could be conducted within the same enclosure.

Plastics could be made by using CO_2 and water as starting materials, which would be transformed into methane and ethylene, and afterwards polymerized into polyethylene. A new production method of ethylene has been developed: it uses an oxidative coupling technique in an electric field. CH_4 and O_2 are used in a redox reaction on a catalytic surface, with temperatures as low as 150 °C[13].

Illustration 8: Glass alley skylight on the Martian surface. Note the tracks from the robot that recently cleaned the glass free of dust. Vacuum insulated glass is used with a thin transparent coating of metal-oxide to block radiative heat loss. A fragile film is applied through sputtering to the interior sides while the exterior panes are hard-coated glass, which is done through a pyrolytic process.

Additive manufacturing, fed by domestic materials, would play an important role in fabricating the agricultural robots, other machine parts and household items. Reliance on this technology will lead to innovations and improvement of precision, printing time, variety of materials used.

The airlocks connecting the interior to the exterior would use waste gas from the atmospheric processing module, to conserve breathable air. The trucks would be hosed down for dust, and the waste water captured in a catch basin, filtered and re-used. Such automated systems are already in existence (ie. designed by Westmatic).

Designing a city *de novo*, non-traditional amenities can be easily added such as a pneumatic delivery system for small items and garbage disposal. The city will necessarily need to be hyper-connected, not only for telecommunications, but would also have an extensive sensor network for homeostatic life-support functions.

Economy

A rich knowledge economy will need to be developed on Mars, and all efforts should be made to forge the most competitive economic system in the solar system. Many space enthusiasts consider asteroid mining and other resources as closing the business case for colonization, however, cosmic distances, and high transport and operational costs will nullify any advantages. Historically, many knowledge-rich and resource-poor economies have come out as winners, one shining example being the post-war Japanese economic miracle. Inversely, examples of countries with rich mining resources and living in misery can easily be thought of. Considerations relating to economic organization for colony development will first be discussed, followed by some ideas on establishing a knowledge economy.

Colonial development-Steep capital investments will initially be necessary, and should be seen as a massive public works program that opens not only physical frontiers, but also knowledge frontiers, especially in the areas of robotics, agriculture and biotechnology. The returns on these investments will not be immediate, and since most private investors look at a short time horizon for returns (except for SpaceX), financing from space-faring governments becomes necessary. After the first manned landings have occurred, it is likely that a perceived race for Mars will start. This will mobilize public opinion and the elites of all space-faring countries, and help marshal the gargantuan government expenditures necessary for geological exploration, technology development, planning, and infrastructure construction. Historical precedents of this process include of late 19th century naval arms races, exploration and colonialism in Africa, early polar exploration, and 20th century nuclear and space races.

European colonies were often financed through trading ventures, organized as joint stock companies chartered by the colonizing state and in which the crown invested both its prestige, and its capital. Plantations and mines were often directly owned by the local colonial state. Trading monopolies and tax privileges were granted by the colonizing state to the local colonial government.

The best way to advance Martian colonial plans is to follow some of these examples, by creating public-private partnerships, where companies and national governments could pool resources. These would fall under the aegis of the Mars Development Authority, a semi-sovereign government entity of ambiguous nature. Its ambiguity would be advantageous because it would be able to manage national rivalries, and at the same time have a measure of independence, which could lead to eventual self-rule. Some of its responsibilities would include the capitalizing of new ventures, and the enforcing of contracts and titles to property. It would organize a competitive procurement process where private companies vie to provide habitat extraction, air, power, telecommunications to the city. Monopolies on the utilities should be avoided to the best extent possible and they should be deregulated to encourage competition. One "tax privilege" the Mars Development Authority should be granted, is to profit from the intellectual property royalties from Martian-related research conducted by national governments.

Colony development will at first necessitate massive amounts of funds, an order of magnitude estimate is of about $ 75 billion over 15 years, about half of the Apollo program, which will be employed for devising the technology and building the initial bases. This appraisal is based on the fact that we are much closer technologically to what is needed for conquering Mars today, than we were for getting the moon in the early 1960s.

The Apollo program created remarkable discoveries in civil, electrical, aeronautical and engineering science, as well as rocketry, making it possibly the greatest scientific and engineering feats of all time. Various estimates for the spinoff Return on Investment (ROI) of the program show that in the long term, the investment was recuperated many times over. It is generally accepted that Research and Development (R&D) have an ROI on the order of 10-20% per year, and 10% will be assumed for the returns that the colony will accrue on all investments. By analogy with the pharmaceutical sector, we can expect about a decade for Martian technology to come to fruition, giving us an idea of the timeline to profitability (see Net Present Value in Illustration 9).

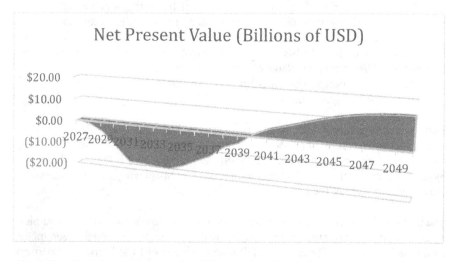

Illustration 9: Order of magnitude of Net Present Value for the Mars Development Authority. This metric is considered more useful than ROI in measuring R&D returns. Assuming public opinion and government mobilization in the mid- to late-2020s, intellectual property profits should overcome costs by around 2040. Cost timeline is copied from the Apollo program.

At a microeconomic level, the prices for groceries, electricity, water, clothing (we can expect a renewal of sewing skills), and consumer goods will be prohibitive.

Staff shortages will be rampant, and medical and educational services will suffer. These problems can be remedied by having an equalization allowance, which would be calculated based on the difference between living costs on Mars and on Earth. Wages would also be a lot higher, and both these financial incentives would help to attract newcomers. As the population increases, and technology and humans adapt to the new environment, living costs will decline. These economic pressures on Martian society are likely to result in technological advances and increased productivity, and the foundations of an affluent knowledge economy.

Knowledge economy-Less than 2% of the population is engaged and research and experimental development, yet it is these research efforts that are at the heart of our modern lifestyle. For any given technique of production, transport or distribution, there are long-run limitations on the growth of productivity, which are technologically determined. The winning of new knowledge is the basis of human civilization.

Innovation is an essential condition of economic progress, and a critical element in the competitive struggle between companies, nations, and soon planets. Innovation is of importance not only for increasing the wealth of nations in the narrow sense of increased prosperity, but also in the more fundamental sense of enabling people to do things which have never been done before. The entire quality of life is changed for the better, and hopefully not for worse. It doesn't mean creating products more efficiently, but inventing lines of products and services that once only existed in the imagination.

The classical example is of radical innovation is the adoption of steam power in the 19th century. It depended on several factors: the fall of the cost per horsepower as engine sizes increased; high pressure engines were developed that required less coal; and improved railway locomotives and their usage increase from 1825 onwards. These trends were mutually reinforcing, widespread adoption of locomotives made the price of coal decrease and so forth. This example illustrates the complete interdependence of technical change and economic progress.

The barren surface of Mars represents the extreme end of processes that humans have unwittingly initiated on Earth: climate change, land degradation, and desertification. Martian technological amelioration, will, by force of necessity, break through the aforementioned growth limitations, and would be fundamental in saving untold millions of future starving and displaced people. These advances could not only revolutionize agriculture and other disciplines (see Table 1), but also fill the Sahara with green troglodytic cities.

New science-related industries flourished in the 20th century, such developments underscore the dependence of advanced economies on the successful use of new technologies, and the extent to which this is partly dependent on indigenous scientific and technological capabilities. Education, research, and experimental development is regarded by many as the basic factors in the process of growth, relegating capital investment to the role of an intermediate factor[14]. The relationship between all the institutions dealing with technology and its dissemination, public or private, is called in the literature a 'national system of innovation'.

Table 1: Areas of Technological Development for Mars

Challenges to Mars Colonisation	Technological Development	Terrestrial Application
Illness due to lack of pharmaceutical infrastructure	Benchtop automated drug synthesis (i.e. "chemputers")	Decentralize and democratize pharmaceuticals
Lack of personnel to tend crops	AI and robotics to autonomously farm	Make labour intensive agriculture cheaper
Limited air, water and food	Closed-loop life support systems	Breakthroughs in waste management
Expensive production of CH_4	Super-efficient biomethane reactors at room temperature	Turn organic wastes into a renewable source of power
No habitat	Autonomous mining vehicles extract habitat	"Dark mines" with no human presence
No or few surgeons	Advanced AI surgery robots using deep learning	Advanced surgery available to larger portions of humanity
Energy is more expensive	Harvesting low-grade heat and other energy resources, conserving energy	Energy-conserving products and industrial processes
Food is expensive and lacks diversity	Cellular agriculture/synthetic agriculture, food science to modify simple products	New food products and cheaper production processes
Illness due to lack of doctors and complex diagnostic labs	Point-of-care diagnostics for all diseases (i.e. immunosignaturing). AI diagnosis.	Revolutionize how medicine is conducted

The importance of national systems of innovation can be illustrated by a historical example. In the 1840s England was in the throes of the industrial revolution, which was partly a consequence of the following elements: strong links between scientists and entrepreneurs; science had become a national institution, encouraged by the state and popularized by local clubs; investment in transport infrastructure (canals and roads, later railways); partnership form of organization enabled inventors to raise capital and collaborate with entrepreneurs etc. By contrast, Germany was decidedly underdeveloped, and this was recognized by the Prussian government.

The Prussian response was to create one of the best technical and educational systems in the world, the *Gewerbe-Institut*. They went on to sneakily acquire British machine tools, reverse-engineered them, and poached machinists. Historians consider these actions and educational advances as one of the main reasons why Germany overtook Britain in the latter half of the 19th century. Still today these institutions constitute the foundation for the superior skills and higher productivity of the German labor force.

The Mars Development Authority should therefore set up an ideal environment within which companies will innovate profitably. If we look at history as our guide, we can see how quickly industrial leadership can shift from one country to another. In the post world war II period, many American multinationals shifted R&D, innovation and dissemination to their European subsidiaries, enriching these countries as a result.

The greatest significance of fundamental research is that it provides a multipurpose general knowledge base on which to build a wide range of scientific and technical services. Every country and planet, without exception requires such a base, even if only on a very small scale. Without it, there cannot be any independent long-term cultural, economic or political development.

One of the main objectives of Mars Development Authority policy for science and technology should be to build and sustain an indigenous R&D capacity. Economists generally argued that public expenditures in support of innovation should be mainly oriented into four areas[14]: 1) Fundamental research, i.e. in universities, 2) Diffusion of generic technologies (i.e. closing the digital divide) 3) Supporting industries that cannot do R&D at the company level (i.e. agriculture) 4) Investment in information services (i.e. libraries, databases etc.). These should be implemented as soon as basic needs have been attended to, by creating the University of Nerio, and by supporting small firms working on agriculture, biotech and robotics.

The ability and initiative of entrepreneurs will create opportunities for profit, which in turn will create a swarm of imitators and improvers to exploit the new opening with a wave of new investment, generating boom conditions. The self-reinforcing events that lead to widespread adoption of steam power had the same characteristics. Implementing a national system of innovation on Mars, combined with the fiery spirit for invention and survival on a forbidding planet, would accelerate this innovation explosion and position le colony for an off-world economic miracle.

Social-Cultural-Political

"Men wanted for hazardous journey. Low wages, bitter cold, long hours of complete darkness. Safe return doubtful. Honour and recognition in event of success."

The world's most famous of ad for Shackelton's Endurance expedition illustrates the difficulties and dangers the first settlers will face. Although risks will drop as the city develops, morale, resilience and wellbeing will be vital in eking out a living. These will be important not only for facing the inherent tumult of the endeavour, but also the intense interest and scrutiny of terrestrial origin from the public and various institutional and national stakeholders.

The over-arching objective would be to maximize the community's wellbeing, defined by the Canadian Index of Wellbeing as "the presence of the highest possible quality of life in its full breadth of expression focused on but not necessarily exclusive to: good living standards, robust health, a sustainable environment, vital communities, an educated populace, balanced time use, high levels of democratic participation, and access to and participation in leisure and culture." Numerous studies have demonstrated the business case for maintaining robust wellbeing: it has been shown to be related to productivity, decreased absenteeism and presenteeism. Small positive changes that increase wellbeing have been shown to produce healthy financial returns[15].

The objectives of the social, cultural and political domains are outlined below, as well as a way to track those objectives, and make adjustments where necessary. It is important not to be too prescriptive, as each community is unique; and the settlement would assuredly be in unique circumstances, and would know best what it requires.

Objective 1: Social Wellbeing-The social domain encompasses the ensemble of social conditions that permit individuals and their communities to thrive and live up to their full potential. The settlement should be able to create opportunities for people to participate in community life and pursue self-realization. Colonists should be able to feel welcomed, safe, and engaged, 24/7, regardless of background or physical ability. Individuals should have access to support facilities and services, daily and during moments of need. People should be engaging socially at formal and informal levels.

Objective 2: Wellbeing related to the Physical Environment- The term "livability" is more concretely related to community wellbeing and is relevant to the design professions. Livability "reflects the wellbeing of a community and comprises the many characteristics that make a location a place where people want to live now and in the future[16]." A livable place is characterized as "safe, attractive, socially cohesive and inclusive, and environmentally sustainable; with affordable and diverse housing linked to employment, education, public open space, local shops, health and community services, and leisure and cultural opportunities; via convenient public transport, walking, and cycling infrastructure." Quality of spaces is more important than quantity of spaces.

Basic physical, social, and emotional needs are likely to be met when people can afford to choose well-designed, quality housing in a convenient geographic setting that accommodates safe living conditions and access to healthy lifestyle options, amenities, and active transportation between home, work, shops, and services. A biophilia plan should be developed and implemented, maximizing human–vegetation interactions for public and common areas. Biophilia refers to human's desire to be with nature.

Objective 3: Economic Wellbeing- The key economic factors that influence population health and socio-economic wellbeing and sustainability include income, housing, labour force activity, and education. When favourable, these factors help people live better, healthier, and more productive lives. People of different income levels should be able to afford a high quality of life. Places for daily purchases, places to work, live, play, study, take transit, should be able to be made within walking distance. Each neighbourhood should be able to offer people economically accessible opportunities to satisfy everyday life needs, including habitat, food, recreation, health care, mobility, and education. The ROI for the development and design of the city should account for ongoing maintenance and operation, and for the economic burden to Martian society. The city needs to support a healthy local economy, specifically a knowledge economy as described above.

Illustration 10: Amphitheater extracted from bedrock for concerts, religious services, lectures or plays. Culture will nurture the mind and provide temporary escape from professional and family duties. The sculptures on the back wall show that art and living space can be made from the same material (Saint-Simon, Cairo).

Objective 4: Cultural Vitality and Wellbeing- Individually and collectively, we depend on expressing ourselves creatively and freely, and nurturing a sense of belonging, delight, and play. The myriad of activities we pursue and enjoy benefit our overall life satisfaction and quality of life. Various forms of human expression help to fully define our ourselves and our sense of wellbeing. Producing art, for example, has been shown to significantly improve psychological resilience[15].

Recreation in settings both formal and informal, is important to individuals' well-balanced life and wellbeing, as well as health. Formal facilities could include structures with dedicated purposes such as an artificial ski hill or volleyball courts (see Illustration 11), and informal could include spaces to play cards, joke around with co-workers, and hold recess-like breaks or gatherings. Evidence also shows that people who regularly engage in cultural activities, such as going to museums, movies, theater, concerts etc., have a longer lifespan[17].

People will need to have access to cultural, recreational and art facilities in their neighbourhoods. A sense of belonging is important, and people need feel included in their communities, connected to their social networks, and engaged in civic and community life, regardless of their background. It is paramount that newcomers would be able to integrate with ease. Quality spaces should be made available for play, creative engagement, frolic, and quiet reflection. A collective sense of stewardship should be present in the planning of these spaces. Learning is an integral part of being human; we should be lifetime students: people of all ages should have opportunities to learn and develop.

Illustration 11: Underground volleyball courts will provide spectacular gameplay in reduced gravity. Ensuring that the population gets enough exercise would be a top priority, to avoid muscle-loss and other adverse physiological effects from living on Mars.

Objective 5: Political Wellbeing- A key finding in the literature is that people need to have opportunities to contribute to their own wellbeing[18]. This is central to the process of being able to flourish and fulfill one's potential. A sense of ownership and stewardship is important with people being able to understand, control, manage, interact with, and transform their environment. Martian citizens should be personally invested in the function and success of the metropolis. Political institutions should incite a sense of collective ownership; their final shape will not be described here, as they should be created in a grassroots manner and not dictated from above as the Mars Development Authority morphs into complete self-government. Decision-makers, stakeholders, and the public at large should have opportunities to meaningfully collaborate with the colony's design professionals from the outset, and in an ongoing manner. A diversity of perspectives, stakeholders, community, and disciplines should be integrated in the beginning, and throughout the life of the settlement.

The 5 objectives presented above could be used as a holistic framework for evaluating community wellbeing that is practical and applicable across time. It could be implemented using standard social sciences metrics such as questionnaires, key informants etc. This could be done by first defining and engaging with the community partners and stakeholders. Then managers, leaders and participants may wish to establish personal or collective targets to move towards the stated objectives. While the targets may vary in definition and scope and may evolve as particular projects unfold, establishing them early in the process will enable the team to focus a conversation and inform the development of options.

Communication throughout the process would be important for success, and a communication plan should be developed. Different designs and decisions should be informed by how they perform relative to the targets and objectives. The outcomes of individual projects, policies or initiatives should be well documented. These records will be significant during the life of the colony, as different aspects are implemented, monitored, and improved upon. Previous experience thus recorded will also be significant for future projects and could be used as a benchmark.

Failed planned cities, such as Brasilia or rigid Chinese mega-projects, constitute a warning, but also an opportunity with the negative example showing us the way forward. The same could be said of planned economies and social institutions. A city is an ecosystem of many people expressing their individual preferences in unplanned and often unpredictable ways. By incorporating feedback from what people want, at the outset, and throughout the life of the metropolis, will lead to the founding of magnificent social and cultural institutions, and the creation of an enticing built environment.

Esthetics

In keeping with 5 socio-political objectives described above, esthetic elements should focus on contributing to the colony's wellbeing. Beauty should be shared by all and not just a wealthy few. Environments that provoke delight and enjoyment through high-quality design features and by connecting users to vegetation impact positively on people's life satisfaction and are associated with psychological health benefits[19]. When people feel good in their space and function well in their environment, community wellbeing is meaningfully improved. Simple yet elegant designs can be caved out of bedrock, and use whitewash, and natural materials such as ceramics and hemp fabrics for decoration (Illustration 12). Plaster, in the form of pure lime, was used as the first decorative architectural element by humanity, as show discoveries at the neolithic village Yftahel in Israel.

Colonial designers should maximize physical and visual connections to vegetation from common areas, and from spaces where people spend several hours a day. One study found indoor plants can prevent fatigue during highly demanding tasks and increase productivity by 8 per cent, for those working in offices that incorporate elements of biophilic design[20]. Alleys covered with vines and potted plants could provide a sense of delight and relaxation (see Illustration 1 and 2) with the plant pots carved *in situ*. Apartment windows could be staggered, so that the view is of plants and not into the neighbours lodging. Domed courtyards, plazas and public parcs will equally be well vegetated.

<u>Illustration 12:</u> The beauty of whitewashed buildings is as old as civilization itself. At the left, a minimalist bedroom has hemp bedding and a bed platform carved from bedrock. At the right, showering and bathing areas are similarly made. Note the paucity of manufactured objects; the shower spout and sink are made of ceramics. Ceramics can also be used as decorative elements. Sunlight is transmitted from the windows and through fiber optics.

<u>Illustration 13:</u> Linear parcs found within the agricultural levels, provide soul-nurturing flora. These paths provide alternative transport corridors aside from the alleyways found above. The greenspaces are enclosed above with plants, to hide the grow lights and provide greenery in several directions. At the left is a bamboo plantation, a plant which could be used for making furniture or other small household objects such as brooms. Bamboo bicycle handlebars are pictured in the foreground.

Natural lighting and ventilation should similarly be maximized in areas where people spend a lot of time. Daylight exposure has been shown to improve wellbeing, quality and amount of sleep, and reduce sick days taken[21]. Ensuring sufficient ventilation is important for alertness and cognitive abililty. Fiber-optic lighting tubes should pipe light to all homes, especially those on the lower levels where sunlight will be the weakest. A biophilic design would use rock faces, plant life, natural light, and other experiences of nature within the habitat (Illustration 12, 13 and 15).

Illustration 14: *In situ* rock amenities where the past informs the future; beauty and minimalism are combined. Sinks and clothes washbasins can all be made in place. To the right, a person is holding an ultrasonic clothes washer that can do most of the washing. Rubbing clothes on the corrugations is only necessary for the most recalcitrant of stains.

One of the constraints of would-be Martian designers is the lack of plentiful resources. There are areas on earth that have the same constraints: deserts, where pre-modern people lived off pastoralism, and where building materials were non-existent. Troglodyte houses in Tunisia are an example of beautiful desert housing that could be directly applicable to Mars, with their sunk whitewashed courtyard carved in rock, and typically contain a palm tree and a few hardy plants. The interiors of these tourist attractions are characterized by decorative rugs, hand women and colored with natural dyes, and by furniture hewn from rock. We can look to the past for design ideas for the future.

We live in a society where supermarkets now have over 50,000 products, and a quarter of people with two-car garages don't have room to park cars inside. Immigrants to mars, even the richest, will have to adopt to a certain extent the tenets of minimalism. An absence of possessions does not equate an absence of happiness, the opposite might even be true (see Illustration 14). By designing frugality into the city and making it affordable, it will maximize the immigration rate, and swiftly achieve "critical mass" for a self-sustaining civilization.

Illustration 15: Expensive exterior domes, and underground linear parcs will provide the greenspaces needed for health and wellbeing. Citizens can commute or exercise in the linear parcs, whereas the aboveground covered parcs are places for meeting, relaxing, or for quiet reflection.

Winning this "New Atlantis" in the heavens, would give life to Francis Bacon's vision of a knowledge society, rife with exploits, a land where "generosity and enlightenment, dignity and splendour, piety and public spirit" are the commonly held qualities of the inhabitants. The Martian adventure need not be for a select few and beset with the utmost of privations and discomforts, but could accommodate a comfortable, delightful lifestyle available to millions.

References:

1) https://www.annualreviews.org/doi/abs/10.1146/annurev-earth-060313-055024
2) https://www.earthdoc.org/content/papers/10.3997/2214-4609.201902585
3) https://ieeexplore.ieee.org/abstract/document/7500775
4) https://www.annualreviews.org/doi/10.1146/annurev.pp.15.060164.002131
5) https://www.sciencedirect.com/science/article/pii/S0094576515004294?via%3Dihub
6) https://pubs.rsc.org/en/content/articlelanding/2019/EE/C9EE01238A#!divAbstract
7) https://pubs.acs.org/doi/abs/10.1021/jp509583c
8) https://agupubs.onlinelibrary.wiley.com/doi/full/10.1029/2007JE002905
9) https://www.liebertpub.com/doi/10.1089/space.2019.0018
10) https://www.sciencedirect.com/science/article/abs/pii/S2214552417300706
11) https://arc.aiaa.org/doi/10.2514/6.2019-4059

12) https://link.springer.com/article/10.1007/s10706-003-3159-3
13) https://pubs.acs.org/doi/abs/10.1021/acs.jpcc.7b08994
14) https://www.taylorfrancis.com/books/economics-industrial-innovation-chris-freeman-luc-soete/10.4324/9780203357637
15) Terrapin Bright Green, LLC, "The Economics of Biophilia"
16) Victoria Competition and Efficiency Commission, A State of Liveability.
17) Organisation for Economic Co-operation and Development, Education at a Glance, 2009
18) Community Wellbeing: A Framework for the Design Professions. Ottawa: The Conference Board of Canada, 2018
19) Steemers, "Architecture for Well-Being and Health."
20) Cooper, Human Spaces Report.
21) Boubekri and others, "Impact of Windows and Daylight Exposure."

14: PHLEGRA PRIME

Michael Behrendt
Mikebehr2@gmail.com

Robert Gitten
robert.gitten@gmail.com

Benjamin Greaves
robert.gitten@gmail.com

Sweeya Tangudu
sweeya2450@gmail.com

Zachary Tolley
ztolley@gmail.com

WELCOME TO PHLEGRA PRIME

Introduction

Phlegra Prime is a vision for a Martian city-state of one million citizens consisting of ten spatially distinct districts located in Phlegra Montes. The city-state will be profitable from day one by relying on existing markets and the radical reduction in the cost of transportation enabled by the SpaceX Starship. Life support will be maintained using a bioregenerative system. Phlegra Prime's GDP is expected to be $63 billion, and the city-state is expected to reach one million citizens 263 years after its founding with a final immigration rate of 120 spaceships per year. The city-state has neighborhood, district, and city-state level governments consisting of representatives chosen by sortition (lottery), in order to ensure equal representation. Work and housing is organized into cooperatives, where participants have voting power and where profits are shared in order to grow a strong local economy.

Layout

Location

Phlegra Montes, located in the Northern hemisphere, in the region of Arcadia Planitia, between 30-52° N and 160-170° E is chosen as our location to build a city-state for one million citizens. This region has access to abundant shallow ice that will be mined for water and further production of oxygen and rocket propellant. It is located at a relatively low latitude, so atmospheric temperatures would be warmer than at the poles and this region would receive ample solar energy. There is diverse topography in this region with vast expanses of flat areas that make favorable landing sites as well as proximity to ridges and thrust-fault structures that suggest a diversity of materials will be found. [Ref 1]

Layout and zoning

This city-state of one million people will be planned as a group of discrete districts with capacities for about 80,000-120,000 citizens each. Each district would have diverse economic and social functions. The economic functions would range among space travel, food production and logistics, mining, power generation, administration and others. The social functions will be museums, convention centers, theaters, large parks, fine dining venues and others. These functions would encourage the state's citizens to move across districts for both economic and social reasons, creating opportunities for business and social interactions.

Each of these districts will have one or more transportation hubs at their center and would be connected by a high-speed mass transit system. Every district will have its own utility systems such as air treatment, water recycling, solar farms, waste management as well as primary agriculture. A district would be able to sustain itself for a few weeks in case the links to other districts are disrupted temporarily. Each district has residences, education facilities, primary healthcare, retail and multi-spiritual centers.

A district's location would be based on geographic proximity to its economic and/or social functions. Citizens will have the freedom to live in any district they prefer. For example, a district based on mining operations would be situated in best proximity to favorable mining sites, but citizens engaged in the mining industry are free to live wherever they wish within Phlegra Prime.

These districts will be built in phases as the state's population increases over the years with the first piece being the inter-district transportation system. While the master plan concept lays down the guidelines for administration and facilities, each district built over time would be different in architectural expression. They will further be guided by the geographic context and economic function, while also being influenced by the evolution of ISRU and construction capabilities during the 263 years of habitation. For example, a district housing the activities for the spaceport and space travel will be the first one to take shape. Hence, a part of this district would have structures brought from Earth and would consist of discrete volumes connected with pressurized tubes and tunnels [Image on cover page]. On the other hand, districts built thereafter will have more evolved construction capabilities including larger spans, and more open spaces feeling like outdoors on Earth.

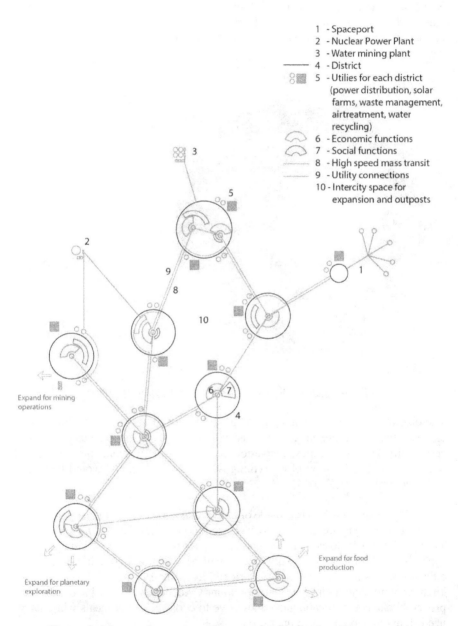

Figure 1: Master plan concept diagram for 10 districts (Not to scale)

As these districts grow in size, it is expected that they will develop sophisticated and differentiated cultures. As an example, if only one district has legal gambling, then time and capitalism will cause it to develop into a state where it resembles the economic structure and recreational draw similar to that of Las Vegas, Nevada. The reason that someone would want to travel to another district is to experience a culture different from the one that all individuals grow up with and experience every day.

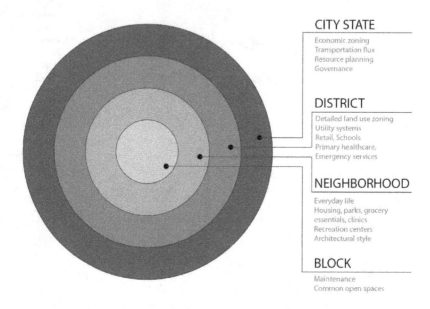

CITY STATE
Economic zoning
Transportation flux
Resource planning
Governance

DISTRICT
Detailed land use zoning
Utility systems
Retail, Schools
Primary healthcare,
Emergency services

NEIGHBORHOOD
Everyday life
Housing, parks, grocery
essentials, clinics
Recreation centers
Architectural style

BLOCK
Maintenance
Common open spaces

Figure 2: Four levels of planning and administration

The districts are further made up of neighborhoods that house about 5000-10,000 citizens. Each of these neighborhoods will have a unifying architectural style, community kitchens, clinics, recreation centers, space for small businesses and retail outlets. The location of everything within a neighborhood would be within walkable limits.

Each of these neighborhoods is made of blocks that will have common open spaces. The neighborhoods are planned to have common cooking and dining facilities. There will be robots to assist in food processing for the neighborhood citizens under supervision of human chefs. This facility will be within walkable limits from the citizen's homes and citizens have the flexibility to dine in any of the community kitchens within their district. The community kitchen will also have adjoining private kitchens that inhabitants can reserve to cook their own meals while using needed raw ingredients from the pantry.

This system of community kitchens will streamline food supply logistics and create minimal wastage of food and packaging materials. It will also encourage inhabitants to convene and interact more frequently. The risk of fires, and the HVAC load for additional exhaust and air supply to maintain homes' interior temperatures will decrease by concentrating food preparation in a common community kitchen.

Food preferences will be monitored to allow for agricultural output to prioritize the wants and needs of the community as well as understand overall nutrition.

Figure 3: A district with agricultural staple crops, food storage, logistics, and recreation as its major economic functions. [Background Image Ref 2]

Structures and Built Forms
Most structures will be a combination of subterranean spaces and superstructures. The subterranean spaces will be naturally shielded from radiation while the spaces above ground will be enclosed with radiation shielding materials such as excavated Mars regolith (for initial settlements), masonry blocks made of Mars regolith, sulfur concrete, polyethylene or ice. The subterranean and superstructures would also be enclosed with temperature shielding layers such as vacuum.

The built environment where citizens live and work will mostly be shielded with less access to views outside or natural light. They will have access to breakout areas that feel like 'outdoors' with large open spaces, vegetation, ambient sunlight, naturalistic landscaping, and even small animals. Windows and glazing for the structures will have multiple layers of glass with operable shutters to protect from radiation when not in use.

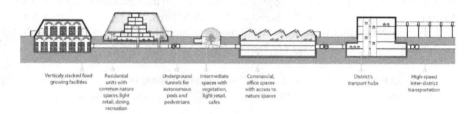

Vertically stacked food | Residential | Underground | Intermediate | Commercial, | District's | High-speed
growing facilities | units with | tunnels for | spaces with | office spaces | tranport hubs | inter-district
 | common nature | autonomous | vegetation, | with access to | | transportation
spaces, light	pods and	light retail,	nature spaces	
retail, dining,	pedestrians	cafes		
recreation				

Figure 4: Section of several common structures. In order from left to right:
agriculture, residential, retail, industrial, transport

Each pressurized space will have a radiation rating. This will differ with the
radiation that can enter through openings in pressurized enclosures and affect the
interior environment. Localized dosimeters will be used to calculate and track the
radiation that each citizen is exposed to throughout the day/week. If an individual
exceeds their exposure limit, the system will notify the citizen and their healthcare
practitioner to medically intervene and make changes to their routine in order to
reduce future exposure.

Transportation
Each pressurized space will have a radiation rating. This will differ with the
radiation that can enter through openings in pressurized enclosures and affect the
interior environment. Localized dosimeters will be used to calculate and track the
radiation that each citizen is exposed to throughout the day/week. If an individual
exceeds their exposure limit, the system will notify the citizen and their healthcare
practitioner to medically intervene and make changes to their routine in order to
reduce future exposure.

Social, Cultural, and Aesthetic Experience
Leisure in Phlegra Prime will encompass the majority of leisure activities available
on Earth. Institutions like libraries, opera and movie theaters, art and history
museums, artificial beaches and waterparks, shopping, restaurants, casinos, and
nightclubs will all be present. Parks will be present in every district, with their
biome, plant and animal life, and weather differentiated allowing citizens to
experience a multitude of Earth climates. The differentiation between the different
districts will encourage citizens to travel throughout the city-state and interact in a
rich social and cultural life. Animal life will be present in close proximity with
humans, from pet animals (such as cats and dogs), wildlife (such as squirrels and
songbirds in park areas), to zoos, containing fauna that will be native born Martians'
only direct interaction with Earth's biosphere. Personal entertainment, such as
books, movies, and video games, will be transmitted from Earth in a digital format.
Expeditions onto the Martian surface will also be encouraged and allow citizens to
enjoy their local environment.

Phlegra Prime is designed to be a pleasant place to live, with park areas complete with high ceilings, natural light, and plant and animal life to alleviate the sense of claustrophobia. Public transportation as well as district streets will be regularly cleaned and sanitized, to improve the perceived quality of public spaces. Pets will be present for necessary companionship, and there will be designated areas to walk or exercise with these creatures. The life support systems will allow each district to have its own unique climate.

ECONOMICS

Near Term Business Base and Initial Investment
In order to ground the concept for Phlegra Prime in reality, the city-state must be assessed as both a near term business case and a long term economic growth project. The guiding philosophy is that the settlement be profitable from day one. Demonstrating that money can be made on Mars, even if marginal, will show that future investment in growing the city-state is worthwhile.

Martian settlement is a highly speculative business venture. Instead of trying to find a "killer app" that justifies investment in Mars, focus must be on how existing terrestrial markets would make use of the revolutionary low transportation costs ($500/kg) enabled by Starship. Assuming an average of 90 kg per passenger, a one way ticket to Mars is approximately $45,000 with a round trip being $63,000. This is comparable to the price of a commercial round trip ticket to the South Pole [Ref 3]. Given the similarity in the cost of transportation, it is reasonable to assume that Earth's Polar Regions are analogous to Mars. Ventures that are profitable in the Antarctic and High Arctic would also most likely be viable at Mars.

Table 1 depicts total economic turnover of two permanent human settlements in the terrestrial Polar Regions, the U.S. Antarctic Program (USAP) in Antarctica and the Norwegian settlement on Svalbard in the High Arctic.

Table 1: Near Term Settlement Market Evaluation Based on Terrestrial Analogs			
Market Segment	Antarctica Turnover ($M)	Svalbard Turnover ($M)	Combined Turnover ($M)
Scientific Research and Education	535	55	590
Tourism	2360	65	2425
Mining	0	62	62
Internal Services	0	130	130
Indirect Subsidy	0	68	68
Total	2895	380	3275

The permanent bases in Antarctica rely on two main sources of revenue, direct government subsidies in the form of a system of scientific research bases and commercial tourism/expeditions. Of interest to note is that revenue from commercial expeditions is almost 4x higher ($2.36B USD, [Ref 3, Ref 4]) than the government funded research bases ($535M USD [Ref 5]). That said, most of the commercial operators make use of runways, fuel depots, navigation aids, and other infrastructure provided by the government research program. This is especially true for operations near the South Pole.

The other examined analog market is Svalbard, the northernmost permanently inhabited settlement on Earth. Whereas Antarctic settlements are permanently occupied bases, the towns of Svalbard are actual human communities complete with children and the elderly. Like Antarctica, research and tourism make up large fractions of Svalbard's economy. However unlike the Antarctic, Svalbard also hosts limited commercial mining for export and a large internal economy, accounting for 34% of the overall turnover [Ref 6]. Like Antarctica there is also significant government subsidy only it is used to artificially lower the cost of living in Svalbard and encourage economic growth. Seeking out similar subsidies either in the form of direct investment or indirect subsidy/tax exemptions will play an important role in closing the business case for an early Martian city-state. The cases of Svalbard and Antarctica show that there is productive economic activity to be done in these extreme hard to get to environments. Even though both regions are somewhat similar to each other, there is still enough of a market to justify activity in both places. Mars as a location that is wholly unique from anything on Earth has scientific and aesthetic value that may allow it to capture a larger market than either Antarctica or Svalbard can on their own. Table 1 depicts a credible upper bound for the current market potential of a Mars settlement, by showing the combined revenue of all activity on both Svalbard and the Antarctic.

Based on these existing markets, the early phases of the settlement will serve two primary markets: tourism and research. From this can be determined the conditions under which an initial settlement of 100 people (the capacity of Starship) would be profitable. The two greatest impacts on profitability are the nonrecurring engineering (NRE) cost to design and build the habitats and the resupply cost of supporting the settlement. The major assumption in this analysis is that the first settlement of 100 will be self-sufficient on the bulk goods of food, water, and oxygen. According to the NASA ALS Baseline Values and Assumptions Document (BVAD), astronauts on long duration missions require 45 kg of essential non-bulk goods such as clothing, toiletries, and medical supplies each year [Ref 7]. A 100 person settlement will therefore need to produce enough revenue to pay for the importation of **9750 kg** of essential supplies every 26 months via Starship at a cost of roughly **$2.25 million** per year. This represents only **0.07%** of the total market evaluation. The majority of the revenue will go to paying back the NRE that went into developing the initial habitat in the first place. This will be a sophisticated first of its kind system capable of harvesting basic resources from the Martian environment and growing enough food to support the settlers and visiting tourists/researchers. Instead of trying to estimate what the NRE would be for this system, a more useful metric is the max allowable NRE cost for the given market conditions. I.e. the maximum cost for the project that would still allow for net return. This analysis assumes a 30 year payback period and a 7% interest rate as this seems to be common for large infrastructure and municipal development projects. For the most pessimistic market case, Svalbard, NRE must be capped at **$5 billion** in order for the venture to return a net profit (about $7one million per year). This does not represent the value of the initial investment in the city-state, but that serves as a conservative upper bound for how much money should be spent on building the first phase of the Phlegra Prime.

Growth and Corporate Governance
Initial profitability is only the first step in building to a settlement of one million. The Earth-facing external economy of tourism, research, and education will pay for the import of crucial supplies but it is the growth of the internal economy that will cause the city-state to scale. This internal growth will only happen if revenue from the external economy is continuously reinvested in the creation of new Martian enterprises instead of going into the pockets of shareholders back on Earth. In order to encourage this the economy of Phlegra Prime will utilize cooperative corporate structure. The structure of each Martian cooperative business is inspired by the highly successful Mondragon industrial group that has operated in the Basque region of Spain since 1956. On the surface a Mondragon cooperative appears nearly identical to any other company with levels of management, shareholders, and a CEO. The main difference is that the shareholders are also the employees who in turn democratically elect the CEO and board. Upon joining a cooperative, employees are required to purchase equity in the venture (this can either be up front or paid off over successive years of employment) which gives them voting rights in the cooperative General Assembly, a body of all employee owners [Ref 8]. The Assembly convenes once per year to review the cooperative's finances and elect new members to the board of directors who serve 4 year terms. Every employee has an equal vote regardless of tenure except the cooperative's founders who get two-times the equity as a reward for starting the business. The two main advantages of this form of democratic leadership is an equitable wage distribution and profit sharing. In their charter Mondragon cooperatives set a maximum pay ratio between the highest and lowest paid employees. Doing this reduces income inequality and incentivizes high pay for lower level workers which in turn leads to the creation of a strong middle class that will grow the Martian internal economy. Traditionally Mondragon has set this at 6:1 [Ref 8] but Martian businesses will enact a pay scale of at least 10:1 in order to incentivize the cultivation of quality managers and executives. 45% of all cooperative profits will be divided up equally amongst the employee owners. This money is put in accounts analogous to 401Ks or HSAs and will help pay for retirement and healthcare. In order to incentivize near term employee performance, employee owners will be able to sell a fraction of their equity back to the co-op during liquidity events that coincide with major work anniversaries. The remaining 55% of the profits will be divided between a reserve fund (45%) and a community fund (10%). The reserve fund is what the cooperative will draw on to purchase new equipment, train/retrain workers, and weather difficult financial periods. The community fund will be used to fund small development and

charity projects in the employee's local community. This profit distribution scheme makes sure that corporate returns will continuously be reinvested in the Martian economy in one form or another.

The initial 100 person settlement outlined above could be run as a single cooperative or as a collection of cooperatives addressing different businesses (tourism, research, education, etc.). Mondragon limits each cooperative business to no more than 500 people in order to encourage communication and democratic decision making. If a cooperative grows past a certain size, it is required to form a new cooperative as a subsidiary. Multiple cooperatives working together to provide a single product or service are called a group. A group operates like any large corporation with the individual co-ops acting like business units. For example a group that makes home appliances might have separate co-ops for different products but also separate co-ops for things like sales, procurement, and marketing. The democratic management structure is repeated with the group having its own General Assembly made up of all the members of all the member co-ops who elect the group CEO and board.

These co-ops are not monoliths. Although they are internally operated differently than standard corporations, they are still business. A mature Martian economy will have multiple co-ops competing in the same market and incentivize people to leave their current co-ops and start new ventures. Doing this will require the establishment of three meta co-ops or what the Mondragon group refers to as Tier 2 co-ops [Ref 8]. These are the basic economic infrastructure required for growth and consist of banking co-ops, insurance co-ops, and education co-ops. Banking co-ops are essentially credit unions that will finance the creation of new cooperative ventures by providing loans. These banks are critical to the success of the cooperative model because they will allow for equity to remain with the cooperative founders and employees instead of being diluted as is the case with venture capital. Insurance co-ops will underwrite other businesses and will even provide health insurance and health care to policy members and their families. A large cooperative group might have its own insurance cooperative. The final type of tier 2 cooperatives are education co-ops. These will consist of secondary, vocational, and collegiate educational institutions that will train and retrain the workforce. These institutions are funded via a combination of tuition and endowments from the reserve funds of other cooperatives. For example, on Earth the Mondragon group currently operates several schools to cultivate talent in everything from machining to business law [Ref 9]. These education co-ops will also engage in R&D on behalf of their sponsor co-ops in order to make sure that their product offerings stay competitive. As tier 2 co-ops all of these institutions will have General Assemblies consisting of both employees and external stakeholders who they interact with. For example the education co-op General Assembly will consist of all faculty, staff, students, and CEOs of the co-ops who sponsor the endowment. The banking co-op will have representation from account holders and borrowers in its Assembly. Unlike most other co-ops the voting power will not be evenly distributed amongst the members of the General Assembly.

The rate of economic growth in the Martian economy will fundamentally be a function of how quickly the population can grow. As described by Zubrin et al. [Ref 10] there will be a virtuous cycle where the labor shortage on unpopulated Mars incentives each new immigrant to maximize their economic productivity. What follows are the results of a simple population growth model that predicts how long it will take for Phlegra Prime to reach a population of one million people. The model assumes a modest birth rate of 11 per 1000 people and a death rate range of 12.4-7932 per 100K people (depending on age) which are comparable to the industrialized world today [Ref 11]. Most of the population growth will be via immigration which is assumed to grow at a rate of **2%** per launch window to Mars. This is a reasonably conservative estimate for the rate of growth in space travel (for context this is about 3x less than the historic growth rate for passenger air travel, [Ref 12]). Given these assumptions, it will take approximately **263 years** for the population of Phlegra Prime to reach one million people.

Vision of a Mature Martian Economy
By the time the population of Phlegra Prime reaches one million, approximately 73% of the population will be of working age. A major benefit of the cooperative paradigm is that it will guarantee near full employment as co-op groups will typically have sufficient reserve capital to retrain members and put them in new ventures instead of laying them off during adverse economic conditions. With one million citizens, Phlegra Prime will need to have an advanced industrial economy that is producing the vast majority of goods and services consumed on Mars. The economy will consist of at least 1500 co-ops (max 500 people in a co-op) ranging from massive industrial and financial conglomerates to small independent ventures like shops and restaurants. Assuming a per-capita GDP that is comparable to the developed world, Martian GDP will be approximately **$63B** which is comparable to the state of Rhode Island or the city of Naples, Italy. Using small economies like these as analogs to the fully developed Martian settlement suggests that the vast majority (~95%) of the economic activity will be internal goods and services [Ref 13].

TECHNICAL DESCRIPTION

Interplanetary Transportation and Spaceport
Based on the population growth model at least 12,000 people will be immigrating to Phlegra Prime each synodic period. For the purpose of an initial feasibility analysis, this assumes that they are transported in a future variant of Starship i.e. a LOX/Methane vehicle with sufficient delta-V to fly back to Earth. Transporting 12,000 people will require 120 Starships, all of which will arrive at settlement within a period of approximately one month (synodic launch window) and which must wait on the Martian surface for 500 days prior to departure. The spaceport, depicted in Figure 5 will consist of ten 85 m diameter launch/landing pads and will handle at least 4 launches and 4 landings per day during the peak of the arrival/departure window. Each pad will be surrounded by a 10 m high berm to prevent Starship from kicking up dust and will be centered in a 1.4 km wide

exclusion zone in order to prevent collateral damage in the event of a launch vehicle failure. LOX/Methane propellant will be stored in underground tank farms and Starships will be stored on an apron between missions. Departing Starships will launch east out over the plains of Arcadia Planitia in order to take advantage of the planet's rotation. The spaceport will be on the eastern edge of the city-state so that launches occur over uninhabited areas.

Figure 5: Phlegra Prime Spaceport [Background Image Ref 14]

Manufacturing on Mars: Bootstrapping to an Industrial Base

Manufacturing on Mars using indigenous resources is the key enabling technology to allow for a sustainable and expanding human presence on the Martian surface. Thankfully, Mars is a world rich in resources and the material inputs for most manufacturing is present on Mars. Industrial infrastructure on Mars can be bootstrapped starting with a relatively small mass of manufacturing capabilities and expanding over time gradually reducing the required input materials from Earth.

90% of the mass of a modern six axis CNC mill can be manufactured on the mill itself with wiring, motors, sensors, electronics, and bearings making up the remainder [Ref 15]. Overall, the initial seed manufacturing capability should be able to replicate 80% of landed mass, allowing a doubling of manufacturing output with 20% of the mass, allowing future imports from Earth to focus on new capabilities. A spiral boot-strapping approach will enable a Mars colony to work its way through industrialization to the point where only high precision goods such as processors and sensors will be imported in relatively few synodic windows.

As a prerequisite to manufacturing, a megawatt class fission reactor will be landed at the future site of Phlegra Prime ahead of the deployment of the first spiral of manufacturing capabilities.

Spiral 1 Capabilities ~50 tons landed
- Regolith Excavator
- Gas Miner
- Hydrogen Reduction Smelter
- Plastic Synthesis plant
- Glass Works
- Wire/Rebar Plant
- Electron Beam Metal Printer
- Six Axis Machining Center
- Contour Crafting Machine

Civil engineering works will likely be the first manufacturing conducted on Mars. Sulfur concretes have comparable or better properties to terrestrial concretes and do not require active curing in Martian ambient conditions [Ref 16]. Sulfur exists on the Martian surface in the mineral Troilite (FeS), which outgasses sulfur when heated to 300C. Regolith can be enriched with 20% sulfur and then heated and deposited like a thermoplastic for form crafting, paving roads and landing pans, and other civil works. The metallic iron from this process can then be magnetically separated for use in the production of metal feedstocks.

Steel can be produced readily on Mars by the reduction of iron oxide, to produce metallic iron which can then be combined with carbon extracted for the Martian atmosphere [Ref 10]. Locally produced steel can then be drawn into wire usable by electron beam metal additive systems. Such systems are limited in scale terrestrially because of the need for printing to be conducted under high vacuum, on the Martian surface the ambient pressures are low enough to enable E-beam welding without a vacuum enclosure [Ref 17]. Basic geometries can be directly printed, with complex and precise components loaded onto a six-axis mill for finish machining.

Plastic synthesis can be readily undertaken by utilizing carbon-monoxide produced from the high temperature electrolysis of atmospheric carbon and hydrogen split from local water to produce syngas. Using non-consumable catalysts imported from Earth, syngas can be used to produce first ethylene and then polyethylene. Polyethene is a high-performance engineering polymer that can be used in applications ranging from habitat pressure vessels to small consumer items. The initial polymers facility will produce HDPE in the form of sheets and uniform particles for selective laser melting. Limited amounts of nylon can also be produced using nitrogen extracted by the air mining system, though this should be limited in applications since its production will be limited by nitrogen extraction which will be in high demand for life support applications. Polymers can also be drawn into fibers to allow the production of textiles and indigenously produced clothing.

The identification of local sources of copper will be a key activity during this period as the production of copper wire for use in electrical distribution and electric motor production will be a highly mass payback capability.

Gas processing during spiral 1 makes use of the Sabatier reaction supplemented with hydrogen and producing methane, and the Bosch reaction which will be used for the production of elemental carbon, for the regeneration of oxygen gas and the production of rocket fuel. The majority of metabolic carbon dioxide in the city will be recovered through bioremediation. Gas filtration systems will be used to separate individual gasses to purities of 95-99.5%. Buffer gas for habitat pressurization will be harvested from trace nitrogen in the Martian atmosphere. Higher purities may be required for breathable nitrogen, and will make use of a second step of pressure swing adsorption [Ref 18].

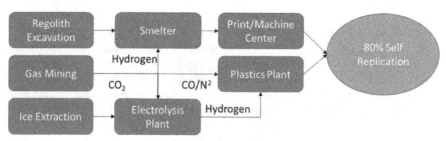

Figure 6: Initial Settlement Industrial Plant

Spiral 2 Capabilities
During the first spiral period the industrial output capabilities of the city-state will undergo several doublings as imported components are added to locally produced metallic and polymer components. The next phase of industrial capability will come about as the colony seeks to move up the value chain and further reduce the mass of goods imported from Earth. The production of metals suitable for aerospace applications require electrolysis and thus significantly more input power than steel. Luckily, nearly all the mass of smelter to enable the production of aluminum and titanium can be manufactured from the boot-strapped industrial base of spiral 1.

Ilmenite (FeTiO3) is common on Mars, and can be reduced to using hydrogen to produce titanium dioxide. Titanium dioxide can then be fed into the Cambridge process to produce metallic titanium, requiring approximately 17 kWh per kilogram of total energy input.

There are known thorium rich regions on Mars as well as indications that Uranium is present on the Martian surface [Ref 18]. At this point sufficient manufacture capability should be present on Mars to enable the indigenous production of reactor components. Indigenous Martian thorium could be bred with imported enriched uranium to jump start a self-sustaining Martian power industry.

Silicon production will also begin during spiral 2. The production of silicon will allow for the production of high-quality optical glasses, but also can be used in CVD furnaces to produce high purity silicon wafers to enable the production of integrated circuits. The production of even simple logic and low-quality memory indigenously would be a significant step towards Martian independence and will be undertaken in this spiral.

Spiral 3 capabilities
The third spiral of Martian industrial development will leverage the previous two spirals to allow Phlegra Prime to approach material independence from Earth, though importation of goods will always be desirable, the city-state will be able to replace any imported good with an indigenous substitution, though in some cases that substitution may be of lower quality.

Spiral 3 will see the importation of the components for extreme ultra-violet (EUV) lithography machines that cannot be produced locally. Doping agents that haven't been yet identified or extracted on the Martian surface may also be imported to allow for the full spectrum production of microprocessors and memory on the Martian surface.

The synthesis and extraction of complex pharmaceuticals will also be expanded in the third spiral. The infrastructure of chemical engineering is relatively straightforward and the physical infrastructure should be possible to manufacture at the beginning of spiral 2. Adapting known synthesis processes to the Martian environment and feedstocks will necessarily be a process of trial and error and the application of human skill. It will likely be the work of a generation to replicate the synthesis processes for common pharmaceuticals on Mars, but that work should have borne its fruits by the end of spiral 3.

Power Generation
The power system will be composed of a baseload fission power system which is supplemented during daylight hours with photovoltaic panels which provide offsetting power during periods of peak demand and a back-up power during periods of reactor maintenance or refueling. Power generation will be distributed so that each district will possess its own reactor and photovoltaic plant designed to produce excess power relative to the needs of the district. This will enable a city-state wide distributed power system which will ensure a significant degree of fault tolerance so that critical life support infrastructure can be powered even in the event of multiple parallel failures.

The initial settlement will use imported solid core reactors with terrestrially sourced fissile fuel. During the buildup period the imported fissile fuel will be used as the starter fuel for molten salt reactors using indigenously sourced thorium in a breed-burn cycle [Ref 19]. If and when uranium ore is identified on the surface an enrichment capability will be established, though 235U based fission systems will be reserved for fuels for mobile systems. Fission fuels for spacecraft sourced outside Earth's gravity well and regulatory reach may prove to be a significant source of revenue for the colony [Ref 20].

Agriculture and Life Support

The agricultural design of Phlegra Prime will balance three things. First, growing crops that are, today, most consumed around the world, as to accommodate immigrants from as many cultures and countries as possible and to provide dietary diversity. On this note, the design will avoid crops that commonly trigger allergic reactions. Second, the agricultural system will focus on crops that already have heritage being grown indoors or in soilless systems due to the present uncertainties in Martian soil plant growth viability. Finally, while the government of Phlegra Prime will not dictate the diets of its citizens, the agriculture system will meet crop growth requirements to provide a million citizens the US Health and Human Services and US Department of Agriculture's caloric dietary guidelines for their age, along with the World Health Organization's "Healthy Diet" guidelines [Ref 21, Ref 22].

All staple crops such as rice, wheat, maize, and cassava will all be grown in a centralized "agriculture" district, 20 square kilometers in size, for ease of monitoring, intervening, and processing [Ref 23]. By using dwarf varieties of crops, this crop growth area will be able to be effectively stacked vertically [Ref 24]. This agriculture district will also produce, handle and distribute protein sources such as lab grown meats, chickens, eggs, and black soldier fly larvae. Each district will have a dedicated space for fresh fruits, vegetables, and herbs to be grown [Ref 25].

The habitat environmental control and life support system (ECLSS) will have five primary functions, CO_2 removal, oxygen supply, temperature/humidity control (THC), trace contaminant control (TCC), and water supply. A bioregenerative life support system (i.e. plants) has long been seen as the answer to meeting these functions because it offers the promise of near resource closure. It may at first seem obvious to tie food production into the bioregen ECLSS, however previous simulations of Martian settlements [Ref 23] suggest that these two systems should be separated. Instead the habitat will make use of a system of algae bioreactors called water walls, currently under development by NASA [Ref 22]. Water walls are an inherently modular and scalable system that can be implemented in any part of the settlement. The elements are a series of bags that can be arrayed to fit any habitat configuration. Some bags use algae to convert CO_2 to oxygen, others absorb humidity out of the air via osmosis, and others take human waste and convert it into fertilizer [Ref 22]. The system has almost no moving parts and elements that wear down can easily be replaced. They also have the benefit of acting as additional radiation shielding. Approximately 2200 kg of material, mostly in the form of water, will be required to support the metabolic needs of each inhabitant [Ref 23].

Although this is more massive than many physio-chemical architectures, the availability of water ISRU at Phlegra Montes enables this architecture to scale.

POLITICAL SYSTEM

Levels of Government
The infrastructure of a space habitat is so fragile that relatively minor civil disturbances and political disagreements will pose a threat to the safety of the entire population. The reality that a small group of people could shut off life support, puncture the habitat, or open the airlocks will create an immense pressure to restrict the actions of the citizenry resulting in tyranny. Responding to these challenges requires the creation of a system of government that empowers all citizens to be part of the political process, make sure that decisions actually reflect the consensus of the body politic, and enshrine individual rights and freedoms above all else.

The government of Phlegra Prime will be divided between three levels, city-state, district, and neighborhood.

Figure 7: Levels of Government

Separation of powers and responsibilities will be such that governance is bottoms up. Most decisions will be delegated to the neighborhood level, encouraging a vibrant participatory democracy. In order for this system to truly represent all inhabitants, citizen rights are automatically extended to anyone who establishes residency within the city-state. When selecting immigrants from Earth, Phlegra Prime will strive to have the demographic makeup that evenly represents Earth's population. If required to achieve this goal, Phlegra Prime will provide, to individuals, merit based scholarships that will subsidize skills training before departure and the cost of the passage to Mars. Phlegra Prime will be open to all of humanity regardless of race, color, religion, national origin, sex, physical or mental disability, or age.

City-State Government
Like many terrestrial governments Phlegra Prime's will consist of three branches, an executive, legislative, and a judiciary with separation of powers and checks and balances:

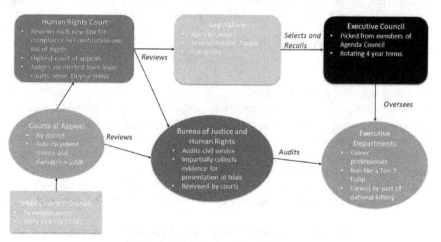

Figure 8: National Government

This three branch system is where the similarities end. Perhaps the biggest departure from current terrestrial governments is the legislature, depicted below in Figure 9. This system is a form of direct democracy directly inspired by the structure of ancient Athens [Ref 26]. Instead of selecting representatives by voting (election) representatives will be selected by lottery (Sortition). The first advantage of this is that it is possible to pick a true statistically representative sample of the population. Statistically, a random sample of at least **384** citizens would represent one million people with a 95% confidence level and 5% error (first past the post voting often has error in excess of 40%, [Ref 27]). This would ensure that the legislature will reflect the citizenry in every metric such as gender, race, disability, age, and political leanings. The second main advantage of Sortition is that it will completely subvert partisanship. Without having to continuously win elections there will be drastically less incentive to engage in partisan jockeying or gridlock. This will be a system that inherently encourages consensus over conflict. The third main advantage is that this system will encourage a strong civic virtue. The knowledge that anyone on the street could at any time be randomly called upon to serve in the settlement's highest office will incentivize citizens to both take an interest in political affairs and to think about how policy directly impacts the people around them.

The lottery will be open to all citizens above the age of eighteen who participate in the civilian service (discussed below). Each year a group of 400 citizens will be picked at random to participate in government. Terms will be staggered so that ⅓ of the legislature is turned over each year (total of 1200 legislators). Winners will be able to decline their slot in government but will also be compensated for taking the time away from their jobs in order to serve. Each representative will serve for a single term after which they will return to the general public. They will be eligible to be selected again but there will also be strict restrictions on what kinds of jobs or changes in employment can happen during their first few years out of office in order to prevent a revolving door.

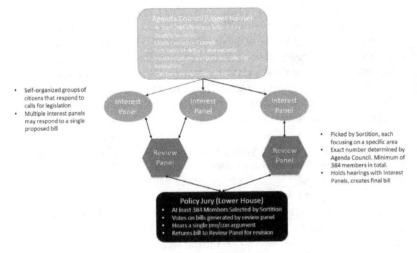

Figure 9: Legislature

Figure 9 depicts the structure of the legislature with its 2 major and 1 minor house. The Upper House will be called the Agenda Council and will be responsible for deciding what laws will need to be drafted/revised during a session and the rules of debate. Legislation will either be initiated by petition or within the Council. The Council will be selected by double Sortition, i.e. a yearly lottery from within the pool of 1200. Once the Agenda Council adds a law to the agenda, interest panels of ordinary citizens will respond with draft legislation. The interest panels will be open to all citizens and there may be several interest panels responding to a single call for legislation. The middle minor house, called Review Panels will be focused working groups that review draft legislation, conduct hearings, and will ultimately create the final bill (analogous to house subcommittees). Each Review Panel will be chartered to only weigh in on a certain area of policy. The exact number and role of each Interest Panel will be determined by the Agenda Council. Members of the Review Panels will be picked via double Sortition from a fraction of the pool of 1200. Once a conference bill leaves the Review Panel it will go to the Policy Jury. The Policy Jury will consist of the remainder of the 1200 who are not in the Agenda Council or in a Review Panel. Their job will simply be to vote yay or nay on legislation. Per rules set in the constitution, the jury will only be allowed to hear one brief pro and brief con argument before voting and the vote is final [Ref 26]. A yay vote will mean that the bill becomes law and the nay vote will mean the bill returns to the Review Panel for revision. All legislation proposed by the Agenda Council will have both a due date for when it must be reviewed by the Policy Jury and a sunset clause for when the law must return to the legislature for revision. Passed laws will automatically be reviewed by the judiciary for compliance with the Constitution and bill of rights, with the Human Rights Court effectively having veto power.

The executive will be led by a 4 person Executive Council who collectively serve as the heads of state. Each member will serve a single 4 year term and is elected on a rotating basis each year by the Agenda Council. All decisions by the Executive Council will be unanimous. The Executive Council will have the authority to appoint/dismiss members of the executive branch. The Agenda Council will at any time have the authority to cast a vote of no confidence and remove all 4 members of the Executive Council from office. After dismissal the Agenda Council will hold a new election to select 4 new members.

The judiciary will be divided between different levels of authority. The majority of crimes committed will be relatively minor, and will be handled by a local tribunal, which will be able to apply justice in a subjective way that better fits the will of the local population. Improperly handled local cases may be appealed to a higher court. Tribunal jurors will be elected by their community, while judges in the court of appeals will be elected by the citizens of an entire district. The highest court, the Human Rights Court, will serve both to consider appeals from lower courts and will review every law passed by the Agenda Council for possible human rights violations. Judges for this court will be nominated by the executive council, and will be approved by the Agenda Council with a 2/3 majority.

Rather than having defendants and prosecutors collect evidence, all evidence will be collected by an impartial government organization called the Bureau of Justice and Human Rights. Trials will consist of a set of evidence agreed upon by these parties, and overseen by an impartial judge. This prevents the emergence of a system in which defendants who are able to pay for more skilled investigators have a better chance of being cleared of charges.

Civilian Service
While Phlegra Prime has no standing military, a period of civilian service after the completion of school will be encouraged. This civilian service will allow individuals to experience work in the context of being part of the essential services of the city-state. These civilian servants may work in healthcare, maintain infrastructure for food production or life support, or even assist in the construction of new buildings. This means that all citizens will work to maintain Phlegra Prime, and work with a diverse set of individuals with different race, age, religion, and political views. In the unlikely event where a military is needed, citizens of a specific age range will be drafted until the conflict is resolved.

Bill of Rights
The city-state's constitution will guarantee a bill of rights, including freedom of religion, assembly, speech, and the press. Citizens also have the right to know all government activities and proceedings. The right to bear arms shall not be infringed. It will also guarantee the judicial rights of protection from warrantless searches, due

process, public trial by jury, and speedy trial. The constitution will also guarantee public transport, body integrity, internet access, housing, privacy, and health care. Citizens also have the right of free movement, protection from indentured servitude, and equal protection under the law. The bill of rights is designed to protect the lives and livelihoods of the citizens, but also to guarantee that they have the ability to lead enjoyable lives in the modern era. This foundation of personal freedom, guaranteed by the constitution and protected by the Court of Human Rights will ensure that Phlegra Prime develops a rich and diverse cultural life that is only possible in a free society.

Education System
The education model will be based on the International Standard Classification of Education (ISCED) 2011 levels of education [Ref 28]. In this system, children under 3 are considered level 0-, ages 3-5 are level 0, ages 6-11 are level 1, ages 12-14 are level 2, ages 15-17 are level 3, and university-aged students are levels 4 through 8, depending on the degree program. Adopting this international model, will allow both students and parents from around the world to effectively integrate into the new educational system. With just over 164 thousand students in the ISCED levels, 0- to 3, Phlegra Prime has a comparable student population to overall population ratio to cities such as Dallas, Texas [Ref 29, Ref 30].

Through thorough analysis of each student's pace and ability to learn topics, curriculum will be tailored to a student's individual learning styles and needs. This will replace conventional teaching strategies with a system more similar to online learning of today. Only a handful of staff members will therefore be able to facilitate the education of hundreds of students, stepping in only when there is an issue with the learning tool.

With around 56,000 students between the ages of 18-25, ISCED levels 4-8 will consist of many universities of around 5,000 students each, with each university having a specialized focus on topics such as arts, technology, science, construction, and humanities. Remote university enrollment from students on Earth will also be a method of financial income for Phlegra Prime.

REFERENCES

1. McEwen A. S., Sutton S. S., Bramson A. M., Byre S., Petersen E. I., Levy J. S., Golombek M. P., Williams N. R., Putzig N. E., 'Phlegra Montes: Candidate Landing Site with Shallow Ice for Human Exploration' Seventh International Conference on Mars Polar Science and Exploration, Available: https://ui.adsabs.harvard.edu/abs/2020LPICo2099.6008M/abstract
2. Wolfe, Jennifer. "MPC Creates Out-of-This-World Effects for 'The Martian' | Animation World Network." *Animation World Network*, 2015, https://www.awn.com/vfxworld/mpc-creates-out-world-effects-martian.

3. Antarctic Logistics and Expeditions, 'Dates and Rates' Available:
    https://antarctic-logistics.com/trip-finder/#view=grid
4. IAATO, 'Data and Statistics' Available: https://iaato.org/information-
    resources/data-statistics/
5. NSF, 'OPP Funding' Available:
    https://www.nsf.gov/about/budget/fy2019/pdf/30_fy2019.pdf
6. Statistics Norway, 'This is Svalbard 2016: What the Figures Say'
    Available: https://www.ssb.no/en/befolkning/artikler-og-
    publikasjoner/_attachment/294354?_ts=15a12de02c0
7. Anderson Molly, S., Ewert Michael K., Keener John, F.. 'Advanced Life
    Support Baseline Values and Assumptions Document' NASA JSC
    Available:
    https://ntrs.nasa.gov/archive/nasa/casi.ntrs.nasa.gov/20180001338.pdf
8. Clamp, Chsistina, A. 'The Evolution of Management in Mondragon
    Cooperatives' Southern New Hampshire University
9. Mondragon 'Mondragon Corporation- 2018' Video Available:
    https://vimeo.com/47333795
10. Zubrin Robert, Wagner Richard 'The Case for Mars: The Plan to Settle the
    Red Planet and Why We Must' Free Press 2011
11. Centers for Disease Control and Prevention, National Center for Health
    Statistics. Underlying Cause of Death 1999-2018 on CDC WONDER
    Online Database, released in 2020. Data are from the Multiple Cause of
    Death Files, 1999-2018, as compiled from data provided by the 57 vital
    statistics jurisdictions through the Vital Statistics Cooperative Program.
    https://wonder.cdc.gov/
12. Lett Joosung J., Lukachko, Stephen P., Waitz, Ian, A., Schaefer, Andreas
    'Historical Trends in Aircraft Performance, Cost, and Emissions'
    Available:
    http://web.mit.edu/aeroastro/sites/waitz/publications/Ann.Rev.reprint.pdf
13. Office of the United States Trade Representative, 'Rhode Island'
    Available: https://ustr.gov/map/state-
    benefits/ri#:~:text=Rhode%20Island%20was%20the%2046,Rhode%20Isl
    and%20GDP%20in%202018.
14. "[38+] NASA Mars Desktop Wallpaper on WallpaperSafari."
    WallpaperSafari, Jan. 2018, https://wallpapersafari.com/nasa-mars-
    desktop-wallpaper/.
15. 4th Space manufacturing; Proceedings of the Fifth Conference. 18 May
    1981 - 21 May 1981. Princeton,NJ,U.S.A https://doi.org/10.2514/6.1981-
    3226
16. Toutjani, H. A., Evans, S., Grugel, R. N., 'Performance of Waterless
    Concrete' Structural Faults and Repair 2010; June 15, 2010 - June 17,
    2010; Edinburgh, Scotland; United Kingdom, Available:
    https://ntrs.nasa.gov/archive/nasa/casi.ntrs.nasa.gov/20100026417.pdf
17. Hooper, William H., Martin Marietta Corporation, 'Final Report: On-Orbit
    Electron Beam Welding Experiment Definition'. Available:
    https://ntrs.nasa.gov/archive/nasa/casi.ntrs.nasa.gov/19900006115.pdf

18. JPL Mars Odyssey: Midlatitude map of Martian Thorium
    https://photojournal.jpl.nasa.gov/catalog/?IDNumber=PIA04257
19. Kerry, Frank G. *Industrial Gas Handbook Gas Separation and Purification*. CRC Press, 2007.
20. Angelo, J. A., & Buden, D. (1985). Space nuclear power. Malabar, FL: Orbit Book.
21. Presidential Memorandum on Space Nuclear Launch Reform
    https://www.whitehouse.gov/presidential-actions/presidential-memorandum-launch-spacecraft-containing-space-nuclear-systems/
22. "Appendix 2. Estimated Calorie Needs per Day, by Age, Sex, and Physical Activity Level - 2015-2020 Dietary Guidelines | Health.Gov." *Home of the Office of Disease Prevention and Health Promotion - Health.Gov*, ODPHP, https://health.gov/our-work/food-nutrition/2015-2020-dietary-guidelines/guidelines/appendix-2/. Accessed 30 June 2020.
23. "Healthy Diet." *WHO | World Health Organization*, 29 Apr. 2020, https://www.who.int/news-room/fact-sheets/detail/healthy-diet.
24. SM, GOMEZ & Montoya, María & Atanbori, John & French, Andrew & Pridmore, Tony. (2019). A low-cost aeroponic phenotyping system for storage root development: unravelling the below-ground secrets of cassava (Manihot esculenta). Plant Methods. 15. 10.1186/s13007-019-0517-6.
25. Cruthirds J, Kloeris V., 'Mathematical Modeling of Food Supply For Long Term Space Missions Using Advanced Life Support' NASA JSC, Available: https://ntrs.nasa.gov/archive/nasa/casi.ntrs.nasa.gov/20030064056.pdf
26. "Sustainable Bananas in Greenhouses: First 'Dutch Bananas' Harvested - WUR." *WUR*, Wageningen University & Research, 2018, https://www.wur.nl/en/newsarticle/Sustainable-bananas-in-greenhouses-first-Dutch-bananas-harvested.htm.
27. Bouricius T.G. 'Democracy Through Multi-Body Sortition: Athenian Lessons for the Modern Day' *Journal of Public Deliberation* Vol 9 Issue 11 Available: https://delibdemjournal.org/articles/abstract/10.16997/jdd.156/
28. CGP Grey 'Why the UK Election Results are the Worst in History' Video Available: https://www.youtube.com/watch?v=r9rGX91rq5I
29. UNESCO Institute for Statistics 'International Standard Classification for Education: ISCED 2011' UNESCO Available: http://uis.unesco.org/sites/default/files/documents/international-standard-classification-of-education-isced-2011-en.pdf
30. United States Census Bureau, 'Quick Facts: Dallas City, Texas' Available: https://www.census.gov/quickfacts/dallascitytexas
31. Dallas Independent School District '2016-2017 Annual Report' Available: https://www.dallasisd.org/Page/27056
32. SpaceX 'Starship' Available: https://www.spacex.com/vehicles/starship/
33. Cohen, Marc M., Flynn, Michael T., Levy, Francois, Mancinelli, Rocco, Matossian, Rene, L., Miller, Jack, Parodi, Jurek, 'Water Walls Life Support Architecture: 2012 NIAC Phase I Final Report' NASA Available:

https://www.nasa.gov/sites/default/files/atoms/files/niac_2012_phasei_fly nn_waterwallsarchitecture_tagged.pdf

34. Do, Sydney, Koki Ho, Samuel Steven Schreiner, Andrew Charles Owens, and Olivier L. de Weck. 'An Independent Assessment of the Technical Feasibility of the Mars One Mission Plan' IAC Available: https://dspace.mit.edu/bitstream/handle/1721.1/90819/de%20Weck-An%20independent.pdf?sequence=1&isAllowed=y

35. Michael T. Flynn, Marc M. Cohen, Renee L.Matossian, Space Architect, Dr. Sherwin Gormly, PhD, Dr. Rocco Mancinelli, PhD, Dr.Jack Miller, PhD, Jurek Parodi and Elysse Grossi 'WATER WALLS ARCHITECTURE: MASSIVELY REDUNDANT AND HIGHLY RELIABLE LIFE SUPPORT FOR LONG DURATION EXPLORATION MISSIONS' Available: https://ntrs.nasa.gov/archive/nasa/casi.ntrs.nasa.gov/20190001191.pdf

36. Sheila Baber, Eva Birtell, Kaixin Cui, Alexa Escalona, Benjamin Greaves, Kevin Kempton, 'CYBELE Advanced Concept of Operations' NASA Game Changing Development Program Available: http://bigidea.nianet.org/wp-content/uploads/2019/10/CYBELE-final-2019-10-18.pdf

15: FOUNDATION

Michel Lamontagne - Designer of Surya
Laurent Gauthier- Lead writer
Jean-Marc Salotti - Albatross designer
Florent Bednarek - Webmaster
Carl Vambert – Real Estate
Olivier Gourdon - Illustrator, designer, Lucie's soul
Rémy Navarro - Designer of Arkadia
Lucie – Special Reporter
michel.residence@gmail.com
laurent.fy.gauthier@hotmail.fr
jean-marc.salotti@ensc.fr
c.vambert@hytoo.fr
bednarekfl@eisti.eu
olivlugdunenses@gmail.com
contact@planetscopemagazine.com

Arkadia City, Foundation, Mars Planet

Lucie's notebook
Sent to Mars for 2 years, our special reporter Lucie will explain her experience on each part relied on the theme.

Arrival at Arkadia. "Of all the questions I asked myself before leaving Earth, the most heady, the simplest and the most important was undoubtedly the one concerning "these other humans": Would these brothers from another planet accept me? Now, I find myself dreading their look. It's silly, but that's what I feel as we approach Mars. In the imminence of a landing on it's crazy spaceport, my heart is racing, thinking of all the intrusive questions that I will have to ask, all these avenues of human civilization that I will have to walk in search of an answer. I will have to understand what is the pride of Arkadians and Suryans, the inhabitants of this strange city state known as Foundation, where each new green plot of 'land' has been won at the price of efforts and ingenuity, bringing life to a planet which was devoid of it. The steam clears from the tarmac. The airlocks opens and a new smell reaches my nostrils. I am on Mars."

GEOGRAPHY OF FOUNDATION

The Foundation City State regroups two main settlement poles and a smattering of small settlements approximately aligned on the 170° West meridian. The two settlements, Arkadia and Surya, often described as the city of water and the city of the sun, are linked by the Meridian Way, a 2,200 km long road that is the backbone of the Foundation city state. Arkadia, located in Arcadia Planitia is the oldest human settlement on Mars. It developed from the first inhabited base on Mars, chosen for easy access to large amounts of water ice. Surya is a younger settlement, located in Nicholson crater near the Equator. The presence of water ice and useful minerals in the nearby Medusae Fossae region guided the choice for this location. The two settlements benefit from a fairly low altitude, respectively -3798 m and -4423 m, providing an atmospheric pressure of about 100 kPa. The Suryan settlement tends towards a greater adaptation to the Martian environment when compared to Arkadia. Surya' investment in solar energy is highlighted by the use of surface greenhouses and space mirrors for food production, rather than the use of underground grow rooms and bioreactors. The architecture is also based on spaces that are more open to the outside, with a more relaxed attitude towards radiation, rather than the initial reliance tunnels and massive shielding found in Arkadia.

AREOPOLITICS

Mars global diplomacy. The first human settlements on Mars were the results of public-private partnerships, whose governance was ensured by consortiums of one or more Earth nations. Their rules were necessarily complex and often poorly adapted to local conditions. The requirements of daily survival on Mars highlighted the need for rapid and collectively accepted decisions, and irresistibly called for the independence of Martians. The natural arrangement of the Martian settlements into cities gave rise to a planet wide civilization of sovereign City States, made of one central city or a coalition of cities. Foundation is the largest and most famous of these coalitions, and to most people on Earth, it is basically Mars.

Mars Congress. At the planetary scale, a consultative body, called The Congress of Mars was quickly established. It emerged to set up and run the Mars Planetary Rescue Force, nicknamed, with a certain irony, the Red Helmets. This organization first administered the contributions of each city (field hospitals, suborbital hoppers, staff) then became the privileged place for discussion on other global themes. The main topic of discussion at this level concerns the terraforming of the planet, which is still in its infancy. A challenge system has been established and is based on bonuses won by the first City-State to achieve an objective linked to the ecopoiesis process. This healthy competition, which some call the "Olympic Games of Mars" is followed throughout the solar system and the subject to a whole betting system. Its winnings are used to finance the prizes and the compensation for prejudices to settlements penalized by the terraforming process. The Congress of Mars also functions as an arbitration court. If coexistence and cooperation are the watchwords of diplomacy on Mars, there is definitively a certain competition for access to resources, and relations between City-States are not free from disputes, and even conflicts.

Use of weapons and the 'Balance of Prudence'. Even if the means to produce weapons are available, there is little incentive to build them and use them. Using a weapon on Mars leads to the risk of rupturing the habitat walls, an act seen as an abject crime against Humanity and the Ecosystem that sustains it and deserving of the most severe punishment. In a similar way that the use of atomic weapons on Earth is restrained by the 'Balance of Terror'. Even if no text explicitly prohibits the use of weapons on Mars, each State knows that endangering a habitat, even in "collateral damage", would lead to all the other States allying against the perpetrator. No State has betrayed this tacit agreement, which is commonly called "Balance of Prudence".

Political organization of Foundation. Each Martian City-State is organized in a different way, with different entities representing the legislative, executive and judicial powers. Foundation operates as a direct democracy. Most public decisions are submitted to referendum by universal suffrage. Votes are electronic and authenticated by *blockchain*. Citizens' solicitations are permanent, most of the votes relate to relatively local decisions and the participation rate is generally low. The discretionary power of political staff is thus very limited, and it is sometimes necessary to resort to the drawing of lots to appoint officers. The Terran lobbies have regularly declared themselves bewildered by the lack of interest of politicians toward their precious advice. The political organization is based on the principle of subsidiarity, aiming to delegate the level of decision as low as possible. The basic level is that of the district, a tightly defined set of habitats serving of the order of 10,000 people, but which is in practice very variable. This district is administered by a mayor and municipal councilors. Each higher echelon has its officers elected by universal suffrage. Unlike on Earth, it's observed that favorite public positions are the lowest and local ones. Subsidiarity is particularly important for the judiciary and executive powers. For example, in Arkadia and Suria, if a crime is committed in a district, the judgement and the sentence are carried out at this level. If the crime concerns two districts, officers of the next echelon are mandated to solve the problem.

The Right to Life Support. The most famous constitutional principle of Foundation is called "the Right to Life Support". Despite its economically liberal culture, the Foundation society implements many costly social protection measures, which are readily accepted because they are subject to the rapid sanction of public control and direct democracy. Education, measures promoting birth rates, medical follow-up for all are considered inalienable public services. As is the Right to Life Support, which includes air, a minimum amount of water, energy and minimal housing. Indeed it is impossible to live outside the system on Mars. This highly decentralized but highly interconnected governance has enabled intense technological development, stimulated by the permanent emulation between districts and City-States. It has led to the selection of organizational methods derived from agile methods, but applied to the development of districts and cities and rule-making. The increase in productivity that resulted from this organizational revolution was comparable in magnitude to the revolutions of Fordism and project management techniques of the Second World War, but this time without the drawbacks of cultural standardization and exclusion from the decision process. These developments were paralleled by high population growth,, stimulated by the birth rate but also by immigration.

This is the Foundation society, very representative of all Martian societies: coalitions of hundreds of small districts, fiercely watching over their independence and uniqueness, but never hesitating to exchange information, and to vote for everything.

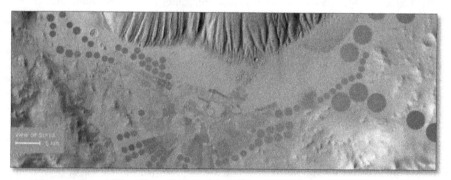

Surya City, Foundation, Mars Planet - Urban planning

TOWN PLANNING

Living on Mars is not without risks, but the design of the life support systems, and its continuous improvement, has reduced the risks to that of a well lived life on Earth. Life on Mars is no longer the one of a crew in an Antarctic station, but the one of a citizen of a comfortable, mid-sized cosmopolitan city. This could be achieved thanks to subsidiarity being taken as the fundamental governance principle. In terms of town planning, this results in maximum autonomy for the district of about 10,000 inhabitants. It has many advantages; it effectively ensures the empowerment of each citizen vis-à-vis the maintenance of the manufactured biological ecosystem that keeps him alive, as well as the sharing of practices between the different districts which leads to healthy competition. In addition, compartmentalization, maximized, ensures that a local accident does not compromise the whole of the city, which will then be able to come to the aid of the stricken district. Thus, each district is as autonomous as possible, both physically and in terms of production, but also in the management of public affairs. The principle of autonomy implies a second one: maximizing the proximity of the different areas necessary for the life of a citizen. Combining residence, work, shops, social and community areas into a single habitat is common practice. This principle combines as well with the fact that agriculture and the maintenance of ecosystems is one of the main activity of the inhabitants of Foundation, since it is the most labor-intensive. It is therefore necessary to reduce the distance between agricultural and urban areas. In this way it is possible to favor pedestrian displacements within a district, The journeys between the districts are made by a system of automated electric vehicles, rolling beneath the pedestrian walkways and through tunnels, integrated in the pressurized enclosures or themselves pressurized, depending on design. Solar energy farms occupy a large area outside of the city (several hundred of square kilometers, depending on local energy mix). They are organized as standardized 'solar islands', in which solar panels surround a central industrial sector, which include all facilities dedicated to solar panel manufacturing and maintenance, alongside with plants dedicated to methane/oxygen energy vector production and storage (for more details, see our article on energy).Finally, the systematic use of the Building Information Modeling (BIM) methodology for the design and operation of the city made it possible to optimize costs thanks to the creation of digital twins. There is room in the human imagination to invent new cities. And what better place than on Mars, where a city must be, by necessity, a mosaic of autonomous villages where each inhabitant, adult and child, is the gardener of the ecosystem that supports them.

Lucie's notebook

The Foundation Space History Museum. "That will be all for us, madam. To answer your concerns, no quarantine will be necessary - the journey is long enough!"It is a graying medical control doctor who concludes her only spoken sentence by waving me away with a gesture. I am allowed out of the glass office with a few drugs in a cellulose bag and a long list of recommendations. Faster than I imagined, after 4 months in weightlessness. It is now 5 days since I landed on Mars and severe leg pains remind me of the kilometers already covered in the streets of Arkadia. The slight gravity of Mars has helped me get used to the vertical station again, but my ungainly gait betrays my origin. My timid little steps and my uncontrolled leaps earn me smiles! I have not yet managed to imitate the long slow and elegant strides of the native Martians.

The local chapter of the Mars Society gave me an interview this morning and provided me with a guide, Adhémar. I have been thoroughly briefed with safety drills and practices and am free to move around. Adhémar offered to accompany me to a speech given by the Minister of Culture on the anniversary of the founding of Arkadia. We have to walk to get there. Again. Arkadians vigorous walking habits relegate the marathon runners of Earth to simple hikers. Everyone walks here. But I expect I'll get used to it soon. The Grand dome of the History Museum finally covers us. It is the most visited attraction on the planet. The speech is held in front of several hundred people. The majesty of the space allows me to listen to a rather inspired and fortunately brief speech (my feet are killing me!). Adhémar shows me some of the dome' treasures. I find myself facing the modules of the first Martian base, preserved intact. It is here, in the center of the city of Arkadia, that the first elements of the colony were set up. Further along, Adhémar, stops in front of a rather incongruous sculpture. For Adhémar, it is the object of a certain emotion: during the early manned tests of the first Martian built space vehicle, a team of astronauts decided to bring back to Arkadia the mannequin Starman at the wheel of his Tesla Roadster. Indeed, a few day after the Falcon Heavy launch, to the question "if I can recover it, can I keep it ?", Elon Musk, founder of SpaceX and owner of the vehicle had answered "Yes". The Martians have apparently taken up the gauntlet.

FOUNDATION SERVICES AND PRODUCTS

Note: G$/yr stands for billions of United States dollars per Earth years.

Tourism (11 G$/ yr). Mars welcome more than 50,000 visitors per synod! Residence offered with premium services:

- For scientists, engineers and others, on missions financed by the Earth.
- Tourists on the experience of a lifetime. Excursion to the surface on emblematic places: Marineris Valleys, the largest canyon in the solar system, Mount Olympus, the largest volcano in the solar system, The Ark of Tartarus Colles, visit of a belt asteroid aboard an Albatross, stellar cruise to Saturn aboard a Starship.
- Adventurers & Pioneers from the whole Solar System,
- Health, Cures for retired people, or anyone else: extension of life expectancy by reducing stress on the heart and arteries by lower gravity.

Real Estate - $ 69 G$/yr

- Licensing: Foundation offers licenses to exploit and explore the riches of Mars. A company is authorized to explore and extract mineral resources; in exchange it has a 30-year property rights and it must participate, via subsidies, to the extension of the city. It can also rent the operating site to other companies.
- Real estate investment: as it is the core job of each Martian settler to build a new world, it is obvious that Mars would be the new Eldorado for construction and real estate financing, compared to a saturated Earth where competition is fierce and profits tiny. Investments in Foundation currently represents a 28% growth share in the solar system property market. Average price: 10 k$/m^2. And remember, it is not a question of demolishing a planet to cover it with concrete, but of greening Mars!

Knowledge Economy - 59G G$/yr

- Education: the Mars Institute Technology (MIT) has designed a training offer focused on a high level of excellence in a number of fields: Agronomy, Sociology and Psychology, Engineering according to specialization, 3D printing, Robotics and AI, Circular Economy and Recycling. MIT has concluded research and development contracts with private companies on Earth.
- R&D and patents: Mars is a harsh place, and inventiveness is a necessity to survive. Innovation is so at the core of Foundation economy. Thanks to its R&D, Foundation draws profile of this for the benefit of Earthlings. It offers patent exploitation licenses for the following sectors: Medicine (cancer treatment, life extension, cardiovascular treatment, procreation assistance and many others, Food industry (GMO, above-ground, soil treatment), Climate and

environmental control, Engineering, Circular economy, Recycling and reuse of materials, Dry extraction of resources, Water Management, Energy.

The example of Earth ISRU: today the technologies developed on Mars to make steel from the regolith are sold under license to Earth and allow the production of high quality steel at a competitive manner directly at the mine site. Foundation collects 5% of royalties on the patent exploitation licenses on steel, this generates for Mars revenues of 40 billion $ / year. The patents related to the highly automated Foundation solar farms are nowadays best-sellers on Earth. All of Foundation university and patent revenues are placed with Bank of Mars. The interest of 5% is used to finance the extension of the colonial projects of Mars.

Space services (5,1 G$/yr):
- Equipment and launch: The Martian space industry is booming. Today it competes with Earth for the supply of space launch and equipment. Currently, Foundation offers the following services: Production and launching of geostationary satellite thanks to the advantage of ΔV compared to Earth, Design and launching of orbital stations, Asteroid drilling and mining equipment, Exploration probes to the outer planets and the asteroid belt, X33 in commercial or government version, Space Internet, Telecommunication equipment.
  Costs to orbit examples: LEO: $ 150/kg // GEO: $ 180/kg // Moon: $ 200/kg // Belt: $ 100/kg
- Supply: Foundation offers an extensive infrastructure for supply contracts for the Moon, orbital stations and the Asteroid Belt. As well as extended trips to the gas giants. Here is a non-exhaustive list of possible supplies: Food, Carbon, nitrogen, hydrogen and other light elements, Consumables and basic necessities, Tools and Machines, some as simple and as essential as toothbrushes, Propellant, Space habitats. As an indication, the annual supply to the Moon, populated by 100,000 people, brings in 1.6 G$/yr only to compensate for life system losses.
- Labor: Foundation provides the Asteroid Belt with the manpower necessary for the development of the mining industry. All trades, from the most intellectual to the most manual are sought after. Mars Services, the umbrella-organization for the management of extra-Martian workers, is responsible for: Providing qualified personnel, Organizing Mars transport - Belt, Providing supplies and consumables, Certifications - Payroll management, Health coverage. There are currently more than 10 000 workers in the Asteroids Belt. And yes, they call themselves Belters.

Rare metals (1.2 G$/ yr). Foundation in collaboration with MarsProof, the most influential Miner' Guild of Mars which is behind the first Martian service station, takes care of extracting, refining and delivering metals essential to your industries. We offer you a diversified catalog of metals, including rare Earth minerals, gold, platinum and copper. We offer a direct route from the belt to Earth. We take a contribution of 5% in materials for Mars and 5% in cash in order to take care of the transportation, and the security of the transporter. Mars also offers deliveries of metals to all human settlements in the solar system.

Entertainment (7.4 G$/yr). Foundation offers various services in the entertainment, games and multimedia sector: annual tourist lotteries, bets on the winning city of the Terraforming prizes (Mars Olympic Games), Sale of broadcasting rights for sporting events, The Grand Rally of the Marineris Valleys. For example, a lottery on Earth is set up with $100 tickets. Every two years, the lottery draws ten winners for a two-year "all inclusive" trip to Mars. The lottery generates a very high margin which is devoted to the development of Mars.

INTERNAL TRADE

Monetary system. Foundation chose to abolish the fiduciary system widely used on Earth. The goal was to gain energy because there was no need to produce physical currency on the spot. During the early development of the economy of Arkadia, a standard system was put in place. This was based on a raw material stored in Foundation with the blockchain (MTC) as a distribution vector. With the development of methane strategic stockpiles for energy storage and as a precursor for industrial activities, methane progressively gained a tremendous importance in the Martian economy. The methane standard enabled the economy of Foundation to remain independent in the face of Earthly fluctuations in the financial markets and is now used in all Martian city states. As a result, one of the most useful organic compound is produced and stored on Mars as would be gold on Earth. Accumulation of organic matter, a major step of Mars terraforming, is at the core of Mars financial system.

Trade Road. Each settlement has developed both a fiercely protected general production capacity and a few specialties for inter settlement and intercity state trade. Arkadia is an industrial center for large scale production of industrial goods using cheap nuclear power. Surya developed its solar industry. The Meridian Way serves as an exchange route between the settlements, creating ties that bind the two parts of the Foundation city state together.

Banking services. Service revenues, patent royalties and immigrant money is placed in a bank on Foundation and is loaned to Earth and Martians with interest. Interest is used to finance colonial works. The bank of Foundation thus became the first space bank. Banking is currently expanding to fund expansion in the belt.

Bonus Malus. Mars has implemented a Bonus / Malus system based on the positions taken by each and each group in order to promote mutual aid and sharing.

ENERGY

Energy is a fundamental subject for Martian societies. It is not the most labor-intensive sector, but certainly the one that attracts the most attention. Some essential points: civilizations on Earth developed from the exploitation of wood from forests, as fuel, but also as building material. It then moved on to the formidable fossil fuel reserves. The Earthlings have thus benefited from a colossal converter of solar energy, the Ecosystem, fruit of billions of years of evolution and stockpiling huge amounts of energy. However, on Mars, Civilization and a mature ecosystem can only emerge from a system of energy production ex nihilo.

Foundation was developed on the basis of a nuclear and solar energy mix. The low density of the atmosphere of Mars makes wind power uneconomical and no significant underground heat source has been found for geothermal energy to be developed on a City-State scale. Nuclear energy offers the possibility of a continuous supply of energy. The preferred technology is that of fast-neutron MSR (Molten Salt Reactor) with in-line fuel reprocessing. However, the absence of usable uranium deposits on Mars makes the planet dependent on natural uranium extracted from the oceans on Earth and imported at each synod. This departure from the rules of autonomy is made out of necessity and has led to the creation of a strategic stockpile of ten years-worth of nuclear fuel. Exploration continues in the hope of finding local uranium that could compete with solar energy. Solar energy on Mars today ensures near complete independence from the Earth, but also between each location. The technology of inkjet-printed perovskite photovoltaic cells dominates the field. They are installed in concentric strips approximately 5 m wide and up to several kilometers long, around central dedicated manufacturing plants. Specialized rovers print them, clean the surface of dust, move the strips when needed and recycle them if they are damaged. Solar energy is sensitive to natural variations in sunshine: daily variations, seasonal variations (induced by the obliquity of the axis of rotation of Mars to which are added variations due to the ellipticity of the orbit of Mars, where the sunshine is more than 40% stronger at perihelion than at aphelion) and dust storms. These regularly plunge the colony into the dark for long periods, with planet wide storms every six years and many smaller ones. Energy supply being the condition sine qua non of survival on Mars, these conditions make the management of energy stocks extremely strategic. To meet the challenge, Foundation uses three generic technologies: batteries for short-term storage, pumped storage hydroelectricity (PSH) for short- and medium-term storage when available, and a fleet of methanox thermal power plants for strategic 'long-term' storage. The two main settlements observe the principles described above, but put into practice different energy mixes. Arkadia has a balanced energy mix of 45% nuclear and 55% solar (in energy produced), while the sunshine in Surya (combined with a fierce desire for independence) has enabled the implementation of a much higher fraction of solar. In addition, in Surya, a massive PSH stores energy (see box on this subject).

Energy consumption on Mars is 1600 GJ / year per person. This may seem huge versus 220 GJ / year per person in North America, or 120 GJ / year / person in Europe but it is, at least in part, an illusion: on Earth, a gigantic fraction of the required solar energy is hidden in agricultural production, which is not included in the balance sheets. On Mars, all the energy is counted and tabulated. Nothing is free

	Arkadia	Surya
Nuclear	17GW	8 GW
Solar	39 GW	53 GW
Batteries	2 GW	15 GW
Thermal (methane)	17 GW	15 GW
PSH	0 GW	5 GW

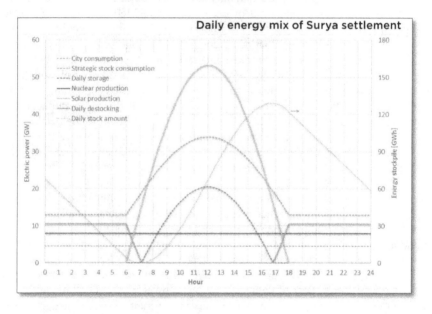

Daily energy mix of Surya settlement

Nicholson PSH. The city of Surya is located at the equator, in Nicholson crater. This crater offers elevation differences of over one thousand meters and has proven to be a prime location for the construction of a massive pumped storage hydroelectric(PSH) station of 5 GW and 200 GWh, partly dedicated to daily storage and partly to seasonal storage. The principle is as follows: Two lakes hold 300 million m3 of water each (4kmx3kmx25m and 3kmx2kmx50m). One is located at the bottom of the crater and the other in a natural hollow near the central summit. They are connected by pipes. To keep the water in a liquid state, the lakes are covered with hanging plastic sheets and maintained at a pressure of 0.1 bar. Thanks to the greenhouse effect and additional light provided by orbital mirrors, the temperature is kept above 5° Celsius. During the day, large fields of solar panels supply electrical energy to reversible turbines which bring the water up from the lower lake to the upper one. Depending on the needs, day and night, the water can run down the pipes and reverse the turbines, driving electrical generators. Thanks to a maximum total flow of 2000 m3/s and a cumulative elevation of more than 1000 meters, the electric power is potentially 5 GW (efficiency of 70%) and the storable energy is around 200 GWh.

Lucie's notebook

Visit to the solar fields. Arkadia is surrounded by hundreds of square kilometers of strips of solar panels whose role is to provide local, abundant and cheap power. These are manufactured and maintained by autonomous factories which tirelessly roam the great expanses of Mars to extend the surface of energy collection. These robots are themselves manufactured in fully automated factories (which they call autofabs). I discover that Adhémar, for his part, works in one of the master factories that manufactures the components of these autofabs. After a 3-day petition and an expensive Hawaiian restaurant, Adhémar announces that he is taking me to see the solar panels. Joy! Leaving early, he finishes his first rotation and picks me up around 11am at the "Domer" station. It's meal time and we'll start with the cafeteria. Like a foreigner recently arrived in Dodge city in the American Far West, I attract attention. I end up exchanging the usual banalities that bring strangers together. While we share a bowl of farmed salmon Poké, several workers ask me, in various ways, how I manage not be afraid, on Earth, without a protective enclosure above my head. I find that my answers remain unconvincing to them. Knowing me to be a journalist, Mayra, Adhémar's partner, pushes him to take me to see one of the roaming factories, something inaccessible for a lambda newcomer as it requires putting on a mechanical pressure suit. I struggle into the tight suit and we take off on a small, wheeled vehicle. No walking this time! We three drive for many kilometers between the long strips, until we reach one of these famous roving factories. I feel very small at the foot of this huge quasi sentient machine, it's bulk enhanced by an enormous roll of film which it carries on his back. A number of subunits wheel around, and the printheads slide back and forth over the strip. I contemplate the immense expanse of panels, while in the distance a wide rolling robot, rather like a linear irrigation machine, cleans the solar panels. Adhémar explains to me that this is how the inhabitants of Mars have managed to drastically reduce the price of energy, to the point that it is this same process, under Martian patent, which is currently implemented in some deserts on Earth. Confronted with this very real scene, an emotion overwhelms me without warning. I understand something essential here: What forests and fields do on Earth by converting sunlight into organic matter and multiplying by seeds is exactly what these replicating robots are doing on Mars: transforming rock into an environment favorable to the emergence of life.

WORK

According to Foundation Law, every citizen is entitled to a Right to Life Support, which includes a breathable atmosphere, housing, access to electricity, water and canteen service. The counterpart of this right is the obligation to perform free civic service. In the territory of the Foundation, work-related activities are divided into two categories: unpaid civic work, intended to produce basic goods and services linked to survival on Mars, and private activity intended to generate the income necessary to buy goods and services. The separation between basic and leisure activities is an important object of democratic debate. Iit varies between districts, however it is systematically observed that civic service includes the following goods and services: maintenance of the atmosphere, management of the water, energy and waste, ecosystem maintenance, health and education - in particular by companionship during civic service. While some neighborhoods refuse to do so, the majority of them also include the production of minimal free social goods for food, clothing, housing and connection - providing a minimal safety net that is unattractive but sufficient in the event of eventual hardships. The rest of the economy is the object of private activities, the supervision of which remains minimal and generally registered in the form of a business. The difficulty and complexity of survival in the non-forgiving environment of Mars, as well as the long work of making the ecosystem (ecopoiesis), fosters an empathetic state of mind and favors complex thinking, allowing this system of dual types of work to keep a balance - which does not prevent heated debates on the location of the tipping point.

The vital systems of Martian cities require consistent and permanent maintenance. In practice, the settlement of Mars requires very large human resources and manpower is more often scarce than the reverse. As a result, children are mobilized very young on ecosystem maintenance activities that are most suitable for them. In this logic of early empowerment of the Martian citizen, the age of majority is fixed at 15 years. The permanent labor shortage for teleworking, piloting drones and artificial intelligence development pushes the Martians to find solutions at home. Foundation and the City States implement an active policy intended to encourage a high birth rate, with in particular the following measures, decided according to the votes of the different districts: free crèches and shared childcare, home support by retirees and family, free education, priority for housing and medicine, or tax reductions. Indentured immigration for a service of two, four or six years, paid by companies is also common. Some telepresence is possible from Earth, but in very limited circumstances due to transmission delays. Automation has increased productivity to unheard heights and made the Martian GDP per individual the highest in the solar system.

Lucie's notebook

The trek to the arch of Tartarus. My first year on Mars is coming to an end and I'm preparing to go on to the other objective of my trip: Surya. Only, I had not expected to leave friends behind, without any hope of seeing them again. Cruel. I embrace Mayra, I breathe the smell of her shoulder to engrave it in my memory and tear myself away from her presence and wave my hand as a farewell. Adhémar accompanies me on this trip and comforts me as best he can. He embarks with me on a kind of night train or a bus, I do not know or care anymore. Throat clenched and eyes clouded, I watch the lights of Arkadia fade away. After a quarter of an hour we reach a convoy of vehicles, mainly trucks, which will follow the track of the Meridian Way to the south. I decide to sleep the first hours.

I wake up as we stop at the Kampe Relay, close to Tartarus Colles, a third of the way to Surya. We get into a smaller vehicle. Its driver, Antonio, smiles at me and says that his job is more pilot than driver. I understand the nuance as soon as we start our trip, we cross the plain at 140 km/h, a dizzying velocity for off road displacements. Along the hills, the suspensions stir in all directions, the huge tires rumble, stones hit the cabin walls; but the cabin itself barely trembles. Definitely an adventure! I did not expect such a journey here. "A trek," corrects Antonio, smiling again. The landscape, which undulates around us, seems to observe us; and succeeds, for a short moment, in making me forget the beloved faces that I left the day before. The elaborate GPS system of Mars places us with precision on our itinerary. We gradually enter a canyon made of ocher, red, grey and orange under a magical salmon pink sky. Then, suddenly restrained by our safety straps, we stop in an opaque cloud that surrounds the rover while Antonio, standing on the brake, signals us by a triumphant "here we are", the presence before us of the famous arch of Tartarus. Unbelievable. "This natural sculpture was detected for the first time on November 15, 2006 by a probe called Mars Reconnaissance Orbiter", Adhémar tells us.

Arch of Tartarus, Mars Planet

We stay here a few hours, Adhémar and I, walking about in our spacesuits, exploring this natural sculpture which attracts some other curious people, such as this young Californian, on a four-year study trip to the University of New Jakarta, a budding city-state located at the foot of Olympus Mons. It is truly surreal to speak through a communicator to a stranger in such a place. Believe me, I am enjoying my luck, to be here, at this time.

TRAVEL, TOURISM

A fuel sector dominated by methane vector. Transportation between the various human settlements on Mars is carried out by land, air or space vehicles. They generally draw their energy from the methanox sector (mixture of methane and oxygen), benefiting from a historic pooling of infrastructure and know-how with energy storage systems and methalox space propulsion. Certain City-States have developed alternative energy vectors based on solid fuels; aluminum-oxygen or magnesium-carbon dioxide (the latter being taken directly from the atmosphere), included in a regeneration cycle, in particular by solar oven. These latter sectors, however, remain in the minority due to the cost of generalizing the infrastructure.

A complete range of vehicles. Land vehicles are of all kinds. For the longest journeys, they have life support systems that make them more like large motorhomes than cars or trucks. They can be all-terrain, but most of them are built for travel on the flat surface of the Meridian Way. In order to maximize safety, vehicles often run in convoys with departures at fixed times. Point-to-point space travel between distant cities on Mars is common. This type of transport is generally carried out by interplanetary vehicles, which are not used between each launch window. The transport companies prefer to operate them in the meantime, and sell a ticket at a reduced price, the vehicle already being amortized by the sale of interplanetary journeys. The recent demand for orbital mirrors for the Surya development has put a strain on this infrastructure. There are a few specialized suborbital vehicles reserved for institutional applications - for example the *Mars Planetary Rescue Force*, with hoppers in each city, able to intervene on the whole planet in less than an hour.

Low intensity planetary transport. Because of the doctrine of self-reliance and autonomy favored by the Martian city states, these are usually not very specialized, which limits the exchange of products and therefore transportation. Most of the transport is intended for the youngest and less autonomous settlements, generally small mining villages. In addition, many prospecting or scientific expeditions crisscross Mars regularly. The movements of people between City-States are also rather limited. The lower costs of transport between locations on Mars, as well as the absence of constraints linked to the launch window between Earth and Mars every two years, in principle make it possible to offer golden bridges to the best professionals of another city. However, rigorous management of human resources and skills is a key point of the long-term survival strategy of each City-State, leading to administrative obstacles and a kind of self-restraint by recruiters.

Tourism, an important user of transportation. Tourism occupies a large place in the local economy. Indeed, Mars is not stingy with natural splendors - who has not dreamed of seeing the sunrise from inside a pressurized chalet built at the edge of the caldera at the top of Olympus Mons? Some major architectural works are counted among the Wonders of the Solar System, among which Foundation offers the Tower of Arkadia and the Botanical Garden of Surya. Tourism is the business of the Martians themselves, but also that of many Earthlings. Some of them are wealthy tourists who have come for two years, usually on a mixed business and leisure trip, and others are retirees who have saved their whole life for this purpose, Mars adds to its marvels the effects of a reduced gravity, than can be beneficial for many cases, with the proper medical supervision. Tourism is not only the affair of rich people: many lotteries on Earth offer prizes of Mars expeditions. These draws are very popular, to the point of occupying today nearly a third of the world market for games of chance on Earth. Some commentators welcome the fact that this sector is being used for a more constructive purpose, namely the greening of Mars. We will not blame them.

AGRICULTURE

Ecosystem management is a part of Mars economy. Agriculture, and more generally the maintenance of ecosystems in Martian districts and settlements is one of the primary activity of Foundation. Its importance also determines in a large part the size of the production of material and energy systems, and if it appears disproportionate by the standards of the Earth' economy, it is due to the fact that on Earth, the ecosystem is self-sufficient and preexisting to civilization. Its work is ''free'' and is not counted in the economy; the closest equivalent to the situation on Mars are, on Earth, the increasing costs of repairing environmental damage.

	Arkadia	Surya
Grow rooms area	50 km^2	32 km^2
Greenhouses area	0 km^2	90 km^2
Bioreactors area	13 km^2	8 km^2

An agricultural mix in evolution. The lower sunshine of Mars, combined with more marked seasonal variations, as well as the long night caused regularly by dust storms, has led to a relatively ineffective historical approach: on the one hand underground grow rooms and breeding chambers with artificial lighting and on the other hand bioreactors, massively producing low price proteins, often used in animal feed (insect mushrooms and fish). The production mix has a relatively low yield. However, it has made it possible to safely produce a large amount of organic matter (including the synthetic methane from the energy sector). The progressive accumulation of this organic matter is indeed the fundamental issue of the development of life on Mars, and a very significant amount of resources and technical developments have been devoted to it (notably in the energy field). The grow rooms are optimized to maximize the use of space. The cultivated species are diverse and mixed: cereals, soybeans, fruit trees, especially for the latter in urban gardens. Livestock mainly concerns fungi and insects, which are fed with bioreactor products, and the insects are then used for fish farming and breeding (fish, poultry, rabbits). Finally, some of the illuminated surfaces of the city are covered with green algae inserts, particularly in Arkadia.

Surya City, Greenhouses

Surya greenhouses. A recent development in Surya, the construction of very large surface greenhouses is an approach more suited to the Martian environment. These greenhouses are highly transparent, using locally developed technologies, and have little or no need for artificial lighting as illumination is enhanced by orbital mirrors. This development, which is less energy intensive but more intensive in terms of land area, has been made possible by the increase in industrial productivity as well as the reduction of the cost for the access to space. Flexible curtains of vacuum insulation reduce nighttime heat losses.

Orbital mirrors. Orbital mirrors are a new development. Thousands of low flying mirrors in highly eccentric orbits focus light onto the Surya greenhouse areas, increasing the light levels to those of the Earth. These light balloons, usually 150 m in diameter but just a few hundred kg in mass, are made of reflective mylar sheets and provided with a minimum station keeping system. They focus a mirror area five times larger than the surface areas they illuminate, to create optimum growing conditions at very low cost. Martians SSTOs have found here a new market for their capabilities.

"Made in Mars" gastronomy. While the consumption of meat from small animals is frequent on Mars, red meat, for example bovine, remains an extremely rare and expensive dish. However, a whole chain of cultured meat production has developed on Mars, which is enjoying growing success today. Some Martian chefs cooking 'artificial' meat have ended up being recognized by Michelin stars and have had a meteoric career on Earth, where the price of animal meat has also experienced a strong surge in cost. Indeed, patent negotiation for this type of product is today a major export issue for Mars.

Lucie's notebook

An evening at the bar. Absorbed by an article in sociology, a persistent ringing breaks through my bubble. Mellibée, a young student with whom I spoke yesterday afternoon, invites me to go out. I have the strong impression that the wisdom of this interruption lies in the benefits of the socialization yet not recommended of alcohol. I finally accept and find myself in a place quite in keeping with the memory I have of a student bar. Mellibée introduces me to her friends and happily advises me on the culinary specialties. I confess to him that I have not yet eaten despite the late hour and his friend Geoffroy makes me roar with laughter by advising me the "mattdams" or "mattdammons", the local name for fries. Obviously, the novel and movie, dating back to beginning of the XXIth century, has persisted until this sparkling generation. The evening's discussions mainly revolve around the exhibition of the controversial "Lollere", an unclassifiable artist whose works show a critical eye on Martian societies and who is starting to gain attention on the global network. Constance hates, Mellibée adores. Nothing can convince either of them, obviously. Geoffroy proposes a game of "flare", a game that looks like darts but takes place in a 20 meter corridor, played with luminescent projectiles. I finally understand that throwing the dart to the edges created a color which apparently brings back points.

Feelings of tiredness and thirst making themselves more insistent, I leave my hosts of the night at an alley which leads to my "sub", cheap apartments installed in the basement of the habitats. The group is meeting me tomorrow to go to the Lollere expo. I do not know yet if I will have time, but this lucid and hopeful youth is beautiful to see.

ARCHITECTURE

A dangerous environment. The natural environment of Mars is deadly for humans and more generally for all living terrestrial species. The atmospheric pressure is a quasi-vacuum, about 1% of the pressure at sea level on Earth. The temperature rarely exceeds 0 ° C in summer, its average is below -50 ° C, and it can approach -100 ° C in winter on the territory of Foundation. The thin atmosphere allows a level of irradiation from solar storms and cosmic rays of over 300 mSv/yr in an unprotected environment, nearly ten times the usually prescribed maximum dosage for workers of the nuclear industry. Architecture thus takes on a particularly critical dimension on Mars, because the buildings must ensure flawless protection of the people and of the ecosystems that they shelter. To this end, Martian architects commonly divide habitats into three spaces, each with its own constraints: the Exterior (deadly, where one only moves in spacesuits or in pressurized vehicles), Living Space (public, enclosed space, containing the ecosystem), and *chez-soi* (private space). The atmospheric pressure of the habitats is generally fixed at 0.7 bar, of which 0.26 kPa is oxygen, the rest being made up of an approximately equal mixture of nitrogen and argon. These values are historic ones and derive from a need to reduce the stresses on the structures, to reduce the work required to produce inert gases while guaranteeing perfect health and physical condition for the inhabitants. The buildings respect historic standards of protection against radiation, fixed during the construction of the first bases on Mars. They allow anyone who wishes to do so to spend their entire life in areas identified as guaranteeing a level of irradiation set at a maximum of <100 mSv accumulated over 5 rolling years, in line with the rules for the protection of nuclear personnel on Earth by the International Commission on Radiological Protection (ICRP). These standards are accompanied by meticulous medical monitoring. However, the strong feedback accumulated by several hundreds of thousands of cumulative lives in this environment has led to the confirmation of the weak effect of low radiation doses, as observed in the Ramsar region in Iran on populations very exposed to a level natural radioactivity (up to 260 mSv / year), almost comparable to that of Mars. In particular, with the exception of outside workers, citizens do not wear a dosimeter, and no one, apart from Terran tourists, loses any sleep over the time spent in open spaces without direct protection. People from the equatorial city of Surya, more accustomed to vast open spaces, are more relaxed in this regard than people from the northern city of Arcadia.

Protective architecture. The construction of buildings on Mars is, to a large extent, reversed with respect to Earth: the stresses on structures produced by internal pressure are much greater than those of gravity: the structures must be built to resist bursting rather than crushing. In Foundation, the pressure structures mainly consist of flexible cylindrical envelopes, intended to ensure gas tightness and resist both hoop and longitudinal stresses. Stainless steel is usually preferred, although other assembly have also had some success. An intensive research effort has led to significant improvements in flexible glass, leading to a material with high transparency, high resistance and low intrinsic energy, the 300M "AresShield". This allows for the creation of spaces that are as open and bright as possible, lighting up dwellings that might otherwise have been suffocating. This design using a continuous pressure envelope removes most of the forces from the foundations. These modules can be joined in many configurations to enlarge the space, usually at the cost of some radiation protection, as can be seen in the larger parks in Surya. The dimensions of these enclosures follow the historical evolution of the Foundation cities, towards more industrial capacity and comfort: the center of Arkadia is made up of Mk1 cylinders of 9 m in diameter, surrounded by Mk2 cylinders of 18 m in diameter. The most recent constructions of Surya reach 36 m to 40 meters in diameter, approaching the practical limits of such constructions. The orientation of the tubes is then a matter of choice, with clearly marked architectural differences between Arkadia, favoring vertical cylinders and Surya that favors horizontal ones. Radiation protection is provided according to available materials and technologies, with the aim of keeping lines of sight as open as possible. It is usually made of regolith; sintered, compressed, bound with various cements. Walls and roofs are thick and heavy, adding their mass to the radiation protection. So light envelopes, enclosing heavy buildings.

Public architecture. Foundation urban planners seek to create open spaces, while breaking the monotony that an entirely artificial environment might, in principle, create. They look for curves when possible by varying the internal environments, and architectural diversity is encouraged by multiple artistic competitions. Over half the average grade space is occupied by vegetation, walkways, squares and parks, maintained by neighborhood associations. Reminders, if such was necessary, of the love of Martians for gardening. The cities are spread over several levels: the most pleasant surface urban spaces bring together dwellings and shops in a vehicle free environment, while the underground spaces hold the grow rooms, light industries, technical and storage areas. This arrangement reduces spatial extension, and therefore promotes walking and health, while ensuring close proximity to open and pleasant spaces. They also make it possible to passively reduce the dose of radiation for much of the working day. Many underground spaces also serve as shelters in the event of an accident compromising the integrity of a pressurized surface enclosure. The regulations require that sealable underground areas must meet the basic needs of the population for a month. Sometimes in difficult but bearable living conditions, while the elements destroyed in the event of a major disaster can be rebuilt and relief organized by. other districts or city states.

Interior architecture. The interiors of the buildings are generally modular, so that the size of the apartments can be quickly adapted to the size of each home. The Martians' way of life is generally more communal than that of the Earthlings: it is not uncommon for sanitary facilities and kitchens to be shared between several homes, although this is more common in the older districts of Arkadia.

Surya City, Botanical Gardens

Lucie's notebook

Visit to the botanical gardens. Some microorganisms live under the ocean at a pressure 400 times greater than the one we live at on dry land. A whale hears our heart beating when we approach it underwater and a dragonfly flies at 100 km/h while spending only 2 Watts and can stand up to 30 Gs". I listen to the words of Mathilda Kampf with the delight of a child enjoying a fairy tale. Jumping up from her desk, Mathilda, a seventy-year-old curator and researcher in aquatic ecosystems, takes me through a series of corridors dotted with fragrant plants and punctuates our walk with an uninterrupted flow of anecdotes. I am swallowed by this gentle tornado and suddenly stop in front of a huge aquarium. On Earth I didn't really care for them, but here it's been 8 months since I've seen water in close proximity in such large quantities. It's a miracle. An aquarium without fish, but still. Mathilda joins me in my enchantment and explains to me that the botanical gardens begin by questioning the visitor about this thing that is life. These seemingly empty aquariums, for example, contain the two things that gave birth to humanity: liquid water and phytoplankton, basis of the oxygen and carbon cycles. "Striking, isn't it?

" I recognize the acuteness of this question and let Mathilda continue on to a crescendo of wisdom that I am not soon to forget: "The concept of this dome goes further than that of a museum and it was not easy to work out, on a purely ethical level, as our initial wishes touched on the principle of ecopoiesis, that is to say, ultimately, succeed in reproducing an ecosystem in its entirety ". Concentrating on the words of my new teacher, I enter a large wooded area. It smells so good! I am overwhelmed by nostalgia for Earth and it's forests. But why? Aren't they already here? Further on, there are clearings and small savannah that we cross through in glass corridor. My host pointing her finger upwards, showing me a light structure high above our heads. We are under a gigantic transparent tent, and the tube keeps us from the inadequate air pressure outside. Inadequate to us, but not to plants, and not for water. A vista onto an immense surface of water open up, the lower reservoir of the Nicholson pump-storage power station. I question Mathilda: "Why do you absolutely want to reproduce the entire Earth ecosystem? You know it's impossible, right? "

"Unfortunately, we felt that we had no choice", retorts Mathilda. Bending down to pick up a small insect on the ground, she puts it up high onto a branch and continues: "if only for the simple reason that Earth needs Mars, just like a computer needs backup. It may seem presumptuous to you, but think about what we can learn from such an approach as the Earth gradually loses its biomass diversity. Water and phytoplankton put man on Earth, man will have put water and phytoplankton on Mars. At best, it will stir up the pride of Earthlings and at worst, it will make a copy that may take several thousand years to make; but it is worth it." Mathilda finishes the visit with me and leaves immediately with a young man in a white coat, popping out of nowhere to drag her away to a conference room. Stunned by information overload, I realize that it is on Mars that I am learning to really understand my own planet.

The tent concept. The tent is a new concept on Mars, using domes and tension structures for the creation of arbitrarily large spaces. The design is based on a flexible membrane made from the new 300M flexible glass, held in a mesh of steel cables. These cables are fixed to foundations that are either driven deep into the Martian soil of ballasted to keep the tent in place. The theoretical maximum height of the tent is a few kilometers, the weight of the cables would then exactly compensate the internal atmospheric pressure, without requiring anchoring. In practice, the height of the tents is dimensioned by the cost of the cables and the pressure inside the tent. A first attempt using very low atmospheric pressure was built successfully for the Nicholson PSHP and may lead, eventually, to having plants growing directly on the planetary surface.

SPORTS

On Mars, sport is important, both because it is necessary to work hard to maintain proper muscle tone and health, and because it offers novelty to Earth. A gravity of 0,38 g changes both endurance and strength, as well as changing some games beyond recognition. Football, in both American and European styles has lost some of its appeal on the red planet, where anyone can kick the ball from one end of the field to the other. So it has been re-invented by the Martians. Vehicular sports have found a niche, exploiting the endless landscape. We have, for example, the motorized trek, inspired by the Paris-Dakar and the 4L Trophy: each year teams leave in all-terrain vehicle of homemade manufacture (approved and properly tested, of course) and embarks on a journey of several thousand of kilometers. This event, broadcast every year on Earth, is appreciated for its adventurous nature, its covering genuinely unexplored spaces, but also because it is a pinnacle of motorsport, where the vehicles must be perfect to adapt to the terrain, have huge endurance and also provide security in the harsh Martian environment.

Other sports, such as climbing, have also experienced a gain in popularity. Indeed the low gravity on Mars has made it possible to exceed the limits of the discipline, and the climbs of the walls of Valles Marineris are impressive. Regular sports activity on Mars exceeds even the well-known dedication to fitness of Australians. Mars, despite all the architecture, is not Earth, and the human body is not perfectly adapted to it. Whether it's in a gym, through team sports or individual sports, and particularly through walking at every opportunity, Martians exercise continuously. A few years ago a special edition of the Olympic Games was held in Arkadia and was an interplanetary success, with a whole new set of possibilities setting new records in practically every field. A strong contrast to the regular Olympics where progress is now practically nonexistent or measured in milliseconds. The dance disciplines, especially aerial dance, have experienced a significant boom and the possibilities given by the low gravity have allowed the creation of ballets of incredible intensity, which have propelled some Martian choreographers to status of interplanetary stars.

Lucie's notebook

The arts. Ranjan stands and raises his hand excitedly; he wants to open up the competition to the other districts. Freina, moderator for the evening, asks him to sit down while asking the assembly to vote. She suggests that we adjourn till the next competition, so that everyone has time to think about it. The 63 people present vote by show of hands. They will keep the competition closed to "outside" for this year. This Thursday is a district meeting and it's not only nice things that the exchanges are about. The formality of the debates, however, usually leads to constructive decisions. Such as a moratorium, decided following the announcement of the construction of the Arkadia tower, which seems questioned at the time when I write these lines, or the creation of a link direct between botany and decorative gardens following the problems of cultivable area in the Vankuist district. Each week, a theme thus preoccupies an assembly and this week it is the competition of bas-reliefs for 6 facades of the district. A meeting to discuss an artistic competition... Don't they have better things to do? It was after this meeting that I became interested in the subject of Art on Mars. What place does it have in a seemingly hyper-utilitarian universe? I then realize, *a posteriori*, the number of exhibited works that I have come across in my walks and the full scale concerts seemingly improvised in public spaces. Unlike Earth, artistic expression is not separated from social and professional life, and it's quite confusing. Schools give the arts a preponderant place, and they show up in CVs and elsewhere. I have the impression, however, that some aspects of this escapes me, like a word missing in a sentence. Adhémar, present at the famous meeting and capturing the wide-eyed interrogation of my puzzled face, explains. On Mars, inventiveness is a question of survival, nothing more, nothing less. To encourage this, Foundation grants the individual the possibility of finding their own mode of expression from an early age. And at all stages of his life, a person on Mars will be able to enhance his existence by traveling in his deep "self" thanks to this medium, in particular. A way of cultivating one's true identity in the service of the whole, "one" first for the "multitude" then. Beyond the words, I feel a deep impact from his speech, a shock wave that resonate in me. For Adhémar all this seems obvious, but for me, it is a quest. By working on oneself, by mixing this loose and mysterious material, one may create by capillarity an acceptable living-togetherness. I find myself doubting the success of this turn of mind on the scale of a city, while having the proof before my eyes that people on Mars seem to know who they are much more than I know myself.

INDUSTRY, ISRU

The main drivers of industry. The allocation between the industrial products made on Mars and those which are imported from Earth is influenced by two factors: the price of interplanetary transport and the Foundation doctrine of autonomy. To summarize, most products whose production price is significantly higher than $ 500 / kg are imported, with the exception of those which appear on the list of critical products. The latter is established by democratic vote on the basis of compromises between the price level, the need for the product and the human resources available for starting a new industry. As a result, the vast majority of the industry in Mars relates to relatively low-tech products when compared to Earth standards: production of primary resources, the construction sector and the manufacture of heavy and strategic equipment. A gradual rise in the technological level has occurred, sometimes for initial reasons of image. In this register, the success of the Foundation's X33M "Albatross" is very telling.

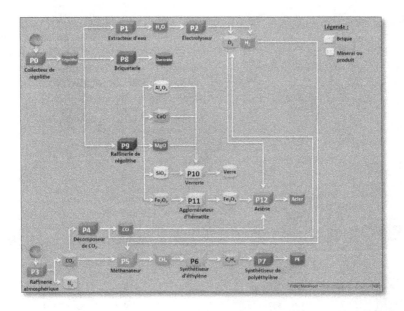

Raw materials. The main raw material produced on Mars is water. For the most part, it is extracted from large surface ice water open-pit mines. This water ice is abundant in Arkadia and also present near Surya, in Medusae Fossae, but with a much higher level of contamination. The treatment of ground up regolith resulting from the water extraction process makes it possible to produce other raw materials: sintered regolith bricks, concrete, structural steel and cables, glass *AresShield*, and other products used in industry. External concrete uses sulfur as a chemical binder, but the availability of some calcium silicate deposits favors the alternate minerals. As a Martian will tell you, nothing smells quite as bad as wet Sulfur concrete. Large atmospheric processing plants separate the main gases from the atmosphere. Carbon dioxide is separated first, by compression and cooling, to be used for the production of propellant using the Sabatier reaction. Alternatively, carbon monoxide production is intended for the reduction of iron oxide into iron and steel. Nitrogen and argon, in approximately equal parts, are used for the atmosphere of the habitats. They are mostly separated out independently and locally from the CO_2 process using an energy efficient adsorption process, as the propellant production no longer provides sufficient inert gases for the build-up atmospheres of the fast-growing habitats. An important part of the nitrogen is used in the production of nitrates intended for fertilizing crops (via, for example, Haber-Bosch reaction and newer plasma-based processes). Water vapor is also recovered, but the production of water by this way is tiny compared to the needs of the colony. A wide variety of plastics are produced from methane. Plastic is however avoided if possible, because its resistance to ultraviolet light is poor its energy cost is significant, and methane often turns out to have more cost-effective uses. The production of raw materials from local resources using principles of In-Situ Resources Utilization (ISRU) has been from the start of the settlements on Mars a rich research vector, which has produced both local successes and processes applicable back on Earth. Some of these initiatives have been able to yield significant dividends, others have been open-sourced, such as the MarsProof ("Tough enough for Mars!") project by the association Federation Open Space Makers and which is at the origin of to the well known miner's Guild of today. Indeed, the label "Earth ISRU" has appeared in reaction to the advantages, demonstrated on Mars, of small locally produced runs of high technology fabricated products using fabs and 3D printers.

Table production volumes (earth years)		
	tonne / yr	PJ / yr
Food	1 100 000	360
Biomass	1 200 000	360
CH4 propellant	72 000	6
Water	75 000 000	40
Structural steel	1 000 000	40
concrete	8 000 000	20
Habitat services	/	120
Solar panels	150 000	150
Others	/	500

A very active construction sector. The nature of Martian housing, light pressure vessels subject to high tension loads and massive construction elements to absorb radiation, creates an important need for construction materials and operations. Combined with the strong growth of the Martian population, this leads to making the construction sector on of the main industrial sector of Mars. Just as the production of solar panels has been streamlined as much as possible, the production of protective enclosures has been the subject of a great optimization: automatic concrete batch plants, stone yards, transporters, metal spinners and fab units. This optimization has always been subject to the respect of the architectural principles aimed at the greatest versatility of modular elements, to allow creativity to express itself at the local level, and to prevent any risk of monotony in an artificial environment.

Recycling, at the heart of human activity on Mars. Recycling of materials is intensive, because Martians are acutely aware of the value of raw materials and of the energy benefits of recycling planned from conception. Don't they drink recycled water and breathe in equally recycled air? An interesting anecdote concerns the choice of stainless steel as the standard grade for habitat manufacturing. This choice was made in line with the choice of the SpaceX Starship hull material, with the aim of recycling end-of-life vehicles. No source of materials has been overlooked to reduce the costs and efforts to build this new world.

The equipment industry. Industry is largely dominated by the needs of the agricultural sector: cultivation and hydroponics equipment, light sources, and more generally the needs related to production, distribution and energy storage. District workshops include small, versatile productions units: plastic, metal or ceramic 3D printing, rapid prototyping and other automated machining equipment, making small series production economically possible. In addition, a substantive activity is related to construction and robot maintenance. The use of open source models is encouraged whenever relevant to reduce engineering costs. A certain emulation exists between makers on Earth to produce equipment usable on Mars, because it is a consecration to see his model selected by many districts of the Red Planet, even for free.

A turning point towards high technology. Industries on Mars are gradually diversifying their production, and gradually reducing the planet's dependence on high technology from Earth. Foundation has recently succeeded in making reliable and deploying a local production of integrated circuits, using Minimal Fab type units. The production of SSTO type interplanetary transport vehicles such as the 'Albatross' series is also a great source of pride, as will be explained in our next article.

X33M, SPACE INDUSTRY MADE IN MARS

Game changing strategy. In the early stages of the settlement, interplanetary transportation was carried out by terrestrial companies such as SpaceX, with their own vehicles. However, the moons of Mars were eventually proven to harbor underground ice, as predicted by Fanale et al. This made for a paradigm shift for Earth-Mars transportation. With industrial production starting in earnest on Mars, local investors created a single stage to orbit (SSTO) rocket, built on Mars to compete with Earth suppliers. The design of the rocket was inspired by the futuristic but ill-fated X33 Lockheed Martin SSTO prototype. A spaceport was built on Phobos to produce propellant to refuel the Martian SSTOs. This extended their range to many destinations in the solar system, including Earth orbits, the lunar orbital station, and the asteroid belt, where they provided the lowest transportations costs for the growing asteroid mining companies. Mars benefits from much lower deltaV requirements than Earth, and can supply consumables to the asteroid belt at a competitive rate.

Length: 21 meters.
Width: 25 meters.
Dry mass: 30 tonnes (with descent shielding)
Main propulsion system: LCH4 and LO2, specific impulse 319s.
Total propellant: 240 tonnes
Payload: 20 tonnes
DV: 5500 m / s

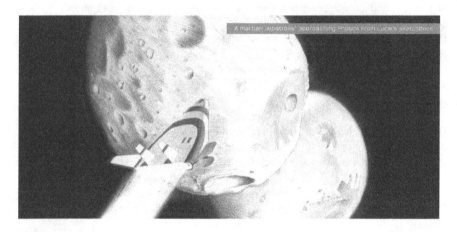

Earth-Mars transportation optimization. For Earth-Mars transportation, numerous Starships (giant vehicles with more than 100 passengers) make the round trips. The stop at Phobos allows an important reduction of costs as well as an improvement of safety. Most manned Starships do not reach Foundation soil anymore, passengers are now transferred in Phobos Station to the winged Albatross SSTO that handles the descent, allowing for less complex, safer and more comfortable landing procedures. Then Starships refuel with methane and oxygen and returns to Earth. As they arrive in Earth orbit with half-filled tanks, only 2 refueling missions are needed instead of 5 for the next interplanetary transit, which cuts the cost by a factor of 2. In addition, thanks to the elimination of the descent and landing on Mars, the payload capability has been increased and the reusability rate has been tripled.

Profitability of the 'Albatross' SSTO vehicles. Martian SSTO such as the Albatross have made Mars the leader for delivering satellites to geostationary orbits. The Albatross SSTO has some important advantages over Earth rockets:
- For the descent to the Martian surface the shape of the vehicle allows for efficient aerobraking in the atmosphere. The landing can be either vertical or horizontal.
- It is a robust and fully reusable rocket. With appropriate maintenance, it can be reused up to 500 times without extensive refurbishment.
- It is a single block. There is no need to assemble different parts for the launch.
- Thanks to the low gravity of Mars, the spaceship is quite small and light. The dry mass is only 30 metric tons. In comparison with Earth rockets, the Albatross can be assembled, moved and set up for launch using much simple equipement.

MARS IDENTITY

The autonomy doctrine is based on the assumption that Mars should be able, as soon as possible, to survive and thrive even in the case of a global disaster on Earth that would jeopardize supplies. The idea of Mars as a backup of the Earth underlies many of the Martians' decisions. This leads to some specific industrial choices, based on risk analysis and mitigation rather than profitability. And it leads to endless public debate, of the sort Martians are so fond of. An overarching Martian identity rose up remarkably fast, tempered by a fierce local pride that feeds the many district's and City States civil life. The people of Foundation are proud to be part of the oldest and largest state on Mars. However, this pride has never deteriorated to nationalism thanks to the shared experience of life on Mars, where every second of your life depends on the work of others, on the careful and attentive maintenance of equipment, systems and ecosystems that must work flawlessly for you and yours to survive. The Martians really are 'All in it together'. The idea of Mars as a steppingstone to space and eventually to the stars, even more than the idea of a backup plan for mankind has proven attractive to millions of Martian settlers. Mars is populated by pioneers and the children of pioneers, people who feel intimately that they are building something new, and something big. Mars' quest to green itself is seen by many as a quest for meaning, the transcendence of Humanity, from a dead world into a vibrantly living one, and the first step to an even grander plan of spreading life throughout the universe.

Lucie's notebook

A new start. "This is my last day on Mars." It took my fingers several minutes before they could type that sentence on my keyboard. 2 years and 9 days crisscrossing the surface of a red pebble floating in space. In a few hours, I will leave everything. Leaving Adhémar is hard. I will also have to leave my gang: Jeremya, Pelemon, Juan, Melibée and Mayra. I can only bring my 40 kg of accumulated belongings, the maximum allowed per person. The images of the next few hours pass before my eyes: Fly up to Skyreach Station on Phobos aboard an "Albator", Then, arriving near the rotating torus at the heart of the base, our little transport will be caught by a capture arm which will dock us next to a huge Starship, waiting patiently for its return journey cargo. We will stay in transit while we trade propellant, food and exchange crews. I will spend the time thinking about Mars and its people. I will look feverishly through one of the portholes at the red planet, looking for a thin shiny line linking two tiny green spots. I will put my fingers on the glass, thinking that I am stroking my lover's face as the vibration caused by the moorings snapping open signals the end of a great adventure. But I haven't left yet, and I am waiting for the taxi at tube 28. For the moment, I smell that odor of air conditioning mixed with enclosed plants that I've grown so used to. This fragile, manufactured air. I know in my heart that the future is playing out here, that human endeavor is taking on another scale with the rebirth of this planet. Mars was once covered with water, someday...
Journalist, it wouldn't be so bad here. The media on Mars are becoming more and more crucial to the communications between the City-States. The next flight to Phobos is in 3 hours... and I think to myself: "There is still time to stay."

Phobos, Terminal One

TEAM, DISCLAIMER & REFERENCES

Many thanks to all the people who contributed to our report: Clarisse, Pierre-Maël, NASASpaceflight.com' discussion boards, MarsProof project and Federation Open Space Makers association.

Some organizations are cited for story convenience in current text (Mars Society, SpaceX, Michelin Guide, Lockheed Martin), but the opinions given are those of the authors and cannot be attributed to these organizations other than by coincidence.

Visit our website for more pictures and references :
www.planetscopemagazine.com

16: VANAHEIM

Shayne Beegadhur
Loughborough University
S.beegadhur-18@student.lboro.ac.uk

Sophia Lee Roberts
King's College London
sophia.lee_roberts@kcl.ac.uk

Theodore Macklin
Imperial College London
Theodore.macklin18@imperial.ac.uk

Andre Nowaczek
University of Cambridge
an545@cam.ac.uk

Caroline Swenson
University of Notre Dame
cswenso2@nd.edu

In 2060 a mission was launched that would change both Mars and Earth forever. It began with an exploration phase; electric tractors and cargo transports were landed a few hundred kilometres north west of Elysium Mons and in Libya Montes, just south of Isidis Planitia. They set about harvesting ice in the north before driving it down towards the equator where they dug vast fields of solar collection troughs. They landed vast power generation modules with the first settlers to arrive; those guys didn't have it easy! Working from temporary buildings, they hooked up hundreds of kilometres worth of piping to the troughs to get those turbines spinning and started to liquify the air for its oxygen and nitrogen. They built the first tunnels through the solar fields and, from the excavated material, the first farms. That's when things really got going. The refineries arrived in time for the first harvest to give them the ethanol and carbon they needed to make the steel and polymer to keep going. From there it's history that first base became Freyja as its population boomed past one thousand and the commercial industry moved in to jumpstart the economy.

With momentum having been built, Heimdall and Delling came a few years later using the experience gained as Freyja continued to expand. Once Freyja had hit 50,000 people and the other two had 20,000 a piece, work started on Eir and Kvasir. After twenty more years, here we are; one million people on Mars.

1: Design Philosophy

In designing the first settlement for one million people on Mars, it was necessary to base engineering and economic decisions on a clear set of design tenets. The fundamental visions for the settlement are:

- Economic and scientific feasibility of design realisation
- Settlement success defined by economic profitability after operating costs
- The highest possible degree of non-reliance from Earth
- High quality of life and scope for individualism

By weighing design options against this set of filters, the final settlement design was ensured to be practical and worthwhile to pursue. Vanaheim has been designed to favour independence from Earth above design complexity and effort exerted, but not above economic sense.

2: Settlement Structure

The one-million-person settlement of Vanaheim would be split into five separate towns, named *Freyja*, *Heimdall*, *Delling*, *Eir*, and *Kvasir*. The settlement and its constituent towns have been named in honour of the chosen site's proximity to the landing site of the Viking 2 mission.

The sites chosen for the five towns lie within Libya Montes, to the south of Isidis Planitia, with an average latitude just north of the equator. The reasons for this choice are manifold. Easy access can be found to Hellas Planitia and the plantias surrounding Elysium Mons, enabling the collection of water ice and mineral prospecting. The equatorial location allows for higher light intensities and lower energy launch from the surface. The mountainous terrain offers higher local mineral concentrations and confers known, predictable prevailing winds to assist with planning for dust storms.

Dividing the population into smaller groups allows for expansion while maintaining a higher quality of life between them. By building each town up so that it can provide a high quality of life for each settler, and then moving onto the next to do the same, we avoid the pitfall encountered throughout history where resources become concentrated in specific areas, leading to a disparity in welfare. A relatively equal quality of life has been shown to lead a higher trust society and increase settlers' happiness overall. Each town will contribute to the overall settlement; the experience of the last town guiding the development of the new. Choosing to develop towns in parallel to one-another offers a faster route to settlement completion, the opportunity to test new techniques, and security for settlers.

Each town has been designed for self-sufficiency however, initial requirements during development will introduce marginal industrial specialisms in each town. These production disparities will drive the internal Martian economy thanks to a bulk cargo transport system. The export economy will rely on both data and physical products exported through the two shared spaceports; Valkyrie and Bifrost.

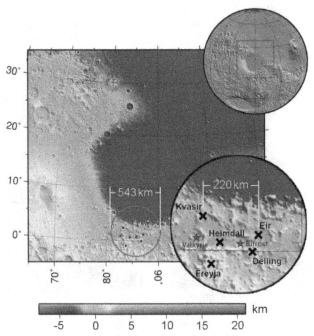

Fig. 1: Elevation map showing locations of towns (black crosses) and spaceports (blue stars) in Vanaheim.

3: Power Generation

Supplying continuous, reliable, and adequate power for any large settlement on Mars must consider numerous environmental, ethical, and economic factors. The long travel time and large haulage expense from Earth makes reliance on it for this key element of infrastructure both dangerous and uneconomical. Mars' greater distance from the sun than Earth means that significantly lower solar intensities are available to be harvested; an average value of only 295W/m² at the chosen location near the equator was used in calculations to acknowledge variations in irradiance throughout the day. Similarly, dust storms, while not forceful, can reduce surface light intensities by upwards of 99%. Nuclear systems are intricate and require a high degree of supervision; manufacturing them on Mars would not be possible until a large, energy rich, permanent settlement was already in place. Initially shipping the infrastructure safely from Earth for the necessary reactor sizes is neither economic nor logistically feasible. Despite the relative abundance of low concentration thorium for nuclear fuel in the Mare Acidalium, a lower supervision, sustainable, and independently expandable system is essential in order to get the settlement to a stage where nuclear energy could be phased in without dependence on Earth.

Solar-thermal energy will provide settlement power until the population expands past three million people, when the choice will be reviewed. This avoids shipping costs or dependence on earth for photovoltaic solutions since all system components for a thermal system can be manufactured solely using Martian materials. The technology involved is well understood, reliable, easily maintainable, and fundamentally cheap thanks to the implementation chosen.

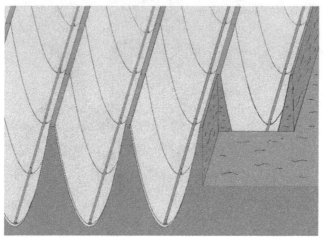

Fig. II: Parabolic solar-thermal power trenches.

Parabolic trenches, 0.75 metres deep by 0.87 metres wide will be ploughed into the Martian soil by Multipurpose Electric Tractors (METs) such that they run parallel to the lines of latitude in large fields next to the settlement locations. The raised material will be formed to extend the parabolic curve (Fig. II). Photopolymer resin mixed with metal particles will be injected into the surface of the trenches, hardened, and polished to a high reflectivity by the forming equipment. Wide (50 meter) spaces will be left in the solar fields every kilometre as access corridors, below which power access tunnels (PATs) will be excavated (see Section 5.1).

A low-carbon ferritic steel pipe, rated up to 250 bars will be located at the parabola focus. The pipes will be supported at Marsquake-relief channels in the trenches. Pressurised water at 210 bars will be superheated to 800°C in the pipes before being partially expanded in a turbine to 56.8 bar, being reheated back to 800°C, and then expanded fully down to 0.3 bar. Ten direct-contact-feedwater-heaters will be utilised. This gives a cycle efficiency of 50.02% with turbines and compressors operating a 97% isentropic efficiency. Heat rejection will occur through a heat exchanger into a higher-pressure water system and used for space heating before heat is dissipated by air-radiators built into all buildings (see Section 5.2). Air heat dissipation is preferred over ground-based heat-exchange due to the risk of melting unknown subsurface ice or other unforeseen geophysical effects.

3.1: Using Sugarcane as a Fuel
Measures must be put in place to provide continuous power during the night and dust storms. As a large infrastructure expenditure has already been put into turbines,

it follows that they should be utilised for all power generation. It is necessary to store energy as heating potential, the most effective form of which is chemical energy. The most practical way to do this is to utilise a fuel crop which can be refined into gas turbine fuel and *SynGas*, as well as reliably remove CO_2 from the settlement's atmosphere. Sugarcane was chosen for this purpose, as up to 5% of sunlight incident upon it may be stored as sugars which may be fermented into ethanol. Under artificial lighting in the optimum wavelength range, up to 10% of energy may be stored as sugars and 40% will be stored in other bio-products. These products allow another 2% of energy to be recovered as ethanol, leaving 23% of input energy to be recovered as SynGas and other flammable gases. The remaining 15% is stored in mulch and charcoal. The charcoal is utilised in the regolith extraction system, and the mulch is dehydrated and is processed into a first stage water filtration medium. Agricultural bio-waste will be pyrolyzed to form viable fuel for the power reclamation.

Town	Freyja	Heimdall	Delling	Eir/Kvasir	Overall
Population	300,000	250,000	200,000	125,000	1,000,000
Individual Power Draw (MW)	78.077	65.064	52.052	32.532	260.258
Industrial Power Draw (MW)	117.116	97.597	78.077	48.798	390.386
Agricultural Power Draw (MW)	321.418	267.848	214.278	133.924	1,071.392
Designed Instantaneous Capacity (MW)	234.231	195.193	156.155	97.597	780.773
Industrial Ethanol Production Power Draw (MW)	60.208	54.188	48.167	36.125	234.813
Solar Power Collection (GW)	5.000	4.204	3.407	2.185	16.981
Solar Trough Field Area (km²)	16.950	14.250	11.550	7.406	57.561
Solar Trough Field Area per Resident (m²)	56.501	57.000	57.748	59.245	57.561

Fig. III: Table showing quantities relating to power generation systems.

Sugarcane will be grown with a reclaimable heat potential equal to 3.2 times the designed instantaneous capacity and agricultural demand. This allows the settlement to operate normally, with the exception of agriculture (see Section 6), during both the night and dust storms for a time equal to the number of days that were available for plant growth or until storage capacity is filled (Section 10.6).

The ethanol, pulverised agricultural charcoal, and viable flammable gases will be used as fuel in a modified Allam cycle gas turbine power plant to supplement solar power, during the night, and during dust storms. The substances other than ethanol will be burnt first during nights before using ethanol. This leaves the more of the more densely storable fuel available for dust storms. The Allam cycle has been modified to compress its working fluid in the liquid CO_2 regime, making it more akin to a supercritical Rankine cycle. The regenerative heat exchange element has been substituted to transfer heat into the working fluid of the water/steam Rankine cycle as an alternative to the solar trenches or in conjunction with them. The modified Allam cycle is operated between pressures of 100 and 68 bar with a maximum temperature of only 865°C. These low maximum temperatures in both the steam and CO_2 turbines allow for the use of traditional polycrystalline turbine blades rather than single-crystal blades. This dramatically reduces the cost of turbines and the onus on the Martian manufacturing industry. The modified Allam cycle has an efficiency of 5.3% on its own, but allows a combined-cycle efficiency of 50.6%, while the combined-heat-and-power energy utilisation is upwards of 95% depending on heat demand. The high temperature and pressure water from the modified Allam cycle can be used as a heat or steam source for industrial processes.

Half-gigawatt combined cycle modules (CCMs) will be manufactured as a partially sealed unit. These will be installed close to the steam heating sites, with turbines in a vertical orientation into holes bored into the earth using the augers designed for drilling farm-tubes (see Section 5.2) in residential buildings. These chambers will then be capped with the CO_2 compressors' radiators to prevent regolith ingress and restrict damage in the event of an explosive failure. The CCMs will be installed near the PATs to be accessed via passageways similar to those for entering residences (see Section 5) to enable convenient access even in adverse weather conditions.

3.2: Future Power Concepts
The solar fields provide a designated land area for energy collection. As Vanaheim expands and becomes more technically capable of producing high quality semiconductors and electronics, these fields can have a rectenna installed above them to receive large intensities of beamed microwave power from solar power satellites. This will remove the need to construct further farm-tubes beyond the need for industrial ethanol as beamed microwaves can penetrate dust storm clouds. This concept is expected to become first feasible and necessary for a population of above one million.

Above a population of 1.2 million, it ceases to be economic to continue to scale up the original system in favour of beamed space solar power. The implementation of this system could be more cheaply and efficiently realised to provide sufficient power for additional immediate consumption and storing an equal amount of energy as ethanol to last the night. This system can allow power to continue to be delivered during dust storms, and so removes the need for long-term ethanol and oxygen storage.

The solar trough fields only cover sufficient area to reliably receive beamed power during dust storms once the area required to supply 1.1 million people has already been installed. Independent nuclear power is expected to become viable and economical at a population of approximately 2 million.

4: Chemosynthesis and Material Production

For an advanced civilisation to be self-sufficient on Mars, efficient closed-loop chemosynthesis and material production methods are essential. This ensures that supply of material to manufacturers is reliable so that the production of goods can occur predictably and to a tight schedule. To do this requires a fresh approach to selecting which organic and inorganic materials and compounds are really necessary for the fastest economic growth. From these base materials, it is possible to diversify the chemicals and materials a colony can produce so that more complex products can be produced on Mars.

Ethanol produced from farmed-sugarcane is the only organic molecule required to produce all organic molecules and polymers needed within the settlement. However, without inorganic chemicals, it would be impossible to create the litany of more exotic compounds necessary for food, medicine and strong polymer production. The greatest difficulty is involved with the extraction and processing of inorganic chemicals from regolith and basalt. As material production becomes more copious and profits are generated, specialist ore mining and production missions can take place for material such as copper. Fig. IV illustrates the method for extraction and production of essential inorganic materials from regolith and basalt that can occur within a single 'refinery' unit. This system takes only waste from ethanol production, small amounts of water, and electricity as its other input. It can extract pure chlorine, titanium, silicon, magnesium, aluminium, and sodium silicate. Iron can be removed from the process, however most will be directed into carefully controlled steel production. A mixed ceramic powder will be left; this will be sintered into usable ceramic products for low performance applications. This production system will be present across all towns in order to grow their economies and expand operations. Towns can therefore offer reductions on Residence-Fees to private industry in order to complete specialist ore extraction missions to obtain materials like copper, nickel, zinc, and eventually thorium.

Although a strong case can be made for the rapid production, consumption and disposal of man-made plastics, it has been energy intensive and a great contributor to the degradation of Earth's ecosystem. Therefore, in Vanaheim, ceramics will become a more dominant material in everyday life; Martian glass will be abundant, and arguably the best quality in the solar system due to the purity of the silicon dioxide it is made from. One product of the mixed ceramic powder will be a 20 mm thick, general purpose plate dubbed 'MonoPlate'. The same ceramic will be formed into a porous medium for plant growth. The production of primary and secondary glasses will be achieved by the beginning of the settlement expansion phase. As Vanaheim grows, products typically packaged using single-used plastic on Earth can use glass or paper instead if possible. Developments in the composition and production methods for Martian glass can be sold back to Earth companies through the Terra branch of MAPD (see Section 10.4).

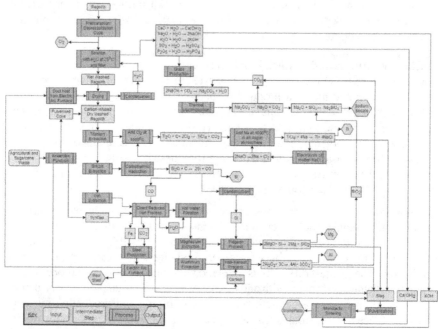

Fig. IV: Chemosynthesis and processes diagram for all refinery units.

5: Habitable Volumes

A core element of all human settlements is immediate access to safe, comfortable, and convenient personal and public spaces. This is much more difficult to achieve on Mars than on Earth; surface temperatures are frequently below survivable, the Martian atmospheric pressure is too low and has no free oxygen content, and yearly radiation dosage can be over 30 times higher than on Earth. Potential solutions to this set of issues must provide not only safe spaces, but comfortable ones, too. In order to utilise the uniqueness of the Martian location, it is also necessary that views of the surface should be easily available in both public and private spaces. A combination of surface buildings and subterranean spaces was chosen to fulfil this set of requirements; subterranean tunnels should be large public spaces, akin to roads or squares, which join the entrances to largely superterranean small, apartment-style buildings. Block living places much less of a burden on resource requirements than designing for individual residences, however this decision could be reviewed in the future with construction method improvements.

5.1: Tunnels:

The subterranean tunnel infrastructure will provide the foundation for each town, with buildings attaching to these vaulted passageways which house utilities facilities, commercial spaces, and parks. Industrial facilities will also follow this structure to provide large, enclosed spaces.

These tunnels will be constructed using a 'cut-under-cover' technique. Material can be broken by explosives, concussive force, or mechanical methods allowing it to be pulled from the ground by cheap, parachute-like slings towed behind the METs. The tractors will excavate material along the line of the tunnels down to a specified depth (typically 20 metres) and with a fixed width of 20 metres. Modern tunnel-boring methods would require complex, expensive, and unadaptable equipment, and so would be unsuitable. Cut-and-cover also allows for excavated material to be stored on the surface at the point of extraction rather than requiring a dirt-movement infrastructure, and it allows flat-walled tunnels to be produced.

Raw tunnel walls will then be impregnated with photopolymer resin to make the volume gas tight, or shallowly sintered to allow impregnation at a later date while providing an adequate degree of gas retention. Locally manufactured steel gantries will then be installed over the excavation and pitoned securely into the surrounding rock. Setting a fixed width for all initial tunnels will reduce the number of different components that must be manufactured and stored. This architectural choice can be reviewed once resources are sufficiently abundant. MonoPlate panels will be installed in aluminium runners attached under the gantries and sealed on their top surface with photopolymer resin. Excavated material can then be piled back and compacted on top of the now-completed ceiling to a depth of 3 metres to help counteract pressure-bearing stresses and provide a good degree of radiation protection in the tunnel. It will be possible to use reinforced glass to replace the MonoPlate panels once materials are available and performance data is available. The surrounding rock can cause either a high rate of heat loss or good insulation depending on the surface temperature. The large thermal mass will, however, mean that changes in the internal temperature will be slow and so can be increased by diverting waste heat from power generation through radiators into the tunnels rather than directly into the surface buildings' air radiators.

Where tunnel walls have been left unimpregnated, it will be possible and safe to carve surface texture and architectural elements into them. This will be preferable for large public areas where aesthetic considerations are of greater importance. Flat walls are likely to be left in industrial areas and the initial constructions. In commercial areas, spaces can be cut into the tunnel walls by conventional methods to increase floor area although large undercuts that extend above 10 metres from the ground should be avoided and load bearing rock structures should not be cut with overhang angles greater than 50°. Trapping a layer of water between this glass and another sheet on the surface will allow for natural light to enter the tunnel where it is aesthetically appropriate, while still providing radiation protection and mechanical advantage. This set of guidelines and the ability to locally dye photopolymer resin allows for a wide range of architectural styles to be commissioned and tunnels to be used for many purposes (see Fig. V).

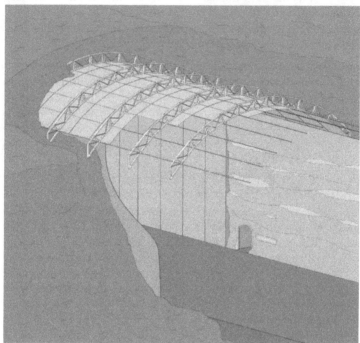

Fig. V: Deconstructed view of tunnel construction and entry door to surface building.

5.2: Surface Structures

To utilise the Martian location, all initial residences will be placed above the surface to provide natural views of Mars. The same three issues of temperature, radiation, and pressure must be actively solved in building design, as shear material bulk ceases to be an economical solution. An innovative implementation of a 'twin-hull' design was chosen to overcome the majority of these dangers and allow utilities routing to be integrated into the architectural design.

Air radiators are needed to dissipate heat from power generation, however the hot water stream (70°C) provides an excellent opportunity for space heating. The buildings were also tasked, therefore with being the radiators for the settlement. In addition to this, water is an excellent radiation absorber as well as being consumed in residences (drinking and washing). These buildings are designed to hold a water layer between a thin outer, and a thicker inner hull, in channels. This gives preference to heat transfer out of the water into the building but predominantly out to the atmosphere. The channels allow temperature to be controlled by changing water flow from the base of the building; ensuring that apartments at the top of the building can be heated as hot as those at the bottom if desired.

As convection from the surface was needed to be steady and predictable, while maintaining a large surface area, an elliptical profile was chosen for buildings. The aspect ratio has been kept below 0.7, with the major axis aligned with the prevailing wind direction to avoid fully-developed vortex shedding, which could cause excessive bending stresses on the building. This shape provides a large internal volume, while using less wall material than a rectangular building; and reduces pressure-related stresses seeking to deform the structure. Setters are expected to prefer large windows, so wall space is valuable. The majority of the building's load will be supported by a hollow, internal structure, providing a large volume to contain an access staircase, lift, and utilities' bus. The remaining central column could then be used for sugarcane farming. This *farm-tube* allows for local oxygen production to improve residential air quality.

The water layer in the building's hull aids with air pressurisation, however it places unusual loads on the outer wall. As this wall must be thin for heat transfer purposes, the building shape is unusual; typically narrowing at its centre. The outer wall thins towards the top of the building as the hydrostatic pressure reduces and it is necessary to have a thinner wall to ensure an approximately constant outer surface temperature as the water cools. The inner wall is designed to have a constant thickness, equal to that required for pressure retention at its highest pressurised point. This choice increases safety, load bearing strength, and ease of construction. All walls have been designed to have a safety factor of six for stresses at infinity in order to allow sufficient leeway for stress concentrations introduced by the large windows included.

With pumping work having already been performed to move hot water to the top of the building, as much water as possible should be stored at low pressure (0.5 bar) in a large tank at the top of each building. This allows each residence to act as a water tower to supply the residences at the bottom of the building and facilities in the tunnel network. Water is also pumped into a smaller, high pressure tank to supply residences on higher floors. Water will be drained from the main, low pressure tank to return to the CCM heat exchangers when water demand is low. Water filtration will occur centrally at the distributor to the wall channels, and again on access to the storage tank. Conducting strips placed into the water will act as baffles, heat conductors to the building's roof, and grounding to neutralise ionised molecules caused by the tank's additional role as the topmost radiation absorbing layer for the building.

Surface buildings will be connected to the subsurface tunnel network by extending the farm-tube, utilities space, and access shaft down to the tunnel depth. This increases the area available for farming and provides space for pumping and services distribution facilities, as well as avoiding the need for airlocks in each building. The farm-tube and access shaft will be connected by a short passageway to one or two tunnels in order to connect the building to the rest of the town (example entrance to a building shown in Fig. V). An inflatable wall system based on NASA's *Echo 2* satellite will be used to seal off the entrance tunnel in the event of a pressure loss. As the location of the breach can be determined, placing deployable walls to either side of the entrance to a building from the tunnel will ensure that residents are not trapped. The likelihood of a breach occurring directly above the entrance to a building is small and mitigable.

Building construction will begin with the use of robotic augers to dig the holes for the farm-tube, utilities space, and access-shaft while surface levelling is performed by METs. Once complete, the hole for the farm-tube will be impregnated with photopolymer resin and used to allow a vertically-crawling, twin industrial arm inspired, material laying construction robot. These arms will heat and project excavated material in thin layers to form the shape of the building. The hot, high velocity rock-powder will sinter itself upon impact with the layer below, forming a structure with lower porosity than conventional ceramic manufacturing methods, high strength, and that does not require a resource-intensive binder. Photopolymer resin will be impregnated into the inner layer of the outer wall and both sides of the inner wall to ensure water and airtightness. The resin will have a pale pigment added for the inner wall layer to improve aesthetics and make the surface more amenable to painting. In this way, all external and internal walls can be constructed quickly and with minimal resources.

The roof will be printed in the same fashion until the printer reaches the edge of the farm-tube and cannot continue further. The robot will be lifted out by a MET crane attachment and moved to a new site. The crane will then plug the farm tube and internal finishing can begin by hand. Pigmented and textured MonoPlate will be used extensively as a substitute for drywall, wood, and stone for internal finishing. The future resident will be offered extensive aesthetic choice over their property's decoration if they are known at the time.

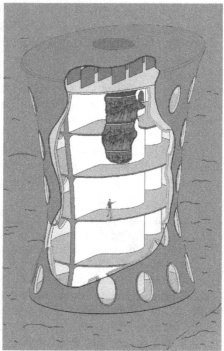

Fig. VI: Cutaway view of a small residential surface building with farm-tube in its centre

The small, example structure shown in Fig. VI has a total height of 27.05 m. The ten apartments in this building shown have a size ranging from 58.9 to 104.9 m² (634 to 1129 sq. ft). It is possible to design smaller or large buildings to house more or fewer people. Ceiling heights were chosen to be high, even at the expense of residential density; those shown are 4.76 m to allow for floor mounted wiring channels, insulation, and utilities routing. This large ceiling height primarily aims to combat claustrophobia in the lower gravity but also allows residents to install a mezzanine floor to increase living space if they prefer.

5.3: Additional Design Elements and Diversification
As residents will spend most of their lives in these built structures, it is critical that they demonstrate variety in order to decrease monotony and environmental claustrophobia. The aspect ratios of surface buildings, their heights, and chord lengths can all be varied within reasonable limits to allow for a wide range of buildings to be designed. The flexibility in tunnel wall design allows for a plethora of aesthetic styles to be investigated by guest architects or artists-in-residence. It is expected that district aesthetics will emerge and both tunnel and building designs may become of local pride in the way that architecture in cities currently does on Earth.

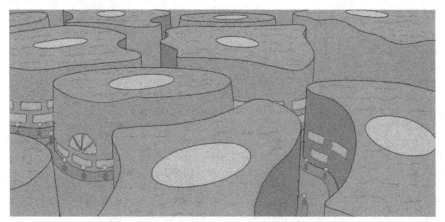

Fig. VII: Example layout of a residential city district with the covered gantries removed.

The ratio of tunnel space to building space will be local to each settlement; with more highly populated towns expected to have a lower ratio since the excavated material from even a short section of tunnel is capable of constructing a large number of buildings. Tunnels will be used where large space spaces and headroom are important, such as for food-crop and pastoral farming. Similarly, towns are expected to incorporate public parks, especially at tunnel intersections where long lines of sight are available in many directions. The capacity for architectural and artistic expression is equal to or even greater than that available in a city on Earth: the Martian sky can be made visible or screens or projections can be used to simulate one, or something else entirely!

The layout of tunnels and buildings is equally flexible; tunnels can be laid densely (although at least 25 m apart) for agricultural and industrial purposes, or more sparsely depending on the need for surface space for buildings. Tunnels should, however, be laid with a minimum radius of curvature of at least 500 m to their centreline; this ensures that no shrapnel from an explosion can travel more than 140 m down the tunnel before being arrested. Many tunnels will have road surfaces for internal transport and pavements for pedestrians within towns. Buildings may be clustered in estates, or may be more uniformly distributed about the city depending on the phase of development. Both styles have benefits and disadvantages, estates enabling larger communities while distributed buildings afford greater privacy. Care should be taken by city-planners to note the effects of turbulence caused by the buildings in the prevailing wind to ensure that heat can be rejected as predicted and also that neighbouring buildings do not significantly block each other's views of Mars.

6: Farming

Sugarcane farming is essential to the energy storage needs of the settlement. A genetically modified sugarcane will be engineered so that its phototropism will dominate over its gravitropism, as well as inferring the necessary efficiency and pest resistance advantages. Sugarcane must have its stem cut at an angle to ensure that the cane can be regrown from the same plant. This is typically a highly manual process, however with the conferred changes to the growth pattern, the hydroponic lighting can be angled to cause the plant to grow at an angle of approximately $60°$ as shown in Fig. VI. A flat cut across the dense, angled crop can then cause the angled cut on the stem, vastly simplifying harvesting. The farm-tube model for sugarcane growth offers the opportunity to make full use of built structures and integrate carbon dioxide scrubbing directly into residences. Growing platforms with 7-metre diameter, will be placed with a 2.85 m height in order to maximise growing volume for the angled sugarcane (effective height of 3.29 m). The average farm-tube will house 20 growing platforms. Using this system means that a harvesting robot can be hoisted up the centre of the farm tube to cut the crop every 89 Sols. These robots can be reused in any farm-tube meaning that the infrastructure investment is small but highly productive. The amount of ethanol to be produced also includes a quota to be delivered for industrial purposes, including for the production of construction photopolymers.

Unfortunately, with the amount of sugarcane needed, extra growing space must be dedicated to farming it beyond that available in residences. In order to capitalise on the already available infrastructure, these overflow farms will also be built as farm-tubes, but in a tight cluster for resource efficiency. These clusters will then be shrouded with a single sintered-dirt wall to save materials and avoid the necessity for a concave wall shape seen with residential buildings. These specialist farm buildings will be placed on the surface by the PATs (Section 3) and connected in a fashion similar to normal residential buildings. This location alongside the PATs ensures fewer heat losses will occur between the CCM heat exchangers and the use of that heat by the sugarcane.

Town	Freyja	Heimdall	Delling	Eir/Kvasir	Overall
Population	300,000	250,000	200,000	125,000	1,000,000
Food Crop Harvested Per Day (km²)	2.6000	2.1667	1.7334	1.0833	8.667
Total Food Crop Farming Area (km²)	222.043	185.036	148.029	92.518	740.143
Energy Storage Ethanol Demand (kg/s)	89.439	75.200	60.960	39.100	303.798
Industrial Ethanol Demand (kg/s)	10.0	9.0	8.0	6.0	39.0
Total Ethanol Storage Volume (10^6 m³)	5.753	4.837	3.921	2.515	19.541
Industrial and Energy Storage Sugarcane Area Harvested per Second (m²/s)	37.266	31.333	25.400	16.292	126.583
Total Sugarcane Growth Area (km²)	294.613	247.707	200.801	128.795	1000.711
Total Number of Farm-tubes	384,732	323,478	262,224	168,192	1,306,818

Fig. VIII: Table to show quantities related to food farming and ethanol production.

The hydroponic element for the food production infrastructure will be isolated from the rest of the settlement in the PATs. This will protect crops from contamination and to allow it to have a different atmospheric composition. Doubling the CO_2 content in the growing environment for crop plants has been found to double water efficiency, which, in an aquaponic system will integrate the farming of fish (salmon, tilapia, trout, yellowtail, anchovy, tuna, red snapper, bluefish, and pollock) and hydroponic food farming; the waste products from the fish are high in nitrates which are needed for crop growth. An aquaponic environment can double the energy density of a crop. Genetically modified crops will aid with land productivity. This CO_2 rich air is then shared by the sugarcane overflow farms and the food plants.

Fig. VIII shows the areas needed for the growth of both sugarcane and food plants. It is clear that, despite stacked farming efforts in the PATs, additional food growing space will be needed as a town reaches its design capacity. Buildings such as those for residences, but with no internal floors, can be constructed along the PATs to provide a growing area of approximately 3 km² each when populated with stacked hydroponic layers between access corridors. These structures can be designed to have a cylindrical profile, rather than elliptical for material efficiency and to simplify designs for future automated systems. Initially these hydroponic buildings will have to be harvested by hand by workers with oxygen masks to counteract the high levels of CO_2 maintained. A given area will be harvested every 83 Sols with a different area harvested each day. A high airflow will be maintained throughout the PAT to homogenise the air content and ensure that new moisture and carbon dioxide reaches the top layers of all farm buildings (see Section 8.1).

As food and sugarcane growth (not including farm-tubes in residences) must be shut down during dust storms, stockpiling and storage of food-stuffs must occur to provide a 12-month buffer on food supply (see Section 10.6). If this supply is unused, stored stock will be cycled back out into the food supply to ensure that stores can be freshened.

A variety of basic crops will be grown to allow for a varied diet and cuisines, this will include wheat, maize, corn, potatoes, oats, rice, and other vegetables. Chickens will also be raised for meat and eggs in the industrial districts of a given town.

Other dietary staples will be synthesised from their base chemical components to produce suitable replacements (e.g. milk). A varied diet not only provides settlers with all the necessary nutrients, but also improve their quality of life by keeping meals interesting.

7: Social Factors
7.1: Political System
Elected *Community Representatives* will gauge the opinion of their constituency (approximately 200 people) on bills that will affect them in order to report back to the local Committee of Experts. They will be paid on an hourly basis for work recorded through a government website portal. They will act as a spokesperson for the community in any challenges to proposed or actioned legislation.

A Local *Committee of Experts* will be made up of professionals and academic experts from different fields in each town. Members will be nominated by others within their sector. They will inform policy making on a full-time basis, and are expected to return to their sector after their four-year term is complete. Campaigning for nominations will be discouraged and treated as poor form. A *Grand Committee* will comprise all members of all five towns to propose Vanaheim-wide legislation.

The residents of each town will elect a *Town Governor* for a five-year term. Direct individual voting prevents the formation of ineffectual political parties. The candidate with the largest proportion of the vote with at least 33.3% will be elected. If no candidate reaches this threshold, a transfer of votes system will be initiated, where the candidate with the least votes is removed from the ballot and voting is repeated.

A former Governor will be elected as *Head of State* responsible for approval of Vanaheim-wide legislation and maintaining settlement cohesion for the same term length. A people's vote of no-confidence any member of this governing body may be initiated by a majority vote in the Grand Committee. In this way, government is beholden to technical expertise and community needs while being protected from nimbyism.

7.2: Education System

Students will enter into an ability-based grading system. At aged 4, children will be tested on their reading, numeracy, and problem-solving ability. Based on their results an impartial algorithm will place them into a suitable grade. Throughout their academic career, they will undergo similar tests in each of their subjects every 3 months, re-determining their ability levels. By grouping

students in this way, teachers will be able to adapt their teaching styles to an attainment level, tailoring learning to the ability of the group. Students are more likely to remain engaged with content, as the pace of a class can be better set and those who struggle would not be under any pressure to memorise content for the sake of passing a standardised exam. This promotes a growth mindset to produce deeper and broader understanding from an early age.

To complement this approach students' educational attainment may be described by the level they achieve in wide-ranging core and optional subjects by the time they leave education. Students will remain in education until the age of 18, or until they reach a minimum attainment level in core subjects. After this, they may decide to continue with academic education or enter the workforce through vocational traineeships.

There will be 3 to 4 schools in each town, supporting a variety of attainment levels. All schools will cater for the traditional core subjects. Schools cannot be expected to provide all possible optional subjects; students will be encouraged to attend classes at other schools where possible. In addition to regular schooling, Vanaheim will provide a number of government funded entrepreneurship grants to encourage the development of new companies and innovation.

7.3: Healthcare

Free physical and mental healthcare will be provided to all Vanaheim citizens. Doctors' remit will extend to keeping settlers healthy not just healing the sick, putting a focus on preventative rather than reactive measures. This will not only improve the welfare of individuals, but also ensure a strong and sustained workforce on Vanaheim. With residents' consent, passive health monitoring systems can be implemented to monitor for temperatures, pathogens in saliva in wastewater, urine components, blood in stool, etc. This data will be processed in each residence and will be flagged to the individual if concerns arise. They will then have the option to raise concerns with medical professionals if they choose.

One large central hospital in each town will carry out all scheduled and emergency procedures. This will allow for medical resources to be concentrated in specific locations meaning that patients have access to the most advanced technology and best doctors; providing them with equal access to high quality medical care regardless of where they live. Medical research may be sold back to Earth as the production aspect of our supply and demand model.

7.4: Culture
Vanaheim will operate logistically on the length of the year on Earth, even if this falls out of sync with the Martian year. However, it will operate on the length of the Martian sol, including a time-slip of 39 minutes 35 seconds. For this reason, leap days will have to be computed for each 'Vanaheim year' to bring it within 12 hours of UTC on Earth, leaving it with either 355, 356, or 357 Martian days in a year. All dates will be displayed alongside with an equivalent date in Sols and Martian years.

Settlers will be coming from different countries and backgrounds on Earth, making Vanaheim a truly multicultural society. In order to maintain and embrace all of these cultures, only two national holidays will be specified each year. Consecutive days at New Year and a two-day holiday on the anniversary of the arrival of the first settlers. To make up for this small number of holidays, the statutory minimum number of Vanaheim-annual paid vacation days will be set at 24.

It is hoped and expected that the Vanaheim will develop its own culture and traditions; the cultural melting-pot will enable and encourage new cross-cultural interactions and relationships. The etymology of culture suggests that to have a culture, is to inhabit a place sufficiently intensely to cultivate it -to be responsible for it. The settlers should be able to identify with their surroundings to a point where they can develop culture specific to the settlement. To facilitate this an office of arts and culture will be responsible for introducing settlers to the unique aspects of their new home and informing them of special milestones and astronomical events.

8: Systems
8.1: Air
All pressurised regions will be designed for 1 atmosphere of pressure, to ensure that humans travelling between Earth and Mars are better accustomed to the atmosphere, reducing the adjustment time needed and ensuring that no long-term health risks are incurred. Instead of using the same composition of air as on Earth, an air mixture containing no noble gases (21% oxygen, 78.96% nitrogen and 0.04% carbon dioxide) will be used except in agricultural areas where higher concentrations of carbon dioxide will be used to improve crop yield. Carbon dioxide is trivial to obtain, as it can be taken directly from the atmosphere, stored and regulated, by autonomous control systems. Excess carbon dioxide from ore refinement and manufacturing can be stored for use in case dust storms risk damage to the normal systems.

Oxygen is easy to obtain in the initial phases because the water demands of the colony during early missions are low enough to justify electrolysis of water and onsite emergency stores composed of liquid oxygen imports. Given the ratio of agricultural land to residential land in each settlement, once farming has begun, there will be sufficient oxygen to meet the needs of the residence, agriculture and industry (see Section 10.6). Indeed, within the PATs, an unusual issue in spaceflight will be encountered; excess oxygen will need to be extracted from the air while preserving nitrogen and introducing CO2. The excess CO2 working fluid from the modified Allam cycle can be injected into the PAT atmosphere to replenish CO2 and moisture within the farming volumes and tunnels. Additionally, the Martian atmosphere can be compressed and injected directly to increase CO2 and nitrogen content.

Nitrogen proves significantly harder; whilst it is the second most abundant gas on Mars, it only makes up 2.3% of the atmosphere. Nitrogen is also very scarce in the Martian soil and in order to extract any useful quantity of the gas requires access to nitrogenous ores. At the start of the settlement's lifecycle, the power requirements associated with collecting nitrogen via these methods are too high. Therefore, the remaining option is for the gas to be shipped to Mars from Earth ahead of the first settlers. Solid nitrogenous compounds would be harmful and require a lot of energy to decompose into gaseous nitrogen, so it would be preferable to send shipments of 100 tonnes of liquid nitrogen at a temperature of 200K and 125 bars. This shipment would arrive just before the first inhabitants move into a new settlement and would cost approximately $50,024,000 per shipment. This quantity of nitrogen is able to support the operation of 25 residential buildings along with the agricultural land required to support this populace.

Air within the PATs and farm buildings will be processed through air liquefaction sites at the CCMs. Here, air will be liquefied by the Linde-Hampson process and solidified carbon dioxide ejected back into the PATs or Martian atmosphere. Oxygen can be evaporated out by a slight increase in temperature before passing it through a heat exchanger to recondense it while evaporating the nitrogen. The cool nitrogen stream can be passed around the CCM to cool its casing or used to cool the cooling water at times where the town's heat demand is low. Excess nitrogen can be extracted and transferred to top up the residential area's atmosphere. The liquid oxygen will be stored near the CCM as its oxidiser for when solar power is unavailable and chemical fuels are being used. Once liquid oxygen storage is full, the liquid oxygen will be stored in readiness for transport to the launch sites, Valkyrie and Bifrost, for use as a propellant. This system can be used directly on the Martian atmosphere to assist with initial acquisition of nitrogen. The rate of nitrogen acquisition is one of the key factors which determines the rate of expansion of a town. This rate of expansion continues until a town reaches "maturity". We define this as a population capacity of 20,000 people as the automation of basic functions is at an adequate level to warrant the development of a new town. At the start of a new town's life cycle, an initial injection of nitrogen is required which may require further imports. However, as new towns are established, previous towns will also provide the initial injection from their nitrogen storage. When Eir and Kvasir are set up, 100% of the nitrogen for the towns will be produced on Mars.

8.2: Water
Water is essential to the survival of humans, so there must be a robust and reliable system in place by the time that the population on Mars reaches one million. During

early missions, the crew are likely to bring the required quantity of water or components necessary to create water, as the water demand would be much lower, limited to activities such as drinking, bathing and agriculture. However, making water from components is not enough to meet the needs for critical activities of a larger settlement. This can be achieved by melting ground ice from the Martian surface, which can exist at latitudes as low down as 35N or 35S. In these regions, ice is located 1-2 meters below the surface. The majority of the material covering the ice is regolith and soil, which would make an excellent proving ground for testing key technologies which will make the automation of excavation, construction and transport possible.

A wave of precursor Multipurpose Electric Tractors (METs) and External Large Cargo Transport Vehicles (*Dustcrawler-C*) will be sent to a region between 35N and 50N before the arrival of humans. The METs would then carry out excavation tasks to uncover and push the ground ice into the Dustcrawler-Cs. All Dustcrawler-Cs have temperature-controlled cabins, which ensure the ground ice is stored and transported in a liquid state. Data gathered from these tasks will be used to influence the design of future METs and transport vehicles so that excavation and transport can operate at the greatest possible efficiency when fully automated. Furthermore, the composition of the ground in this location is ideal to test the basic operation of the METs before autonomous excavation and construction of the power trenches. At the end of the mission, the precursor METs will remain at the site and the Dustcrawler-Cs will return to Freyja. The long journey will test the necessary ground vehicle technologies we require for long distance autonomous transport on a roadless highway (active suspension, electric powertrains to travel at speeds of 100km/h, energy recovery systems, dust protection systems etc).

Early exploration and science missions, in addition to ongoing developments from Earth research, will improve the efficiency of water recycling systems. Therefore, the water deficit percentage is expected to decrease with time so that the frequency of water excavation missions can decrease. Through the analysis of water consumption and use of predictive algorithms, ice excavation missions can be planned to avoid water deficits. These missions will include sending new METs to the sites, which will be designed to remain operational at the site for a period of 15-16 months from the beginning of the Spring Equinox, and transport vehicles to collect the ground ice. After the operational period, the METs will shut down, avoiding the dust storm season. With time, ice excavation missions will be pushed further north in order to access ground ice which is closer to the surface. Although this will increase transport time, it provides access to a larger volume of ice, which will be easier to extract. And by this point, the transport network is expected to have the necessary reliability and capacity to bring the ice to the towns. This ensures that the growth and maintenance of the water supply will be more than sufficient, allowing Vanaheim to thrive.

8.3: Sewage
Sewage will be stored and treated on the Martian surface for ease of access during upgrading and maintenance procedures. Furthermore, it is cheaper to set up a superterranean pipe network as there are no excavation costs. Each town has its own

sewage system rather than sharing a centralised sewage plant to reduce the risk and cost that comes with long stretches of pipe networks. As a large proportion of waste will come from the superterranean housing, a superterranean sewage network also reduces the amount of work done to resupply the houses with potable water after filtration. As well as waste from residential sectors, waste from power generation, material extraction, manufacturing and a number of other sectors will also contribute to the total sewage.

The primary stage of purification is to filter the entire sewage mass through mulch which is a waste product for power generation. This removes large solid debris and begins the introduction of Proteobacteria to assist with the second stage. Past the mulch filter is a stage of progressively finer filters to ensure that the second stage receives only liquid waste. In the second stage, wastewater undergoes active aeration with an oxygen supply of 3mg/L. The population density of bacterial colonies in the wastewater increases and breaks down organic matter at significantly higher rates whilst requiring a lower volume to process the same mass of wastewater compared to passive aeration. The active aeration process is repeated over several cycles and once complete, the water will be injected with high pressure oxygen to force it through a fine mesh to remove any fine particulates. Following ventilation of the excess oxygen in the stream and chlorination to sterilise the water, it is re-entered into a town's water system.

9: Transport
Transport has enabled the growth and development of economies and civilisations on Earth by creating trade networks and generating income. With transport comes a high set-up cost associated with vehicle manufacturing, infrastructure development, maintenance etc. In the colonisation of Mars, transport spans a wide range of routes and vehicles such as subterranean and superterranean ground vehicles which require roads and rockets which require spaceports. Therefore, steps must be taken to keep transport costs on, to and from Mars as low as possible.

9.1: Ground vehicles
For the rapid and reliable transport vehicle production, the process will be contracted to existing Earth automotive designers and manufacturers, and Vanaheim will provide the infrastructure required for manufacturing these vehicles. Companies submit proposals and based upon cost, design and production plan, proposals are shortlisted. The winning company for each contract will be given access to the state-funded Martian consultation firm, MAPD to further optimise their operations on Mars. Following the fulfilment of the contracts, companies will have the option to purchase and own the production facility to continue their operation on Mars (see Fig. IX).

9.2: Road Infrastructure
Large highways between industrialised cities have consistently proven to be expensive feats of civil engineering. Whilst the long stretches of road provide better vehicle performance, the initial production and maintenance cost associated with large highway systems is not justifiable for a one-million-person civilisation. Instead, the Yggdrasil Adaptive Roadless Network (YARN), a network of non-

paved transport routes connecting settlements, spaceports and industrial cities together, has been developed. The routes within YARN may change location and architecture based upon the status of Mars, allowing for an adaptive transport network which can operate indefatigably in any weather condition or state of repair.

Fig. IX: A concept image of a cargo and crew convoy leaving Bifrost.

Contract	Description
Internal transport bikes	Easy mechanical vehicle which can be mass produced easily and sold as a commodity or operated on a rideshare scheme.
Internal transport automobiles	All electric fully autonomous automobile with 4-6 passenger capacity designed for rideshare or personal use.
External transport automobiles (Sandstorm)	Ethanol-powered entry-ignition-engine/electric hybrid fully autonomous vehicles with 4-6 passenger capacity. High speed and comfortable transport. Designed to have mixed-terrain capability.
External large transport automobiles (Dustcrawler)	CO_2 entrained oxygen-ethanol entry ignition engine/ electric hybrid fully autonomous vehicles with 96-108 passenger capacity. Provides transport between industrial regions, cities and spaceports. Fitted with a reserve life support system and emergency power generation systems. Designed to have mixed-terrain capability.
External large cargo transport automobiles (Dustcrawler-C)	CO_2 entrained oxygen-ethanol entry ignition engine/ electric hybrid fully autonomous vehicles with a 400 cubic meter internal volume. Provides transport between industrial regions, cities and spaceports. Fitted with emergency power generation systems. Designed to have mixed-terrain capability.

Fig. X: Table to show proposed ground vehicle contracts for private industry.

The autonomous network of vehicles coupled with the Martian Positioning System allows us to identify general trends in optimum transport routes in terms not only of transit time but also safety. Furthermore, we required all external vehicles to have mixed-terrain capability in order to traverse the Martian landscape and add flexibility to the topology of YARN. As well as this, the mixed-terrain capability allows us to push back the eventual date when we will need paved highways. These methods will ensure that YARN will provide the best performance possible whilst our economy can grow further. YARN is a necessity to allow for the most profitable expansion of our colony. The money saved on highway construction and maintenance along with the profits generated by transport ensures that the government will hold all the power when it comes to awarding transport infrastructure upgrade contracts. Thus, we avoid an all too common situation on Earth where highway construction and maintenance projects are toyed with as political bargaining chips rather than a basic necessity for a very large population. Within towns, however, a formal road infrastructure will be built. This will allow settlers to navigate through the tunnels quickly, using fully autonomous automobiles. Unlike around the YARN, towns will have a significant number of pedestrians, so by designating footpaths and roads settlers can be kept safe and accidents can be better avoided.

9.3: Spaceports

There will be two spaceports servicing the settlement for redundancy and efficiency. The first to be constructed, Valkyrie, will primarily serve the western cluster (Freyja, Heimdall, Kvasir) while Bifrost will serve the Eastern Cluster (Delling, Eir). Each spaceport has 5 landing pads, this covers regular imports with some spare pads. Instead of having a central terminal, the mobile service tower for a pad can act as a temporary staging area if necessary. This means the entire spaceport is decentralised and expanding capacity is as simple as constructing a new landing pad and service tower for it.

Each pad at a spaceport is assigned to a ship for the entirety of its stay planet-side. Once landed, a service tower will be moved into position to allow unloading and subsequent loading and refuelling of the ship: much like passenger boarding bridges at airports. This gives leeway to ships on the accuracy of their landing and is more effective than using a crawler to transport the entire ship somewhere else for unloading. Cargo and passengers will be transferred to the relevant ground transportation and return via the YARN. Any outgoing cargo and passengers can be brought to the ships in the same way.

10: Economics
10.1: Centralised Vanaheim Bank (CVB)

Vanaheim will have its own banking system, where residents can receive interest on capital used to invest in projects to further grow the settlement's economy. High transaction fees from Earth bank accounts will be implemented, in order to encourage settlers to transfer money into new accounts on Mars. They will be able to access and maintain their own accounts in much the same way as pre-existing Earth accounts.

The CVB will issue the *Martian Æsir* (symbol: Æ, code: MAE), which will be a floating digital fiat currency supported through credit cards, debit cards, stored-value cards and other typical forms of contactless payment. Digital currencies reduce operating costs, increase efficiency, and make the currency more versatile. The MAE will also have incredible seigniorage and provide unparalleled control over the economy allowing us to keep up with the rapid growth rate without making the currency scarce.

10.2: Exchanges

There will be two exchanges operating out of Vanaheim: the Martian Home Exchange (MHEX) and the Martian Public Exchange (MPEX). These exchanges aim to leverage Vanaheim's unique position as the first large settlement on Mars to establish it as the planet's financial centre while also giving it a presence on Earth. These exchanges will make profit through the traditional avenues of transaction, listing, and data fees. The market data for these exchanges is likely to be in high demand. The listing fee will be waived initially to help establish the exchanges.

The MHEX is like most exchanges on Earth and allows local businesses (or local divisions of companies) to list themselves for trading. The MHEX also allows commodities trading on Martian commodities, such as sugarcane. This will not only attract financial institutions to Vanaheim, but will incentivise setting up comprehensive local divisions for serious trading as the time delay from Earth represents a significant disadvantage for Earth based traders on the MHEX.

The MPEX is a public exchange aimed at Earth traders. It is modelled after the Investors Exchange (IEX) and will use the Earth-Mars time delay instead of an artificial delay like the IEX uses. Large financial institutions gain advantages over the average trader by using High Frequency Trading algorithms and having private connections that come from computers extremely close to the exchange. The delay will help level the playing field and is proven to help stabilise the market overall. The MPEX will not be limited to trading Martian instruments. The MPEX will have lower than standard transaction fees to encourage trading by anyone. Conversely, it will have a more substantial yearly listing fee, reflecting its increased stability and to compensate for the lower transaction fee.

10.3: Taxation and Government Expenditure

Instead of a revenue-based tax model, Vanaheim will run on a volume-based *residence-fee* model. Both settlers and corporations will be charged a residence-fee based on the volume in cubic meters of property which they own. These fees will include a basic allocation of water and power for each settler and company, as well as the right to live in Vanaheim and benefit from its community. Additional water and power will be charged for at a rate linked to the ethanol stock price on the MHEX. The standard rate for building space will start at the Aesir equivalent of $43.40 per m³, while subterranean space will be charged at $34.95 per m³ (USD as of 30/06/20). Building space will entitle residents to 11.5 kWh electricity and 192 litres of water per m³ per Vanaheim year. Tunnel space will allow for 15.0 kWh and

249 litres of water per m³ per Vanaheim year. Additional electricity will be set and charged at $0.25 per kWh and water at $0.0125 per litre during storm-ready conditions when ethanol prices will be most stable. Prices throughout the year will then vary with ethanol stock price.

Settlers who take state roles in elementary occupations (as described by the International Standard Classification of Occupations) and unoccupied critical roles may receive discounted residence-fees as part of their contract. The degree of the discount will depend on the job and the circumstances of employment, e.g. a doctor may be offered a 10% reduction when moving to a different town or 30% if they move to Vanaheim from Earth, while an agricultural worker may be offered a fixed 80% fee reduction. These waivers enable settlers in lower paid, state jobs to enjoy a higher quality of life for the same salary and encourage people to take these roles in order to move to Mars.

Department	Budget($m)	%
Payment of Public Sector Employees	8,750	31.25
Education (inc. research and higher ed.)	6,440	23
Healthcare	3,080	11
Social Services (pensions, disability, etc)	2,100	7.5
Emergency Fund	1,400	5
Residence Fee Rebate	1,400	5
Maintenance	1,260	4.5
Miscellaneous	980	3.5
Emergency Response Expenditure	980	3.5
Import Subsidies	840	3
Expansion Budget	770	2.75
Total	28,000	100

Fig. XI: A table to show budget expenditure.

As a result of this system, the government of Vanaheim will not profit off of increased company revenues, it will benefit from the growth of successful companies. In this way, companies will find greater reward in increasing efficiency than on Earth. It is hoped that this will aid with the movement of industry into Vanaheim.

Initial estimates suggest that one in eight people will work in a state-funded capacity. Based on an expected income from residence-fees of $28 billion, the budget is allocated accordingly.

10.4: Martian Acceleration & Protection Department (MAPD)

Most companies on Earth are well optimised for operation on Earth, but when the opportunity arises to establish operations on Mars, billions of dollars may be lost in research, development and production of essential systems. MAPD would be a division of the Martian government, designed to optimise the transport, establishment and operation of private companies and governmental organisations on Mars. MAPD would be composed of experts in Martian STEM, law and economics who could provide consultation to clients on Earth and Mars so that they could run as efficiently as possible. An organisation partnering would be initially introduced to an Earth branch of MAPD, in order to identify how the cost of their core operations and transportation of business assets can be reduced during the transition to Mars. Meanwhile, the Martian branch will work to provide a package to allow the establishment and operation of the partner organisation on Mars. This package includes cost for construction, operation and maintenance from existing private Martian companies. The cost of this package would be a similar price as a package offered by government companies. To ensure this, the Vanaheim government will provide Residence-Fee rebates to the private companies to enable them to operate profitably while offering their services at lower rates.

Since the notion of colonising Mars became popular, experts have always claimed that large profits were possible through Intellectual Property (IP). However, no legitimate method had been proposed which wouldn't result in private firms gaining enormous power by transactions and litigation around IP. Therefore, MAPD offers reliable protection for partner organisations that have set up businesses on Mars. The IP would be held in the name of the owning organisation with MAPD acting as the only conduit for the transmission of this information between Mars and Earth. Organisations may have genuine concerns regarding the reliability of this transmission architecture.

It is important for MAPD to ensure that information is not given to the wrong people on Earth. Fig. XII details the MAPD IP transmission architecture. On top of the organisation's encryption, the MAPD-developed Meiosis Duplication Satellite (MeDuSa) Protocol is applied to the Martian Data Packet. Similar to biological meiosis, the Encrypted IP data and Decryption Key are separated into millions of tiny portions of data. One portion of Encrypted IP data is randomly packaged with a portion of the Decryption Key. This process is duplicated so that an A and B set which are completely different from one another and both sets are uploaded to the Mars Satellite Network and transmitted via laser communication.

MeDuSa data sets are protected from interception via Quantum Data Packet Transmission (QDPT) encryption which is a form of mistrustful quantum encryption optimised for long-distance transmission. The Earth Satellite Network subsequently decrypts the incoming data using Quantum Data Packet Receival (QDPR) decryption (inverse of QDPT). The data packets are bundled together as they are decrypted and run through the MAPD-developed Data Retrieval Protocol. This protocol opens all the data packets and groups all the Encrypted IP data and Decryption Key data in separate sets. An AI Supercomputer runs through both sets and looks for overlap between A and B data. By finding one overlap, the

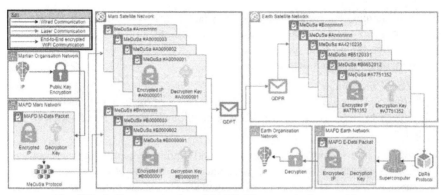

Fig. XII: A diagram to show MAPD IP transmission architecture.

Supercomputer can reconstruct both copies of the Encrypted IP and Decryption Key. The final check is to ensure that the A set exactly matches the B set after which both sets are sent to the organisation who then can access the raw IP data and any remnant of the information within the MAPD transmission architecture is destroyed. Throughout the entire transmission process, MAPD will never be able to decode, access or see the raw IP data so partner organisations can rest assured that the risk of espionage or leaked information is reduced whilst the profitability of the intellectual property is preserved.

10.5: Required Industrial Facilities
Where equipment and facilities are needed for the construction of Vanaheim, those facilities will be operated as a government-owned company. These companies, with excess capacity not given over to state operations, will offer their services to other customers on Mars and off-world. With a high barrier to entry for product profitability, only manufactured goods, and construction services will prove sources of government profit outside of the Vanaheim ecosystem.

Government companies will be broken into supply and production companies. Production companies will include a construction company, core chemicals provider, regolith refining, turbine blade production, advanced metal and ceramic forming. Supply companies will be dedicated to the operation of water filtration and storage, sugarcane fermentation, farming, food production, air processing, oxygen management, and ethanol management. Government supply companies will be expanded in capacity to fulfil expansion needs.

The government owned production companies will offer the best value services of any provider on Mars as all investments made into their equipment and facilities will be written off into the initial cost of Vanaheim. Government projects utilising government production companies work from them will be inexpensive, as employee wages are already covered in the settlement budget. During waves of expansion, however, government production companies will be largely dedicated to the needs of settlement growth rather than supply of industrial components or buildings. Government company capacity will be limited to grow only to meet the maximum needs of designed settlement growth. This ensures that private companies will have demand for production services from industrial operators in Vanaheim. However, a competitive natural market rate will develop during consolidation periods when government producers are able to contribute to fulfilling industrial and individual consumer demand.

10.6: *Oxy-Ethanomics*

Oxy-Ethanomics is a portmanteau of 'the economics of oxygen and ethanol' which will have a big impact on life in Vanaheim until alternative power sources are phased in.

Solar energy is stored as ethanol, biomass, and liquid oxygen in order to continue Vanaheim's operations even in the absence of sunlight. As described above, biomass and ethanol will be produced using energy available during the day able to continue settlement operations and crop growth during the night. Sufficient energy will also be stored to ensure that settlement operations, and growth of sugarcane in farm-tubes within the settlement can be continued during dust storms. This dust storm contingency stores masses of ethanol and liquid capable of fulfilling this requirement for up to a period of one Vanaheim year. The stores of oxygen and ethanol to provide power during dust storms will be called *storm-tanks*, despite this, they will be integrated into unified tank systems containing volumes used for other purposes also. The total stored volume will be split between two equally sized many-tank systems. The stored volumes of the ethanol and oxygen are listed in Fig. VIII and Fig. XIII respectively.

The power consumption of food and sugarcane farming is the greatest of any function in Vanaheim. To produce and store enough ethanol to maintain these productions during dust storms would be, in the former case, unfeasible and, in the latter case impossible. This result of this decision is that stockpiling of non-perishable and stabilisable food-stuffs is necessary to supply food for during dust storms. However, as it is not feasible to produce sizable excesses of food to put into this stockpile, it will be a necessary expenditure to initially fill it. This will cost $1.609 billion. In the event that this supply is used up, it will be necessary to refill it to full. This is budgeted for as part of the emergency fund.

While the storm-tanks are filling, the ethanol rate available to industry will be the industrial ethanol flow rates listed in Fig. VIII. These values can be increased if industrial demand is sufficient. Once the storm-tanks are full, however, the considerable ethanol and oxygen rates dedicated to filling them can be repurposed for industrial and expansion uses. This will allow a fall in the price of ethanol; however, oxygen price will be unaffected for reasons stated below. A proportion of this ethanol will be dedicated to the production of construction photopolymer by the government core chemicals producer until a stockpile able to support one Vanaheim year of settlement expansion is filled and current expansion demands met. This ensures that, even after a 12-month dust storm, expansion may occur as soon as adverse conditions lift.

Town	Freyja	Heimdall	Delling	Eir/Kvasir	Overall
Population	300,000	250,000	200,000	125,000	1,000,000
O_2 Production mass flow rate (kg/s)	3,364.71	2,829.01	2,293.30	1,470.93	11,428.88
Human Consumption O_2 mass flow rate (kg/s)	29.16	24.30	19.44	12.15	97.20
Storm Pneumatic O_2 (kg/s)	30.00	30.00	30.00	30.00	150.00
Energy Storage O_2 (kg/s)	165.79	138.16	110.52	69.08	552.62
O_2 for Industrial processes (kg/s)	40.87	36.78	32.70	24.52	159.39
O_2 for expansion (kg/s)	1.50	1.25	1.00	0.63	5.00
O_2 for Ground Vehicle Fleet (kg/s)	1.50	1.50	1.50	1.50	7.50
O_2 for Agricultural Waste Combustion (kg/s)	31.17	25.97	20.78	12.99	103.89
O_2 used for Wastewater Aeration (kg/s)	278.80	232.34	185.87	116.17	929.34
O_2 Storage for Rocketry (kg/s)	5.00	5.00	5.00	5.00	25.00
Experimentation O_2 Storage (kg/s)	5.00	5.00	5.00	5.00	25.00
O_2 Excess mass flow rate (kg/s)	2,775.92	2,328.71	1,881.49	1,193.90	9,373.94
Max Storage Capacity (10^6 m^3)	5.42	4.66	3.89	2.74	19.45

Fig. XIII: A table to show oxygen usage throughout Vanaheim.

Excess ethanol can also be supplied as a rocket fuel or cracked to methane for this purpose. A specialist *Mars Ascent Vehicle* research project will be supported by the government to utilise ethanol and liquid oxygen as propellants to potentially reduce launch costs paid to third parties. The supply of fuel and oxidiser to third party carriers will also provide income. If excess ethanol is being produced is still beyond usage capacity, the power diverted to sugarcane production can be reduced to slow growth times and so production rate.

Oxygen will be produced passively by plants' photosynthesis and then be harvested by air liquefaction. With the significant amount of farming necessary to produce food and ethanol, oxygen production is in sizable excess even when accounting for storm-tank filling and processes that require it. This ensures that oxygen prices, except during storm-conditions when agriculture is shut off, will be low. For this reason, oxygen will be evaporated to cool CCM casings in closed vesicles in order to increase its pressure. This high-pressure gas can be used as a pneumatic fluid for industrial and maintenance processes where appropriate. Examples of applications include its use to blow dust out of solar trenches and to actuate industrial robots. Excess oxygen will either be burnt with unstorable ethanol to return CO_2 to the atmosphere and recapture water.

As dust storms prove such a large impediment on settlement growth and profitability for resident companies, a government project will be initiated as an ongoing priority to collect Martian weather data. This, like on Earth, will allow for meteorological predictions to be made about the formation of local dust storms. As a local dust storm has a 33% chance to grow to a global dust storm, methods to quell formation of dust storms at formation will be investigated as a matter of urgency. Characterisation of storms that will affect Vanaheim will allow better preparation for refilling of food for any storm that lasts longer than the stockpile.

10.7: Trade

Whilst Vanaheim will be self-sufficient when it comes to the production of basic goods, the journey to this point requires frequent imports from Earth. After the success of early missions, initial imports will consist of the systems required for Vanaheim to create its own life support systems and contingency supplies. Following the rise of Martian material extraction and manufacturing, the nature of imports shifted to more *shortfall supplies*: commodities that Vanaheim needs but cannot make yet. Within these shortfall supplies are also emergency food reserves which will be ordered in preparation of dust storm season.

Given the reliance on Earth, exporting goods and services produced on Mars to be sold back on Earth. As government companies will have limited growth to entice private companies to Vanaheim, the majority of exports will continue to consist of advanced mechanical components. However, the abundance of ethanol will enable the production of complex metal alkoxides which typically retail for well over $200/kg. Furthermore, transportation of the raw materials to produce metal alkoxides is less expensive than on Earth given that each town has a healthy supply of the necessary raw materials. This means that rather than transporting materials over hundreds or even thousands of kilometres, it will take place over only tens of kilometres. These benefits will help to entice chemical manufacturers to Vanaheim in order to produce these molecules at a cheaper rate compared to Earth.

However, with all these benefits, the government cannot focus on only trade, so as private industries begin to populate Vanaheim, they will be responsible for their own launch costs for imports and exports. Launch pads can be rented for a fee of $250/hr to cover maintenance and operation of the pads. Fuel will be sold to private launch contractors with the expectation that every private launch will require fuel from Vanaheim.

Import Asset	Total
Life support establishment Systems	$6,169,640,000
Contingency Supplies	$1,023,235,000
Shortfall Supplies	$4,741,500,000
Total import cost from establishment to one million people	$11,934,375,000

Fig. XIV: A table to show import assets.

10.8 Beyond

Vanaheim aims to be self-sufficient and will expand its goals as it grows to sufficient numbers. There are a number of projects that funding will be dedicated to after the settlement reaches certain milestones. The most prominent of which are the Nuclear Power Programme (NPP), the Export Improvement Programme (EIP), and Solar Power Satellites (SPS).

The NPP encompasses the long-term goal of using nuclear power as one of the settlement's main power generation sources. Along with SPS, this will mark the transition to a consistent flow of power while freeing up ethanol for new industrial partners. The SPS will re-use the solar trenches as receivers. The contracts for the projects will be awarded to private companies.

The EIP aims to develop a Martian Ascent Vehicle (MAV) to launch a company's exports and necessary fuel to large, in-orbit cargo transporters, like *Starship*. By using MAVs, the cost of exporting goods is reduced because the large surface-to-surface vehicle will not have to perform a landing burn so refuelling costs are decreased. In the long term, trade expenses will decrease once the MAVs are in operation. During the EIP, contracts will be awarded by the Vanaheim government to private industry as a means to help the companies grow.

Vanaheim is dedicated to improving the livelihood and quality of life of anyone who wishes the further the mission for self-sufficiency. But one should never forget to look beyond what is possible today. And by no means can one anticipate that this journey will go off without a hitch; there will be mistakes. However, the culture on Vanaheim is designed to improve where we have failed in the past shall embolden the children of Mars do not make the same mistakes a before when expanding humanity's reach beyond Mars.

Bibliography

"NASA Mars Rovers Braving Severe Dust Storms." *NASA*, NASA, 20 July 2007, www.jpl.nasa.gov/ news/news.php?feature=1421.

Brandenburg, John E., and Paul Murad. "Evidence for Past, Massive, Nuclear Explosions on Mars, and Its Relationship to Fermi's Paradox and The Cydonian Hypothesis." *Aiaa Space 2016*, 2016, doi: 10.2514/6.2016-5529.

"Climate Change: Surface Ice at High Latitudes." *Planet Mars*, pp. 173–175., doi: 10.1007/978-0-387-48927-8_55.

Helliwell, John F., Richard Layard, Jeffrey Sachs, and Jan-Emmanuel De Neve, eds. 2020. World Happiness Report 2020. New York: Sustainable Development Solutions Network

Allam, Rodney, et al. "Demonstration of the Allam Cycle: An Update on the Development Status of a High Efficiency Supercritical Carbon Dioxide Power Process Employing Full Carbon Capture." *Energy Procedia*, Elsevier, 18 Aug. 2017, www.sciencedirect.com/science/article/pii/ S187661021731932X.

Carbon Dioxide, National Institute of Standards and Technology, webbook.nist.gov/cgi/inchi? ID=C124389&Mask=4.

17: THE DESIGN OF CITY STATES ON MARS, A VISION FROM SGAC

**Nissem Abdeljelil, Emma Baltide, Ahmed Baraka, Beenish Batul,
Sujab Bhusal, Lauren Church, Andrew Foster, Chaitanya Krishna Gopal,
Supun Gunarathne, Ben Hayward, Umang Jain,
Vignesh Kottayam Viswanathan, Gracio Joyal Lobo, Bijaya Luitel,
Pavithra Manghaipathy, Rigel Kent Lorico Marco, Aleksandar Nikolov,
Simone Paternostro, Bruno Pavletić, Swaraj Sagar Pradhan,
Harlee Quizzagan, Kristi Ray, Ekaterina Seltikova,
Krittanon Sirorattanakul, Ankita Vashishtha, Alina Vizireanu,
Nataly Yauricasa Cárdenas**
Space Generation Advisory Council
PG-coleads@spacegeneration.org

1   INTRODUCTION

The nascent and unprecedented nature of setting up an off-world colony for a population size of 1 million humans presents unique challenges towards the political, social, cultural and technological considerations. The approach taken here is one of "lessons-learned" from millennia of anthropological studies and understanding the workings of civilizations from past and present. Ideas of a democratic social and political fabric from various interpretations have been explored to perceive a feasible fit for a 'new world order' on Mars, while still having the system compatible enough to engage with Earth amicably. Here, the citizens of this Martian colony are in the process of having already detached themselves from their Earth-rooted identity, and they are focused on building and empowering a local Martian identity. The colony is considered a functioning state which is not seeking to serve Earth as its sole benefactor, but rather as a community that works towards its own interests and goals. The ideological decisions of this work are pivoted on concepts of self-reliance, self-governance and sustainability for the Martian people. Here, we assume that the founders of the colony elaborated on ideas of collectivism, rule of law and general trust to foster cooperation towards scientific and technological goals[1]. We think pluralism, in this case, can be justified using a common adversary on Mars being the natural inhabitability of the Planet, where more cooperative systems have a better probability of flourishing than others. Again, we assume the availability of active dispute resolutions and rules of engagement with one another, applied universally based on the concept of general trust and rule of law[2]. By owing to technological advancements in artificial intelligence (AI), machine learning and empathetic technology, most administrative and management tasks encompassed by traditional governmental institutions would be offloaded to and driven by artificial, cognitive administrators[3]. Some of these ideas are readily used on Earth in 2020 and are being explored as more 'smart cities' are transformed. However, being a bit more conservative, strongly empathetic functions such as punitive judiciary and other ethical resolutions would be reserved to human interpretation. The colony would launch highly innovative solutions to solve the most challenging problems of living on Mars. This sustainable economy would not just utilise the benefits of technology, but also be a leader in innovation, research and development. The Mars City State would be established as a nurturing ground for technologies that would be inconceivable to develop elsewhere. This amalgamation of human cultural sensitivity and technological prowess may well define the Martian experience. And most importantly, this colony would be the answer to the age-old question: *Is there life on Mars?* Yes, and it is flourishing with human empathy, and a nexus of technology!

2   LIFE ON MARS

The concept of aesthetics is defined as "A set of principles concerned with the nature and appreciation of beauty"[4]. It may seem at first glance that there is little room for such concepts in this challenging and extremely technically driven environment. Still people, especially in such remote, dangerous, and potentially cramped surroundings, need to maintain contact with their humanity. It is therefore vital to ensure an appropriate aesthetic mix is built into the design of the settlement and of the society that inhabits it.

2.1   How to Live a Happy Life on Mars

We can transport people to and from the Red Planet and provide them with the necessities of life whilst they are on the surface. Looking at a hierarchical needs' perspective, it is clear that fulfilment relies on the ability of society not only to provide the basics but to promote self-actualization and transcendence through, for example, "devotion to an ideal (truth, art) and experiences with nature and aesthetics"[5]. Incorporating effective aesthetic elements is, therefore, a key part of settlement design.

2.1.1   Aesthetics Drivers & Challenges

Drivers (such as psychological and societal necessity, challenges) are the shaping force for aesthetics. For long term viability of the settlement in human terms, the concepts of attraction and retention need to be considered. People will by necessity live in an enclosed environment, where resources and space are at a premium. *Attraction:* How to make sure Mars looks like a desirable location to potential settlers and tourists from Earth. *Retention:* To ensure residents and visitors are motivated to stay on Mars, bearing in mind that for a 1 million inhabitant city state, the population may well not all be explorers who will give up everything and embrace hardship. It may be advisable to introduce the concept of an "aesthetics budget" – a certain amount of space and energy per capita is designed into the settlement for these aspects – such as public spaces, parks, viewing galleries, outdoor environment simulation, sports areas, running water features, illustrated in Figure 1.

Figure 1: Artistic Rendition of Public Spaces, Part of the Concept of Aesthetics Budget

Technology is the key enabler for a life on Mars, but it does not have to be seen only as a means to survive on the planet. The same technology can make life enjoyable for the settlers: 3D-printing can support the customization of the interior design of the habitats, helping to give the inhabitants strong feelings of comfort and attachment to the habitat by living in a proper home; high-definition flat screens or even holographic technology can completely transform the private and common spaces, simulating relaxing environments and scenarios. The final objective is to make life on Mars as good as on Earth or even better in terms of quality of life.

2.1.2   Unique & Salient Aspects

Living on Mars would present opportunities to incorporate unique aspects into the aesthetic mix. Unique scenery, experiences, technologies, architecture, cultural aspects would create opportunities not yet possible on Earth. Huge windows in canyon walls, unique architecture, crater cities, indoor and simulated outdoor spaces and parks, monuments and cultural items could all contribute to a unique Martian aesthetic. Many major civilisations began in geographical areas that have some visual similarities to the Martian surface, and analogues exist in some extreme terrestrial communities, for example environmental conditions in Antarctica have given rise to innovative and distinctive architectures[6]. Later, Section 5.1 elaborates the settlement concept envisioned in this Martian colony, including settlements proposed in Martian lava tubes and planned settlement expansion in Martian craters, the latter shown here in Figure 2. More views of the settlement can be seen in the Appendi.

Figure 2: Bird's Eye View of the Crater Cities and Biodome Structure, Including Launch Pads

3   GOVERNANCE ON MARS

The function of an incumbent government in the colony would be very decentralized and mostly executive. A hybrid system of governance based on direct democracy, assisted with the AI-administration is seen as a suitable fit.

3.1   Governance System

The driving principles of such a government would be focused towards technological advancement and innovative capacity-building. The system should be able to foster competitive markets, while providing its citizens with social safety nets in parts of life that would be considered essential and intellectual. In addition to the collectivistic ideals, the political system would also be responsible for rendering personal freedoms to all its citizens, allowing access to certain fundamental rights such as Right to life, Right to self-determination, Right to liberty, Right to due process of law, Right to privacy, Right to freedom of thought, Right to freedom of religion, Right to freedom of expression, and Right to education. These would not only aim to provide equality to everyone, regardless of their beliefs or origins, but also function as a balancing factor preventing delusions of aggressive technological breakthrough at the expense of its constituent citizens. Right to life will specifically ensure the access to essentials like water and oxygen for every citizen. The colony will be inclusive in its execution of the aforementioned driving principles, considering the diversity of the population will be derived from the existing social and cultural make-up of humans from Earth. Further, new cultural identities amongst the population will also be tolerated and encouraged, keeping in mind the inherent long-term detachment from Earth. Liberal and social democratic ideas will form the foundation of the society. The government, however, will also play a role in regulation of industries to a certain extent, and exercise some form of egalitarian use of resources.

3.1.1 Democratic Institutions & Elections

The direct-democracy idea will be a core feature of the governance on Mars in the colony. Along with their rights, their responsibilities towards the functioning of the colony will also be disseminated amongst the citizens. Universal suffrage will allow all adults, regardless of their background and function, to vote in political elections and other decisions of collective importance. Participation in politics and governance will be encouraged as a second nature to all citizens. Adult citizens will have democratic rights to vote for certain elected officials, who would hold executive powers in a Mars Council, administering the collective will of the colony. These council members would have fixed term limits to avoid politics as a lifetime career. The citizens would also vote for judicial nominations to oversee any dispute that would be outside the scope of the AI-administration. The elections to the Mars Council will not employ "First Past the Post" or "Electoral College" voting. Voting would rather be executed by "Ranked Voting", where citizens instead of voting for one representative would vote their preferential order, creating a more representative voting system.

3.1.2 Hierarchical Definition in the Government

The Mars Council, in summary, would be a democratic institution and instrument consisting of arbitrary members who would not pass decrees, but rather discuss and debate policy issues publicly and present back arguments to the citizens to vote on the final laws or amendments adopted by the colony and subsequently by the Mars Council. The judiciary, in this case, will uphold the Constitution, both in human and AI interpreted forms. The citizens would also have access to public referendums, where they could usher in debates and discussions on topics of relevance in the Council and then have it put up for a vote in the colony. This instrument is important to make sure the Council does not aggregate exclusive power at any time, and always is kept in check by the population. The same instrument is applied to the appointment of these Council Members. People's initiatives will allow people to make alterations to the Constitution. For all these instruments to be invoked, a certain number of signatures of approval must be collected, in a defined period of time to uphold the legitimacy of the same[7]. The direct democracy system would be labour intensive and be much slower than other systems that have more concentration of power, but in the long run it would be a more fair and equal system, with the best possible representation of its population's ideals. To prevent abuse, in any case, a "Two-thirds Majority" system would be applied for quasi-unanimity on decisions.

3.1.3 Dependency on Technology

The dependence on AI for administration and maintenance of the colony would require data collection from sensor suits and biological markers around the settlement. Even democratic participation of the population would occur through their personal devices, and not always in-person. Data security and access will be a strong focus on the design of such a system to protect citizen's right to privacy and ethical use of all the data. The AI-administration would also include grassroots level law enforcement and overall functional and operational management of the colony. All the data collected would, for most purposes, be agnostic of the identity of the citizen, aside from areas where identity-attachment would be absolutely necessary to maintain the integrity of the data. Most data would be used to analyse and optimize the functioning of the colony.

3.2   Government Interactions

This section discusses the interactions of the government with its citizens and along with extern-Martian entities such as the Earth, the Moon and Asteroid Belt legal subjects. Representation of its people is one of the primary functions of the Mars Council on policy and intergovernmental matters.

3.2.1   Interaction with the Population

As stated in the previous sections, it is the responsibility of the government to ensure the will of the people in its jurisdiction is administered to the greatest and most efficient extent possible. This is done via decentralized and independent institutions embedded in the political fabric of the colony. These institutions have legislative, executive and judicial responsibilities, completing the three pillars of a democratic system. Firstly, **Legislative Responsibilities**. The Mars Council is responsible for making the various avenues of public discourse and decision available to the population. Once the procedural public voting on various facets has revealed a decision, the Mars Council will print this result into law, and communicate and coordinate the same with the executive and judicial arms of the colony. The Mars Council shall also assess the viability of all results "after-the-fact" to ensure compatibility with the guiding principles of the colony. In such a way, while being directly responsible to the population, the Mars Council is also inherently accountable to the judiciary and executive branches. This reporting feedback loop will help foster ideas of general trust among the population, and the inherent checks-and-balances shall also provide the population with an independent audit of the government's performance. Secondly, **Executive & Judicial Responsibilities**. For the most part, executive and other law enforcement functions will be conducted via the AI-administration, overseen by human counterparts. However, since the laws on Mars will be directly derived from the population, it is only fair that they be interpreted and used as a tool for dispute resolution by their human counterparts in the judicial branches of the government. Hence, the judiciary, on all its levels will be benched by Martians elected and appointed by the citizens of the colony. This hybrid between the AI-driven executive branch and human judiciary will, again, provide an inherent system of checks-and-balances between the two branches. Aside from law enforcement, other public offices that allow certification and licensing services, along with tax collection etc. will be completely administered by automation and AI-administration.

3.2.2 Interaction with External Influencers

Diplomacy will be one of the most important tools for the sustainability of the Martian colony envisioned here. As mentioned in previous chapters, the colony is envisioned as a centre for technological excellence, which would also count as one of the most valuable exports of the colony. Using trade as a tool for diplomacy will be a standard protocol for engagement with external influencers such as the Earth, the Moon and Asteroid Belt legal subjects. On the subject of trade, Mars will seek to use technology and tourism as primary exports, and in turn also accept minimal imports, even if unnecessary in the largely self-sustaining economy, to serve peaceful bilateral and multilateral relations. Specifically, for tourism, any transient population coming into Mars will be subject to local laws and particular jurisdictions, however allowing for extraplanetary extradition where necessary. Mars will maintain close diplomatic ties and aim to garner a neutral stand in the political arena when engaging with the aforementioned external influencers. This will also serve as a primary tool for deterrence against external aggression. Firstly, **Treaties & Summits**. Primary tools to support peace and dispute resolution for Mars would be maintaining a policy of neutrality in all multilateral treaties and summits. Mars will not, however, deliver preferential treatment for recourse to any other legal subject other than its own population. Secondly, **Protectionism & Military**. Although Mars will stand neutral to all external influencers, the government will employ deterrence capabilities outside of diplomacy and other forms of soft power. Mars will have a passive military capability to protect its territory both, on surface and its neighbouring jurisdiction in outer space. The sole function of such installations would be to protect Martian citizens and Mars' vital in-space assets from external aggression. Mars will not employ aggressive force to any external entity as a tool for dispute resolution.

4   SOCIETY & ECONOMY ON MARS

An economic system is built on a mesh of labour, exchange, and consumption. One million people colony structured to be independent from Earth needs an efficient and expandable economy. The colony effectively achieves this if it optimises its sectors from the point of inception. Not later.

4.1   Economic Model

The unique combination of factors on Mars could create unique opportunities with the need for prioritisation of capacity building. This should include basic skills and focus on emerging in-demand skills, and should be iterative. As the colony grows, it may well be the ground to pioneer the hybrid AI-governance, interplanetary economists and interstellar research. Later, it would incorporate planetary defence, population research and cutting-edge communication technologies.

Iterative means that the colony should "continuously" work on programs that would describe Mars colonists as pioneers of each emerging industry, before it even emerges. Capacity building thus, would be highly predictive, enabled through a carefully structured human development system that can mirror the Finnish education system, where social welfare includes rights to education, scaled down financial burden, highly tailored curriculum, and an infrastructure where everybody is onboard. In emerging industries, the first player would always have a head-start unmatched. If the colony becomes the first premium provider, new players may have a hard time entering the market. For this, the Martian economy should immediately invest in new innovations, always seeking to challenge limits and improve. The colony should invest in capital goods and systems that can be treated as such, instrumental to their main objective. These include capacity building, non-redundant and non-burdensome regulations on entry, or licensing to encourage entrepreneurial participation. A reward system for participation should be thought of instead of one that remediates government spending. The amount of primary economic drivers coupled with government investment may lead both the general and specialized research sector. Balanced economic growth to foster competition and innovation represents the key to a thriving Martian independent colony.

4.1.1   Revenue Generators

The Mars economy has a high level of complexity and would be dynamically developed, its variety of markets implying total collaboration between its services. These will mainly be based around sectors and technologies which underpin industries, such as mining, biotechnology, constructions, agriculture, manufacturing and machinery, as well as research and development (R&D), AI, education and culture and further welfare programs. A prosperous economic society would require a skilled and highly educated workforce which could take a vision through all development success phases. Space mining might well represent one of the core industries for export, as detailed in the trade report, mining minerals that lie in the belt of asteroids between Mars and Jupiter being estimated at a wealth equivalent to about $100 Billion for every individual on Earth[8]. The high-quality mineral ores such as iron, nickel, cobalt and platinum could lead to a total transformation of the market. Predominantly, these would lead to the total transformation of the Mars Cities, given such minerals would play an important role into Martian construction and technological sectors. The education sector, both on Earth and on Mars, would have major breakthrough programs in setting the new entrepreneurial mindset for the next generations, being supported with private funding schemes. Elon Musk, founder and CEO SpaceX, has solely funded the Ad Astra school of 30 students, allowing for $475.000 per year in 2014 and 2015 of his personal funds to be directed towards an innovative educational concept. Considering this type of project at international level, coupled with University-level interest in research conducted on the Mars soil for geological studies, the educational sector will play a major role for Mars - Earth interplanetary relations. Strong partnerships could foster the knowledge and interpersonal skills of future generations, to be able to act collectively towards the good of humanity.

Research & Development

The severe environment on Mars will be an obstacle for many human endeavours but could also present a serious advantage in some fields. The lack of air, light, sound pollution and the absence of space debris from decommissioned satellites in Mars orbit for example, could be an attractive reason for building radio telescope arrays and observatories on the surface of the planet, or could stimulate the positioning of space-based telescopes, with possible revenue of up to $1 billion[9] and $8 billion[10] respectively. Considering that the manufacturing of the equipment will be mostly done on Mars with Mars and Asteroid Belt resources, such commissioning from Earth could be a serious bolster for the Martian economy. In our vision, the economy of the Martian colony resembles that of a highly-developed country on Earth of similar size and economic features – Singapore, with approx. $106,000 GDP PPP/capita for 2020 estimated[11]. Adequate R&D activities would play a crucial role in overcoming the restraining factors of the Martian environment and for securing the sustainability of the Martian economy. Considering the possible threats for the sustainable development of the economy much more significant than these on Earth, we chose the highest value for R&D spending on Earth as the most conservative value to aim for on Mars – Israel's 4.9% of GDP for 2018[12]. Extrapolating these values for the Martian colony of 1 million people and using a scaling factor of 1, we expect a GDP for the Martian economy of approximately $106 billion and a total of $5.19 billion expenditures on R&D. Another important factor for the further development and growth are the Foreign Direct Investments (FDI) from the public and the private sectors based on Earth. The global public sector accounted for $44.5 billion on space activities for civil programs in 2018[13]. With the establishment of the Martian economy as a scientific outpost we expect the main flows of funds to be directed to three key areas - space science & exploration, satellite communications & navigation, space security. These alone could be totalling to $18.5 billion[13]. In view of an average 7:1 rate of economic return on each dollar invested in NASA in the form of spinoff technologies, products, services and jobs over the years, we estimate a cumulative result of $129.5 billion in continuous economic output in the long term[14]. The opportunities for scientific R&D on Mars and its favourable position next to the resources-rich Asteroid Belt would attract FDI from private companies in sectors like mining, biotechnology & pharmacy, as well as production, 3D-Printing, electronics, or consumer goods. With the predicted depletion of mineral resources on Earth, we expect the redirection of activities and funding from Earth to Mars and the Asteroid Belt. Selecting some of the most innovative companies in the world[15] in these sectors as benchmarks to do so, Rio Tinto, Johnson & Johnson and 3M among others, and taking into account their current R&D expenditures similar to the expected for Mars, due to the aforementioned reasons, we estimate a total R&D expenditure from the private sector of approximately $13.93 billion and a combined return on investment of $27.9 billion for these sectors, Table 1 for reference.

Table 1: Estimated R&D Expenditures and Return on Investment Rates for Investments from Earth-based Public and Private Sector (based on calculations from herein sources[13, 16, 17, 18])

Foreign Direct Investment (FDI)	R&D expenditures [billions]	Investments	Return on investments (ROI)	
			[%]	/ [billions]
Public sector				
Space sciences and exploration	$ 7			
Satellite communications and navigations	$ 9,4			
Space security	$ 2,1			
Total	$ 18.5			
Private sector				
Mining	$ 50,67	$ 84	9.54	/ $ 8
Biotechnology and pharmaceuticals	$ 11,36	$ 104	14.7	/ $ 15,3
Production	$ 1,9	$ 32	14.45	/ $ 4,6
Total	$ 13,93	$ 220		$ 27,9
Total FDI and private sector	*$ 32.43*			

Transit population

Visiting the Martian colony as a tourist or as part of a job assignment it's also something to take into account. An interesting possibility is using Mars as an acclimatization base towards longer missions in a zero-gravity environment in space. As the distribution of resources on Mars will require careful forecasting and having in mind restraining factors such as travel costs, medical concerns for both incoming and local people among others, a limitation in incoming people is required to about 10% of the local population. The announcement of the concept of the Starship Interplanetary Vehicle by the private company SpaceX, presents an inspiring opportunity for the democratisation and commercialisation of interplanetary human space flight. With a payload configuration option for crew or cargo and based on estimated values from Elon Musk[19], and The Mars Society[20], we have calculated a value of $50 million for a crew round trip and $70 million for a cargo round trip per Starship, as detailed in Table 2. Here, data are estimated assuming a launch window every 26 months and a travel time of about 80 days. Using the world average for total contribution of tourism to GDP at 10.2%[21] we get approximately $10,812 billion general economic output for the case of the Martian colony, or approx. $108,000 per incoming traveller/tourists or professionals.

Table 2: Estimated Cost Per Transfer

	Starship Earth-to-Mars		Starship Mars-to-Earth	
Capacity crew or cargo	100 people	100 tons	100 people	100 tons
Cost per person or metric ton	$500,000	$500,000	free	$200,00
Total cost	**$50 million**	**$50 million**	**free**	**$20 million**

4.1.2   Trade – Import & Export

Assuming that the Martian colony is mostly self-sufficient, the viability of its economy would be determined by the scale of the trade relations with Earth, the bases on the Moon and space stations in orbit. A promising economic model envisions a maximum utilisation of resources on Mars and the Asteroid Belt for securing its autonomy. In most public reports on a Martian colony, the settlement depends on Earth as the main market for its exports. While this might be true at the beginning, growing the colony to 1 million people and beyond would require substantial and constant inflow of funds to secure further development of the colony. In our vision, Mars is positioned as the main outpost for human exploration of the Solar system and deep space, producing the required space infrastructure by itself. Natural resources on Earth are becoming increasingly scarce and the costs for their extraction is expected to rise. With the expected peak in extraction of mineral resources on Earth, the prospect of mining in space may become more and more appealing. Due to Earth needs of precious metals and ores, the Martian colony could set up itself in an outpost for mining and trading. Despite this concept, it seems more reasonable to utilise these resources in situ for the production of machines, infrastructure, fuel and more. The current advancement in automatization of the mining operations on Earth points to the future possibility of a fully robotic, automated process of capturing an asteroid, extracting the resources (minerals, metals, water) and transforming them into end-products via 3D-printing. The most obvious source of physical export revenue from Mars in the beginning, methane as a propellant for example, can offer stable income for the colony. The United Launch Alliance offers currently \$1,000 per kg. for propellant delivered to L1[22] which translates to \$100 million in revenue per Starship /100 tons capacity. With estimated cost for a Mars to Earth cargo transfer of \$20 million (Table 2) we get \$80 million in net income per Starship /extracting cost of methane on Mars not included. The unique position of Mars within the Solar system next to this resources-rich region will turn the colony not only into an important manufacturing base and the main supplier of water, minerals and metals for the human settlements in the entire Solar system, but also into a central refuelling station for interplanetary transport vehicles.

4.1.3   Marketing Rights, Labels & Entertainment Sector

With the Martian colony being well developed and autonomous from Earth, we expect revenue from the export of goods, technologies and services labelled "*Made on Mars*". Possible income source could be the sale, transferring or licensing of Martian labels and trademarks to third parties. While the amount of the revenue from these operations cannot be currently estimated we assume that it may represent a small but steady income generator for the colony. Therefore, the Martian colony would be of interest not only for the governments and private companies but would capture the attention of the general public too. Thus, attracting investments from media networks, television providers and the movie industry could take the revenue towards \$200 million for programming costs alone[23].

4.1.4   Social Welfare Programs

The role of welfare programs and whether it is instrumental in economic development is long debated. Central to these are Nordic countries where social expenditures and economic growth are global highest. The Mars colony can learn from their model, despite deviating from it in terms of priority. Key characteristics of these models are however adoptable in Mars. This includes: i) Mars' model should be clearly defined in terms of overall objective instead of specific policies, and ii) the model should be regularly reformed accounting for changes in social behaviours and economic state. For the herein model, we propose to revise the original Lampman analysis where social strategy is based on the cost-benefit analysis of two categories: non-quantifiable and quantifiable[24]. Instead, Mars planning should recategorise it as consumption vs investment, recommended by the author. This is more viable for Mars because organic economic activity is still underway and so investments weigh more and should be forecasted into the analysis, instead of a) being combined with other general quantifiable costs, or b) treated as unquantifiable. In this regard, the best strategy for Mars' social expenditure allocation can be patterned. The key is to perform cost-benefit analysis every phase to re-allocate budget to the sector that is deemed timely. The first settlements will definitely focus on housing and survival but as laid out in the blueprint for a scalable economy, the one million colony should later shift to expenditure focused on capacity development. Further on, having a focus on investment-status spending would leverage temporal factors for sustainability of the colony, from 1 million and beyond. Given the highly sophisticated technological infrastructure of the Martian society, a consistent, data-driven optimization is feasible. For the million economy, it's important to determine the key social policy areas. First, **Basic Life Support System**. Manpower represents the pinnacle of the economic engine in a million Mars colony. It is crucial to provide basic survival necessities such as access to water and oxygen, energy and radiation shielding, but as discussed, this would just be the initial major expenditure. Long-term expenditure should be on services that have extended durability and is not in-cash, as this welfare expenditure is treated as investment. By spanning several generations of benefactors, expenditure would involve sponsored construction of lava tubes, general and specialized research on energy systems and more. Second, **Human Capacity Development**. The colony is characterized by state-sponsored access to skills enhancement, professional development, and basic education. Here, born-Mars citizens have access to basic to advanced education which focuses on skills acquisition. Mirroring the Finnish system, the educational institutions are not overtly standardized but instead leverages the plethora of data to allow a tailored curriculum, giving the educational administrators flexibility to revise sets. The state plans for the shift to increase budget allocation for this sector as the colony grows. Developing a demand-ready population, the state would provide financial assistance for education evaluated by family size and economic capacity, to bridge the gap between wealth disparity. As the colony is also a centre for innovation, rapid change in the required skills and solutions to labour inefficiencies would drive unwanted dismissals. To augment long-term educational systems, the state would also provide in-service (not in-cash) assistance for professional development on forecasted emerging in-demand skills. Third, **Healthcare**. Universal healthcare is expected to be a strong backbone for all social, psychological, political, and economic aspects of the colony. The unfamiliar Mars environment may well pose challenges in addressing medical emergencies ranging from physical injuries all the way to psychosocial distress. This state-sponsored access will include basic medical care such as preliminary tests for diseases/detrimental ailments, mental health support through free consultations and discounted access to group activities promoting healthy interaction, access to basic pharmaceutical grade drugs for recurring medical issues on Mars, and free access to contraceptives. As resources are highly optimized, reproduction and choice to start a family on Mars should be assessed through a state-sponsored institution with focus on understanding the population dynamics and help the couple evaluate their situation given medical conditions and financial capacity. Forth, **Old Age and Survivor's Pension**. Unlike Earth, the colony would slowly shift to allocating budget to this sector later as the colony grows. This is because the blueprint for the colony is to be built from a scalable labour force focused on innovating and developing. However, there would inevitably be growth in this sector of the population and to encourage the society from the onset of their Mars citizenship, a promising pension plan should be in place. Hence, the pace at which budget becomes bigger here should be carefully considered. Unlike the first three, this type of state investment is determined much later but needs to be promoted even before the sector grows to encourage healthy contributions to benefit from this sponsorship.

4.1.5   Economic Expansion Strategies

Room for expansion is wide for the Mars colony, especially if it plans and sustains wise investments. The colony must therefore strive to pioneer new ventures before, during, and after reaching a million. Sustainability of the million-colony requires massive expenditure, one that can only be amassed over time. Export should not be the initial primary goal for the colony's still developing economy[25]. The export-driven economy should stem from a mature active local economic activity. This means aggressive support for establishing groundworks such as advanced capacity and efficacious governance. This should not be fixated on merely increasing export activities but should be the sustained living standard which should be viewed as the primary driver of economic growth. Regulations regarding business, trade, and economic expansion should be simple and open but highly optimized[26]. The colony's strategies to attract inward investment is crucial. Positioning itself as a technology hub in the interplanetary economy will be reinforced with compelling state investment in welfare systems and attractive incentives for new players both local and incoming to Mars.

4.1.6   Economic Regulations

Economic freedom is characterized by many points that the Mars colony needs to have fully mastered to sustain the million size. The economy should adopt policies that embrace diverse and to some extent, even competing strategies[22]. Full economic freedom may be guided by individualistic philosophies which might pose a risk to a colony where more collective reasoning is needed. Hence, Mars can adopt a hybrid system. Regulations are still needed since a fully autonomous economy can be very inconsistent and largely unstable. However, the colony should focus on expansion policies that are just moderation mechanisms. Upon establishing the colony as an innovation hub, expansive schemes should be drafted such that it predicts how external trade will advance so that it has policies that are not economically compressive and instead adaptive. Economic freedom entails regulatory efficiency to safeguard 'freedom' in business, labour, and monetary matters. Deep bureaucracy and redundant regulations are unsustainable for a growing Mars colony. Productivity always starts with easy minimum requirements for participating, to create new ventures faster, causing faster competition, driving faster innovation, viz. economies of scale. After this layer should be policies that are not a burden to decision-making and pricing strategies. Instead, one that encourages participants (or allows/ rewards players) to/who have long-term business plans. Attempts to police pricing and ownership are likely to lead to a fluctuating environment that can cause government desperation and unresearched plans or band-aid policies. Labour freedom is crucial. Since sustainability and efficiency are central, Martian businesses should be free in both labour arrangements and dismissals. Boosting productivity means redundancy should be easily addressed. Thus, the colony needs to treat human capital development as the baseline focus. All these strategies only hold if the monetary freedom of the colony's economic activities is backed by a stable currency and whose contracts are priced by the market, not controlled. The colony's direct democracy is essential. The colony's banking, financial system and monetary policies are the main players.

4.1.7   Banking & Fiscal System

Central to monetary freedom is an independent central bank[27]. Aside from ensuring no price-control systems, the Martian bank will spearhead a highly integrated structure as it will accommodate an interplanetary economy. A design for a Space Bank may be befitting. This will facilitate an interplanetary financial system that designs loans, reserve strategies, and financial services for the Mars colony's government, corporations, and other private entities. One of the main aspects of the first Space Bank will be to allow an extra-terrestrial entity such as those on Mars to raise equity and scalable debt instruments in any market within the economic system which is defined as including existing Earth markets. An access point for interplanetary investors will be in place. Such a specialized entity is required as the economics of outer space will no longer be confined on Earth, but across a system of basecamps, colonies, and sub-economies who may have their own government or presiding entity independent from Earth nations. The dynamics will need to be purposeful and tailored, and can kick off with the million Mars colony, transforming itself to be the backbone for its currency and general financial system. The colony's social welfare programs and the financial system are acutely braided. The currency must therefore be in a form that is secure, sustainable, and efficient. An obvious form for this would be cryptocurrency as it can be quintessential for a smart system linked across finance and non-finance sectors[28]. Blockchain technology offers undeniable security and efficiency in any extra-terrestrial economy where resources are managed with paper currency treated as costly. Blockchain solves this along with numerous auxiliary non-finance benefits. The ability to transfer data securely across a network can be the pillar for a lot of innovations. The chain can be developed into a computer network that can communicate securely over an intricate network, or across colonies through space-assets, or even extending into deep space network[29,30].

4.2   Demographics

The building blocks of the Mars colonization is foreseen by futurists such as Michio Kaku to be taking root in the next couple of decades, 2030s to 2070s. At that point, human's extra-terrestrial exploration would have been supported largely by well-matured technologies on Earth. Hence, the goal to advance in being an interplanetary species would be ripe. For this, a Martian colony characterized as an industrial-driven society is deemed appropriate. In this setting, the majority of the 1 million Martians are seen as progeny of some initial settlers (such as the first 1,000 Martians). From these initial settlers, the society has been carefully monitored and expanded into a society imperative to supporting the objective. Population control would have to be set in place to achieve this. In the upcoming sections, such a strategy would be expounded upon.

The 1 million industrial society would be composed of a majority of production designers, technological innovators, and visionary engineers. It will also be composed of leading scientists with a similar segment size of service-oriented crews! There's also a peripheral population of non-contributing citizens who are Martian retirees who have served their purpose for the majority of their lifetime, or are citizens contributing mildly into sectors such as creativity, finance, health, etc. for a narrow time of their aged life.

4.2.1   Social Structure in the Colony

Humans are used to structure, especially for a band that operates with a unitary goal. Most proponents of the Mars colonization look at it as a strategy of mere survival, an escape of an impending extinction. As laid out above, the 1 million society is not for such purposes, but instead for augmenting advancement. Interestingly, such a position can be viewed through a cultural lens as a form of man's cultural invention. As such, there are several things that need to be taken into account when designing the urban environment and the social structure of the colony. How society is shaped is largely impacted by psychological and social science. In this regard, the Mars colony should be designed as futuristic but as close to our human reality as possible. This is to make sure that individual psychological and collective social stability is maintained. Our human realities for the context of Mars includes our tendencies and undeniable preferences as described below:

- Human curiosity is natural and so learning and growth should always be integrated.
- There's an ever-present tendency to always aggregate in groups and usually for the sake of group resilience against adversities, to derive

livelihood and generate connected experiences, and to establish institutions
as backbone for systematized collective affairs.

- There's an innate commitment to aesthetics and spiritual needs, in
  whatever form each prefers.
- Evident form of human identity as multi-layered identities that needs to be
  both maintained and dissolved or reshaped for social cohesion and for
  preventing social fragmentation.

These four archetypes should be the focal points of deliberation on structuring the
colony. The last point is particularly difficult but is crucial. The colony will be born
out of a conglomerate of settlers of diverse culture, belief systems, and personalities.
Despite having a clear goal as a colony, it should not implement constricting
mechanisms to force cohesion or ignore individualistic philosophies. Instead, it
should focus on allowing an almost exact reflection of human society we already
have here. As such, the democratic governance will be central here. Participation is
encouraged and it is so as the collective thought of survival is a priority. However,
smaller group experiences are also promoted to maintain companionship and
distinctiveness as factions within the colony.

4.2.2   Importance of Maintaining Identity

The dynamism of our behaviour should not be quashed. Being a centre of
innovation, creativity is imperative which will only come from freedom of self-
expression. Furthermore, the absence of an idea that you can personally identify
with robs humans of reasons to connect or to learn and grow, it discourages the
colonists to think outside the box, as a unitary belief system will grow to make
creativity dormant. Therefore, in a Mars colony, the habitat and the urban
environment should present both spaces that can be personal and communal. Living
habitats inside communal complexes are likely to be beneficial for all and in the
long-term. Instead, Mars habitats or pods should be structured where small groups
of people such as families, close friends, or circles sharing personal belief systems
can live, distinct from other subgroups. A large dome where everyone lives in
compartmentalized spaces will be a strong friction to normal human dynamics.
With pods/habitats characterized by some form of physical separation, identities
and privacy is still maintained. On top of this, a structural form that encourages
social cohesion or community participation should also be in place. Despite each
habitat defined as separate, tubes connecting these habitats, providing channels for
intergroup interactions should still be present. A social experience will be the
driving force of the colony to achieving the objective, and a collective belief system,
layered on top of the personal ideologies will be the structure. Aside from
connective tubes between habitats, a central hub in the middle of the urban colony
should also be present. These hubs include universities, parks, entertainment hubs,
health and personal care facilities, and the government hub. These shall be central
to the overall plan as it encourages subgroups distributed outside it to come and
cohere through group experiences, shared in a communal space.

4.2.3   Cultural & Social Impacts

As the colony is composed of diverse culture and belief systems, living in a colony in a seemingly harsh environment such as Mars will be challenging at first. As is normal with human dynamics, conflicting aspects of human nature, especially in the desire for self-identity will be provoking. As such, a collective colony culture, one that is hopefully mastered by the human race by then can mitigate this. We can see progressive movements now and these are crucial to sustaining the million Mars colony and should be adopted. Interestingly, the manner by which the colonists grow from initial settlements to the million size can actually be the key to this. When the first humans will be sent to Mars, it's definitely representative of several nations of different cultures and beliefs. As initial settlements are particularly challenging, this diverse group will be blended to work for a common goal: survival. If the colony grows from such a culture, a dynamic of cooperation for the good of the group, then the million colony will gradually grow out of a culture of natural cooperative dynamics.

5   TECHNOLOGY ON MARS

5.1   Settlement Concept

This section is aimed to discuss important considerations for choosing a settlement site and how Mars City can look like.

5.1.1   Locations

The prominent settlement location on Mars is Hellas Planitia region, the deepest basin consisting of the topographically lowest parts on the Martian surface attenuating 50% of radiation[31] and correspondingly having the highest atmospheric pressure[32]. Three approximate lava tube locations are identified in Hadriacus Mons (HM) region in Hellas Planitia, shown in Figure 3. To calculate the approximate number of people, which can be placed in these lava tubes, few assumptions are made: 1) height of all lava tubes is considered as 10 m and the length of lava tubes for the 2nd and 3rd locations is considered as 1000 m due to absence of strong gravitational pull; 2) 1 settlement will occupy 75 m^2. Taking these assumptions, we estimate that 3 lava tubes combined can hold about 25,000 people maximum. Lava tube settlements can be utilized for research labs as an additional protection of the sensitive experimental equipment from solar wind and Galactic Cosmic Radiation.

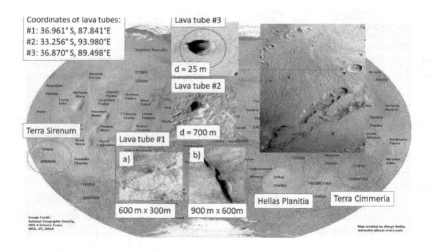

To accommodate the remaining 975,000 people, we need to choose the location that would permit the constant increase in size as we plan that 1 million people is not a limit. On the Martian surface, some areas contain crustal (or remnant) magnetism, such as Terra Cimmeria and Terra Sirenum[33]. In these regions, design of crater city can be proposed, the settlements can be built by furrowing in crater walls as shown in Figure 4, additional radiation shielding can be built by covering the crater roof with a transparent polyethylene membrane or other lightweight shielding material just like pressurized bio dome structures. Further cities extension can be done by buil   rater structure. The   oximity to the Ma

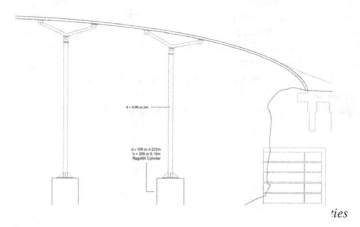

ies

5.1.2   Radiation Shielding, Monitoring & Dosimetry

As it was mentioned above, we propose to use a polyethylene cover above a tube and in the case of the surface settlements - an ice dome. For a settlement in Terra Cimmeria and Terra Sirenum, a 40 cm thick ice layer is sufficient to bring gamma rays intensity below the UK legislative limit of 20 mSv/year[34]. For other areas, we would need a 70 cm thick protective ice layer. However, there will be colonists that are at risk of significantly higher amounts of cosmic radiation such as during mining, exploration, and more. Individuals with such professions will have to monitor their exposure and to ensure that they are under a limit of 0.5 Sv/year as recommended by NASA using a Personal Radiation Detector[35]. In addition, dosimeters capable of measuring gamma rays and neutrons should be placed and maintained in public areas to track radiation level on the surface. A dedicated department will monitor these dosages in real-time for the purposes of early detection of breaches in radiation protection systems.

5.1.3   City Plan Distribution

Figure 5: Schematic Polycentric Proposal of Linear Development[36]

The Martian Lava Tubes settlement would reflect a linear city design running along the main transport/supply in tight strips of development. With efficient public transport, the fact that the linear city has no central core leads to more equal access to space and services. Aesthetically, placing green spaces outside the tight cylinder to urban spaces will give a more balanced living experience to residents and improve the quality of life. A sample of the linear city space distribution is illustrated in Figure 5 as a superimposition onto an image from Tufek-Memisevic et al, 2015[36]. For the settlement on the surface (Terra Cimmeria and Terra Sirenum craters), we will use a radial city design, such as those proposed by Emil Jacob for the World Civic Center. The Martian development would concentrate with industrial and governmental buildings in the inner circle, followed by commercial, recreational and research areas, and after the residential area in the rim of the crater, illustrated in Figure 6. Both city plans make sure to settle all the areas to have equal access to resources with every resident living "downtown" in the heart of all activities and services and in both layouts for a balanced society without clustering and even distribution of resources to eliminate societal divisions and instil a sense of community and unity.

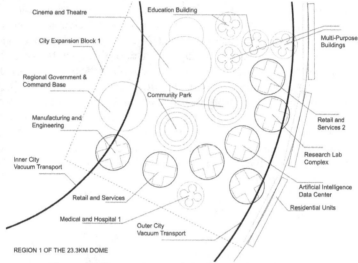

Figure 6: Section View of the Planned Crater Cities

5.2   Satellite Systems

Three satellite systems are already placed around Mars orbit and are well established. These are: MConnect (Mars Communication System), MGuide (Mars Navigation System), MEye (Mars Remote Sensing System) & Near Mars Object Detection System.

5.2.1   Communication System (MConnect)

MConnect, shown in Figure 7, is a Mars-wide wireless network, consisting of a constellation of 60 low orbit small satellites communication systems with complete coverage of the Martian surface. The constellation is at an orbital altitude of 780 km, consists of 10 orbital planes, with 6 satellites in each plane. MConnect will use the combination of RF and optical communication systems. RF link is used for uplink and downlink transmission between satellites and Mars ground stations. Optical ISL is used between two satellites that utilizes a direct detection system for deep space communication between Earth and Mars satellites. The MConnect is used for communication between users sharing services such as tele-education, telemedicine, communication between drones for delivery, virtual reality, virtual explorations, facilitating high-definition virtual tourism visits, communication between transportation systems, IoT connectivity and machine learning.

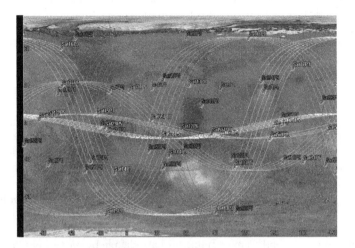

5.2.2   Navigation System (MGuide)

The MGuide provides 24hr continuous navigation services on Mars throughout the Martian surface as well as within Hellas Planitia Lava Tubes. The MGuide, as shown in Figure 8, consists of a constellation of 18 navigation satellites that are placed in 6 circular orbital planes, having 3 satellites in each plane. The constellation is placed at the orbital height of 9,477 km, having an accuracy of up to 1 mm.

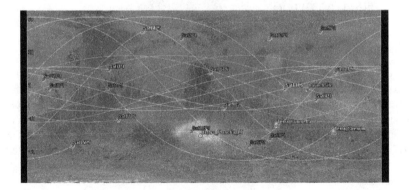

The advanced satellite technology used in the MGuide autonomously does the on-board corrections for the depth of the Lava Tubes to provide accurate navigation services to the settlements. The accuracy and coverage of MGuide is not affected by atmospheric conditions, buildings, infrastructure and vegetation. The receivers used at the Mars city have atomic clocks that do not require continuous time corrections. MGuide is used by the Mars transportation network within the city and outside and also by launch and landing site at the equator. The MGuide is used by drones to monitor remote locations, for disaster management and for delivery services.

5.2.3   Remote Sensing System (MEye)

The MEye platform is space borne and consists of 6 satellites that are mainly monitoring habitat, equatorial and polar regions and are placed in polar orbits having eccentricity of 0.3 and a semimajor axis of 5,000 km to have constant sun illumination. MEye is equipped with some of the following sensors[37] to monitor Martian condition.

1. **Mars Airborne Synthetic Aperture Radar** penetrates through Martian clouds, storms and collects data.
2. **High Spectral Resolution Lidar** aims at advancing Martian society's ability to predict and manage future climate change by providing relevant quantification and understanding of the carbon cycle.
3. **Martian Atmospheric Infrared Sounder** measures air temperature, humidity, clouds, and surface temperature.
4. **Mars Land Vegetation and Ice Sensor** carries out surveys on the Maritain ice poles, also monitoring land and vegetation.
5. **The Advanced Topographic Laser Altimeter System** helps in measuring vegetation canopy height to help estimate biomass amounts and how changes in the Martian atmosphere.
6. **Martian Airborne Visible/Infrared Imaging Spectrometer** helps deriving vegetation classification, mineral mapping, and ice cover.
7. **High-Resolution Grating Spectrometer** studies atmospheric carbon dioxide (CO_2) from space.

5.2.4   Near Mars Object Detection System

The key priorities for an asteroid detection system include a space-based infrared mission to initially characterize asteroids, which is supplemented by a ground-based visible and near-infrared station to further chart the asteroid's parameters of interest. The satellite-based detection system will have a mirror aperture of 50 cm, observing in the infrared and will be placed at L1 heliocentric orbit for Mars. These specifications are based on NEOCam[38]. The ground-based system will be located at Elysium Mons (24.8E24.8N, 146.9E). Satellite-based Asteroid Redirecting System will include:

1. **Kinetic Impacting System** is used to deflect the asteroid
2. **Nuclear Detonation System** - The detonation of nuclear weapons on the asteroid surface
3. **Hypervelocity Asteroid Intercept Vehicle Hybrid System** includes 2 spacecraft, a mother spacecraft, and a detonator spacecraft[39].

4.   **Gravitational Traction Deflection Method** - In this process, a spacecraft will orbit the asteroid, slowly deflecting its trajectory[40, 41, 42].

5.3   Ground Systems & Facilities

This section explores essential ground facilities and transportation in our Martian colony. We start with exploring essential facilities, follow up transportation and next-generation technologies for mining and artificial intelligence.

5.3.1   Environmental Control & Life Support System (ECLSS)

Since humans cannot survive in the harsh Martian environment, a successful colonization will require a robust Environmental Control and Life Support System (ECLSS). The system is semi-centralized, composed of 2,500 independent nodes. Each node can provide water, oxygen, pressurized air, and disposing wastes for 120 households of 4 people per household. If a node runs into technical issues, 5 nearest nodes will then act as backup supply centres. Regarding water supply, Martian ice caps are the major sources of water with an average thickness of 3 km. To extract water from the poles, rovers will be sent directly on the chosen site. Rovers will emit microwaves to heat the ice and transform it into vapor, which is collected and condensed and stored into liquid form. A system of pipeline and reservoirs is put in place to transport water from the poles to the main reservoirs at the different locations of the settlements[43]. In terms of oxygen production, since atmospheric concentration of oxygen is limited to only 0.16%, we need to produce oxygen from other resources. It is estimated that a human under normal activity requires 0.84 kg/day to meet the metabolic load[44]. For a population of 1 million, the numbers come to 840 metric tons of oxygen consumed per day. The Sabatier process of atmospheric carbon dioxide in conjunction with an electrolysis of water produces 4.5 tons of oxygen per 1,000 kg of hydrogen feedstock consumed per reaction[45]. The hydrogen requirement can be met by electrolyzing the water available on Mars such as the polar icecaps and in the regolith. In addition to that, another prospective method to mass-produce oxygen is to treat the perchlorates present abundantly in the Martian regolith with microbial enzyme reduction. It is theoretically projected to produce 9.1 tons of oxygen per ton of regolith, thus leading to ca. 93 bioreactor units with reactor volume of a metric ton to satisfy the daily requirements[46].

5.3.2   Medical Equipment & Medical Aspects

Major medical events require planning, including specialist medical care and surgical capabilities since there will be no medical evacuations to Earth and all aspects of healthcare will need to be handled on Mars. A risk analysis for emergency incidents in a general population is approximately 0.06 events per person per year, meaning that for a 1 million population we should be prepared to handle 60,000 medical emergencies per year[47]. Medical complications consist in all possible medical problems experienced on Earth plus those specific to space travel. These include spacecraft depressurization leading to barotrauma, arterial gas embolisms, decompression illness, and vaporization of water into soft tissues[48]. Telesurgery capabilities are already advancing on Earth, with the first Transatlantic robotic telesurgery being reported as early as 2001[49]. Robotic surgery could be utilised between habitats on the Martian surface, as there would be no communication delay between these; a surgeon in one habitat controlling a robot performing a surgery in another. All medical records will be in electronic format. All instrumentation will have artificial intelligence that is linked to the electronic medical record[50]. For our 1 million population, there will be 3,000 hospital beds distributed proportionally between colonies[51].

5.3.3   Transportation Within & Outside the Cities

Since lava tubes are elongated in shape, the best transportation system to use would be a combination of rail system and electrical vehicles, illustrated here in Figure 9. Hyperloop systems like the one in development by SpaceX will be used. Rails will be placed above the ground so that we can take an advantage of lower atmospheric air pressure on Mars. With magnetic levitation technology, the train can travel as fast as 1,280 km/hr making this a convenient method of transportation[52]. For transportation over a large distance, we suggest the use of innovative solutions through the development of gas hoppers. A hopping vehicle is cheap, efficient, and sustainable. The power source is the radioisotope that is placed in the centre of the vehicle. With radiation generated, it can be used to heat the methane propellant that will be exhausted through a nozzle and as a result will provide thrust to the vehicle. As the vehicle moves, methane will be collected from the Martian atmosphere. Consequently, 4 kg of isotopes are required for each 300 kg hopper[53].

Figure 9: Illustration of the Lava Tube City, Including Hyperloop-like Rail System

5.3.4   Transportation Between Earth & Mars

Current systems mainly use a Hohmann Transfer manoeuvre to move spacecraft between orbits, since it is an efficient way to minimize fuel consumption. However, this type of manoeuvre foresees the use of a very long coasting trajectory between the starting and ending orbit. The new concept involves accelerations for half of the trajectory, then rotate 180 degrees and decelerate until the desired insertion velocity is reached. It is projected to reduce travel time from 6 months to a few weeks[54]. Besides, the use of docking stations located at the Lagrange points allows the interplanetary system to be lighter[55].

5.3.5   Mining Technologies (Ground & Asteroids)

The asteroid belt in our solar system contains ~8% metal-rich (M type) asteroids and 75% volatile-rich carbonaceous (C type) asteroids. To mine these materials, we need to place processing facilities between the orbits to enable regular access. Refined and advanced robotic technologies will be needed for the asteroid to be mined successfully[56]. These rely on the propulsion technology which will be able to catch and de-orbit the asteroid, returning it to a safer orbit. Different methods are available for mining including drilling, microwave heating or newer technologies specifically created for applications on these surfaces[57,58].

5.3.6   Automation

Based on present-day technologies, we envision that automation would be able to guide numerous aspects in the Mars city-state, including, but not limited, to being utilized to perform recurring tasks that humans can be fatigued by, process and run e-governance tasks, automate life support systems monitoring and maintenance, provide autonomous travel and logistics between the habitats and facilities, allow for data-driven medical research and drug compound optimization. In addition, planetary defence, food growth rates, environmental aspects and others will also be monitored and simulated by AI enabling humans to only analyse a pre-run set of data for less cluttering. All the habitats, medical systems, information systems shall be run with IoT enabling a smooth transition of activities[59,60].

5.4   Resource Production and Management

Besides essential facilities, we also need to consider how we can sustainably generate resources required for our colony. This section explores production of different resources from food and electricity to pharmaceutical products.

5.4.1   Food

Martian soil has some of the required nutrients to cultivate crops and due to lower gravitational force on Mars, growing crops require less amount of water as it will drain away at a slower rate. In our settlement, we will grow half of our crops on the ground using innovative farming technology known as "aquaponics". It refers to any system that combines conventional aquaculture (raising aquatic animals such as fish, crayfish and prawn) with hydroponics (cultivating plants in water) in a symbiotic environment[61]. The excretions from the animals raised are fed to a hydroponics system where nitrifying bacteria convert excretions into nutrients for plants. Then, the water is recirculated back to the aquaculture system. Through aquaponics methods, we can grow fish, crops, and simultaneously optimize the water usage. We require 95% less water and 90% less land for farming by aquaponics[61]. Since aquaponics have limitations, we also supply extra nutritional or pharmaceutical needs through the technique of 3D-Printing[62].

5.4.2   Power

To sustain the colony of 1 million people, we need about 6,500 MW of power. Power is more sensitive on Mars than Earth. If power is lost and ECLSS fails completely, humans cannot survive. As a result, we develop a rigorous backup system for power including a reserve equivalent of 30-day. To reduce risks, we obtain power from three different methods.

1. **Geothermal plants** produce 77% of the power. We will divide them up into 50 separate plants. Each plant has 100 wells. Compressed carbon dioxide in a fluid form injected at a rate of 100 kg/s will be used to harvest geothermal energy[43]. Regions with highest geothermal gradients on Mars are located near the lava tubes, making this power source convenient based on our settlement location.
2. **Nuclear plants** produce 20% of the power. We assume that our technology advances enough for us to develop nuclear fusion. It is a clean energy that can produce large amounts of power. We will have 4 reactors with only 2 reactors running at a given time
3. **Renewable biomass generation** produces 3% of the power. Human wastes will be used as an input to the plant. For wastes coming from 1 million people, we can generate 300 MW (see Section 5.4.4).

5.4.3   Raw Resources

The presence of basaltic rocks in the Martian crust under Hellas Planitia is an ideal source of construction materials required for infrastructure development for the settlement. Further analysis of the rocks has found the presence of iron oxide (~18%) in the southern highland while those in the northern plains contain silicates and are composed of andesitic mineral[63]. Martian regolith possesses volatiles mainly frozen water (~5%) in the form of hydrated phyllosilicates, iron oxides and other hydrous phases such as in clay and zeolites which can be mined and processed for ECLSS or can serve as a propellant source under cryogenic conditions[64]. The presence of silicates in the regolith also serves as a raw material to manufacture glass after removing the ferrous impurities. The substantial presence of hydrogen in the Martian surface and polar ice caps can be used to produce plastics and other complex compounds to satisfy the requirement of feedstock for additive manufacturing. The presence of nickel-iron alloy (20 to 25 % by weight) in the form of enstatite chondrites can be extracted and purified using Monds process or chemical vapour deposition on S-type and M-type asteroids in near asteroids orbits[65]. These deposits can be mined and transported for further processing to export purified nickel and iron metals. The rich presence of carbon dioxide in the

Martian atmosphere can be used to produce primary consumables like oxygen, water and methane.

5.4.4   Waste Management

Biogas is an effective method to treat human faeces waste. A dedicated compact plant can be built for every 20,000-50,000 people for ease in operation and avoiding large pipelines for long distances. Taking an average human excretion of 250 gm/day per person, detailed calculation shows about 6,350 kWh/day electricity can be generated from the resulting methane[66]. Urine contains about 91% water and treating it can recover up to 80%. For a colony of 1 million people, we expect to treat and reuse 655,200 litres of water per day[67]. Melting technique can be deployed on Mars easily. This technique requires fuel or electricity as a power source to run. The plants can be run once a month to conserve energy. Approximately 2 kg of biohazardous wastes per day are created by each patient. These biohazardous waste can be either inactivated (ionizing radiation), decomposed (melted if plastic, glass or metallic material) and recycled, or incinerated.

5.4.5   Pharmaceutical Products

A state-owned pharmaceutical R&D department should be created for (a) studying and testing potential therapeutic molecules under Mars conditions, (b) development and validation of drugs, (c) development and validation of drug manufacturing techniques suitable to the Martian environment, and (d) temporary production of essential drugs until sufficient growth of the city. About 22,500 biologically active compounds can be extracted from microorganisms and about half a million can be obtained from plants[68]. However, despite such biological resources offering a large variety of active compounds, the production yield for most of these molecules is still low when compared to chemical synthesis. In the future, genetic engineering technology will allow the optimization of an important number of plants, fungi and bacteria in order to suit astronauts' needs for therapeutic molecules during space travel.

5.4.6   Manufacturing

The widespread use and availability of 3D-Printing would likely have accelerated by the time a city-state is established on Mars, considering the amount of contemporary usage. Therefore, we envision that 3D-Printing will be used in order to design and create everything from small tools to large houses[69]. Also, 3D-Printing will require resources like plastic and metals. Plastics can be made synthetically, while metals will need to be imported until land-based surveys on Mars are able to detect metal ores on its surface, or nearby metal-rich asteroids are mined.

6   ACKNOWLEDGEMENTS

The authors have equally contributed to this work. The authors gratefully acknowledge the support of the Space Generation Advisory Council (SGAC). The SGAC Team is made of international members, primarily students and young professionals of 18-35 years old. The participants would like to thank the mentorship received from the Project Group Leaders of SGAC: Ilaria Cinelli, Antonino Salmeri, José Ferreira, Maren Hülsmann, Anne-Marlene Rüede, Daria Stepanova, Alex Linossier, Rochelle Velho, Anthony Yuen, Smiriti Srivastava, Georgia Maria Kalogirou. Special thanks go to Stefan Siarov from Valispace, for his helpful advice, and Rigel Kent Marco, Motion Graphics Artist - Philippines, for his direct support in presenting the Aesthetics of the colony. Authors would like to thank The Mars Society's initial challenge of 1000 people colony because all presented reports made for one Thousand can lead to one Million in the generations to come. Finally, the authors would also like to thank all the SGAC members who have been able to dedicate even just a few hours towards the project.

7   APPENDICES

A.   View of the Planned Crater Cities

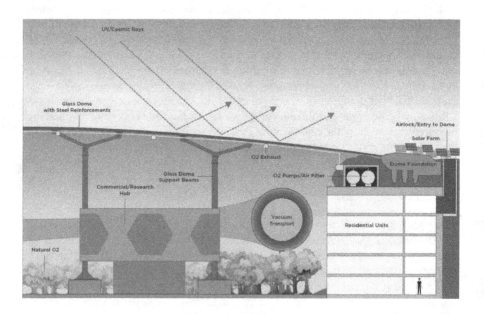

B. Schematic Views of the Residential Spaces Designed for the Citizens

8 REFERENCES

[1] Van Lange, P., et al., 2017. *Trust in Social Dilemmas*. NY: Oxford Univ. Press.
[2] Bruhns, S. and Haqq-Misra, J., 2016. A Pragmatic Approach to Sovereignty on Mars. *Space Policy*, 38, pp.57-63.
[3] Casares, A., 2018. The brain of the future and the viability of democratic governance: The role of artificial intelligence, cognitive machines, and viable systems. *Futures*, 103, pp.5-16.

[4] Lexico, 2020. word definitions [online] Available at: https://www.lexico.com/definition/aesthetics [Accessed 23rd June 2020].

[5] Koltko-Rivera, M.E., 2006. Rediscovering the later version of Maslow's hierarchy of needs: Self-transcendence and opportunities for theory, research, and unification. Review of general psychology, 10(4), pp.302-317.

[6] Nytimes.com, 2020. The Coolest Architecture on Earth Is in Antarctica. [online] Available at: <https://www.nytimes.com/2020/01/06/science/antarctica-architecture.html> [Accessed 24 June 2020].

[7] Frey, Bruno S., 1994. Direct Democracy: Politico-Economic Lessons from Swiss Experience. *The American Economic Review*, vol. 84, no. 2, pp.338–342.

[8] CNBC, 2018. Space mining could become a real thing - and it could be worth trillions [online] Available at: <https://www.cnbc.com/2018/05/15/mining-asteroids-could-be-worth-trillions-of-dollars.html> [Accessed 10 June 2020].

[9] Cartlidge, E., 2019. *Square Kilometre Array hit with further cost hike and delay.* [online] Available at:<https://physicsworld.com/a/square-kilometre-array-hit-with-further-cost-hike-and-delay/> [Accessed 10 June 2020].

[10] Goddard Space Flight Centre, NASA, Technical FAQ on a variety of mission issues, aspects and capabilities. [online] Available at: <https://www.jwst.nasa.gov/content/forScientists/faqScientists.html#cost>. [Accessed 10 June 2020].

[11] International Monetary Fund, 2019. *World Economic Outlook Database.* [online] Available at: <https://www.imf.org/external/pubs/ft/weo/2019/02/weo data/weorept.aspx?pr.x=69&pr.y=6&sy=2017&ey=2024&scsm=1&ssd=1&sort=country&ds=.&br=1&c=576&s=PPPPC&grp=0&a=> [Accessed 14 June 2020].

[12] OECD Data, 2020. [online] Available at: <https://data.oecd.org/israel.htm> [Accessed 30 November 2020].

[13] Seminari, S., 2019. *Global government space budgets continue multiyear rebound.* [online] Available at: <https://spacenews.com/op-ed-global-government-space-budgets-continues-multiyear-rebound/> [Accessed 05 June 2020].

[14] Comstock, Douglas A. and Lockney, Daniel P. And Glass, Colemann, 2011. *A Sustainable Method for Quantifying the Benefits of NASA Technology Transfer.* [online] Available at: <https://spinoff.nasa.gov/pdf/AIAA%202011%20Quantifying%20Spinoff%20Benefits.pdf> [Accessed 03 June 2020].

[15] Boston Consulting Group, 2019. *The Most Innovative Companies 2019; The Rise of AI, Platforms, and Ecosystems.* [online] Available at: <https://www.bcg.com/publications/collections/most-innovative-companies-2019-artificial-intelligence-platforms-ecosystems.aspx?linkId=65059629&redir=true> [Accessed 10 June 2020].

[16] Investing.com. 2020. Equities. [online] Available at: <https://www.investing.com/equities/rio-tinto-plc-exch-income-statement> [Accessed 10 June 2020].

[17] Investing.com. 2020. Equities. [online] Available at: <https://www.investing.com/equities/johnson-johnson-income-statement> [Accessed 10 June 2020].

[18]Investing.com. 2020. Equities. [online] Available at: <https://www.investing.com/equities/3m-co-income-statement> [Accessed 10 June 2020].

[19] Wall, M., 2019. *Tickets to Mars Will Eventually Cost Less Than $500,000, Elon Musk Says.* [online] Available at: <https://www.space.com/elon-musk-spacex-mars-mission-price.html> [Accessed 12 June 2020].

[20] The Mars Society, 2020. *Mars City State Design Competition Announced.* [online] Available at: <https://www.marssociety.org/news/2020/02/11/mars-city-state-design-competition-announced/> [Accessed 28 May 2020].

[21] Pascariu, G.C. and Ibanescu, B.C., 2018. *Determinants and Implications of the Tourism Multiplier Effect in EU Economies. Towards a Core-Periphery Patter. Amfiteatru Economic.* [online] Available at: <https://search.proquest.com/openview/7824d6e3dc91c4a4011343095c4f7f3c/1?pq-origsite=gscholar&cbl=1926338> [Accessed 16 June 2020].

[22] David, L., 2016. *Inside ULA's Plan to Have 1,000 People Working in Space by 2045.* [online] Available at: <https://www.space.com/33297-satellite-refueling-business-proposal-ula.html> [Accessed 04 June 2020].

[23] Peter Hamilton Consultants, Inc, 2010. *What do Discovery's 'Big Four' Networks Pay for Programming.* [online] Available at: <https://www.documentarytelevision.com/sweet-spots/what-are-networks-paying-for-programming/ > [Accessed 11 June 2020].

[24] Institute for Research on Poverty, n.d. Social Welfare Spending and Its Effects on Growth: Another Look At The Lampman Analysis. [online] University of Wisconsin-Madison. Available at: <https://www.irp.wisc.edu/publications/focus/pdfs/foc123g.pdf> [Accessed 22 June 2020].

[25] Mayer, J., 2013. Towards More Balanced Growth Strategies in Developing Countries: Issues Related to Market Size, Trade Balances and Purchasing Power. In: United Nations Conference on Trade and Development. [online] Geneva: United Nations Conference on Trade and Development, p.18. Available at: < https://unctad.org/en/PublicationsLibrary/osgdp20134_en.pdf > [Accessed 17 June 2020].

[26] Miller, T., et al., 2020. Economic Freedom: Policies for Lasting Progress and Prosperity I 2020 Index of Economic Freedom Book. [online] Available at: <https://www.heritage.org/index/book/chapter-2?version=81> [Accessed 19 June 2020].

[27] John Hopkins University, 2020. Why Do Societies Need Independent Central Banks?. [online] Riksbank, pp.2-4. Available at: <http://archive.riksbank.se/Documents/Rapporter/POV/2016/2016_3/er_2016_3%20Why%20do%20societies%20need%20independent%20central%20banks%20by%20Jon%20Faust.pdf> [Accessed 21 June 2020].

[28] Wei, J., 2017. RNCP: A Resilient Networking and Computing Paradigm for NASA Space Exploration.

[29] Center for Space and Policy Strategy, 2020. Blockchain In the Space Sector. [online] El Segundo: Center for Space Policy and Strategy, pp.6-12. Available at: < https://aerospace.org/sites/default/files/2020-03/Jones_Blockchain_03052020.pdf > [Accessed 14 June 2020].

[30] Brathwaite, R., 2018. How Blockchain Can Benefit Space Exploration (Op-Ed). [Blog] Space.com, Available at: <https://www.space.com/41003-blockchain-benefits-space-exploration.html > [Accessed 19 June 2020].

[31] Paris, A. et al., 2019. 1 Prospective Lava Tubes at Hellas Planitia -Leveraging Volcanic Features on Mars to Provide Crewed Missions Protection from Radiation. *Journal of the Washington Academy of Sciences*, 1–16.

[32] Voelker, M., et al., 2017. Grid-mapping Hellas Planitia, Mars–Insights into distribution, evolution and geomorphology of (Peri)-glacial, fluvial and lacustrine landforms in Mars' deepest basin. Planetary and Space Science, 145, pp.49-70.

[33] Langlais, B., & Quesnel, Y., 2008. New perspectives on Mars' crustal magnetic field. *Comptes Rendus - Geoscience, 340*(12), pp. 791–800.

[34] *The Ionising Radiations Regulations*, 2017. Available at: <http://www. legislation.gov.uk/uksi/2017/1075/schedule/3/made> [Accessed: 23 June 2020].

[35] Pradeep Kumar, K. A. *et al.*, 2020 'Advances in gamma radiation detection systems for Emergency Radiation Monitoring', *Nuclear Engineering and Technology*. Korean Nuclear Society. https://doi.org/10.1016/j.net.2020.03.014

[36] Tufek-Memisevic, et al.,2015. A linear city development under contemporary determinants.

[37] NASA, 2020. *Remote Sensors,* online available at: https://earthdata.nasa.gov/learn/remote-sensors. [Accessed on 14 June 2020]

[38] NASA Jet Propulsion Laboratory, 2020. *NEOCam Finding asteroids before they find us.* [online] Available at: <https://neocam.ipac.caltech.edu/> [Accessed 24 June 2020].

[39] NASA, 2020. An Innovative Solution to NASA's NEO Impact Threat Mitigation Grand Challenge and Flight Validation Mission Architecture Development. [online] Available at: <https://www.nasa.gov/directorates/spacetech/niac/2012_phaseII_fellows_wie.html> [Accesssed 24 June 2020].

[40] Olympio, J., 2010. Optimal Control of Gravity-Tractor Spacecraft for Asteroid Deflection. Journal of Guidance, Control, and Dynamics, 33(3), pp.823-833.

[41] Fahnestock, E. and Scheeres, D., 2008. Dynamical Characterization and Stabilization of Large Gravity-Tractor Designs. Journal of Guidance, Control, and Dynamics, 31(3), pp.501-521.

[42] Ketema, Y., 2017. Asteroid deflection using a spacecraft in restricted keplerian motion. Acta Astronautica, 136, pp.64-79.

[43] Badescu, V., 2009. Mars: Prospective Energy and Material Resources, Springer.

[44] Ewert, et al., 2003. Advanced Life Support Requirements Document. [online] Available at: <http://www.marsjournal.org/contents/2006/0005/files/Lange2003.pdf > [Accessed 20 Jun. 2020].

[45] Williams, R. D. and Shaw, M. J. 2005. A Crewed Mission to Mars. [online] nssdc.gsfc.nasa.gov. Available at: <https://nssdc.gsfc.nasa.gov/planetary/mars/marssurf.html> [Accessed Jun. 2020].

[46] Davila, A.F., et al., 2013. Perchlorate on Mars: a chemical hazard and a resource for humans. International Journal of Astrobiology, 12(4), pp.321–325.

[47] Summers, R.L., et al., 2005. Emergencies in space. Ann. Emerg. Med. 46(2), pp.177–184. https://doi.org/10.1016/j.annemergmed.2005.02.010

[48] Hodkinson, P.D., et al., 2017. An overview of space medicine. BJA: British Journal of Anaesthesia, 119(suppl_1), pp. i143-i153.

[49] Marescaux, J., et al., 2001. Transatlantic robot-assisted telesurgery. Nature, 413(6854), pp.379-380.

[50] Cermack, M., 2006. Monitoring and telemedicine support in remote environments and in human space flight. BJA: British Journal of Anaesthesia, 97(1), pp.107-114.

[51] The World Bank Group, 2020. Hospital beds (per 1,000 people) [online]. Available at: <https://data.worldbank.org/indicator/SH.MED.BEDS.ZS> [Accessed 20 June 2020).

[52] Williams, M., 2016. Musk says Hyperloop could work on Mars… maybe even better. [online] Universe Today. Available at: <https://www.universetoday.com/127356/hyperloop-on-mars/> [Accessed 24 June 2020].

[53] Williams, H. R., et al., 2010. A Mars hopping vehicle propelled by a radioisotope thermal rocket: thermofluid design and materials selection, Proceedings of the Royal Society A, 467, pp.1290-1309.

[54] International Space University Space Studies Programme, 2019. Final report of the Team Project Fast Transit: Mars and beyond. Available at: <https://isulibrary.isunet.edu/doc_num.php?explnum_id=1658> [Accessed 24 June 2020].

[55] Vashishtha, A. and Mehrotrab P., 2012. Committee on Space Research (COSPAR) 2012: Manned Deep Space Mission - Challenges and Future Design. Asteroid mining, 2020. [online] Available at <http://web.mit.edu/12.000/www/m2016/finalwebsite/solutions/asteroids.html?TB_iframe=true&width=921.6&height=921.6>. [Accessed 26 June 2020]

[56] Shaw, S., 2012. Asteroid Mining.[online] Available at: <http://www.astronomysource.com/tag/m-type-asteroids/> [Accessed 06 June 2020].

[57] Shane D. Ross, 2001. Near-Earth Asteroid Mining. [Accessed 26 June 2020]

[58] Andreas M. Heina, et al., 2020. A Techno-Economic Analysis of Asteroid Mining [Accessed 26 June 2020]

[59] Subair, S. and Thron, C. eds., 2020. Implementations and Applications of Machine Learning (Vol. 782). Springer Nature.

[60] NASA Science Mars Exploration Program, 2017. [online] Available at <https://mars.nasa.gov/news/2884/ai-will-prepare-robots-for-the-unknown/> [Accessed 24 June 2020]

[61] Wilson, G., 2008. The future of aquaponics looks bright at Brooks. [online] Greenhouse Canada. Available at: <https://www.greenhousecanada.com/the-future-of-aquaponics-looks-bright-at-brooks-965/> [Accessed June 2020].

[62] Lempert, P., 2018. 3D-printed food could help humans get to Mars. [online] <https://www.winsightgrocerybusiness.com/technology/3d-printed-food-could-help-humans-get-mars> [Accessed June 2020].

[63] McSween, H.Y., et al., 2009. Elemental composition of the Martian crust. Science, 324, p.736.

[64] Muttik, N., et al., 2014. Inventory of H2O in the Ancient Martian Regolith from Northwest Africa 7034: The Important Role of Fe Oxides. Geophysical Research Letters, 41(23), pp.8235–8244.

[65] Lewis John, S. and Hutson M, L., 1993. Asteroidal Resource Opportunities Suggested by Meteorite data. Resources of Near-Earth Space. Tucson: Univ. of Arizona Press, pp.523–542.

[66] Andriani, D., Wresta, A., Saepudin, A. and Prawara, B., 2015. A review of recycling of human excreta to energy through biogas generation: Indonesia case. Energy Procedia, 68, pp.219-225.

[67] Vogt, Gl. L., et al., 2009. Waster limitation management and recycling design challenge. [online] NASA. Available at: <https://www.nasa.gov/pdf/396719main_WLMR_Educator_Guide.pdf> [Accessed June 2020].

[68] Demain, A.L. and Sanchez, S., 2009. Microbial drug discovery: 80 years of progress. The Journal of antibiotics, 62(1), pp.5-16.

[69] Monsi C. Roman, et al., 2018. Centennial Challenges Program Update: From Humanoids to 3D-Printing Houses on Mars, How the Public Can Advance Technologies for NASA and the Nation. [online] Available at: <https://ntrs.nasa.gov/archive/nasa/casi.ntrs.nasa.gov/20180006610.pdf> [Accessed 24 June 2020].

18: THE ENGINEERING, ECONOMICS AND ETHICS OF A MARTIAN CITY-STATE

R. Mahoney, A. Bryant, S. Dunn, T. Green, T. Mew, T. Renton, T. Sandlin, C. Simmich & K. Raynor
Southern Cross Innovations, Australia
robert.mahoney24@gmail.com, abryant143@hotmail.com,
stewart.dunn@live.com, green.thomas@icloud.com, tmew91@gmail.com,
renton.tim@gmail.com, sandlin.trevor@gmail.com, Deurlaus@gmail.com,
katrina.eve08@gmail.com

ABSTRACT

Populating Mars is a problem grounded by engineering, and economics. Building on our previous work, "The Engineering Requirements for a Large Martian Population and their Implications", we have endeavoured to solve both problems. Using Mars's abundant natural resources, it is possible to build structures resembling domed Colosseums, each housing nearly one thousand people. Over the course of sixty years, thousands of these can easily be built from the Martian soil, though more concentrated sources of materials can be exploited.

The final city, resembling a collection of 'superblocks' not unlike Barcelona or Buenos Aires, would cover approximately 42 km^2, with another 230 km^2 dedicated to high intensity farming. Energy would be provided from locally sourced thorium; water would be processed from mined ice. Strategic metals - specifically copper - would be mined from ancient seabeds, in particular ones exposed to volcanism in the past. Given their prevalence in copper ores, precious metals extracted from the mining process would make an ideal export, further highlighting the need to locate copper ore deposits on Mars. An extensive network of heavy-haul rail would connect the city to its satellite mining sites and other, nearby city-states with an effective range of over 1,000 km. After this, sub-orbital rocket propulsion becomes more economical.

Economically, the city-state will be heavily dependent on asteroid mining activities, both in the belt and on near-Earth asteroids, as its proximity to these would allow for a competitive advantage over Earth. This would not only provide the bulk of the city's economic activities, but such mining operations would also help reduce the cost of travel to Mars from Earth, providing a triangle-of-trade and a positive feedback loop that drives down prices and drives up economic activity. When combined with exports of licenced technologies, deuterium, radiologically modified plants, precious metals, and a number of novel 'Made-on-Mars' goods, the city state will be able to achieve a robust economy, with a GDP measured in the hundreds of billions. The equal of a developed city in America or Europe here on Earth.

Politically, socially and culturally, the city would be a fusion of state control and fierce individualism, with the requirements for living on Mars necessitating the city provide a large number of guarantees and services, but the need for a dynamic economic engine and the belief in the new frontier pushing people to individualism and liberty. Fortunately, these forces will be balanced by a highly educated population and a culture of community fostered by a shared responsibly to maintain their homes and habitats. Though life would be tough, the high wages and adventurous nature of living and working on Mars would no doubt attract millions of people, truly opening up a new branch of Homo sapiens, and more importantly, the entire inner solar system for humanity.

1. TECHNICAL DESIGN

1.1. Resources and Production

The air and soil of Mars can be used as the basis for most construction required for a Martian settlement, being rich in oxygen, carbon, silicon, iron, and aluminium. A summary of both can be found below in Table 1 and Table 2. As has been previously shown [1], using refining methods similar to those used in sand mining on Earth, these materials can be readily extracted with existing technologies. Despite these abundant resources, the Martian soil does lack several key strategic materials, specifically copper, which will need to be actively sought out. In addition to this, deposits of thorium and precious metals will need to be found for use in the city, and for export back to Earth. Finally, while it is certainly possible to extract all the necessary steel, aluminium and silicon needed for the city from Martian soil, it may be advisable to find more concentrated sources of iron or aluminium in order to improve the efficiencies of production and reduce costs. To this end, Martian resources will be divided into two groups: General Resources; and Strategic Resources.

Table 1: Martian Atmospheric Composition Percent by Weight [2]

CO_2	N_2	Ar	Trace
95.66	2.30	1.77	0.27

Table 2: Average Martian Soil Composition by Weight [3]

SiO_2	Fe_2O_3	Al_2O_3	MgO	CaO	SO_3
46.00	17.85	7.75	6.88	6.05	6.40

Na_2O	TiO_2	Cl	K_2O	Trace	
2.20	0.85	0.70	0.23	5.09	

1.1.1. General Resources

General resources will be collected, refined, and processed at a complex referred to as the *Manufactorum*. This will be a series of interconnected buildings that will take in Martian regolith, air, and water, and turn out steel, aluminium, magnesium, fertilizer, plastics, brick, glass, and all other day to day products. To produce enough steel to build all the required habitats, roughly 2,700 tonnes of regolith will need to be collected per day. While this may seem large, it is dwarfed by mining operations here on Earth. This will be autonomously collected using oversized versions of the RASSOR [4], dubbed the RASSOR XX, as they are two orders of magnitude bigger. Each one will have approximately 5,000 kg mass and be capable of collecting 60,000 kg of regolith per day. Such an approach will allow the city to begin and grow anywhere on Mars, allowing access to more valuable, strategic resources.

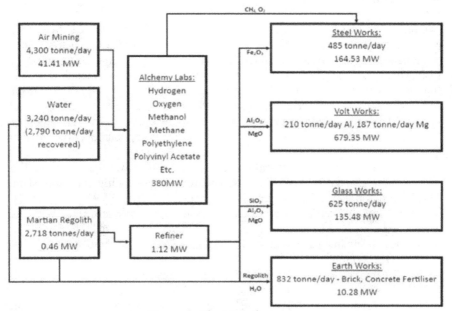

Figure 1: The Manufactorum

Table 3: Total Manufacturing Mass, Energy and Cost

	Mass	Energy	Cost
Regolith Harvesting	830,000 kg	460 kW	$415,000,000
Air Mining	690,000 kg	41,410 kW	$345,000,000
Refining	450,000 kg	1,186,440 kW	$226,000,000
Production	45,000,000 kg	413,510 kW	$22,743,000,000

1.1.2. Strategic Resources

Strategic Resources are those that cannot be found in the regolith, such as copper, thorium and rare-earth metals. On Earth, ore for such materials tend to form in areas that once had ancient underwater volcanism, referred to as volcanogenic massive sulphide ore deposits, which represent a significant source of Earth's copper, zinc, lead, and iron ores, with cobalt, tin, barium, tellurium, gallium and germanium as co- or by-products. As such, the city will need to be located near to areas that likely saw such activity in the past, such as the edges of the Tharsis Bulge and Eridania Basin [5]. This is of specific interest as such geologic regions tend to typically have high concentrations of other strategic resources, like sulphides, gold, silver, and precious metals. In most cases, noble metals like gold and platinum are by-products of copper and nickel refining, being extracted from 'anode mud' left over from the refining process. Once a sufficient supply of water has been found, the next two strategic resources to find will be thorium or uranium for power, and copper for electrical systems and rare-earth metals.

1.2.   Power

1.2.1.   Generation

The primary means of power generation for the Mars city-state will be via a series of generation IV Thorium Molten Salt Reactors (MSR). These reactors use a combination of thorium and uranium 238 in a closed loop dual heat exchanger system to drive a steam turbine. An overview of the power system, including energy storage systems can be found below in Figure 2.

Mars has multiple large surface deposits of thorium, the largest of which being at ~45°N, 30°W [6]. A combination of easy access to fuels and high-power output per kilogram enables thorium based MSRs to be a viable energy source for the Martian city-state.

An estimated 2.5 GWe of electrical power is required to power the city-state. The majority of this, about 1.5 GWe, is consumed by the everyday life of the city-state. This is based on terrestrial consumptions of approximately 1.5 kW per capita [7], while an additional 1 GWe is allocated for industry and infrastructure. With the current target of 1 GWe per MSR, 3 total power plants will be required with at least one of these being in close proximity to the primary industrial complex.

1.2.2.   Distribution

Power distribution for the city-state will be managed via an underground high voltage distribution network in the range of 275 kV. The distribution network will be managed primarily via two concentric rings housed in large service tunnels, with interconnects at multiple locations to provide redundancy if part of the network must be taken down for maintenance or repair. Switching yards to reduce the distribution voltage can then be strategically placed and tied into the HV distribution network as required.

1.2.3.   Storage

In addition to power generation, power storage will be employed so excess energy can be consumed when required. Thermal Energy Storage (TES) is the selected storage method, with the specific technology being molten silicon storage, which can store more than 1 MWh(t)/m^3. This technology has been selected due to the abundance of silicon on the surface of Mars and the high energy densities achieved. The molten silicon is heated via a combination of electrically powered joule heating and solar thermal heating which is thermally insulated for storage until required. The stored thermal energy can then be used to generate power in two distinct ways:

1.   The high temperature silicon can be fed through heat exchangers which in turn drive a steam turbine for thermal storage units located near the MSRs; and,

2.   High efficiency thermo-photovoltaic (TPV) panels can be used to directly generate electricity as the molten silicon is pumped past them.

The electrical energy recoverable from the TPV panels is in the range of 200 - 450 kWh(e)/m^3 [8].

In addition to the high temperature energy storage, the output water from the MSR turbine heat exchangers will be used as a low temperature energy storage system. This heat storage will not be used to generate electricity but will be used to assist in temperature control for the city-state. This water will be pumped around habitats in proximity to the MSRs within a closed loop, non-potable water system which will help to regulate the habitat temperatures.

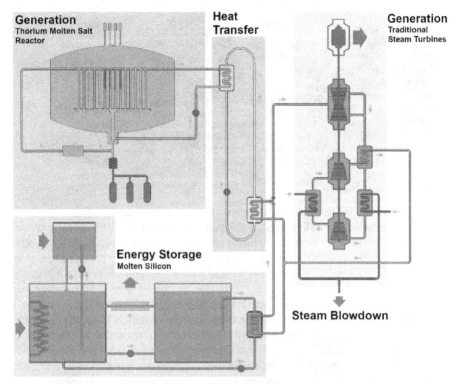

Figure 2: Power Generation Overview

1.3.   Habitation

In order to simplify the design of habitation on Mars, the first step is to reduce the internal pressure as low as possible. While it would be simplest to have a pure oxygen atmosphere at low pressure, this would be inadvisable as flammability is proportional to percent oxygen content, and not amount of oxygen present [9]. To this end, an atmospheric composition of 50 kPa and 30% oxygen was selected to reduce engineering requirements, while still having flammability properties similar to those on Earth. This will be maintained at 10 - 20°C to reduce thermal losses.

Using this atmospheric composition, it is possible to design a habitat capable of being built primarily from brick, glass, and steel, and that can house a sizable portion of the city's population. However, such a design should not be restricted to geographic features, which may limit the city's location and ability to scale. Additionally, flexibility in design, enables precincts to be expanded and repurposed as necessary, with little variation to the basic configuration. To this end, an example of the design selected can be found below in Figure 3.

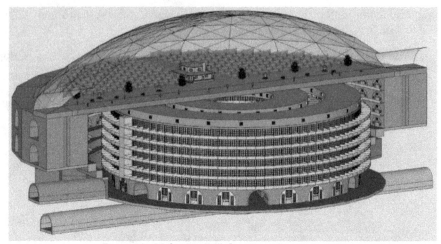

Figure 3: Habitation Colosseum Cross-section

The above design is 180 meters in diameter, with an inner and outer brick wall providing protection, radiation shielding and anchoring. Between these two walls is a layer of steel plate, 24 mm thick. This forms the walls of the pressure vessel and will see a maximum stress of 160 MPa, well below the yield stress of 250 MPa, and fatigue stress of 205 MPa making this design robust even by non-aerospace standards. These will be protected from corrosion with sacrificial anodes of magnesium or aluminium. Within the dome is a pit 14 meters deep, allowing for a number of internal structures to be built. There is also a 'second floor' on top of the inner walls and internal structures, rated to carry a layer of soil 2 meters deep. This allows for the construction of large, open social spaces such as parks or recreational farmland, while also further reducing radiation exposure to the inhabitants within. Figure 3 shows an example of a Colosseum dedicated to habitation, with an internal apartment block housing people in the centre, as well as in the outer walls for a total population of 950. Each person would have a personal space of 45 m². These structures will then be connected to each other either underground through tunnels dug with road-headers, or above-ground, on raised walkways linking dome-to-dome through expansive boulevards. Each of these structures will require almost 2,000 tonnes of steel. Manual labour on the surface will be needed for the construction of the steel walls and the dome, welding together mass-produced sections of steel. The foundation, brick walls and excavation will be constructed automatically. Once the external features have been constructed, the Colosseum can then be pressurised, and further construction can be completed in a 'shirt-sleeves' environment. Base on previous designs, it is estimated that 11,000 tonnes of construction equipment will need to be imported over the total development time of the city-state.

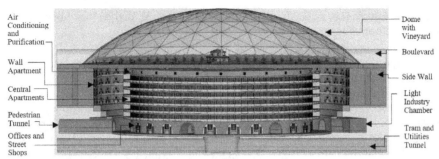

Figure 4: Typical Habitat Layout

1.4.   Emergency Management

Attracting, retaining, and sustaining a population of one million people in the city will require robust emergency management policies and procedures to provide public safety and reassurance. In addition to this, given the hostile environment of Mars and the relative inaccessibility for first responders and aid relief to be freighted from Earth within a reasonable timeframe, it is evident that emergency management and recovery will need to be self-sufficient. Further, the size of the population on Mars would limit the possibility of an effective mass-scale evacuation during an emergency. All emergency management will therefore need to be executed within the confines of the city.

Designing utilities that may act independently of one another can assist in ensuring critical infrastructure is not compromised during an emergency — such as additional water storage in each habitat and back-up power plants and power supplies. Further to this, an Emergency Management Plan may be designed with citizens grouped into either emergency or essential services to ensure end-to-end support in combating an emergency. An example of such a plan was developed as part of this project, the Emergency Mars Evacuation Program [10].

Emergency services would comprise law enforcement, medical personnel, firefighters, and a specialised State Emergency Service that comprises volunteer and paid personnel trained in flood, vertical rescue, chemical and radiological events, solar and Mars storm response, and environments within the habitat experiencing atmospheric loss. Each emergency service would act as the combat agency and incident controller for its respective emergency, be it terrorism, civil unrest, pandemic, fire, or other events.

Meanwhile, essential services would comprise staff and specialised citizens who are trained in maintaining utilities during the course of an emergency, and may include the Departments of Public Works, Utilities and Transport, as well as specialised civil agencies such as IT teams for maintaining communications throughout an emergency incident. Private sector organisations may also be legislated as essential services for the purposes of emergency management and recovery.

1.5.   Town Planning

While establishing a city-state on Mars presents several challenges and opportunities that are different to those experienced-on Earth, several common issues remain. Modern cities have typically grown organically from small towns near key features (i.e. rivers, harbours, crossroads, etc). With the crucial benefits of warmer temperatures, year-round sunlight and lower energy requirements for orbital launches, the Mars city-state will prioritise a large, clear equatorial sites near to strategic resources instead of specific natural features like craters and canyons.

Designing the city from the beginning to house a million inhabitants is a rare opportunity that permits the establishment of a clear expansion plan. The city will begin a close but safe distance from an original landing site and expand along a core rail line towards a future city centre. With current technologies, the city will be able to be largely decentralised, with residential apartments, commercial spaces and clean and quiet industries mixed. As detailed in Section 1.3, the city will be built within large steel cylindrical habitats, each with a domed roof and an excavated interior of varying depth. The upper habitat floors, right under the domes will feature common use parkland, luxury crops, orchards of trees, or sports fields. Habitat floors near the original ground level will feature four to eight floors of residential, shopfronts and common use areas, with light-rail connections to adjacent habitats. Lower levels will primarily be used for storage, offices, and light industry, with the deepest habitats featuring terminals for long-distance trains.

With approximately 1000 habitats required to house the one million inhabitants, the habitats will be closely packed in a honeycomb layout. This will reduce heat losses, minimise material required for connecting infrastructure, and create more spacious atmospheres for inhabitants travelling between habitats. The habitats will be arranged in 'superblocks' with large vibrant avenues running through busier commerce, transport and entertainment focused habitats. Filling the large areas between the avenue habitats will be clusters of quieter habitats, featuring more space for residential, office and small industry zones. Cosy connecting lanes will meander through these habitats and back to the avenues, empowering inhabitants to walk, use personal mobility devices, ride bicycles or move small cargoes with light delivery vehicles, as shown below in Figure 5. By enabling interconnections at various elevations, the city vastly decreases journey lengths and elevator use when compared to the two-dimensional town planning prominent on Earth.

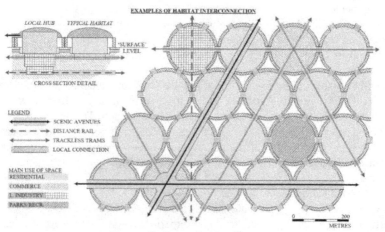

Figure 5: Example City Layout

Large fields of agricultural modules, as featured in Section 1.7, will be constructed around the designated city centre. With their modular design, the greenhouses will be connected in long North-South rows, maximising sunlight exposure, with new modules added as needed. The fields of greenhouses will be interspersed with a grid of rail and utility corridors, to transport automated farming equipment, nutrient sludge, goods, and maintenance personnel.

Heavier industries may be considered unsuitable for co-locating within the city habitat modules, due to excess noise, equipment size or safety reasons. These heavier industries will be located within dedicated habitat modules further from the city or in deep hollowed-out caverns. Trains or tunnels will remain the predominant transport mechanism for personnel, goods, and infrastructure to these heavier industries with specialised access and design as required.

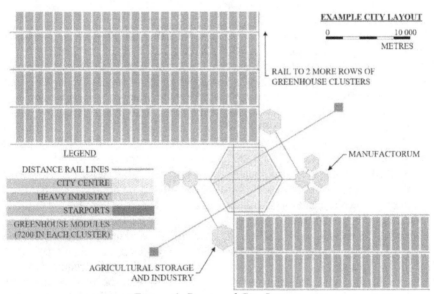

Figure 6: Proposed City Layout

1.6.   Transport

Rail systems have long been used on Earth as a cost-effective means for mass transport of both passengers and freight and has long been demonstrated to vastly outweigh the benefits of road-based transport. To this end, a rail-based system was selected for long-distance and heavy-haul transport, with a rocket-based transport system to be used at the point when rail becomes cost-ineffective, or at the Rail-to-Rocket transition point.

Rail systems continue to be constructed and expanded on Earth, with a significant number of major cities having a light-rail (tram) system for mass transit of the population between city zones. This system can be replicated on Mars for effective transport of the population within the city, and as the city-state is being designed from scratch, the system has been planned into the design to maximise efficiency.

1.6.1.   Long Distance Transport
1.6.1.1. Rail Transport Costs
Estimates for costs of bulk materials to be transported by rail have been calculated using Australian freight costs as a benchmark, with the unit being $/'000 net tonne kilometre (ntkm). The calculated cost for bulk transport is $25/'000 ntkm. This cost includes fuel, vehicle wear and tear, maintenance and consumables (e.g. lubricants, brake shoes etc) and is dependent on distance. A driverless system will be used; similar to iron ore trains in the Pilbara region of Western Australia; on a standard-gauge ballasted track design. Methane and oxygen will power the rolling stock as oppose to electrification, to further reduce costs, and hopper type wagons will be used to transport bulk materials, like ice and copper ore. A containerised system will be used for other materials which will improve haulage while ensuring supply-chaining from localised environments.

1.6.1.2. Rocket transport costs
The rocket transport costs have been calculated using the SpaceX Starship as a benchmark [11]. Each rocket has an estimated cargo carrying capacity of 330 tonne if refuelling can be conducted at the destination point. At a cost of $13 million per launch, this implies a cost of $40/kg to anywhere on the Martian surface. Note that this cost is relatively independent of distance, with roughly the same cost per payload for a 1,000 km journey as one to the other side of the planet. For more details on the selected rocket system are in section 2.1.2.

1.6.1.3. Rail-to-Rocket Transition
Using a nominal payload of 330 tonne, the transition from rail to rocket becomes economically viable at approximately 1,600 km from the city-state. However, significant loading & unloading infrastructure will be required. A standardised container between rail and rocket freight will ensure that costs and time are reduced when freighting goods at interchange and logistic centres. In addition to this, bulk materials to be transported by rail can be efficiently loaded using relatively simple load out facilities (e.g. ice, copper ore etc.) using a rail-loop for freighting. Rocket launch facilities and similarly complex infrastructure will also be required at star ports for loading and unloading of rockets. Rockets may be used to transport commercial freight between city-states and transport of specialist materials, while bulk materials should be sourced within the 1,600 km nominal effective radius of the city-state.

1.6.2.   Intra-City Transport
1.6.2.1. Passenger
Due to the planned close proximity of the Colosseums and their layered design, the primary method for passenger transport within the city is trackless electric trams operating on an overhead wire-free network, similar in concept to the Autonomous Rail Rapid Transit system developed by CRRC (see Figure 7). The use of an overhead wire free system negates the requirement to have overhead wire structures, simplifying construction and improving city aesthetics. Charging occurs with some overhead wire structures at each station, similar to some sections of the Newcastle & Sydney City Light Rail networks in NSW, Australia. The use of trackless trams is advantageous as it involves significantly less construction, materials, maintenance requirements and improves aesthetics.

Figure 7: CRRC ART Trackless Tram System [12]

1.6.2.2. *Freight*

Freight will be transported by autonomous driverless electric trains running in tunnels between industrial and commercial nodes beneath the city-state, powered by an overhead solid conductor bar. Solid conductor bars are advantageous over traditional overhead wire systems as they require less complexity in construction & maintenance (e.g. removal of tensioning requirements, rewiring etc.). The tunnels will be constructed with insulation & floating slab track design (e.g. similar to the Sydney Metro system) to reduce noise & vibration to occupied spaces above.

1.7. Food and Agriculture

Previous work investigated the provision of both energy and nutritional requirements for 1000 Martian colonists, and while those requirements discussed in that report do not change with scale, the methods used to fulfil them will (1). It had been proposed that a vegan diet be used on Mars in order to minimise food security risk. As the colony grows, substantial and sustained food surpluses will be generated. This will allow a relaxing of the vegan diet, by incorporating animals that are able to thrive on the non-edible portions of the crops grown, most likely insects, fish and crustaceans. While it's not impossible to raise traditional poultry and livestock on Mars, the requirement to grow forage crops or forgo food for human consumption to provide for them will make them too expensive for the vast majority of the populace.

High intensity plant factory agriculture was previously recommended due to its resilience to the Martian seasons and dust storms; however, its power consumption severely limits the growth of the city-state. Due to this, as the city-state develops, it will transition to making use of the natural sunlight on the Martian surface. Calculations based on Appelbaum and Flood [13], indicate that on average there is enough energy in the sunlight to maintain a greenhouse temperature of >15°C without supplemental heating at latitudes up to 45°, provided there is sufficient thermal mass. Crops in higher latitudes may require additional insulation, thermal reservoirs, or alternative heat sources. Approximately 230 square kilometres of cropland, about the area of Manhattan and Brooklyn combined, will be required to feed the city, and provide storage surpluses.

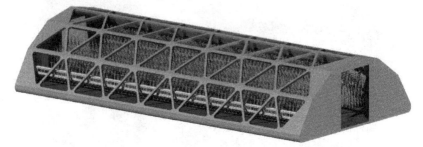

Figure 8: Artist's Conception of a Greenhouse Module

Modular greenhouses will be constructed with Martian materials, primarily steel and glass. A likely format will be cylindrical segments, see Figure 8, which will be laid end to end in rows. This allows power, data and water to be distributed to all greenhouses with a standard interface. Roads between each pair of greenhouse rows and regular cross paths allows easy vehicular access to all greenhouses. Transparent sections of the greenhouses will be double glazed to reduce heat loss and to allow for automated leak detection. Geodesic triangles will likely be the preferred glazing, due to ease of manufacture.

Relying on the sun comes with risks, particularly due to the risk of months long dust storms that will prevent the growing of crops. Multiple mitigation strategies will be used. Most obviously is food storage, with extensive granaries required to hold surplus food. This may require altering the diet to include a higher proportion of grains and other long-term storable foodstuffs. Growing crops on Earth is constrained by local seasons, but Martian agriculture is entirely within climate-controlled greenhouses, so crops can be sown, grown and harvested on a rolling cycle, such that there is always a crop ready to harvest. In the case of a catastrophic dust storm, the crops that are ready will be harvested, while those that are close to harvest may be moved to the plant factories, remaining from the earlier days of the colony. Crops that are too young to survive to harvest can be left to freeze and the biomass later used as feed for edible insects or as substrate for mushroom farming. Plant factories will also be used for seedlings to restock greenhouses after dust storms or other large-scale crop losses.

One of the primary limiting factors of plant growth is the level of bio-available nitrogen. While the nitrogen cycle within the city-state will be closed, modern water treatment involves nitrification, which reduces bio-available nitrogen. Nitrogen fertilizer will be required to maintain food production. Ammonium Nitrate, a common terrestrial fertiliser can be produced with an entirely exothermic reaction chain, if supplied with hydrogen, nitrogen, oxygen, and water. Ammonium Nitrate is also a commercial explosive, which may have application in Martian or asteroid mining. Another key element crucial to farming that may be in short supply is phosphorus. If a sufficient supply cannot be found on the surface, it may need to be imported from Earth or mined from carbonaceous chondrite asteroids. Interestingly, Mars's moons, Phobos and Deimos, are both carbonaceous chondrites and may serve as viable sources of phosphorus.

1.8.   Water

The Martian city-state will be well developed, and as such will require a resilient and efficient water transportation, recycle and reuse network. The water demands of the agricultural, and industrial segments of the city-state have been redeveloped due to changes in efficiency scale, but the fundamental requirement for municipal water is unchanged from what was previously determined, averaging 174 L/d/person (1). This increase means that the municipal demand is the highest consumer of water. The industrial and agricultural demand has been provided in Sections 1.1 and 1.7. It has been assumed that 99% of the water cycle can be successfully recaptured and treated, due to the learnings of operating on Mars for several decades improving the overall efficiency of water retention. Water consumed by the Manufactorum will not be returned to the water system as it will either be reused, or used to store various chemical solutions and pollutants. This will have the added benefit of keeping heavy metals away from the municipal and agricultural water network. As the water load for the Manufactorum is small, this will not be a significant drain on resources. If further reuse is required, the polluted water can be boiled off and reclaimed through condensation. The demands of the Martian city-state are summarised in Table 4 below.

Table 4: Water Demand and Recovery

	Need (m³/d)	Immediate (m³/d)	Recovery (m³/d)	Net Intake (m³/d)
Civic	174,000	172,260	31,320	33,060
Agriculture	23,000	22,770	-31,320*	-31,090
Industrial	450	0		450
			Net Intake	2,420

* Recovered from RO Waste Stream

The net intake per day for the colony is 2,420 m³/d, however a total of 3,025 m³/d will be required to be mined due to a 75% efficiency in the intake RO treatment plant. As before, this demand will have to be sourced ahead of colony development in order to generate a sufficient reservoir of water to sustain the colony in case of mechanical failure or significant water loss. This means that a higher rate of mining vs actual demand would be preferable to fill reservoirs as they are constructed. This demand is the minimum required to support 1,000,000 citizens and internal industrial demand. This does not include the use of water resources for commercial ventures considered further in Section 0.

1.8.1.   Water Mining on Mars

Due to the increase in daily water demand, the number of effective water sources is restricted further. The daily intake demand means that 'rodriquez well' or rodwell water extraction techniques, a method of pumping steam into ice to melt and extract a cavity of water, are no longer practical, as the number of rodwells would be excessive, and the constant replacement and relocation of the rodwell infrastructure would prove taxing and manpower heavy. While these techniques would be suitable for isolated mining or research domes, on a larger scale, more efficiency is required. The two remaining options are drilling for the Martian hydrosphere or mining the subsurface and glacier ice. In the last few years, there have been several developments in the hunt for the Martian hydrosphere. The first is the discovery of the potential liquid water lake beneath the south pole of Mars [14]. This is the first indication of a body of liquid water existing on Mars, and while the South Pole is not an ideal landing site, it opens the potential for further subsurface water discoveries. The second development is an analysis [15] of the recurring slope linae. Using a research methodology based on desert biomes on Earth, there is evidence of a potential hydrosphere as shallow as 750 m beneath the Martian surface which is creating the slope linae.

This is below the current penetration of the MARSIS and SHARAD probes, so verification of this theory is not currently possible, and the Martian hydrosphere remains tantalisingly hidden. Exploitation of the hydrosphere has been discussed previously and remains well within terrestrial technologies. The other option is the physical mining of subsurface and glacier ice. The total intake of water per year is 1,100,000 m^3, which is approximately on par with an industrial mine. Physical extraction would be best focused on extracting large blocks or chunks of ice and transporting them back to the city in a frozen state, for further processing there. The hardness of ice at Martian surface temperatures of -70°Celsius, is 6 on the Mohs hardness scale, approximately equivalent to orthoclase materials. This would put a significant wear & tear on physical mining equipment. Localised heating on the cutting tools, via electrical resistance would raise the temperature of the ice around the cutters, lowering the resistance of the ice, allowing carving to take place. This could then be loaded onto trains and transported back to the city-state for melting and treatment, requiring roughly 200 MW of power to meet demand.

1.8.2.   Water Storage

Due to the increase in volume of the proposed city-state domes, a significant volume of water can be contained within the habitats, using even a small subsection for storage. Installing a tank 2 m in height and 120 m in diameter in the under-section of the habitats, could provide each habitat with up to 22,500 m^3. This storage volume could be optimised into the walls and subterranean sections of each habitat as practicable. Utilising this volume of storage in 1,000 of the habitats provides a 100 day buffer for the entire colony, however the total volume will likely be smaller. The advantage of this arrangement is that additional heating for water storage will not be required. This will provide sufficient local reserves for breakdown or calamity, and a breach of any one dome will not create a significant reduction in water storage.

1.8.3.   Wastewater Treatment and Recycling

The treatment processes of the colony's water and wastewater will be identical to the initial colony of 1000 people (1), but on a larger scale. Instead of using skid mounted treatment units, wastewater and advanced water treatment plants will be required. Expansion will require the construction of larger facilities within the new domes, and breeding of additional bacteria within the existing reactors. An analysis of standard Reverse Osmosis (RO) and Wastewater Treatment Plant facilities within Brisbane, indicate that a 50,000 m^3/d plant that treats municipal waste into drinking water could be constructed within one dome, as depicted in Figure 9.

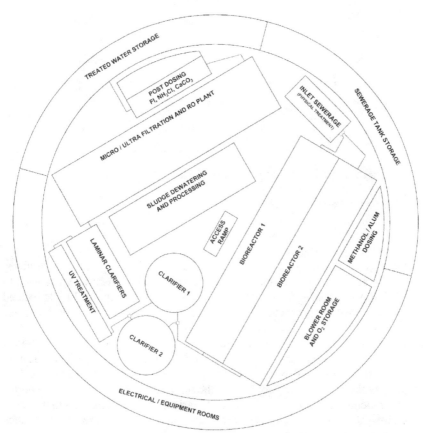

Figure 9: Layout of typical water treatment dome of 50,000 m³/d

Based on the treatment demand, four such domes would be required for the entire city-state. Each dome would contain two trains but would not have complete redundancy. As such, it is suggested that up to six domes would be constructed, to allow for train shutdowns and redundancy without impacting the ability to treat and reuse water. These domes could be co-located for ease of sludge transportation or spread out amongst the other domes to provide localised treatment of wastewater. Due to the need to transport sludge away from the plants to the agricultural facilities, the domes should be located on the outskirts of the colony, near transportation routes and the agricultural zones. Nitrogen and CO_2 are produced by the process and will have to be captured for reuse in fertiliser or disposal. H_2S is also produced by aging sewage and would need to be captured in the inlet works as it is toxic to humans. Each dome would require 29 MW of power and 17 t of oxygen per day to operate. Additional dosing of Alum, Methanol, Ferric Chloride, Chloramine, Calcium Carbonate, and Fluoride would be required for the operation of the plant. Agricultural water will be mixed with the municipal water prior to treatment and will be removed from the cycle after the treated water is hardened, but not treated with chloramine and fluoride. A full diagram of the water and wastewater network is shown in Figure 10.

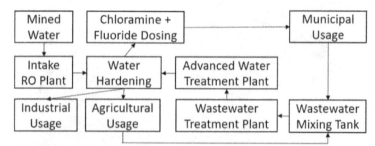

Figure 10: Water Transportation, Treatment and Reuse Network

2.   ECONOMICS

2.1.   City-State Development
2.1.1.   Location on Mars

The location of the city-state is going to be dependent on several variables, specifically access to water, thorium, and areas of high historic volcanism to provide rich ore deposits. In addition to this, the city-state will need to be located close to the equator to provide better orbital access, and better conditions for growing food. This may clash with the need for water, which is more plentiful in the north; however, a robust heavy-haul rail network will allow for a number of 'satellite sites' over 1,000 km away. Based on this and Figure 11, the ideal location for the city-state will be 30°N, 60°W, near to Tempe Terra or Chryse Planitia. This will allow ready access to ice deposits to the north, thorium deposits to the north-east, and mineral ore deposits in Tharsis and Marianas outflow channels to the south and south-west. This is also close enough to the equator to reduce seasonal variance in sunlight and to allow for ready access to Low Mars Orbit.

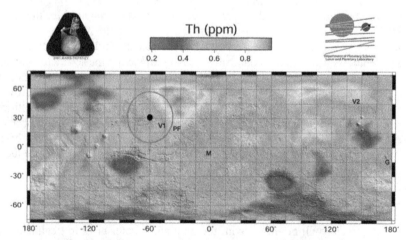

Figure 11: Maps of Thorium Concentration (6) with the City's Effective Radius (1,600 km)

2.1.2.   Interplanetary Shipping

The economic growth of the city-state will be heavily dependent on shipping cost not only to and from Earth, but to other locations in the inner solar system. If a base line rocket design is considered and combined with assumptions of $500/kg from Earth to Mars, and $200/kg from Mars to Earth, the costs of shipping to other locations in the inner solar system can be determined. To this end, the following rocket design based on the SpaceX Starship was chosen. This was done as the Starship is the only rocket in development that is capable of the variety of missions needed. While this design may not be ideally optimised for all locations, it does provide a useful baseline to illustrate the relative costs of shipping through the inner solar system. The one main variance from the current SpaceX design is that the Interplanetary Rocket will have a maximum crew of 50 people, due to volume constraints rather than payload constraints. From this, shipping costs to a number of key locations throughout the inner solar system can be determined, as summarised in Table 7 and Table 8 [16].

Table 5: Interplanetary Rocket Specifications

Dry Mass	Wet Mass	ISP	Payload to LEO	Aerobraking
120,000 kg	1,500,000 kg	370 s	150,000 kg	Yes

Table 6: Interplanetary Rocket Capacity

Capacity	Operational Life	Manufacture Rate	Fleet Size
50 people	30 years	35 per year	1100

Table 7: Shipping Rate - Earth

Location	Delta (m/s)	Travel Time (days)	Refuel Flights	Cost per Mission ($,000s)	Payload (kg)	Cost per Mass ($/kg)
Low Earth Orbit	-	-	-	$8,333	150,000	$55.56
Mars*	6,000	183	9	$75,000	150,000	$500
Moon	6,400	3	9	$75,000	134,780	$556
Phobos	5,100	183	9	$75,000	245,220	$306
Deimos	4,800	183	9	$75,000	276,870	$271
Eros	8,467	253	9	$75,000	23,720	$3,162
Vesta	8,756	399	9	$75,000	12,660	$5,924
Tycho Station	8,994	534	9	$75,000	4,190	$17,873
Ceres	9,477	472	9	$75,000	Cannot Ship to this Location	
Juno	9,528	453	9	$75,000	Cannot Ship to this Location	
Pallas Shipyards	9,558	472	9	$75,000	Cannot Ship to this Location	
Eugenia	9,654	463	9	$75,000	Cannot Ship to this Location	

Table 8: Shipping Rates - Mars

Location	Delta (m/s)	Travel Time (days)	Refuel Flights	Cost per Mission ($,000s)	Payload (kg)	Cost per Mass ($/kg)
Low Mars Orbit	4,100	-	-	$13,154	361,790	$36.36
Earth*	7,000	183	1**	$13,154	65,770	$200.00
Moon	5,200	183	5	$65,768	235,250	$279.57
Phobos	5,000	0	1**	$13,153	255,480	$51.49
Deimos	5,300	0	1**	$13,153	225,540	$58.32
Eros	5,962	544	5	$65,768	167,650	$392.30
Vesta	3,943	880	5	$65,768	383,210	$171.63
Tycho Station	4,041	493	5	$65,768	369,730	$177.88
Ceres	4,950	554	5	$65,768	260,720	$252.26
Juno	4,963	572	5	$65,768	259,350	$253.59
Pallas Shipyards	5,041	575	5	$65,768	251,240	$261.77
Eugenia	5,102	563	5	$65,768	245,020	$268.42

*Location used as baseline
** Direct Launch assumed due to operational simplicity

2.1.3.   City Growth

Assuming the beginning of the city is after the establishment of a 1,000 person 'pilot' site, it is estimated that the growth of the city-state to its 1,000,000 person population will take approximately 60 years. This is mainly due to the primary source of population growth being immigration, with births making up a relatively small portion of the new population. This growth is limited by Earth-Mars synodic periods, and by the estimated carrying capacity of the Interplanetary Rocket, which has been limited to 50 people due to volume constraints, their operational life, and their manufacturing rate. It should be noted that the fleet size and manufacturing rate are specifically for servicing one single city-state on Mars. A larger fleet and higher manufacturing rates are likely, though they will be utilised elsewhere.

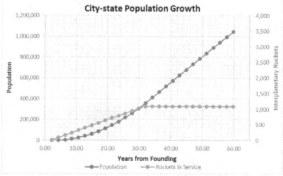

Figure 12: City-state Population Growth since Founding Year

2.1.4. Development Costs

The primary development costs for the city will come from the initial imports to begin large-scale construction, and from the wages of the construction crews. If 5% of the population over the 60 years are construction crews, and they are paid an estimated wage of $100,000 per year, combined with the import masses shown below in Table 9, the total cost to construct the City over 60 years will be approximately $260 billion. It should be noted that this economic model does not consider inflation, amortisation, or industry growth, as these are difficult to forecast over half a century in advance. As a result, the analysis here will represent a 'snapshot' in time, once the city hits the 1,000,000 mark.

Table 9: City Development Costs over 60 years

Category	Cost
Production Equipment Imports	$85,425,440,000
Construction Equipment Imports	$5,509,620,000
Power Plant Equipment Imports	$42,480,000,000
Farming Equipment Imports	$300,000,000
Water Purification Equipment Imports	$28,000,000,000
Transport Equipment Imports	$1,200,000,000
Construction Wages	$56,666,400,000
Growth (20%)	$43,916,290,000
Total Colony Cost	**$263,497,750,000**

2.1.5. Required Gross Domestic Produce

Using the above cost of $260 billion and the final population number of 1,000,000 people, it is now possible to determine the average cost of moving to Mars once the 1,000,000 mark has been reached. If the cost of construction is paid for by immigrants, then buying an apartment on Mars and 'citizenship' in the city will cost $260,000 per person. If we also assume the cost of traveling to Mars is $250,000 then the total cost of immigrating to Mars is ~$510,000. If this is financed via a loan at 3% with a 10% deposit, to be paid back over 20 years, an immigrant will need a deposit ~$51,000 and would have a yearly mortgage payment of ~$40,000. To finance this, and still have enough income for taxes, expenses and savings, the city will have an average wage of ~$100,000. With the required population, this implies a GDP ~$100 billion for the city to survive. This number is roughly in line with a major Western city with a similar population today. Here in lies the major incentive for a person to move to Mars. Financing such a move with a loan makes it attainable, while the high wages and sense of adventure will make it desirable.

2.2. Revenue

Several different revenue streams have been identified to generate income for the city. It should be noted that while a number of other goods and services will be provided within the city, for example financial, legal or catering services, these will all need to be paid for by an external source, making the city essentially a contained economy. As a primary example, land sales and immigration may pay off the construction of the city in an immediate way, but the debt they generate will then need to be financed by money coming in from Earth, either for a direct good that is exported back, or for a service provided elsewhere in the Solar System.

2.2.1.   Asteroid Mining Services

The primary source of income for the city will be providing goods and services to Asteroid Belt mining operations. As can be seen above in Table 7 and Table 8, shipping from Mars to the Belt can be up to two orders of magnitude cheaper than from Earth, and in some cases it is not even possible to reach some locations given the assumed rocket performance. It is worth noting that technological developments would lower these costs and make more locations accessible from Earth, but these developments would also make it cheaper still to ship from Mars. The values noted above illustrate the advantage of Mars clearly, though they are not absolute.

It is highly unlikely that Asteroid Mining will be completely automated. Given the lengthy telecommunications delay, operation from Earth or Mars will not be tenable. In addition, moving asteroids to low Earth or low Mars orbit would be extremely inefficient. Therefore, it is highly likely that future mining operations will be based off a mobile mining rig that will be located within several light seconds of a mining target, acting as a home base. This mining rig which would also act as a refiner, boosting the yield of payloads returned to Earth. These mining rigs would need to be able to produce their own air, water, and fuel at a minimum, but everything else would be a potential export provided by Mars.

Exports to cislunar space would also be possible but would be less common given that Earth is so much closer. They would need to be something that cannot be attained on the Moon, would be needed consistently and not worth the cost of quick delivery, like computer components. A good example would be carbon compounds, nitrogen, and fertilizers for green houses. These would be consistently needed, unavailable on the Moon and not worth the cost of 3 day delivery.

Finally, two other locations need to be discussed, Phobos and Deimos. Both are likely captured carbonaceous chondrites and are already likely mining targets, as they potentially contain water, carbon compounds and phosphorus for farming back on Mars. Shipping to these locations would be very cheap, and they would likely have very large populations either on their surface, or in nearby stations. They would likely require new supplies of food, parts, and people daily from the surface of Mars.

Given the high number of potential export locations, the likely consistency of demand, and the high value of exports to these locations, it is anticipated that providing asteroid mining services will represent up to 54.4% of the city-state's economy, or almost $75 billion annually. While this seems high, dependency on trade partners is not uncommon, for example, China represent 35.5% of exports from Australia. It should also be noted that while this has been analysed in a monolithic sense, this represents exports to dozens of different locations, across several different economic sectors. Mars will be dependent on a Solar System wide, or 'Solarised' economy. A summary of potential exports to different locations can be found in Table 10.

Table 10: Export List

Export Destination	Potential Products	
Asteroid Mining Rigs	• Power Equipment (Solar Panels, Nuclear Reactors, Nuclear Fuel) • Food, Spices, Flavouring • Fertilizers, micro-greenhouse components • Engines	• Repair/Replacement Components • Feedstock (plastics, metals etc.) • Drones, Drone Parts • Personal effects • Crew Rotations
Phobos, Deimos	• Foods • Repair/Replacement Components • Habitats	• Crew Rotations • Test Components
Moon, Cislunar Space	• Carbon Compounds • Spices, Flavouring	• Fertilizers, micro-greenhouse components

2.2.2.   Technology Licenses

Patents have been assumed to be one of the potential ways for the Martian city-state to earn an income. Traditionally, patents have been designed to both protect intellectual property (IP) and exploit technological development for commercial gain. However, the commercialization of patents is an ongoing problem and of the 2.1 million active patents in 2014, 95% failed to be licensed or commercialized, and do not provide any income [17]. Part of the issue is how patent law is expressed, with lawsuits becoming a standard part of patent negotiation and exploitation. As such, exploitation of Martian IPs, particularly where external users play a part, could be hampered by ongoing legal activity or similar technological development on Earth. In addition, should state actors decide to deliberately ignore Martian patents, the reliance would be on existing legal systems to enforce Martian IP. While historical trends tend to favour resolution in favour of the IP holder, and patent law is improving, there may be attempts to claim Martian IP by various actors on Earth.

Despite these potential risks, one of the potential Martian IP spheres that could be commercialised successfully is digital goods and patents that protect methods and processes. Digital IP is easier to commercialise as the concepts can be proven to be radically different from existing programs. Other potential sources of digital IP could include automation and industrial programs, Martian gaming studios, shows, streaming services, and education programs. While some of these would have heavily contested markets, the size of the markets, and the novelty for Martian designed products would be an attractor for a portion of the market. There are several companies that trade in digital assets and earn significant profit from their IP. An example of a digital IP service provider is Siemens Digital, who have an annual profit of $16 billion. This is an appreciable fraction of the required Martian economy, and only incorporates digitization and automation software. As such, there is a market for several digital based providers in the Martian economy. Based on this, it is possible for Martian program licenses to generate a conservative $10 billion of revenue per year.

The other option is for unique Martian industrial process IP. These could include advanced chemical, refining or recycling technologies. IP could be protected using Martian design engineers, and appropriate information classification. Similar systems are used in controllers of existing industrial process IP on Earth. Due to contractual limitations, and earning reporting being totalised, the earnings of these IP holders are harder to judge but could be negotiated depending on the rarity of the process in question. As the Martian city-state would be developing unique technologies and processes, there is a potential for this exploitation of technical knowledge, assuming the risks of commercialisation can be overcome.

To assess the value that could be generated from such licensing, several key industries that will be vital on Mars and a significantly large currently on Earth were considered. These industries included the metallurgical, plastic, renewables, petrochemical/synthetic fuels, recycling, robotics, chemical fertiliser, and vegan industries. These represent a combined $5.7 trillion of value. If a small average market penetration is assumed, 2%, and of that, a revenue of 15% can be generated off the licencing agreements, then roughly $18.5 billion could be made per year. Combined with the revenue from industrial programs, then technology licenses have the potential to generate up to $28.5 billion per year.

2.2.3.   Shore Leave and Tourism

Mars is ill suited for mass tourism from Earth. The costs of transporting a person will be on the order of $200,000 - $300,000 with minimum round-trip times on the order of 2 years. Only the extremely wealthy, capably of affording roughly $500,000 per year, or $1,000,000 per trip to Mars would be able to afford to go. Notwithstanding costs associated with tourism, opportunities exist for some tourists to come and stay for that period.

A better source of 'tourists', however, would be asteroid miners from the Belt returning for 'shore leave'. Given the need to have humans present outlined in Section 2.2.1, the high wages such a job would earn, the small numbers of opportunities to spend their wealth on rotation and the ease of reaching Mars over Earth, it is likely that many operations would either encourage people to take holidays on Mars as opposed to Earth, or draw on Martians as miners.

In addition to fresher food and better facilities, Mars's natural gravity would provide more comfort and could even act as a 'rehabilitation' spot before returning to Earth. Mars would also have relatively open spaces when compared to an asteroid mining station, where facilities will likely be comparable to the International Space Station, though perhaps larger. While Mars would not be as expansive and open as Earth, it would certainly be an improvement. If 5% of the City's population are transient tourists willing to spend up to $500,000 per year, then an approximate revenue of $26 billion per year can be generated.

2.2.4.   Precious Metals Exports

As discussed in Section 1.1, copper cannot be found easily in the Martian surface and concentrated deposits will need to be sought out. While this represents a restriction on where the city can be located, it also represents an opportunity. This is due to the fact that copper deposits often incorporate or are found adjacent to rare metals. As an example, iron oxide copper gold deposits can contain up to 4,000 million tonnes of ore, with grades up to 5% copper and 3 grams per tonne of gold. Additionally, the Olympic Dam mine in Australia is the world's fourth largest copper mine, the largest uranium deposit in the world, and produces a small quantity of gold.

Precious metals represent a reliable, if small, potential for exports back to Earth. Gold, platinum, iridium, palladium and silver all have values well over $200 per kilogram and have large global demand. Given it is unlikely that the city will have production levels even as high as 10% of global demand due to limited manpower, the risk of flooding the market and massively depreciating values is minimal. If the city is capable of at least 2% global production of the above metals, then over $3.5 billion in yearly revenue can be generated.

In addition to this, precious metals such as platinum may also be found as a by-product of copper production. Given the city's current requirements, an additional $500 million dollars of additional revenue could be generated.

2.2.5.   Land Sales

While the majority of the land reclaimed from the surface will be used for residential or city related industrial processes, there will certainly be demand from Earth to purchase land and use it for a variety of different needs. From government agencies to entertainment companies, there will be several opportunities to rent or sell land for independent use. Once the 1,000,000 population mark has been reached, roughly 60 Colosseums could be built per year. Roughly 50 of these will be needed to house the anticipated immigrants coming from Earth. The remaining 10 will be available for other uses. At a cost of $220,000,000 to build a Colosseum, land could be sold at a cost of $7,000 per m^2, roughly the equivalent of land in Hamburg or Dusseldorf, Germany. This allows for a potential revenue of $2.2 billion per year.

2.2.6.   Luxury Goods

Goods are considered luxuries because they are rare or hard to make. Wine and whiskey can easily be sold for over $1,000 per bottle if they are unique enough. Such products made on Mars would almost certainly have distinct flavours from those produced on Earth. As an example, the Malbec grape was disfavoured for production in France, but produced far better flavours when transplanted to Argentina, and is now a staple of the country's wine industry. Transplanting to not only a different climate and soil, but a different gravity will no doubt produce unique and interesting effects that would be in high demand with 'Martian Reds' potentially being a highly sought-after brand.

In addition, given the high cost of jewellery, paying a small premium for Martian gold with Rubies from the Red Planet would certainly be a viable way of generating revenue. The sale of luxury goods, as well as Martian wine and whiskey should be able to generate a revenue of over $1.3 billion.

2.2.7.   Heavy Water Exports

Given the large amount of water required by the colony, a prime export back to Earth is heavy water. Heavy water is roughly 9 times as abundant on Mars and as of writing sells for $925/kg [18]. The electrolysis process for producing hydrogen could be easily modified to preferentially increase the concentration of heavy water in the remaining fluid. Given the estimated concentration of 1 in 1,284 deuterium to hydrogen, this represents a yearly production of 885 tonnes of heavy water, worth $818 million yearly. The water used in this would have to be mined and processed in addition to that required by the colony for normal operations, but these activities could be scaled as required to support the export industry.

2.3. Economic Summary

Given the cost of moving to Mars, a minimum GDP for the city of $100 billion was determined. However, from the above revenue streams, it is possible to achieve this and more, with an estimated total GDP of $137 billion. The ability to command a higher salary coupled with the exotic nature of the city would incentivise immigration to Mars.

Corporations, however, will require a business case to both set-up operations on Mars and recruit for high-income positions. The aforementioned primary industries could present the business cases required to attract and retain high-income workers, with economic activities on the ground multiplying that wealth.

To encourage companies to come to Mars, and to this city in particular, the use of special economic zones with little to no taxation, similar to such zones in China or Vietnam, would help provide the incentive. Instead, the city would collect revenue from companies through renting of services, land leases, value-added taxations and a progressive income tax on employees. Following the economic model of Singapore, the city would be able to generate a high GDP relative to its population with business-friendly policies, but with strong protections for the necessities of its citizens. By having an effective average tax rate of 22% of GDP, the city will have an annual budget of $30 billion. This will allow the city to provide healthcare, education, and a universal basic income of $12,000 per person, as well as paying for infrastructure and services throughout the city.

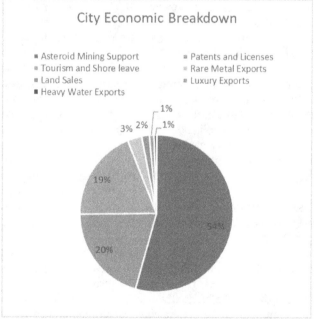

Figure 13: City Economic Breakdown

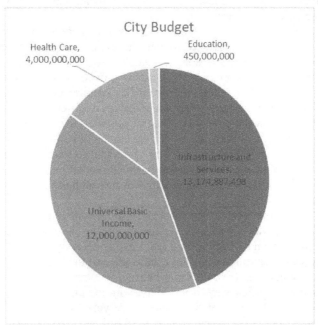

Figure 14: City Budget

3.   POLITICAL

3.1.   City Political Structure

The Martian city-state will be a fusion of free enterprise and centralised economies, not dissimilar to Singapore. Due to the dangers of living on Mars, all citizens will need access to shelter, food, water, electricity, communication infrastructure, healthcare, and education. In addition to this, a basic income of $12,000 would also be a boon to the economy, by providing a safety-net to allow people the freedom of expression and experimentation, granting more innovation and higher rates of productivity. This would not be enough to live comfortably; $12,000 per year is the current poverty line in the United States, but it would allow people to live, providing them with more freedom and economic power. The city would also provide basic resources for construction as well, such as steel, brick and aluminium, however given that these are produced from the soil and the air, they would be incentivised to produce these as cheaply as possible. Not only due to risk of competition from private corporations, but also from other city-states.

To maintain a business-friendly environment and to limit corruption, an official Chamber of Commerce would be created as an official branch of the government, with representatives being voted into roles by the local employees of a company. This would be balanced by a Chamber of Representatives, representing the citizens and workers of the City. Any and all elections would remain publicly financed out of a legislated stipend, with private donations forbidden to mitigate the risk of conflicts of interests forming for serving members. This set-up is to democratise corporate influence on the democratic processes, allowing industries' voices to be heard, but in an official and transparent capacity.

Voting will operate through a Single Transferable Vote system, which is a proportional voting system designed to achieve or closely approach proportional representation through voters ranking candidates in multi-seat organizations or constituencies. This will minimise the negative effects of multiple parties and gerrymandering. In addition to this, voting will be a requirement of all citizen of the city-state, with all new arrivals being automatically registered, and a small fine being levied for those who do not vote. This is again done to streamline the system, making it harder to tamper with. To reduce the burden of voting on the Citizens, it will be done via a secure electronic system. The right to free expression and the freedom from discrimination will need to be enshrined in the laws of the city, in order to ensure that the worst of human impulses stay on Earth, and not in the new world that could be created.

Finally, to maintain the habitats and ensure a sense of community, all citizens will be required to work within the city's maintenance teams, no matter their status or regular job. This is vital as maintenance will literally be everyone's responsibility, and everyone will need to have an active knowledge of the life support and safety systems of the colony and would serve both as training and employment. This would be done in a similar way to jury duty, with the citizens in question being required to give 2 - 3 weeks of their time every year to working on randomly assigned maintenance tasks. Such programs of national service are widely used through-out Europe and Asia.

3.2.   Legality and Laws
3.2.1.   Current Legal Framework

Article II of the *Outer Space Treaty 1967* (UN) (OST) states that '[o]uter space, including the Moon and other celestial bodies, is not subject to national appropriation by claim of sovereignty, by means of use or occupation, or by any other means.'

It should be noted that the Treaty does not prevent countries from establishing facilities, but from appropriation of celestial bodies through the establishment of facilities. Where a country establishes a settlement and claims sovereignty on the area either adjoining the settlement, or below the settlement, these claims would be voided pursuant to Article II of the OST. However, sovereignty within the settlement would survive, as it does for nation's satellites, ships within international waters, and landing craft and probes that have previously visited Mars. To that end, the settlement would remain within the scope of legislative control by its respective country, but no claims of sovereignty on the surface of Mars would be applicable. This is important to note, as anywhere on Mars a human will have to be inside of a 'facility', be it a Colosseum, habitat, rover or spacesuit and therefore always subject to the legislative control of its respective country.

3.2.2.   Required Changes

The OST, alongside the *Treaty Banning Nuclear Weapon Tests in the Atmosphere, in Outer Space and Under Water 1963* (UN) was drafted within the context of the cold war and was intended to mitigate militaristic endeavours in Space. However, given the size of the population and plans to make the settlement self-sustaining, consideration should be given to amendment of the OST to accommodate the settlement making a viable claim for independent state, and even nationhood, due to latency in communications both to and from Earth as well as supply-chaining of resources, emergency services, judiciary and law enforcement.

Such a scheme for semi-independence as a self-governing territory; or, total independence as a self-governing nation, would allow for effective law making, and a co-ordinated response to emergencies that may arise, which require an immediate response by the Martian population. An independent settlement may continue to be regulated and assisted by Earth-based nations through recognition and admittance by various international organisations such as the World Trade Organisation (WTO), and the United Nations and enforcement of their respective treaties.

Advantages would also arise for the Martian city-state to be admitted into the WTO through preventing anti-dumping measures being placed on the Martian city-state by Earth-based nations, as well as unreasonable tariffs that may unduly burden exportable goods from Mars, while improving the Settlement's overall GDP.

Meanwhile, admission into the United Nations would entitle the Martian city-state to judicial forums for mediation through the International Court of Justice, as well as recognition of intellectual property rights through specialised agencies such as the World Intellectual Property Organisation.

Political orientation on Mars would foreseeably be a republic given the lack of pre-existing land title or peerage that may support either a constitutional monarchy or a similar system. Notwithstanding this, there also remains the risk that the colony may also default to an autocratic process. However, were this to occur, it is reasonably foreseeable that international bodies may compel the city-state to hold transparent elections where an autocratic administration does not provide effective governance to support either the safety, independence or welfare of the citizens through either a one-party system or a junta.

3.2.3.   Law Enforcement & Criminal Activity

Analogous population groups on Earth that could aid in predictive modelling of offences per 100,000 people for the Martian city-state are non-existent. Research stations in Antarctica such as Australia's Mawson station or the USA's McMurdo station possess a hostile environment similar to that of Mars in terms of temperature, however, remain relatively habitable. Further to this, these remote stations possess populations of less than 1,000 people, all of whom are either Government, Military or University personnel alongside pre-screened contractors with specialist qualifications.

The inability of predictive modelling of offences per 100,000 people from an analogous Earth-based population group therefore mitigates any effective intelligence-led policing strategies to mitigate criminal activities on Mars. Notwithstanding this, it is reasonably foreseeable that given the harsh environmental factors and unreliable supply-chains to Earth would develop a strong community culture of inter-dependence and collaboration throughout the settlement, as civil disobedience may have disproportionately severe results from unethical behaviour.

3.2.4.   Regulating Crimes of Unjust Enrichment

To that end, it remains plausible that offence rates would remain relatively low comparable to Earth-based populations of a similar size as community involvement in collaborative survival would mitigate predatory practices ordinarily associated with either public law such as criminal law, or private law such as torts. Moreover, given strong community association, it is also reasonably foreseeable that offences commonly associated with unjust enrichment such as: larceny; theft; fraud; or, misleading and deceptive conduct would be strongly resisted through deterrence by community standards.

In addition to this, organised criminal activity would foreseeably be difficult to establish, maintain or expand within the settlement given strong community ties due to the hostile environment of Mars and a sense of collaborative identity. If organised criminal groups were to operate, it is reasonably foreseeable that community responses would act as mediating factors that may frustrate criminal activities, through a variety of responses including a prodigious number of informants to law enforcement, availability of key witnesses who would cooperate with law enforcement, as well as individual responses to organised criminal activity such as citizen arrest for summary offences and civic resistance by those who are exploited, such as through protection racketeering. It is also unlikely that organised criminal activity would obtain the cultural and sub-cultural support ordinarily associated with Earth-based criminal organisations through artistic works that lionise organised criminal membership such as through fashion, music, film or books.

3.2.5.   Regulating crimes of passion

Notwithstanding the above factors to act as effective deterrence for crimes of unjust enrichment on either the individual or organisational level, it is also reasonably foreseeable that the harsh environment of Mars would have the consequence of exposing pre-existing behavioural problems in persons as well as increasing the level of stress, anxiety or mental fatigue in the general population. In such circumstances, persons acting beyond the scope of community standards, either due to a behavioural problem or an environmental factor, may engage in offences ordinarily associated with crimes of passion. These crimes span a variety of indictable and summary offences as well as private suit, and include: murder; voluntary or involuntary manslaughter; grievous bodily harm; actual bodily harm; indecent assault; sexual assault; affray; assault; domestic violence; elder or child abuse; or, the tort of battery.

Intelligence-led policing may assist with the mitigation of these offences and apprehension of offenders who commit these offences through community consultation, community outreach, welfare checks and having access to social services to provide support for distressed communities and individuals who are at-risk of committing an offence, or being a victim of an offence. It is also reasonably foreseeable that incorporating mental health, psychological and sociological education into law enforcement training may also assist in identifying at-risk offenders and directing would-be offenders to government services for further assistance.

3.2.6. Criminal & Civil Justice System

Given distance and latency issues between Earth and Mars, holding a video or virtual courtroom for an alleged offender for an Earth-based court would be unviable. Notwithstanding this, it may be reasonable to have an Earth-based appellate high court hear cases in the public interest that have exhausted the local and district court systems on Mars.

Furthermore, the establishment of a Mars-based professional law society to ensure minimum standards and regulate billable hours may also be advantageous in ensuring access to justice, as well as encouraging parties to seek alternative dispute resolution strategies such as mediation and tribunal hearings.

4. SOCIETY AND CULTURE

The fundamental bases of the society will be established at the time of the first Martian settlement, but it will change with the growing population and adapt to events as they occur. English as the current primary language of international trade is likely to become the default language of Mars, but if initial settlements are made by a single nation or a common-language coalition this may not be the case.

Demographics in a Martian city will feature an even gender split, but likely be skewed in age compared to a terrestrial city. Once births are allowed, they will quickly become commonplace, but the primary driver of population growth will be immigration. As positions will be limited, a strong selection bias to those skilled in required areas will be applied, this combined with the increased radiation risk to children will result in most of the immigrants being at least in their mid-twenties, having completed university or trade qualifications. Similarly, most people are likely to return for the comforts of Earth in their retirement. As such Mars will likely feature a large proportion of highly educated people in their economically productive years.

It would be impossible for the inherent danger of the environment to not have impacts on the society as a whole. The entire city is an ecosystem that relies on proper functioning for survival. Failure of any part can result in deadly consequences ranging from individuals to the entire city, on timescales from seconds (explosive depressurisation), to months long starvation due to crop failure. Constant cooperation from all settlers will be required, and a strong sense that the community comes before the individual will likely be organically fostered, as the city must survive for its citizens to live.

To prevent the formation of an underclass of people only performing menial or monotonous tasks, duties will be distributed throughout the entire population, likely on a rotating roster not dissimilar to Jury Duty. These community service tasks will be undertaken in groups comprised of people from all levels and lines of work, to maximise community engagement.

Community building activities will be encouraged, such as sunrise Tai Chi in the park; while others will be taken from Earth customs and maintain connection to the terrestrial calendar, like Birthdays. New festivals and cultural events tied to the Martian calendar may also be established, such as major historical events like Settlement Day, Mariner 4 Day, Mapc 2 Day, Mapc 3 Day and Sojourner Day;

astronomical events like equinoxes, solstices and syzygies; and Martian cultural events like Earth Arrival Festival and Earth Launch Festival.

Much as an individual needs to put the community ahead of themselves to ensure survival, the city needs to make the physical and mental wellbeing of its citizens its highest priority. To that end, aspects of the city environment will be inherently detrimental to well-being and efforts will need to be made to counter these issues with good design. Working through Maslow's hierarchy of needs, it can be seen that the city must provide at all levels.

Physiological needs like air, water, food and shelter are supplied by the basic life support systems. Leisure, both socially and independently, will also be promoted through natural environments and social hangouts away from stressors as part of the design philosophy of the city.

Natural light is not strictly a need, but it is an important biological signal for circadian rhythms which will cause stress if disrupted. The benefits of sunlight need to be tempered with the increased radiation hazard on Mars. For workers who would be in artificial light all day, taking a break during the day to get some sun may be advisable and this should be factored into working schedules. Similarly work that would otherwise be high exposure may need to be done around twilight or at night.

Safety needs like health, employment and security can be supported by system and physical design. Healthcare will be provided by the city, as the city's incentives are aligned with the individuals, that is maximal health for a minimal cost with a strong focus on prevention rather than treatment. Health care will not just be limited to physical either, regular mental health check-ins with a therapist will help identify and resolve issues. Due to the inherently limited immigration and fixed pool of workers between launch windows, employment will be secure. Employment security coupled with the constant growth will mean that workers on projects that are cancelled will be immediately cross trained into new roles. Security should not be much at risk due to the focus on wellbeing for all settlers, but design can improve perceptions of security, including by ensuring sufficient personal space in both home and work environments, allowing personalisation of spaces, providing a smooth privacy gradient from public to private spaces and sightlines that allow for natural surveillance [19].

Delays in communication range from 4 to 24 minutes, which prevents synchronous (live) communications, so all communications will be pre-recorded. This will adversely affect the sense of connection to Earth. It will be important to engage new settlers into the Martian community and help them feel at home. Work and community service teams will be a part of establishing social networks but becoming integrated into established groups will also be key. The city will have many groups to facilitate this engagement, ranging from sporting teams and craft groups to literary and philosophy clubs. Escape spaces and communal eateries will also allow the chance encounters that friendships or romantic relationships develop from.

Esteem needs like a sense of accomplishment and freedom will be challenging. Some freedoms will be inherently restricted by the nature of the city, such as the inability to return to Earth or work schedules that require constant manning. Where possible flexibility will be given to the settlers to ensure they have a sense of control.

Accomplishment will be readily available, as there will be an unending list of challenges to overcome, to grow and develop the city, as well as support new ventures in the asteroid belt.

Self-actualisation will be actively supported and is expected to come in a myriad of forms but will come with new challenges. Access to education or instruction from Earth will be available, but due to the communications time lag, feedback will be delayed, requiring a greater degree of self-awareness and reliance. Creative pursuits, such as art and music will also be encouraged, and unique Martian styles will develop due to material scarcities (paints, pigments) or environmental factors (lower pressure affects acoustics).

Engaging the populace to customise and decorate the habitats will improve their sense of belonging, as well as provide much visual aesthetics. Over time, district styles will develop in different areas, which will aid with way-finding through the city.

5.   AESTHETICS

Significant research has been conducted on Earth into the review of environmental psychology of built environments for human wellbeing in constrained environments (e.g. Antarctic research bases and the International Space Station) (19). In addition to town and community planning, the city requires sufficient design to be considered aesthetically pleasing and to foster positive mental health. As on Earth, the design of each site within the city will still require the skills of specialists such as architects and psychologists to continually improve the look and feel of the overall city. Critical aspects of human wellbeing affected by the built environment include a feeling of choice, private and personalised spaces, a sense of place, and self-expression (19).

Compared to a life on Earth, the Martian citizens may not have as much freedom of choice with regards to their immediate environment. This feeling of choice will instead need to be provided by encouraging personalisation of internal and external layouts and appearances of their residences and businesses. While structural components will be modular, a sufficient selection will exist such that a huge variety of overall forms and functions are available. Materials common to Earth, such as wood, may not be available, but the versatile options of steel, glass, aluminium, bricks and tiles will be plentiful. The colours of which will be plentiful through addition of trace elements (i.e. stained glass) or glazing (mosaics). These materials will lend themselves well to a seamless blend of brick expressionism and contemporary architecture, with the lower gravity of Mars permitting more radical forms than those achievable on Earth. Alternative use of material, colour and even street art will be encouraged in order to create ever changing, vibrant spaces that pleasantly distract from the sterile Martian surface and the clinical or industrial objectives of the city.

It will be important to ensure that citizens do not feel crowded due to interactions with other inhabitants or due to the enclosed nature of the city. The 'structure within a structure' design of residential habitats creates spacious volumes that allow individuals to subconsciously 'scan' their neighbourhoods from their apartments, providing a sense of security and awareness, while improving community protection. Spaces dedicated to commercial or industrial activities will generally be located deeper beneath the city and hence will embrace the subterranean aesthetic, while incorporating comforting space, lighting and functionality.

Areas subject to high volumes of human traffic will be large and streamlined, with atriums and small gardens that distract from or break up the sound and sight of the crowds. Parkland, spiritual and peaceful places, located far from thoroughfares, will combine water, flora, insulated materials and shielded layouts to diffuse sound and form tranquil pockets for relaxation and restoration. The habitats dedicated to parkland could even feature mega-flora and water features that take advantage of the lower gravity, to create truly unique artificial landscapes that surpass the best of those seen on Earth (e.g. Singapore's Changi Airport).

Figure 15: Singapore's Changi Airport (Left) Cold War Bunker Data Centre (Right) [20]

The upper habitat levels, directly under the domes, will house these varied and extensive natural spaces under a view of Earth and the stars above. The natural daylight will cascade through the domes, onto the streetscapes of the habitats below via special purpose light wells that filter the radiation and close during the harshest radiation events. Finally, viewing outlooks at the summits of the habitat domes or specifically constructed observation towers will provide an unprecedented vista of the achievement of humankind, the view of the city itself. The lush tapestry of vegetation protected beneath an array of thousands of glass habitat domes and farm modules, sprawling across the ceaseless red plains of Mars will be a picture of humankind's intelligence and cooperation that inspires the dreams of many generations to come.

6.   SUMMARY

As has been shown, the problem of populating Mars is one that can be overcome, both from an engineering standpoint and an economic one. The design presented above is a scalable solution, using technologies that can be pioneered with small 1,000 person settlements then scaled up, as necessary. Given the costs of labour and the importation of construction equipment, it is estimated that the construction of the city will cost $260 billion dollars. This will enable immigration to Mars for roughly the same price as buying a house on Earth. As a result, wages will need to be high, however, a 1,000,000 person city-state implies a robust inner solar system economy, and as such Mars's unique position relative to the Asteroid Belt will allow an export economy to develop, giving the Red Planet a key position in a solar system wide, or solarised economy. Supplemented by several other revenue streams, the city will be able to achieve a GDP of $137 billion. This will allow for a robust safety net for its citizens, while economic development zones will incentivise businesses to invest. By selling utilities and renting space to these companies, taxes can be kept low, and a robust, rigorous democratic process will ensure the effective use of these resources, creating a unique polity in the solar system, and a second planet for human to thrive on. Politically, socially and culturally, the city will by necessity be a fusion of fierce collectivism and fierce individualism, with the requirements for living on Mars necessitating the city provide for its citizens, but the need for a robust economy requiring a free-market. These two philosophies will be balanced by a robust democracy with programs designed to encourage a sense of community, rather than enforce it. Though life will be tough, the high wages and adventurous nature of living and working on Mars would no doubt attract millions of people, truly opening up a new branch of Homo sapiens, and more importantly, the entire inner solar system for humanity.

7.   REFERENCES

1. **Mahoney, Robert, et al.** The Engineering Requirements for A Large Martian Population and Their Implications. [ed.] Frank Crossman. *Mars Colonies: Plans for Settling the Red Planet.* Lakewood, Colorado : Polaris Books, 2019.

2. **Nasa.** Mars Fact Sheet. [Online] 2018. [Cited: 19 03 2019.] https://en.wikipedia.org/wiki/Atmosphere_of_Mars#Observations_and_measurement_from _Earth.

3. *Mineralogic and Compositional Properties of Martian Soil and Dust: Results from the Mars Pathfinder.* **Bell, J. F., et al.** E1, 2000, Journal of Geophysical Research, Vol. 105, pp. 1721-1755.

4. **Mueller, Robert P., et al.** *Regolith Advanced Surface System Operations Robot (RASSOR).* s.l. : NASA, 2013.

5. *Ancient hydrothermal seafloor deposits in Eridania basin on Mars.* **Michalski, J., et al.** 15978, 2017, Nature Communications, Vol. 8.

6. **The University of Arizona.** Gamma Ray Spectrometer. [Online] 2008. [Cited: 15 05 2020.] https://grs.lpl.arizona.edu/latestresults.jsp?lrid=32.

7. **CIA.** The World Factbook. [Online] 2016. [Cited: 10 06 2020.] https://www.cia.gov/library/publications/the-world-factbook/rankorder/2233rank.html.

8. *Ultra high temperature latent heat energy storage and thermophotovoltaic energy conversion.* **Datas, Alejandro, et al.** 2016, Energy, Vol. 107, pp. 542-549.

9. *Pressure Effects on Oxygen Concentration Flammability Thresholds on Polymeric Materials for Aerospace Applications.* **Hirsch, David, Williams, Jim and Beeson, Harlod.** 1, 2008, Journal of Testing and Evaluation, Vol. 36, pp. 69-72.

10. **Sandlin, Trevor.** Emergency Mars Evacuation Program (EMEP) Packet. Hawaii : Southern Cross Innovations, 2020.

11. **SpaceX.** Starship. [Online] 2019. [Cited: 30 06 2020.] https://www.spacex.com/vehicles/starship/.

12. **CRRC.** The world's first ART demonstration line runs today. [Online] 2018. [Cited: 2 June 2020.] https://www.crrcgc.cc/en/g7389/s13996/t292853.aspx.

13. **Appelbaum, Joseph and Flood, Dennis J.** *Solar Radiation on Mars.* Lewis Research Center : NASA, 1989.

14. *Radar evidence of subglacial liquid water on Mars.* **R. Orosei, S. E. Lauro, E. Pettinelli, A. Cicchetti, M. Coradini, B. Cosciotti, F. Di Paolo,E. Flamini, E. Mattei, M. Pajola, F. Soldovieri.** 6401, 2018, Science, Vol. 361, pp. 490-493.

15. *A deep groundwater origin for recurring slope lineae on Mars.* **Abotalib, A.Z., Hegg E.** 2019, Nature Geoscience, Vol. 12, pp. 235-241.

16. **Chung, Winchell.** Atomic Rocket - Mission Table. [Online] 2020. [Cited: 12 06 2020] http://www.projectrho.com/public_html/rocket/appmissiontable.php.

17. **Fisher, Daniel.** The Real Patent Crisis is Stifling Innovation. [Online] 2014. [Cited: 0 06 2020.] https://www.forbes.com/sites/danielfisher/2014/06/18/13633/.

18. **Cambridge Isotope Laboratories, Inc. .** DEUTERIUM OXIDE. [Online] 2020. [Cited: 29 06 2020.] https://shop.isotope.com/productdetails.aspx?itemno=DLM-4-99-PK

19. **Donoghue, Matthew.** *Urban Design Guidelines for Human Wellbeing in Martian Settlements.* s.l. : University of Washington, 2016.

20. **Wikimedia.** Wikipedia. [Online] 2020. [Cited: 30 06 2020.] https://en.wikipedia.org/wiki/Singapore_Changi_Airport.

19: KOROLEV CRATER SPECIAL ADMINISTRATIVE REGION

Alex Sharp
CTO, OrionVM
alex@asharp.id.au

Saleem Ameen
University of Tasmania
saleem.ameen@utas.edu.au

John Walker
Massachusetts Institute of Technology
jhwalker@mit.edu

ABSTRACT

The Korolev Crater Special Administrative Region (KCSAR) is a special administrative region of the United States of America established in the 2060s. It is located within the Korolev Crater, an ancient, ice-covered impact crater, which houses one million residents in dome-like structures that are constructed into the ice. The city is largely self-sufficient: it is powered by an integral design nuclear reactor; it generates water using a novel TSSE wastewater treatment approach; and it applies a range of farming methods to meet its basic needs. It operates as an essential autarky, producing the vast majority of its economic resources, such as bulk chemicals and raw materials, in situ. To obtain the few resources it cannot produce itself, it exports a small range of goods and services in which it has a comparative advantage. Its political system operates as a delegative democracy and the government takes an interventionist stance in selecting for KCSAR colony settlers; in encouraging fertility; in providing incentives for important behaviours, such as a Universal Basic Income; and in directly regulating emerging natural monopolies. Perhaps most importantly, it takes a role in shaping the cultural development of KCSAR by investing in local initiatives and implementing an architectural style that celebrates human achievement. The future of KCSAR is bright. In the coming years, a substantial geoscaping operation will be undertaken to melt the frozen crater and create an environment that can support a new ecosystem, lush with life, and reminiscent of life on Earth.

1. INTRODUCTION

Since the very first images of Mars, taken by Mariner 4 in the mid 1960s, became public, the red planet has made its way into the public consciousness, and has ignited a yearning for the human race to become a space-faring, interplanetary civilization. A planet, so very like our own, lies within our sights, but just beyond our reach. In just the past few decades, significant advances in engineering and more nuanced understandings of social structures have provided a platform to examine the ways that we can extend the upper-limit of humanity's existence, particularly in the face of a number of looming existential threats. The result has inspired a revolutionary idea - that of a human colony on Mars. This idea, now, more than ever before, seems possible within our lifetimes. As technology continues to evolve at an alarming rate, addressing many of the challenges that have previously limited our ability to move deeper into space; the need to articulate an implementation strategy that presents a unified, tangible, and sustainable vision for what a Mars colony could look like, across the myriad paradigms of society, is necessary.

In this paper we seek to concretize the dream for a Mars colony into a simple outline of the central processes and systems that we believe will enable a substantial colony of one million people to not only survive, but flourish and thrive on Mars. To provide a substantive analysis, we limit ourselves to dealing mostly with the processes that we think are most unique to the Martian environment that cannot simply be transplanted directly from modern societies on Earth, as well as those that are unique to the location of the initial settlement that we have selected, which is based on the Korolev Crater.

The Korolev Crater is an ancient ice-covered impact crater at 73°N, 163°E on Mars. The Crater was selected because it contains around 2000 km3 of water ice, which provides an abundant supply of water and serves as a simple and cheaply implemented heat sink and radiation shield. The state itself, known as the Korolev Crater Special Administrative Region ('KCSAR'), is established in the 2060s, as a special administrative region of the United States of America. KCSAR is designed as a subglacial city, housed within interconnected domes constructed into the ice, as seen in figure 1, and small domes that sit atop it that offer a means of egress to the Martian surface. It extends in a crescent shape from the northernmost point of the crater along its inner rim.

The structure of this paper has been divided into three key sections. The first explores the technical challenges associated with supporting life on Mars, and proposes a number of novel solutions to supplying many of the essential components of life, such as generating power, creating a suitable atmosphere, accessing potable water, producing food, manufacturing bulk materials, mining raw materials, building consumables, and supplying internet. The second explores social engineering, presenting our vision for the social, economic, political, and cultural makeup of KCSAR - the key drivers of any modern society. Finally, we end this paper with a glimpse into the future of KCSAR - a self-sufficient, innovative, and vibrant society. While life on Mars is initially difficult, we envision KCSAR to evolve beyond its humble origins and become a society that is the envy of Earth's residents.

2. TECHNICAL DESIGN

Due to the presence of high transport costs from Earth and the inherent risks of delays in the transport of critical imports; for human life to survive and ultimately flourish on KCSAR, the colony must be highly self-sufficient and operate as a close to autarkic entity. To achieve this, cost-efficient manufacturing processes of essential bulk materials are critical. In this section, we outline the engineering that underpins these processes, first dealing with necessities for human life, namely, power, air, water, and food; then addressing economic necessities, by focusing on bulk materials, mining, consumables and the internet. Within each of these sectoral subsections, we introduce the structure and scale of resource dependence, before analysing the key production pathways (chemical, engineering, manufacturing) needed to achieve essential self sufficiency. We deal finally with the endogenous formation of a market for Martian products, by using a representative example from chemical manufacturing.

Figure 1: Artist's rendition of the dome under the ice.

2.1. Absolute Necessities Of Life

Given the resource scarcity, the systems for producing a number of the essential necessities proposed here are designed to operate synergistically with one another. This maximises productivity while enabling a futures market to arise from the multiple production processes with distinct production curves. The resulting overall system, therefore, is robust, efficient, and responsive to any market turbulence.

2.1.1. Power and Thermal Design

We begin with power consumption and generation, as all subsequent processes depend upon it. KCSAR requires on average ≈1GWe (with an upper limit of ≈3GWe) to sustain life for one million people. KCSAR's complete power generation model can be described as an interconnected system that encompasses thermal and electric power generated via two different processes: (1) a nuclear base-load plant that produces 12GWth power output from an integral fast-spectrum two-fluid molten-salt reactor operating at 750°C, of which, 4GWe power is obtained from a series of in situ mass-producible low pressure turbines; and (2) a cryogenic peaker plant that produces 4GWe power output derived from a cryogenic energy storage system ('cryobattery') that stores energy in the form of LN_2, with capacity of 96GWh, providing the energy requirements to meet peak demand. As seen in figure 2, this two-way system maximises the productivity of the power generated, by interacting with other energy producing processes. For example, during periods of peak demand, the risk to grid instability arising from network overload is ameliorated through the usage of a cryobattery and during periods of minimal demand, the excess thermal output can be redistributed for other uses, such as powering the chemical market to increase the production of bulk goods, or recharging the cryobattery. This system relies crucially on the presence of a sufficient heat sink, due to the heat density of the molten salt reactor design, and the particularities of the Martian context. The absence of large bodies of water and a thick atmosphere make thermal rejection particularly problematic. Opportunistically, however, KCSAR is located on water ice, which provides the necessary heat sink needed for the long-term viability of the system.

The reactor itself is an integral design made up of standard high temperature materials, and houses a single crucible that contains the entire assembly. It uses Hastalloy N disposable fuel pins, which are filled with a fuel salt mixture that comprises ^{239}Pu and ^{238}U. For this system, the pins are the only superalloys that are required to be produced and shipped from Earth. Since it is a fast spectrum reactor, there is no requirement to isotropically separate lithium for the FLiBe mixed fuel salt to maintain acceptable neutronics and it can be made considerably smaller and somewhat simpler due to the lack of a moderator. It does, however, still require active moderation via control rods. By using this dual two fluid-contained pin design, we can separate the fuel to simplify the reactor maintenance and construction, and therefore avoid the complications arising from (1) the potential condensation of fission products, and (2) the production of Xenon and Krypton. Additionally, as a liquid salt, it can be burned more completely as it lacks the intercalation limits that plagues more traditional solid MOX (mixed oxide) fuels. The chloride based coolant salt is then wrapped in a Thorium salt 'blanket', allowing it to be bred for ^{233}U via ^{233}Pa for use in nuclear thermal vehicles, along with Kilopower style stationary reactors for use in mining colonies and the like. Additionally, ^{242m1}Am, produced as a fission product, may be used to fuel smaller reactor designs [1]. By operating under low pressure Argon on Mars, the usual corrosion problems associated with water or oxygen impurities in the fuel salt are eliminated. Overall, the reactor is designed to run at a 100% capacity factor, with redundant heat exchangers and turbines, and at $\approx$100% load factor between the electrical and thermal energy demands on the system, matching demand via a smart grid system [2].

The thermal energy generated by the reactor is converted to electricity via low pressure turbines. The benefit of using these turbines over standard high-pressure turbines is that they do not require precision manufacturing or advanced metallurgy, allowing them to be mass-produced in situ, on Mars. Inputs into the thermal distribution system are a combination of the rejected heat of the reactor fluids themselves, absorption chillers, and waste heat from chemical industrial plants, which are then used to generate various output loops at chemically useful temperatures, as well as a final rejection loop for any heat that is too low grade for industry, which is used to melt ice and for district heating. The combined heat and power system (driven largely by the reactor's excess thermal energy), produces 'plant heat 'that can be used within various chemical plants to develop other essential goods such as food (refer to section 2.1.4) and fuel (refer to section 2.2.1), while also providing the necessary heat needed for natural ecosystems to thrive (e.g. by facilitating the growth of cold water fish and algae). Furthermore, waste heat from the reactor is used to form large inlets of liquid water by heating the ice (an essential resource for life, that is discussed at great lengths in section 2.1.3) and to address the challenge of cost-efficient chilling on Mars. By using a series of absorption chillers, such as a LiBr absorption chiller that produces water at 2°C, or an ammonia absorption chiller that produces a chilled methanol loop at -30°C, the waste heat has useful properties that allow for the chilling of objects below the temperature of the inlet.

Finally, any excess energy produced is stored within a cryogenic battery [3] made up of LN_2, that is used as needed to meet peak energy demands. LN_2 is heated by industrial waste heat to (1) form high pressure N_2 that turns the locally produced low pressure turbines, and (2) provides the means to both operate the LN_2 turbopump and pressurize the LN_2 tanks. To ensure that the cryogenic tanks remain chilled, a novel approach to radiative cooling is used. A solar absorber that is made up of an undoped Ge film, below an epitaxial GeO_2 scratch resistant layer, is used to create a shading mechanism for the black body radiative cooler through a ZnSe window. An anodized black-body radiator in a dewar then acts as a heat exchanger for He, which enters through the bottom and is cooled to create the cold loop that chills the cryogenic tanks. The chilled He can then be used to either liquefy boiled off gasses, or cool liquid gas as required. Given that there is a thin atmosphere on

Mars, it is relatively simple to build larger vacuum chambers to support this system as needed [4]. This thinner atmosphere also radiates less IR at night as a black body, aiding in system efficiency. Some of the key benefits of using cryogenic batteries are: (a) they provide a sacrificial LN_2 blanket that reduces the loss associated with the relatively valuable LO_2 and CH_4; (b) economically, it creates elasticity in the supply of electrical energy, by ensuring demand for electricity from the reactor and offering the capacity to redistribute it as needed; and (c) since there is a market for both reactor heat and power that can be interconverted, the fungibility of these resources softens the demand curve for each, ensuring a stable market equilibrium.

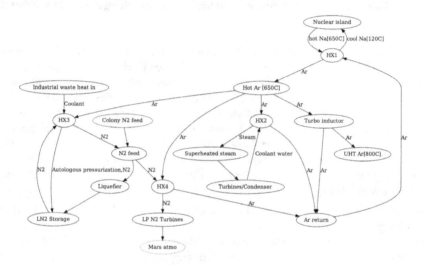

Figure 2: Reactor thermal distribution diagram. Colours: blue are fluid inputs, green are fluid outputs, acronyms: HX=heat exchanger, UHT=ultra high temperature, LP=low pressure.

While one may assume that the bitter cold of the Martian pole would be problematic for a human settlement, it is important to note that most of the colony exists within domes below the glacier that are designed with striking similarities to Igloos. This enables them to take advantage of (1) the insulating properties of ice and (2) the large thermal mass of the glacier, in order to maintain a reasonable internal temperature. Furthermore, at the surface, even though the temperature itself may be below -100°C, the low pressure environment reduces convection losses, while the CO_2 atmosphere reduces conduction losses (compared to a N_2-based environment). As a frame of reference, this means that the engineering problem of heat conservation is actually comparable to the equivalent of much warmer temperatures on Earth [5].

2.1.2. Air and Atmosphere

Next, of primary concern is the supply of the chemicals that form the basis of survival: air and water. In this section, we explore the atmospheric conditions required for life on Mars, dealing first with oxygen, the foundation of human life. By far, the easiest method to produce oxygen on Mars is via photosynthesis. However, as the supply of oxygen from photosynthesis is inelastic (as farms take time to build) and unreliable, the quantity required must be forecasted ex-ante and more than optimal quantities supplied. To mitigate these issues, cryogenic oxygen that is kept cold via the cryobattery's LN_2 that we introduced in section 2.1.1, offers a short term solution to stabilize the supply. In the medium term, or where farming is impractical, a Zn/S/I thermochemical cycle [6] is proposed to produce oxygen from water at a reasonable cost, using a turboinductor heat pump to efficiently provide optimal operating temperature [7, 8]. The reactor would use an electrochemical Bunsen stage [9], with a modular decomposer akin to [10], to save on complexity and cost. Similar designs would also be used to produce O2 from CO_2 by omitting the HI decomposition stage, making them suitable for other tasks, such as life support on mining outposts, powered by kilopower style power plants. A further benefit to Zn/S/I systems is that they both produce the syngas needed for the direct reduction of iron, and consume the water/CO_2 generated from the reduction. The interrelationship of these processes used to generate atmospheric conditions are displayed in 3.

Given that KCSAR's atmosphere is primarily made up of low pressure Oxygen and Argon (Argox), the photosynthetic farms can capitalise on a CO_2 enriched atmosphere, which can produce an enriched Argox fraction when processed, by recycling the CO_2 back into the farms. This Argox fraction underpins the city's atmosphere, supplemented as needed using Air Handling Units (AHU) which maintain total pressure, along with things like temperature, the partial pressures of O_2/CO_2, and humidity. To ensure that excess atmosphere does not pose a serious threat during events such as a fire, a reject line is used; and any water produced via dehumidification is rejected via a drain line. Another potentially lethal threat on Mars is the issue of dust, which can lead to health problems such as chronic silicosis. To address this issue, cyclonic filtration with a second electrostatic stage, is employed as a scalable way to remove small particles. Provisions are also made for potential atmospheric failures. Respirators that enclose the ears and eyes are maintained at regular intervals to ensure that citizens have enough time to return to safety if they are unable to access a full pressure suit. Atmospheric monitors and alarms are built into each AHU that operates if pressure falls to a dangerous level, or if the partial pressure of oxygen or CO_2 moves outside of safe bounds.

The potent odorant Ethanethiol is added to the emergency habitat pressurisation lines, to provide a warning for underpressure and hypoxic situations. While above the ice, depressurization due to pinhole leaks is a major concern, under the ice, the primary hazard is hypoxia or toxic gas buildup due to fire or a chemical leak that rapidly replaces the atmosphere with high pressure, unbreathable gas. An overpressure system within the AHU's allows for them to rapidly vent out the excess gas, preventing the problems associated with overpressure.

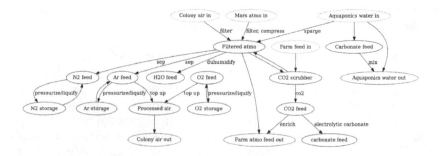

Figure 3: Flow of atmosphere between settlement, farms and industry. Colors: blue are inputs, green are outputs.

2.1.3. Water

Extending on our treatment of air, access to liquid water is equally critical, and, once again, the value proposition of settling on the Korolev crater presents itself. Since KCSAR is built into water ice, it is surrounded by an accessible water source if heat energy is available. To heat the ice, KCSAR capitalises on the heat exchangers from the reactor proposed in section 2.1.1. The high temperature output from the reactor produces waste heat that leads to the formation of local ponds, which, over time, transforms into an inlet of accessible liquid water, particularly as the heat dissipates through the surrounding ice walls. While the lake water is not initially potable, a novel two-stage temperature swing solvent extraction (TSSE) process [11], provides an effective solution for the bulk recycling of water. In the first stage, the solvent Diisopropylamine (DIPA) is heated to absorb water from the raw water inlet until it is saturated. It is then cooled so that the DIPA rejects excess water and forms an aqueous layer that can be removed [12]. In the second stage, Diethyl ether (Eth) is used to extract the remaining DIPA from the extracted water. Given that Eth is mostly immiscible with water, and has a low boiling point and high vapour pressure, it is easily extracted. Finally, the water is treated to maximise its safety using: (1) activated charcoal filtering, (2) NaOCl is added as a disinfectant, and (3) a buffered mixture of common ions to ensure that (a) the pH of the water remains slightly alkaline, (b) metal pipes passivate properly, and (c) the leeched metal ions are insoluble. Minerals are also added to improve nutritional outcomes, such as NaF for dental health.

TSSE is an attractive candidate process because (1) it only requires simple mechanical equipment, and (2) it runs on low grade waste heat, at standard pressure, and with mild reaction conditions. This is preferred to more standard reverse osmosis systems that rely on high precision semi-permeable membranes that are complex to manufacture, and that require energy intensive high pressure pumps. Moreover, this TSSE process can internally recycle the low-grade waste heat using a recuperator, which makes it highly energy efficient, and only requires chemicals and materials that are easy to produce in situ. This ensures that the system is easily scalable as new settlements are created [11]. Particularly as the population grows, recycled wastewater is therefore the most efficient primary source of water and only supplemented by the energy-costly process of converting large blocks of ice into water where necessary. A relatively simple recycling process is used for 'greywater', produced from sources such as showers on Mars. Greywater does not contain harmful biological organisms, and therefore, the water can be passed through a simple filter stage and fed back into the aquaponic loop that we describe in section 2.1.4. 'Blackwater', by contrast, does contain harmful organisms that need to be removed, but also contains nitrates and phosphates that make it a good fertiliser, which is eventually used in farms. Thus, by diluting blackwater with greywater, and using NaOCl and thermal processing to sterilise it, it is also capable of being utilised in the aquaponics loop. This overarching synergistic water treatment process is laid out in figure 4.

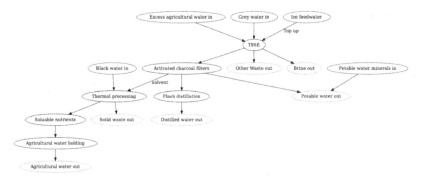

Figure 4: Settlement's artificial water cycle. Colors: blue are inputs, green are outputs.

2.1.4. Food

For an isolated community such as KCSAR, maintaining food security is vital. Producing a sufficient quantity of cost-efficient, nutrient-diverse foods, while maintaining efficient supply lines, is a prerequisite for the sustenance of life. In this section, we outline the artificial ecosystem that provides a varied diet for the citizens of KCSAR, building from the simplest (and most cost-efficient) food sources, to the most complex luxury goods. Given the complexity of maintaining food security, we have divided the treatment of this section into two sequential parts, answering the vital questions of: (1) What chemical processes are needed to enable the production of food? (2) How then, do we mass produce the food in a sustainable way?

The Chemical Pathways for Food Production

Edible industrial products are an extremely cheap food source, and underpin food security, even when there are disruptions in the supply of other nutrient sources. They serve a dual purpose, forming inputs for the biochemical industry discussed in section 2.2.1 once they are processed. For example, edible glucose is converted into plastic and protein into ammonia. These edible industrial products are primarily produced in ditch-style farms and grass farms. Ditch-style farming involves injecting water into an earthen dam and covering it in a layer of oleic acid to protect it from the low pressure / temperature of the surface, so that optimal growth conditions are maintained. Figure 5 shows the depth of oleic acid needed to enable growth. Below the red line, no economically useful growth is possible. Between the red and orange lines, many forms of algae can be grown in useful quantities. The two green lines show the range of optimal growing conditions for Spirulina, a useful cyanobacteria that has adapted to very warm water environments. Within this range, it grows rapidly, producing nutrient rich biomass that feeds both people and animals. The cost-efficiency of producing candidate algae is routinely reviewed to create an appropriate environment for rapid growth. This is achieved, for example, by rotating the crop, or adjusting the depth of the oleic acid. Since oleic acid is produced from biomass and is needed for biomass production, this forms a feedback loop that limits the growth of farming over time

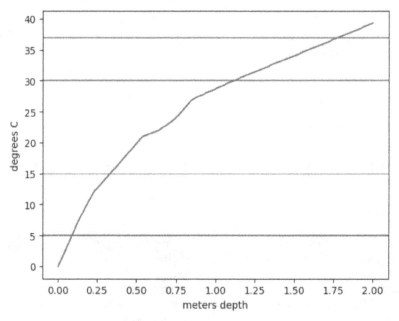

Figure 5: Depth of oleic acid vs. maximum water temperature

The primary obstacle to ditch-style farming is the method of gas perfusion. There is no viable natural path to enable CO_2 and N_2 from the Martian atmosphere to reach the aqueous layer for biosynthesis. Hence, perfusion stations are used, replacing consumed dissolved gasses such as CO_2 and possibly N_2, and then extracting the dissolved O_2.

Mushroom farming is used to catabolise otherwise indigestible biomass into nutrients for future farming. This removes insoluble products that cannot be ground into a colloid and fed into hydroponic systems. Mushroom farming requires O_2 and waste biomass that can be seeded with spores to form edible mushrooms.

Given that nitrogen is critical to facilitating plant growth across the farms, either atmospheric nitrogen, or the artificial urea derived from the Haber process described in section 2.2.1, will be used as a nitrogen source. While the N_2 found within the Martian atmosphere is cheaper to use and requires less processing, its primary drawback is that it results in slower growth. Artificial urea, on the other hand, is more effective, but subject to market fluctuations in its industrial price.

The Sustainable Mass Production of Food

Whilst building robotic greenhouses that can grow standard crops is a daunting technical challenge, automated farming of light grasses and particularly Miscanthus Giganteus ('elephant grass') have been reliably implemented. On KCSAR, this procedure is replicated to produce a cost-efficient source of industrial glucose. Small modular greenhouse sections are fused together and filled with low pressure CO_2-enhanced atmosphere. Quantum dots are employed as a layer on top of the greenhouse membrane to increase quantum efficiency by downconverting light that is unsuitable for the grass, such as UV light, into light that is most effective for photosynthesis [13]. The growth beds resemble traditional hydroponic growth beds and are made from processed, crushed regolith. These are robotically mowed, with the farmed grass cracked into glucose by cellulase enzymes. Using a similar approach, hydroponic gardens will be used to grow more traditional foods including vegetables, grains and spices. Due to the complexity of harvesting, hydroponic gardens require laborers and a shirt-sleeve environment, making their output substantially more expensive.

Aquaponic farming is also used to supplement the Martian diet with fish and other aquatic lifeforms. Aquaponic farms are contained within large bioplastic boxes with an Oleic acid cap, placed in segmented areas cut into the ice. The box acts as insulation, keeping the water warm and separating out the nutrient-rich (particularly phosphate-rich) water that accelerates the growth of cyanobacteria and algae, from the diluted water. As these organisms are at the bottom of an artificial food chain, the speed of their growth is highly correlated with the supply of fish of various kinds, that are a staple of the relatively wealthy.

While relatively expensive, meat is also sufficiently cheap to supplement the diets of most households with varying frequency. Battery poultry, rabbits and similar livestock are the primary source of meat as they may be grown relatively cost-efficiently, rapidly and in confined spaces, and they are farmed for eggs and meat. A substantial amount of farmed meat is processed, with cellulose 'padding 'added (such as in chicken nuggets) to make it more affordable for consumers.

Yeast, farmed in bioreactors, is used to grow bulk substitute animal proteins, such as albumin from eggs and caseins from milk, from industrial feedstock. This allows for the creation of cheaper substitute products, such as artificial egg whites, as opposed to far more expensive chicken eggs. These are also mixed with processed meats or fish to form more complete processed foods, or even added to semi-natural pathways to create, for example, real cheese out of artificial milk [14]. Similarly, using artificial gluten and eggs enables the colony to bake bread and make pasta, which would otherwise be infeasible, thereby allowing for a more varied diet.

In the highest price bracket, are the imported foods from Earth that are infeasible to replicate on Mars such as chocolate and coffee. Where possible, these foods are produced using a mixture of local and imported ingredients. For example, chocolate is produced from combining artificial oils and bulk glucose produced on Mars with cocoa that is imported from Earth. To improve food flavour in a relatively inexpensive way, some spices are engineered chemically (such as simple esters) and biologically (such as limonene, linalool, among others), which makes Martian food more palatable to those familiar with traditional foods consumed on Earth. Spices that cannot be produced through these methods are imported from Earth. The change in the relative costs of ingredients provides the impetus for culinary innovation and the growth of a distinctive Martian cuisine.

While these diverse farming practices meet most of citizens 'nutritional needs, some nutrient deficiencies arise because foods that are high in particular nutrients are difficult or expensive to produce. To address the possibility of widespread deficiencies and resulting health impacts, many foods are fortified with artificial sources of Vitamin C, Vitamin D and Iodine.

2.2. Economic Necessities Of Life

In the previous section, we outlined the basic processes needed to sustain human life on Mars. In this section, we analyse the processes necessary to sustain a stable, working economy that can maintain those basic processes and provide for human consumer 'wants 'as well as needs. As before, cost-efficiency and stability are the primary motivation for selecting particular production and maintenance processes.

2.2.1. Bulk Materials

The next most important chemical pathways are those that are economically essential for survival - the production of plastics and other bulk products. For simplicity, we have compiled a list of useful bulk products that are producible on Mars in table 1. A key consideration in the robust production of bulk materials is that there are multiple pathways with different production functions. As an example, producing Aniline in a bioreactor requires very little in the way of capital outlay and so is easily scalable, but has a higher marginal cost per ton than building a dedicated plant to aminate phenol. A market that includes both, then, along with a robust futures market in industrial products, is able to produce cheaply, while scaling rapidly, in order to robustly meet market demand.

Another key consideration is that while some biological processes are directly useful, such as the production of bioplastics, Penicillin, Vitamin B12 and the like, there are many that are not. In these cases, we can augment existing biological processes with industrial chemistry to produce useful products. For example acetaminophen can be produced from biosynthetic Quinone, Nylon from synthetic

Adipic acid [15], Rubber from ethanol [16] (via isobutene), Epoxy from phenol, THF from processed plant sugars [17], among many others. The symbiosis between biological and chemical systems is the bedrock of the KCSAR modern industrial economy.

Product	Production Process
Benzene and its derivatives	BXT [18], along with Phenol [19], Quinone [20], Aniline [21], among others [22] are produced from biomass, either directly through digestion, or via catalytic conversion of biological precursors.
Ammonia	This is produced through the combination of a catabolic degradation of protein [23] and a low pressure advanced haber reactor [24] utilising plant heat.
Ethene	This is produced via the dehydration of bioethanol, steam cracking of hydrocarbons, or directly from syngas through a Fischer Tropsch style catalytic process
Sulfuric / Phosphoric Acid	Generated from acid gases produced in the metal refining process observed in section 2.2.2.
Nitric Acid	This is produced by the oxidation of ammonia.
Methane	This is produced through the sabatier [25] process using reactor plant heat, along with bioalkane cracking [26, 27].
Hydrogen	This is produced from water via the Zn/S/I cycle, via steam reforming of methane [28], as well as via ammonia decomposition. Storage of hydrogen chemically then allows for the non-cryogenic storage of bulk hydrogen.
Biomass	Biomass is mass produced in farms and in bioreactors. Specifically using Methanotrophs (*Methylococcus capsulatus* et al) fed on methane/methanol.
Methanol / Alkanes	These are produced either chemically from Syngas via the Fischer Tropsch process or other catalytic processes and plant heat, or biologically from biomass.
Syngas (H_2, CO)	Syngas is produced via the Zn/S/I cycle (as seen in 2.1.2) as well as through the gasification of Biomass.
Bioplastics	PHB, PLA and other bioplastics are produced from biomass.
CO_2	CO_2 is scrubbed from both the Martian atmosphere as well as artificial atmospheres through amine or ionic liquid scrubbing [29, 30] as seen in figure 3.
NaOH, NaOCl, Cl_2	These are produced in a standard chloralkali plant.

Table 1: Production processes involved in a range of bulk materials.

2.2.2. Raw Materials and the Mining Process

There are a number of raw materials, such as iron and steel, that are required to build and sustain the economy of KCSAR, which can be acquired through mining. Mining bases on Mars need to be small and relatively self-sufficient, since transporting oxygen and water is not economically viable. As such, mining operations use a Zn/S/I oxygen generator, powered by a mass-produced kilopower-style [31] reactor. Water and CO_2 are extracted from the air using ionic liquids, which become inputs to the Zn/S/I plant and the water recycling plant. The water recycling plant takes in recycled waste water, extracted via the TSSE process detailed in section 2.1.3, and rejects concentrated brine as waste. This process is sufficiently efficient to allow for metabolic water production and imported foods to meet the water needs of the small base.

Figure 6: Artist's rendition of the automated mines.

The majority of the mining operation is performed by semi-autonomous robots that are mass-produced using interchangeable parts found on KCSAR, as illustrated in 6. There is some oversight required by mining technicians, whose role it is to (a) ensure that the robots complete the desired tasks, without engaging in dangerous or counterproductive behaviours, (b) issue override commands to redirect robots in the event that they are incapable of completing the task, and (c) provide maintenance and repair support for the robots. The primary form of transport for personnel and cargo between the main settlement and the outlying mining settlement are autonomous trucks. However, trains are also used within areas that exhibit elevated congestion. To support the mining colonies, cables and pipes traverse alongside the roads to nearby settlements, so that there is access to power and copper or optical fiber for telecommunications.

While there are several resources that can be mined to support the Martian economy, this paper specifically focuses on the production of iron and steel, due to the fact that they are necessary to sustain a variety of industries on Mars, such as construction and manufacturing. Since oxygen is expensive to produce on Mars, the standard oxygen-intensive iron production method is prohibitive from a cost perspective. Therefore, instead, we propose that a water-splitting Zn/S/I cycle be used to produce syngas, which is used to reduce iron oxides to sponge iron powder.

Direct reduction iron is implemented as described in [32] using syngas regenerated via a Zn/S/I cycle reactor. The iron ore will be ground into a fine powder and injected into a H_2-filled ceramic reaction vessel that is kept at 800°C. At that temperature, H_2 and CO reacts with iron oxides to form Fe, CO_2 and H_2O, while converting sulfides and phosphides back into their corresponding acid gasses. Once the output syngas stream is cooled and processed, the product gasses can be easily extracted. The acid gases can then be further processed to form sulfuric and phosphoric acid for industry and as critical components in biomass formation. The cold syngas stream is then reheated by a recuperator and reintroduced to the reaction chamber. While the syngas volume is reduced by the reaction, it will be replaced through syngas regenerated from the generated CO_2 and H_2O. The majority of the syngas is thus recycled, with small amounts of H_2O and CO_2 periodically added to replace losses.

The iron produced is a sponge iron powder that is not immediately suitable for construction, due to silica contamination. Therefore, the sponge iron powder is placed in a furnace with carbon and other metals to form bulk steel, which can then be cast into various steel products. An important example of such a steel product is heavy presses. Presses can take in large sets of sheet metal and hydraulically press them to fit molds, enabling the creation of very large, precisely manufactured, load-bearing parts, that are otherwise impractical to produce. The presses can then be used to construct things such as industrial machines, curved panels of domes, and support beams. These manufactured parts may be used in the construction sector or in the production of capital equipment, contributing to the productivity of other Martian industries. Similarly, the iron powder is also refined through iron carbonyl [33] to form pure iron powder, which is suitable for laser or electron beam 3d printing. Electron beam metal printing is relatively practical on Mars, as the lower atmospheric pressure makes the construction and maintenance of large pressure chambers simpler, allowing for the creation of larger parts.

2.2.3. Consumables and Factories

One of the pertinent challenges that has pervaded much of our discussion of KCSAR is its reliance on domestic manufacturing due to the high cost of imports. Given the availability of several bulk and raw materials, as reflected in sections 2.2.1 and 2.2.2, it is possible to develop and manufacture a range of specialised goods and consumables directly on Mars through the industrial sector - a key component of the Martian economy. To put this in context, importing complete objects such as computers is uneconomical, however, importing specific parts such as bare silicon wafers is relatively cheap, and they can then be cost-efficiently transformed into packaged silicon chips, which are crucial for many processes. To support the development of bespoke capital and consumer goods, capitalising on 3d printing and other additive manufacturing processes is important. However, it is worth noting that current 3d printing technology is prohibitively expensive and slow at scale, which forces KCSAR to rely primarily on more traditional manufacturing systems for bulk items, at least until the technology further develops. Electrical Discharge Machining is used to produce injection molds for the mass production of high precision plastic parts made from bioplastics, such as PHB. Since factory systems are only partially automated [34], a sizable labor force is required to produce goods for the local population, and as a result, factory workers make up a significant proportion of the population.

2.2.4. Internet

The internet-based economy will be a key sector, both as an enabler of the export economy and in meeting domestic demand for communications and access to a forum for creating and sharing cultural products. The provision of stable, affordable internet service is therefore crucial. Long range lasers are used for packet forwarding at the link layer, from an HEO outer shell of the Starlink constellation on earth and a similar constellation in HMO. Both constellations form Autonomous Systems, extending the route advertisements of Earth and Mars to each other. Using the OSI model, this is then the L3 of the Martian internet.

The key issue is that, while packets can be routed and sent, the time delay prevents the use of higher level protocols. As such, KCSAR's ISPs provide L4/L7 proxy services, similar to a traditional VPN service, that proxies connections on Earth via a proxy that understands the time delay. Existing CDN technology, such as NAPs are then used to allow large scale content providers (such as YouTube) to deploy content on Mars, with application-level changes that adjust for the increased spooling delays. An example of such a change is that movies or content created by artists that a user has 'subscribed 'to are uploaded automatically, making them accessible from a local data center. If an individual clicks on a link that is not cached, they will be brought to the holding page, while the page is spooled from Earth and added to the local cache. Once added, it can be accessed almost instantly. ISPs provide caching proxies for most users but for security reasons, many companies run their own, renting capacity from Earth.

2.3. A Market For Martian Products

In this section, we briefly outline the endogenous formation of a stable market for the products outlined in section 2 thus far. We take the bulk products market as an important representative example, with a focus on the market for oxygen, to show how projected market demand drives the production of an appropriate range of goods in the most cost-efficient ways possible. We propose that the success of KCSAR's markets derives from its focus on maintaining different production pathways for the same goods, thereby enabling the creation of a futures market that stabilises supply chains while optimising for the lowest cost production method at any particular time.

We consider a simplified model for the production of oxygen to show how the presence of two production pathways offers both stability and efficiency - photosynthesis in farms, and electrolysis. Even with robotic construction, the fixed start up costs associated with acquiring land and converting into a functioning farm are substantial. Once the farm is up and running, however, the per unit cost of oxygen is relatively cheap but it is slow to produce and there may be substantial variability in yield. By contrast, while a basic electrolyzer has a low cost per unit, the high cost of electricity makes the per unit cost of oxygen substantially higher than that produced through photosynthesis but it can be produced rapidly to meet consumer demand. As such, in this simplified example, the stable component, which forms the majority of oxygen demand, is produced via the farm method financed through a long-term business loan. The final volatile component of demand is met through the use of electrolyzers, which can ensure that consumer demand is met, albeit at a higher marginal price.

The mechanism that efficiently allocates which production method is used is underpinned by the bulk goods market. The bulk goods market is based on two types of contracts: immediate sales contracts, which form the spot market, and future sales contracts, which underpin the futures market. The spot market enables instantaneous trades for consumers with fluctuating needs or producers that need to rapidly vary their productive output but levies a premium associated with the higher cost of rapid production (from our earlier example, oxygen is produced through electrolysis to supply the spot market). The futures market, by contrast, enables suppliers and consumers to lock in long term production plans, minimising production costs but limiting their flexibility (by producing oxygen through farming). While individual consumers struggle to project their individual demand, limiting the effectiveness of the futures market, financial intermediaries, generally banks, forecast aggregate demand and supply of bulk materials and operate as representative consumers. If their financial models predict a shortfall in the supply of methane, for example, these intermediaries buy up futures contracts at an above market price and store the methane in space rented in the cryobattery. When the shortfall arrives, they then sell their methane stockpile, stabilising the market supply and helping to stabilise the long term price of commodities.

3. LIFE ON MARS: SOCIETY AND ECONOMY

Having outlined the technical processes needed to sustain human life on Mars in section 2, in this section we lay out the defining characteristics of KCSAR's social and economic makeup. Throughout, we integrate commentary on the ways in which KCSAR's political, cultural and aesthetic makeup contribute to its vibrant society.

3.1. Governance

The KCSAR political system embodies the most direct feasible form of democracy, a delegative democracy. It has been argued that one of the greatest benefits of a direct democracy, is that it acts as a constraint on the actions of political elites, forcing them to represent the interests of the populace at large and to adopt cooperative rather than confrontational strategies [35]. The delegative democracy of KCSAR is governed by four distinct branches, with government powers separated between the executive, legislative, judicial, and an auditor branch.

Procedural laws substantially limit the powers of each of the branches of government, enforcing oversight from each branch upon the others and requiring that for any substantial change in policy or exercise of non-standard powers, the issue is put to a ranked choice vote by all citizens. Additionally, where matters of judicial interpretation or issues raised by citizens receive a sufficient amount of public support through citizen-led petitions, they are put to a vote. Thereby, citizens directly determine government decisions.

These procedural arrangements are unpopular on Earth primarily because of the prohibitive time cost associated with voting on such a broad range of issues. In KCSAR, this cost is limited through a system of vote delegation by which citizens may delegate their voting right to other individuals or between individuals based on the subject area of the vote. For example, a citizen may delegate their military decisions to person X and economic decisions to person Y. These delegations may be changed at any time and are supported using cryptographic primitives such as homomorphic encryption and linkable ring signatures, allowing for both anonymous voting and cryptographic security [36].

To incentivize the emergence of delegates and enable them to publicize their voting positions, the government provides a stipend proportionate to the number of votes a delegate receives. Strict campaign financing laws prevent the use of external funds or exercising undue influence to promote a viewpoint. To ensure that decisions are data and logic-led and that a full range of opinions voting options are represented, 'index voters' - robot delegates that vote on specific issues which follow simple 'if then 'logical statements or rely on more advanced techniques from machine learning are employed. The data and logic used in their decision-making is open to the public and auditable.

3.2. The Malthusian Trap

The Malthusian Trap emerges where the average individual within an economy produces only as much as they require to live over their life cycle. Modern credit systems reframe the traditional problem by offering emerging economies the ability to finance the 'importation 'of a high total factor productivity by utilising capital-intensive productive technologies. The viability of the KCSAR economy thus turns on its ability to sustain a high enough economic output to exceed the combined costs of its existing debt burden (interest and loan repayments) and the costs of supporting its citizens. Hence, maintaining a high GDP per capita is the primary concern of

KCSAR's emerging economy and its social design prioritises the maximization of GDP per capita, while supporting a more equitable distribution of resources than that found on Earth. The relatively equitable distribution of resources is important in ensuring that all citizens are provided for and live a life without poverty, helping to enhance social stability. It also assists in the generation of higher domestic demand, as people living on lower incomes tend to consume a greater proportion of their incomes. The larger domestic demand enables higher productivity gains through economies of scale. [37]

3.3. Immigration

Having highly productive citizens is a key priority of KCSAR. High demand to be part of Earth's first extra-planetary colony will provide the opportunity to select for ideal candidates using an AI-based actuarial model of a citizen's lifetime production alongside an AI-based forecast of the skills needed by the colony. As re-skilling on Mars has a substantial time and monetary cost, having a diverse range of essential skills (detailed in 3.8) represented among the first settlers is a precondition for the success of the colony and avoiding critical skill shortages remains a key target of the education system. However, the cost of travelling to KCSAR (USD 500k approx.) makes it difficult for otherwise ideal candidates, who are young, well-educated and therefore have limited wealth, to immigrate. The government subsidizes or underwrites a portion of these individuals 'ticket prices, enabling them to access commercial loans by reducing the risk to banks of their default. Friends and families of family members are also encouraged to pool collateral to further reduce the risk associated with commercial loans, thereby making them more available and reducing individual interest rates. Thus, government interventions provide the opportunity for Earth's 'best and brightest' to join KCSAR.

3.4. Fertility

In the presence of extremely high immigration costs, for an isolated colony to maintain its population, the fertility rate must remain above the replacement rate (approx. 2.1 children per person capable of bearing children). In most developed nations, the fertility rate has been rapidly declining and the direct importation of the culture surrounding fertility in those nations to KCSAR would result in demographic collapse. The incentivization of child-rearing is therefore a first order concern.

In KCSAR, direct economic incentives, are offered to guardians proportional to the number of children for which they care and these have been shown to improve fertility outcomes [38, 39]. For example, a Universal Basic Income ('UBI') is offered for each minor for which a guardian cares. Additionally, non-cash incentives such as additional leave and guardian allowances are offered. These incentives are routinely adjusted to maintain a stable fertility rate and help contribute to a culture that celebrates caring for large families while continuing to productively contribute to KCSAR society.

As the cost of raising children is substantial in both direct costs (such as providing food and atmosphere for the child) and indirect costs (such as the loss of productive labor of parents) KCSAR is concerned with the 'quality 'of children - their capacity to contribute to GDP - as well as their quantity. 'Embryonic selection 'is used to select for the most fit potential children. Gametes are extracted from parents and fused to form blastocysts [40]. A cell is taken from each blastocyst and its genome is fully sequenced using a nanopore sequencer. The full genome sequence is then processed to generate a polymorphism map that is analysed using artificial intelligence to determine a 'fitness 'index score. To estimate this fitness function requires the search space be both explored and exploited simultaneously, posing a 'one-armed bandit 'problem. Therefore, we use a regret minimising strategy to estimate the value to both individuals and society of particular phenotypes.

The KCSAR also faces a phenotypic diversity bottleneck created by the small size of its population. Studies have shown that a lower limit on a genetic bottleneck is in the tens of thousands, with higher estimates for K selective species like humans that have few offspring but vast resources available for them. As the population must thrive while adapting to the hostile Martian environment, genetic diversity must be thought of as a resource with risk amelioration benefits even for a population as high as one million. Therefore, frozen gametes from screened and compensated individuals are imported to the KCSAR en masse. The transport costs are relatively low due to the light weight of the cargo.

3.5. Demographics

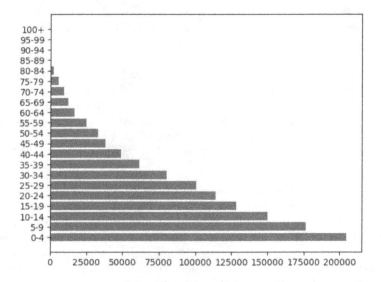

Figure 7: Implied demographic pyramid

To project KCSAR's population demographics, we scale demographics from Gambia, which has a similar fertility rate (3.5 children/woman) and population growth rate (40%) [41]. In figure 7 we approximate a demographic pyramid for KCSAR.

Young children and the elderly are not part of the labor force and are provided for by the government. As the cost of providing for non-productive individuals is much higher on Mars than Earth, the proportion of people who are part of the working population must be relatively high. Assuming various cutoffs for the working age population, we project the occupations of citizens in figure 8. It shows that the age at which young people enter the workforce has a dramatic impact on the ratio of working age people to others in a society with such a high reproductive rate. While the proportion of the population entering retirement is low, children engaged in schooling are a substantial burden on KCSAR.

3.6. Education

Optimising the educational process and limiting the amount of time spent in education is one of the key methods to increase the size of the working population. With the advent of the information age, ever more efficient educational methodologies have emerged. KSCAR's educational facilities rely on developments in AI-assisted learning such as in optimising curriculum to the needs and abilities of individual students and in adapting examination difficulty to the level of students. Students are grouped together by ability, rather than age group and provided access to pre-recorded lectures, teaching assistants and learning materials from a variety of sources, enabling them to take ownership of maximising the speed of their learning experience. Curricula are simplified and specified to the greatest extent possible so that students are prepared to enter an occupation that reflects both their preferences and ability. Particularly for children, methods that focalize 'learning through playing' and student enjoyment are preferred to foster a culture that values and enjoys the process of learning and work.

3.7. Work

The majority of the workforce are engaged in key sectors laid out in KCSAR's technical design and the largest proportion are engaged as factory workers. KCSAR does not receive natural sunlight, so rather than relying on traditional day-night shift working cycles, KCSAR workers work to an eight-hour shift cycle that enables continuous work. To maximise the working population proportion, full employment is sustained through a federal jobs guarantee. All able-bodied workers are provided work at below market wages, enabling otherwise unemployed labor for government operations and ventures. Workers are also encouraged to engage in education to reskill to meet the needs of a dynamic labor market and are provided significant subsidies.

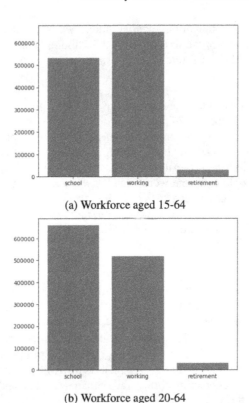

(a) Workforce aged 15-64

(b) Workforce aged 20-64

Figure 8: Impact of working age cutoff on size of labour force

3.8. Critical Services

Much as the mass production of goods is needed to fulfil the material requirements of the settlement, ensuring that the critical services are adequately staffed is crucially important for the settlement to function cohesively. Critical services range from the roles that require significant education, such as doctors and chemical engineers, through to traditional roles in the trades, such as electricians and plumbers. To ensure that there is a reasonable distribution of human capital across key service sectors, the Department of Labor is responsible for monitoring the economy and predicting the required number of each type of critical service worker. This process is facilitated by machine learning models that analyse the movements of the labor market as well as the supply of human capital from educational institutions. When these models predict that there is a shortfall in supply, policy levers are used to incentivize more citizens to become critical service workers.

3.9. Retirement

KCSAR follows the Singaporean model for retirement saving. Each employee pays 10% of their gross income to purchase units of the KCSAR Sovereign Wealth Fund ('KCWF'). The KCWF is a trust, managed by the government, which invests in the equity and debt of local businesses. The value of the KCWF grows at the roughly average rate of the market on KCSAR. On retirement, workers sell down their units to fund a pension. For most workers, this is a sufficient pension but the government contributes to pensions that do not meet a specified living wage cutoff. The sovereign wealth fund receives proceeds from the sale of land, mineral rights, patents generated by public universities and other public goods.

3.10. Wages

All KC citizens receive a Universal Basic Income ('UBI'), which is a cash payment indexed to the cost of a basket of essential goods and services. The provision of a UBI produces a society in which the basic needs of all citizens are met, which has four primary benefits:

1. On Mars, where space and air have material costs, unemployment and homelessness have even larger social burdens. A UBI provides a security net, helping to tide people over and thereby reducing the number of people who are initially part of cyclical or frictional unemployment from becoming long-term unemployed.

2. Studies have shown that there is a strong association between relative poverty and crime. In a Mars colony, where crime is relatively difficult and expensive to prosecute, making sure that all individuals have access to a living wage may be a cost-saving measure [42].

3. It provides support to unpaid care workers and individuals who are not engaged in the formal workforce. This in turn relieves pressure on public services that provide care to ill and differently abled people as well as incentivizing the growth of families.

4. It simplifies the provision of transfer payments to citizens by eliminating the need for secondary transfer payments.

The primary objections to a UBI are that it reduces motivation to work and that it perpetuates inequity, as the most wealthy individuals receive the same benefit as the poorest. Incentives to work still exist with a UBI as the receipt of a wage above the UBI award rate is associated with far greater access to luxury goods and particularly preferred foods such as fish and meat. Moreover, a UBI eliminates the structure of adverse incentives that may arise from other social security programs. Traditional social security programs may offer negative net wages for marginal increases in hours worked, whereas the return from work strictly exceeds the additional taxation burden under a UBI system. In KCSAR, the UBI also has a positive net impact on the equitable distribution of income as it is accompanied by a progressive taxation system such that for anyone of middle class or above, the total value of tax paid exceeds the value of the UBI.

While there is no economy-wide, government-mandated 'minimum wage' in the KCSAR, unions are formed within each major industry that then demand appropriate working conditions and remuneration for members. In guaranteeing that the returns of corporations are distributed to workers (rather than absorbed entirely by capitalists), labor unions serve an important role in allocating resources to the population most likely to spend them and encourage further growth, as the Marginal Propensity to Consume ('MPC') of workers is higher than that of wealthier share-holders. These labor unions will operate as delegative institutions. As companies grow larger, the importance of their unions also grows and unions will take on a direct oversight role as members of the company board.

3.11. Taxation

The taxation mix broadly aims to incentivize taxpayers to engage in the most efficient economic activities (maximising GDP per capita) whilst remaining progressive (maintaining a minimum degree of inequitability). The largest component of the taxation mix is a simple income tax, that is highly progressive and ensures that the net effect of the UBI is positive for low income earners but negative for higher ones. Additionally, a substantial inheritance tax will be utilised to reduce wealth inequality and encourage social mobility. Lower wealth inequality is associated with higher capital mobility, as wealth is negatively correlated with MPC and higher social mobility likely improves labor engagement and provides incentives for hard work and innovation [43].

Taxation is also used as a lever to encourage productive economic behaviors. Capital gains taxes are levied on the increased value of capital goods with active asset rollover exemptions that allow for deferral or reduction of tax burden if the profits generated from the sale of a capital asset are reinvested in new capital assets. This creates an incentive for individuals to reinvest returns, helping to improve liquidity in the capital market and grow the economy as a whole. Land value taxes are levied on the unimproved value of the real estate of the colony (e.g. on shops, apartments etc). This creates an incentive for land to be used for its most efficient purpose and thereby encourages the endogenous formation of regional specialisation. For example, if an area of land were an industrial zone, and it would be more productive for that land to be used for the creation of a shopping area, the tax provides an economic incentive for the industrial zone to relocate, likely to areas that already have a mass of industrial sites, where firms can benefit from the economies of scale of regional specialisation.

3.12. Property Market

Within the city of KCSAR, land is granted to private individuals for a fixed duration and may be traded, mortgaged and rented. As each new section of the city is built, the 'land 'inside is sectioned off for land deeds with a fixed ownership term, which is its design refurbishment lifetime (around 30 to 50 years into the future). On the expiry of the grant, the land is returned to the government, refurbished, redivided and reallocated. This model leads to land growing cheaper over time, making it more accessible to first home buyers. Additionally, mortgages are easily bundled together into collateralized mortgage obligations and are traded between banks, providing a substantial source of liquidity for the short term loan market.

3.13. Regulation Of Natural Monopolies

Many of KCSAR's essential markets are natural monopolies in which the cost function of mass production is more highly correlated with the number of distinct designs produced, rather than the number of items that each produces. In particular, regulated modules are the building blocks of the multitude of designs required to service the colony's needs. As these must be produced at scale, through an Original Design Manufacturing model, to minimise costs, they form a natural monopoly. To make these monopolistic industries profitable, while limiting the undue usage of monopoly power, a regulated utility corporation model is used. Candidate organisations are offered monopolitic control over a sector as well as cheap loans provided through the KCWF in exchange for significant public oversight. The charters offered to these organisations issue board seats that are managed by the community through delegative democracy. The audit branch is given additional powers to inspect company documents, ensure compliance and to provide the public a means of redress. This oversight also ensures that these corporations are extremely stable, making them ideal candidates for conservative investors.

3.14. Law and Order

On Mars, the cost of incarceration is prohibitive, as the cost of space and providing for individual citizens are extremely high. KCSAR implements a rehabilitative rather than retributive model of justice that aims to limit the incentives for crime and to prevent the emergence of environments that may encourage radicalisation. It prioritises early intervention programs that identify and satisfy the needs of citizens experiencing hardship. The political and economic system outlined above also ensures that every individual has a say in government decision-making as well as access to meaningful work and the promise of a bright future. This removes some of the key incentives around which larger scale structural crime might develop [44]. Policies and monitoring are also implemented to prevent ghettoization, and thereby structural disprivilege that may lead to structural crime.

3.15. Leased Rockets

An increasingly mature market for leased rockets is pivotal in the reduction of costs associated with interplanetary travel and for the viability of interplanetary trade. A leasing market for reusable rockets enables relatively small companies involved in interplanetary travel and commerce to operate, by reducing the otherwise extremely high startup costs. As rockets are extremely expensive purchases and proper actuarial models to determine the risk associated with the asset did not yet exist, an insurance policy for the potential shortfall was created to ensure the profitability of investing in rocket ownership. The insurance policy relies on the usage of catastrophe bonds and government intervention through targeted reinsurance programs that restrict the unhedged risk to the insurer to standard commercial risks (a Credit Default Swap). This enables the repackaging and reselling of debt with the same credit rating as the insurer and thereby the on-selling of leases on the wholesale debt market, refilling the initial fund and including extra profit on the lease. As the market continues to mature and the asset risk is better understood, liquidity continues to improve.

3.16. Significance and Composition of Trade

As an isolated colony, KCSAR remains dependent on key imports from Earth including pharmaceuticals; chemical reagents that are prohibitively expensive or impossible to produce on mars; silicon dies; and other consumer and capital goods. As the cost of transport is prohibitively high at $500/kg, these goods are imported in their most weight efficient form. For example, bulk pharmaceuticals are transported as active ingredients and undergo final transformation on Mars. To incentivize importation of capital goods to improve firm productivity, the KCWF is also used to subsidise the import of capital goods.

To purchase these crucial imports, KCSAR must produce a sufficient quantity of exports. The prohibitive costs of transporting goods to Earth ($200/kg) requires that only products with extremely high value-add to weight ratios be exported. These products include:

1. Precious materials are mined from asteroids near Mars by 'Honeybee-stye ' automated robotic mining [45]. Modelling of the price elasticity of demand for gold, platinum and other precious metals indicates that significant quantities may be exported before oversaturation of supply makes it uneconomical. The movement of mined goods and intra-colony trade is facilitated by 'nuclear hoppers' relatively cheap reusable single stage nuclear thermal rockets that use Martian CO_2 as propellant [33].

2. Fuel mined from comets is exported to refueling stations around the solar system. These stations enable spacecraft that would only able to reach Low Earth Orbit to reach Mars and the outer solar system via an on orbit refueling stage.

3. Creative and productive intellectual property produced on Mars serves as a key export. In particular, on Mars, there is a greater imperative to semiautomate or completely automate tasks, producing a comparative advantage in the creation of some AI systems, designs and datasets. Where applications for efficiency gains in Earth-bound processes are found, these assets are exported with close to zero marginal cost.

4. The sale of debt and equity will also serve as a key 'export 'that will finance ventures on Mars, helping to stabilize the balance of trade. In particular capital

intensive projects such as mining ventures and infrastructure projects will require external financing which will be recorded as a credit on the capital account.

5. Mars will also have a competitive advantage in the building and operation of telescopes, probes and landers that are to be launched to the far reaches of the solar system. For example, the building of large scale mirrors is far easier on Mars due to its low gravity and access to a vacuum. Some peaks on Mars, such as Olympus Mons also serve as a good location for deep field astronomy.

6. Mars is an ideal candidate for large scale nuclear production, so isotropic extraction from spent fuel will be a significant source of revenue. Nuclear isotopes such as ^{238}Pu, ^{90}Sr, ^{241}Am and ^{60}Co are extremely expensive on Earth but would be relatively easy to produce on Mars. In particular, ^{238}Pu is a major component in radio-thermal generators, used in deep space probes that are likely to be built nearby. Extraction of fission product rare earth elements also provides a non-terrestrial source of these indispensable elements.

7. In the future, it's likely that more uses for zero and low gravity, large vacuum environments and large ultra-cold environments will emerge. It is already clear that some types of research and some types of production are either only possible or substantially easier in these environments. Leasing out space in KC and providing labor to conduct experiments and create these goods may therefore constitute a substantial source of revenue. Examples of these include Quantum Physics research, the production of ultra-low loss ZBLAN optic fiber [46], the production of various alloys of reactive elements [47], and 3d printing of organs [48]. As the price of trade falls, as discussed in 3.17, there will be more accessible opportunities to capitalise on the comparative advantage of these environments accessible in the KCSAR.

3.17. The Mechanics Of Trade

The exchange of packages through the interplanetary postal services is the basis for trade between KCSAR and earth-bound parties. The KCSAR is a member of the United Postal Union (UPU) and therefore part of the standard postal infrastructure between member states. The distribution network relies on 'multimodal' routing, by which packages are carried by the most cost (or time, depending on the cargo) efficient mixture of trucks, trains, boats, planes and cargo rockets, which form the final step in transport from Earth to Mars. Cargo is transported in standardized containers, allowing for the economies of scale created through standardized packing and handling that have been realised on earth to dramatically reduce the cost of transporting goods to Mars [49].

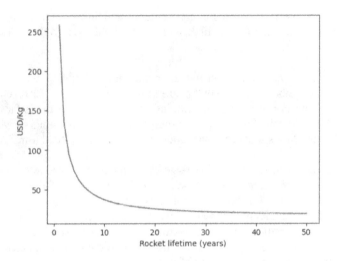

Figure 9: Launch price to mars vs. reusable rocket lifetime

As Mars is crucially reliant on imports of some necessities, the economic viability of the transportation system that underpins trade is a primary concern. The key results of a sensitivity analysis, shown in figures 9, 10, 11 and 12, indicate that the primary mechanism for reducing the transport cost for delivery of freight to Mars is increasing the lifespan and reusability of leased rockets (at least to the 10-year mark). We take the partial derivative of price with respect to a vector of potentially explanatory input variables to provide an indication of how much a change in each factor would impact price. While the rocket lifespan effect dominates, secondary factors such as reducing launch costs (by, for example, connecting the launch complex on Earth to a natural gas pipeline) and reducing the cost of rockets would have significant impact. The technological innovation and reduction in cost of factors of production necessary to change these key factors is most easily stimulated through economies of scale. To enable these economies of scale, a significant market must be generated to enable producers to benefit from scaling up production facilities and investing in research and development.

Two mechanisms are used to generate stable demand for interplanetary trade and thereby capitalize on scale economies: government intervention in its futures market and competition regulation. A futures market for interplanetary trade emerged endogenously, helping to stabilize the demand for and price of traded goods and thereby enabling small and marginal producers to reduce their exposure to liquidity risk and enter the market. It also encouraged more efficient allocation of resources over time. The government intervened by providing a demand floor for interplanetary trade by consistently buying futures for freight transport and using them to transport mail, capital and other government materials. This provided the stable demand requisite for the realization of economies of scale, which, in turn, reduced the cost of transport, increased the quantity of trade traffic, and the broadened the composition of trade as lower prices make more trades profitable.

This increased and broader trade base then stimulated economic growth both on both Earth and Mars.

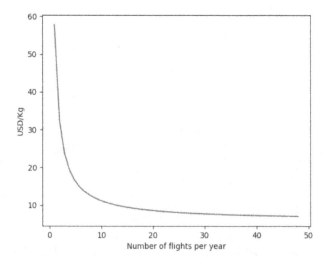

Figure 10: Launch price to mars vs. number of flights per year at fixed 10 year lifetime

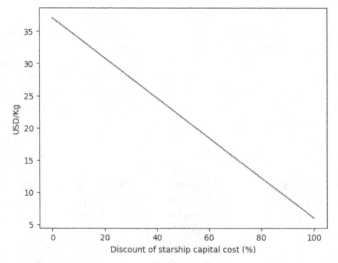

Figure 11: Launch price to mars vs. discounted capital cost

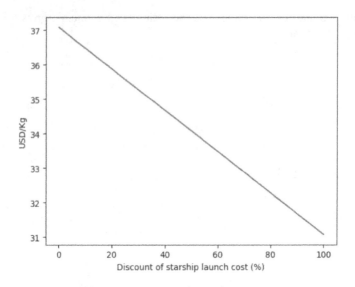

Figure 12: Launch price to mars vs. discounted launch cost

An independent watchdog monitors, reports and provides recommendations to the government on non-competitive behaviors in the rocket production industry. Due to the substantial start-up costs and competitive advantage associated with holding intellectual property in the industry for producing and maintaining rockets. The emergence of vertically integrated firms and firms with substantial market (monopolistic) power is therefore likely. Through the use of the leased rocket model, nonintegrated rocket production companies can easily support an ecosystem of launch providers if they are encouraged to through substantive oversight. A substantially competitive market will keep the costs of production low and encourage innovation amongst competing firms.

3.18. Culture And Architecture

Given that the residents of KCSAR face extraordinary labor and resource scarcity and live in closely contained quarters, it is critical its emerging culture is one that: (a) cultivates a passion for work that is motivated by a desire to advance the collective interests of the colony, (b) values close communal relationships, (c) fosters and applauds innovative thinking, (d) promotes an appreciation and love for both knowledge acquisition, and knowledge sharing, through a life-long pursuit of education, and (e) celebrates large families. To develop these cultural tenets, many of the societal superstructures (e.g. the design of shared spaces, immigration selection criteria, human resource allocation, food manufacturing, among others), are implemented in a way that aligns with and encourages this cultural ethos.

Perhaps the most powerful example of such cultural 'nudges' is in the aesthetic and architectural design of KCSAR's city. While on the surface, the architecture appears to be dominated by function over form with small pressure domes being the norm, as seen in figure 1; it is underground where the culture is brought to life with a sophisticated style that captures the cultural significance of the various spaces, and the cultural essence of the society as a whole. For example, industrial areas are of the brutalist style, made up of blocky, cheaply mass-produced segments, reflecting the role of these areas as the functional drivers of KCSAR's economy. Government buildings, by contrast, use a formal neoclassical style as a reminder that the people, and their government are the living heart of KCSAR's culture, while emphasising the government's role as servant to its people. In recognition of the cultural aspects of human achievement and our need to be in spaces that encourage unity and collaboration, while simultaneously reminding the people of KCSAR that life extends beyond the confines of factory labour; the important common social areas such as opera houses, and theatres provide the most compelling of designs. Using the design tenets of Parametricism, these buildings are 3d printed into unique and visually striking shapes that capture the artistic essence of the space. With values of openness, communication, and collaboration at the epicentre of KCSAR's cultural epoch, common spaces are specifically built with a design dogma that aims to reflect these ideals through the thematic concepts of 'green space' and 'blue space'. In practice, this means that there is an emphasis on local gardens, plants, and water features, along with more striking transparent plastic segment walls that look up into the ice or subglacial lakes. Fountains and other natural water features are used to dual effect. On the one hand, they enhance the natural aesthetics of the environment, while on the other hand, they functionally aid in maintaining the atmosphere. Lights under the transparent segment also illuminate the environment, while providing the stimulation needed for the seeded photosensitive marine life.

Two extremely unique pieces of architecture in KCSAR that really capture its values, are (1) the government mandated and government-subsidised cultural centre that exists at the focal point of all domes, and (2) the underwater beach (figure 13) that is designed to give residents a place to unwind after a challenging day at work.

The cultural center is made up of three concentric circles defined by an open plan aesthetic that encourages social engagement. In the innermost circle exists a place where community members are provided with a range of amenities. These include free access to the internet, luxurious recreational seating, local food delicacies. Here, they are also encouraged to engage in prize-giving activities that foster intellectual expression, such as debates and hackathons. To reach the innermost circle, community members make their way across the two outermost circles, which are designed to share and celebrate local achievements. For example, local artists and hospitality workers are encouraged to express themselves in the outermost circle, with public displays of original music, food, and art. Government grants are available for the creation of such works. In the middle circle, virtual presentations adorn the walls, showcasing the accomplishments of local scientists, innovators, and business champions. In this way, a cultural ethos that simultaneously perpetuates artistic expression, intellectual inquiry, originality, innovation, and collaboration is encouraged.

Figure 13: Artist's rendition of the underwater beach

The underwater beach is a dome segment that is set under the ice, such that the water pressure on the outside of the segment is equal to the air pressure inside the segment. This then allows for small openings to be made through the dome, which facilitates the free movement of marine life between a small artificial pond in the sand and the wider sub-glacial lake beyond. The segment itself has a transparent roof and wall, overlooking the lake itself. Consequently, this almost transforms into an artificial subtropical biome, with a base that is mostly filled with crushed regolith as sand, overhead lights, and waste heat from the reactor, which makes it an ideal home for a teeming artificial ecosystem. The vista of the surface through the ice is illuminated by a set of floodlights, providing a stunning view that makes the seating around it a popular space for KCSAR citizens to rest and unwind after a hard day's work.

It is also worth noting that the government takes a more direct role in encouraging socially beneficial behaviours, by providing economic incentives for individuals to engage in things like childrearing and education (as detailed in sections 3.4 and 3.6), as well as providing educational campaigns such as TV infotainments that reinforce certain cultural tenets (i.e. the importance of communal engagement). Since living and working in the difficult close contact conditions of a Mars colony has ramifications for people's mental health and life satisfaction, these programs encourage people to form tight-nit relationships with their close neighbours and provide information about how to form supportive relationships that improve mental health outcomes. Self expression through the arts and recreational activities are also highly encouraged and subsidised through government grants.

Figure 14: Artist's rendition of the future terraformed crater

4. THE FUTURE OF KCSAR

To this point, we have laid out the bare bones social and economic processes needed to support a colony of one million people facing pressing existential risk. We turn now, by way of conclusion, to look at what the future will hold.

Over the coming years, a substantial geoscaping operation will be implemented to make the Crater habitable for humans. A central aluminium space frame-style construction with a series of support pillars will hold up a cover, consisting of a thin PHB or other bioplastic layer and a clear layer of lipid-derived wax, acting as a counterweight to the air pressure beneath and as thermal insulation. This greenhouse structure will be filled with the potent greenhouse gas, SF_6, until it raises the Crater's temperature above $0°C$. The Crater's ice layer will then melt, forming a stable glacial lake as the Albedo drops. Evaporation of water vapour from the lake then occurs, enhancing the atmospheric pressure within the Crater beyond the Armstrong limit. O_2, N_2 and Ar will then be pumped into the atmosphere to make it habitable for human and other air-breathing life.

The lake will be seeded with nutrients, including metal trace nutrients, sulfates and phosphates to make it habitable for plants and algae. Azolla and algae will be added to produce food and oxygen, whilst fixing nitrogen from the atmosphere. Once it is habitable, the lake will be seeded with fish that will become an important food source for the KCSAR colonists. The upper barrier will form artificial rain clouds as it cools warm moist air that condenses into water and falls as rain. Animals will be able to walk freely about the surface and the lake will be green with life. An artist's rendition of this end state, the dream of the Korolev Crater, is depicted in figure 14.

ACKNOWLEDGMENTS

Thank you to Epifanio Pereira @id_epifanio for the artwork throughout. Thanks to Giovanni D'urso, for comments and suggestions. To the Research Science Institute (RSI) and the BLUEsat UNSW team, for their inspirational experiences. To those in the space industry, for opening the door to the solar system. And finally to the people of the Mars society, for hosting the competition and far more importantly, for steadfastly refusing to let the dream of interplanetary human civilisation die.

MORE INFORMATION

Due to space constraints, references and other materials may be viewed using the QR codes found below (primary, secondary).

20: NEXUS AURORA - MARS CITY STATE DESIGN

**Adrian Moisă, David W. Noble, Henry Manelski, Koen Kegel,
Kim von Däniken, Sean Wessels, Mitchell Farnsworth, Mae Elizabeth,
Cameron Rough, Sam Ross, Dr Orion Lawlor, Il Ho Cho,
Olivier Rinfrent, Daniel Kim, Felix M. Müellner**
and 50 more Nexus Aurora contributors found
at: nexusaurora.com/report-authors.pdf

nexus.aurora.sm@gmail.com

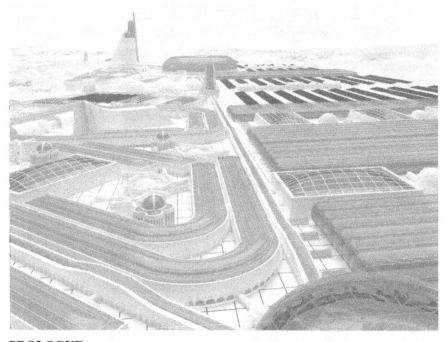

PROLOGUE

Mars' thin atmosphere makes dawn fast. As the sunlight sweeps across the eastern slopes of Hellas Planitia, the glimmer of Orbital-1 visible overhead, banks of lights across the city fade up in a warm glow. The weather outside is cold and crisp with a medium wind, and the climate control creates a corresponding chilly breeze down the streets.

As clocks hit 7:30 Standard, alarms begin to buzz across the habitat. The household murmurs into life. Zach Ellis is first up, his commute demanding an early start. His wife, Vala Zubrinova, works inside Willow Node itself, so is afforded the luxury of a morning shower. The two children, Jasmine and Alexei, lope down to the kitchen.

Breakfast is different for all family members - Zach prefers seeded rye toast, the children have cereal with soya milk and Vala treats herself to a smoked salmon sandwich. Alexei finishes the milk with his cereal, and shouts to the voice assistant to add some to a shopping list.

Zach leaves the house first, donning his work uniform and heading out onto the habitat central street. The buildings are fairly old, so lack the decorative flourish seen on newer buildings around the city, but the facades are painted in a charming range of pastel colors. Despite the amber glow of the lights, the street is lit with flickering patterns from the meters of shielding water overhead.

Vala hops onto a packed metro car (standing room only at rush hour, this line serves over 10 thousand people in Willow Node) and is whisked away towards the central hub. She makes a quick change at the ring interchange to the Northern Branch, onto a train full of workers also heading to the Mars United Biopharma plant.

Zach also changes at the ring interchange, but he swaps to Eastern Branch and a train headed for the Node main rail station. Once there, he takes a familiar route to the northward departures and through a series of airlocks into a low, rounded passenger pod. Bang on time, the entry hatch closes, and a robotic crane lifts the pod onto the bed of a waiting train. With only the faint vibration of an electric motor and the rumble of hardened steel tracks, the train pulls away from Willow Node.

Jasmine and Alexei cycle past the ring interchange and through pedestrian tunnels, which lead them to the Node's central dome. In keeping with the Node's name, the designers have chosen to fill the open spaces between the buildings in the dome with willow tree-surrounded ponds. The two children scurry off the trains and join the throng of students heading into the school complex.

Vala works at Mars United Biopharma as a production line supervisor and quality control checker. Vala's work is easy, monitoring purity of the product at various stages of production with frequent sampling, and trimming inefficiencies out of the robot schedule. The primary output of the MUB plant here is salicylic acid, which gives Willow Node its name.

Zach works outside his home Node as an automation engineer at the vast steel rolling facility in Geim Crater - the city's industrial center. He disembarks the train near Geim Crater and joins the crowds of engineers and operators heading through pedestrian tunnels into the crater complex. Zach splits his day between supervising the semi-autonomous robots on the mill floor, and working in the control room on advanced automation software for the same robots - part of the never-ending project to reduce the load on human operators.

Back in Willow Node, Alexei and Jasmine are in classes. Alexei is 9 years old, so he is still learning from the standard curriculum. Morning is classroom teaching, and after lunch he takes part in a depressurization drill. After school finishes, he joins his friends at the rock-climbing club, walking to a climbing gym a few minutes from the school. Jasmine is 22, and hopes to progress to Further Learning next year after her capstone project concludes. Morning classes are a mix of compulsory and optional side modules. But Jasmine's real passion is geology, so she spends the afternoon working on detailed plans for her geological research trip this summer. Like all native Martians, she ever seeks to push the boundaries of human exploration. Her research trip will be a bold undertaking, but one that she is ready for.

After the working day concludes, all the members of the family return to their apartment in the habitat on the outskirts of Willow Node. Jasmine and Alexei help Vala in the kitchen, preparing a meal of tuna steak and boiled potatoes, while Zach tidies the living room. The meal lacks expensive imported spices, but Vala is practiced at working with the herbs available from the grow boxes on the windowsill. After dinner, Alexei runs from the table early, eager to call with friends across the city. Jasmine heads off to her room to work on a lab report, as Zach and Vala settle down on the sofa to watch some imported Earth sitcoms.

This isn't a typical family on Mars - just an example of a single family living in, more or less, happiness and tranquility. Just one of a quarter-million families living in their quarter-million homes, who can call the red planet "home."

CONCEPT

To cut free the shackles of our limited capacity for thought and drive humanity towards a shining future, where unhindered technological progress is coupled with considerate, human-centric design: this is our optimistic vision of societal progress - this is the grand goal of Nexus Aurora. The plans for a Martian colonization program with a flourishing colony of one million people by the year 2100 encapsulates this vision. It is our garden city full of open vaulted spaces, natural light and community.

In order to build a city of one million, we must do more than live off the land. We must use Martian resources to build an industrial base to rival terrestrial nations. Quarries, algae photoreactors, farms and atmospheric compressors providing feedstock to vast rolling mills, additive manufacturing systems and chemical reactors: we account for all these systems and more, powered by Martian-made nuclear reactors.

Building designs must adapt to withstand the pressure, radiation and temperature extremes of the Martian surface, while simultaneously enabling happiness and fulfillment for the inhabitants by providing airy, light-filled community spaces. We reject the concept of cities built of tunnels and underground habitations. Nexus Aurora has wide boulevards and urban parks made possible with pressure vessels of basalt-fiber reinforced plastics (BFRP), utilizing the powerful radiation-shielding effect of water to protect our residents.

Such a city would be expensive to build, as a huge amount of industrial equipment (and a huge number of people) would need to be transported from Earth. As with any financially significant endeavor, Nexus Aurora will not be profitable in its early decades. A thoroughly planned economy balances income from tourism, exports of construction materials and services to other Mars colonies with the continual high cost of imports. A democratic government will ensure stability and freedom for the population.

This 20-page report represents less than 1% of the technical documents that comprise the comprehensive plan of Nexus Aurora. Our plan is a phased approach that grows the Martian colony from a tiny outpost of 20 Earth-dependent colonists into a self-sufficient settlement, expands into a formidable industrial base, and finally to become a thriving and autonomous nation.

Nexus Aurora is not tied to any one nation or organization. It draws on the spaceflight legacy of NASA and ESA, the technical experience of SpaceX and Mitsubishi, and the industrial base of Germany and Japan. Only by opening up both space itself and the design of our future in it can we embolden humanity to move beyond our current limits. Thus, we present:

1   BUILDING NEXUS AURORA

1.1   Preparation

The requirements of building a large permanent settlement on Mars are very different to those of short-term missions. Our number one priority for site selection was availability of water in the form of glaciers. Water forms the backbone of industry on Mars, so the typical landing sites in the low northern latitudes are simply inadequate for our purposes. Instead, we are forced to consider relatively high latitudes.

The second priority is easy access to a range of minerals. Without accessible and rich deposits of metals, we will have considerable difficulty building a city. Remote sensing from satellites, particularly the MRO, is invaluable for broad-stroke analysis but cannot detect the kinds of small deposits that can be targeted by mining operations. For that, we must turn to geology and areology, and it tells us that the best indicator of varied, rich deposits is a history of fluvial and volcanic activity.

Based on these primary criteria, and a whole raft of others too extensive to discuss here, we eventually selected a landing site on the edge of Dao Vallis, situated on the eastern rim of Hellas Planitia.

The advantages of such a site are considerable, beyond the geology. The landing zone is 4.5 km below Mars datum, which offers higher atmospheric pressure, more height through which to aerobrake, and better protection against radiation. Our landing site is also only 600 km from Talas Crater: one of the lowest elevation areas on Mars and a prime candidate in the search for life.

However, this site is not without its difficulties. The southern latitude, combined with Mars' orbital eccentricity, makes the winters of Hellas cold and dark.

The thicker atmosphere also means harsher dust storms. Despite these barriers, Dao Vallis is the chosen site of Nexus Aurora, and we have plans to overcome these challenges.

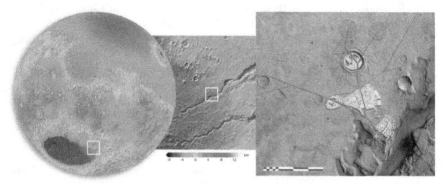

Locality plan

The selection of a landing site is but the first step - we must know every centimeter of the landing zone and the surrounding regions before we set foot on the surface. An extensive satellite network, both to provide the first connectivity for the colony, and to extensively survey the landing site, is required. A ring of satellites in areostationary orbit, communicating with Earth in the EHF band, will provide a stable communication line for robotic operation. Deployed simultaneously to these is a fleet of low polar orbit observation satellites, gathering high-resolution imagery of the landing zone (in both visible and various EM bands, for better remote mineralogical analysis). The deployment of this constellation is anticipated for the 2024 transfer window.

However, satellites can only take us so far. To precisely locate and analyze mineral veins on the surface, survey robots are required. These robots will accompany the first landed cargo ships at the landing zone and represent a radical departure from the NASA doctrine of valuable, high-redundancy rovers. Instead, a Starship will deploy dozens or even hundreds of small, lightweight surveyors onto the landscape. These will spread out in a swarm, providing a wealth of data about the underlying geology and geography of the landing zone in the form of elevation maps and core samples. This will be invaluable for the laying out of habitats, mines, scientific outposts, and transport lines as the city swells beyond the confines of the initial landing zone.

1.2   Buildings
Building on Mars requires a radical rethinking of terrestrial ideas for what a building should look like, how it should function, and how it should be built. We have identified four key ways in which building on Mars is significantly different to Earth, with solutions visible in the concepts of every building.
1.   **Material availability** - reducing manufacturing energy in the later stages is vital for reducing power consumption, and restrictions on material

    manufacturing on Mars severely limit potential building materials and techniques.

2. **Pressure** - to maintain Earth-like pressures inside buildings of any practical size, the walls must withstand enormous tensile hoop stress. This poses the biggest practical limitation on building size and form.
3. **Radiation** - despite popular misconceptions, Mars is not rendered utterly inhospitable by cosmic radiation[1]. However, to avoid long-term health concerns, appropriate shielding must be provided to reduce exposure by 80% in order to fall within terrestrial safety limits.
4. **Temperature** - the southerly latitude of Nexus Aurora means we can expect rapid radiative heat loss[2]. This challenge is reversed in later years of the project when effectively rejecting heat from industry becomes a major challenge.

In the first years of Nexus Aurora, all building materials must be shipped from Earth, and ease of construction and assembly is critical. To this end, we have developed a standardized modular structure, based around a 6 m diameter cylindrical pressure vessel with steel end-domes and reinforcing rings. This modularity is useful - modules can be split into two levels for work and habitation, or kept in a single high-ceilinged space for workshops and machinery. The interchangeability of the end-domes allows modules to be rearranged on the surface, creating vastly complex spaces from a few simple building blocks. These modules are so versatile that we expect them to continue to be reused even as the city grows thousandfold, such as in remote outposts and small industrial sites.

We have determined that burying these modules, or building underground in general, is significantly worse than structures on the surface (protected from radiation by a rammed-soil brick arch, sufficient to bring exposure below the 50 mSv annual limit for radiation workers in the US[3]). Underground structures require tunneling equipment to be operated with associated costs in mass, time, and energy, and the subsurface of Mars has sufficient unknowns to make such an endeavor very risky. In comparison, brick vaults can be built autonomously, provide superior thermal insulation, allow easy maintenance and repair, and create a shielded space for both outdoor work and running utilities.

The interior of the modules will be simplistic, drawing on the wealth of existing development work[4] for the design of short-term Mars bases. Emphasis will be placed on creating both personal spaces for citizens, and small-group community space to facilitate group cohesion.

Constructing buildings in-situ makes use of an abundant material at our landing site - basalt in the form of drawn fiber, a material with incredible tensile strength, which can be made into an airtight composite with a thin layer of imported plastic. Basalt fiber reinforced plastic composite[5] (BFRP) is an incredibly well-suited material for our purposes - strong enough to make pressure vessels tens or even hundreds of meters in diameter, very low import weight, and production energy a fraction that of traditional materials like sintered bricks and steel. Weaving basalt fibers into a net-like structure to reinforce clear polymers (at a fraction of the cost of glassmaking) allows for vast greenhouses.

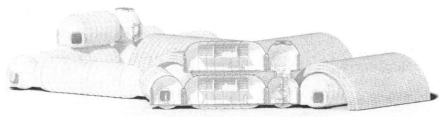

Mars Base made of 6m Modules

Detailed landing site analysis will provide information for selecting the best place to locate the initial settlement - the landing area contains a sizable collapsed lava tube[6], which (structural integrity allowing) would provide an excellent large volume to hold dozens of habitation and industrial modules with protection from weather and radiation.

1.3   Necessities of Life

Power production in the first stages of colonization is provided almost entirely by solar energy, backed up with lithium-ion batteries. This is possible due to the power system's ease of deployment, excellent power density, and decentralized nature, which provides built-in redundancy against failure. We have determined that ideal power cells are thin-film CIGS cells[7]. These can be deployed as roll-out blankets by robotic systems to provide power to the pre-colonization equipment or fixed-tilt arrays on carbon-fiber frames with a combined density of 100 g/m2. The complete design weighs just 120 tons, which includes a 6 MW-average output grid including solar arrays, batteries, transmission equipment, and emergency backup methalox generators. This provides the basis for the life support of the first few hundred colonists, along with power for early farming and industry.

Power demands will increase soon after initial colonization. Expanding the power system is simple; more arrays and batteries can simply be added at will. Battery capacity is designed to ensure continuity of power to vital systems and will not be used to sustain industrial operations, so large capacity is not required.

We have also explored the use of Small Modular Reactors, particularly the Westinghouse eVinci[8] and other candidates for the Megapower programme. These reactors would be connected to aero-derivative turbines[9] for an expected output of 5 MWe and 3 MWth at 90 t shipping weight. However, such a design is dependent on these reactor programmes coming to fruition and favorable geopolitics for launching tons of nuclear fuel, so we do not intend to rely on imported reactors.

The atmosphere for the colony is provided from two sources; the first is compressed and separated Martian atmosphere, and the second is hydrolyzed water. The CO_2 from the atmosphere can be repurposed for the farming industry. Oxygen regeneration will be provided primarily by plants, with LiOH and KOH scrubbers[10] providing backup regeneration. A comfortable atmosphere will be maintained at 80 kPa for superior redundancy across the city. This will be achieved with a decentralized system of heaters, both electrical and district heating with industrial heat. Humidifiers and dehumidifiers will also be utilized to maintain atmosphere.

The location of our landing site is only a few kilometers from a known glacier with over a cubic kilometer of ice, which exceeds requirements for water extraction. The most effective extraction technique identified is the Rodriguez well[11], which is already used for water extraction from glaciers on Earth and can produce 37 L/day/kW.

To prevent the need for excessive purification, we plan to have separate processing for greywater, which is reused in toilet flushing, farming, etc., and blackwater, which is treated as potentially hazardous and will be processed with Earth-similar standards. At larger scales, processed waste will be used as industrial feedstock. Methane from anaerobic decomposition and nitrogen-rich compounds are both valuable products, rather than waste.

Unlike the hyper-optimized spacesuits designed by NASA, suits for long-term use in Nexus Aurora should focus on ease of manufacture and repair along with simplicity of use with minimal training and practice. The EVA suit created by the design group must function for 8 hours on the surface, with all the comforts and safety systems expected from a spacesuit. Easy use of a range of tools is also required.

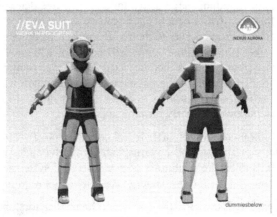

EVA Suit

In order to meet these criteria, a partial pressure suit[12] is optimal, and it will be similar to those worn by fighter jet pilots, providing pressure to the skin through corset-style threading. Thermal regulation and protective outer layers are worn over this pressure-providing suit.

When working indoors in structures with lower safety factors, the extra layer of safety provided by a rapidly-deployable egress suit is required. A breathable atmosphere is provided by a zip-up hood that can be worn comfortably behind the head when not in use, and can provide 5 minutes of breathable air to facilitate evacuation.

These two pieces of uniform, EVA suit for surface operations and IVA suit for frequent indoor use, provide the colonists with function, comfort, and safety outside housing habitats.

An essential part of the construction and growth of Nexus Aurora is the near-omnipresence of autonomous and semi-autonomous robotic systems. Robotic systems allow the automation of menial, arduous, or dangerous jobs. In order to improve interoperability and reduce the number of discrete systems that need to be transported to Mars, a modular robotic platform will be developed. This allows a common drivetrain and control system to be attached to a wide range of tools and equipment systems, with quick interchangeability provided by surface workers or even other robotic systems. Some of the requirements that have been explored, with the inputs of design teams across the Nexus Aurora project, include mobile cranes, brickmaking and bricklaying robots, logistics robots for the transport of cargo both indoors and outdoors, and humanoid avatar robots to allow for remote maintenance and construction.

1.4   Production and Industry

Nexus Aurora cannot rely on Earth imports to build the vast majority of Martian structures due to the steep shipping costs. Therefore, an industrial base for producing construction materials is essential.

Mining in the early days of the colony is an entirely mechanical process, with backhoe and bulldozer excavators moving regolith from targeted areas to industrial sites. As demands for minerals increase and diversify, more traditional methods are required: open-pit quarrying and in-situ leaching. The former will provide the tens of thousands of tons of basalt[13], iron ore and soil needed per year, and will require blasting using perchlorate-based explosives[14] manufactured on Mars. The latter is more appropriate for small high-concentration deposits of specialist minerals such as copper, sulfur, and uranium. The study of sediments along Dao Vallis will provide a low-cost solution to finding mineral deposits if they exist anywhere along the ancient riverbed.

Basalt Fiber Reinforced Plastic Farming Modules

Farming serves two roles in Nexus Aurora: not only does it feed the colony, but it also provides an essential industrial feedstock. Plants can fix carbon dioxide into complex molecules with efficiency far above any artificial process - we intend to make full use of this.

The basic structure of our large farms is simple. It has joined rows of 5-20 m diameter BFRP tubes up to 1000 m long, with the lower segment filled with either processed soil (to remove perchlorates) or water. These tubes are wide enough for large-scale machinery to operate inside.

The nature of producing greenhouses on this scale leads to an interesting and rarely-considered conclusion. Building greenhouses requires between 200 and 600 kWh per square meter. Artificially lighting and heating the same area to maximize yields requires 800 kWh per year[15]. Therefore, provided industry can keep up, it is energetically cheaper in the long run to simply build more greenhouse area than resort to high-density, high-energy farming techniques like stacked hydroponic basins or artificial lighting.

The downside of this strategy is that a huge farm area is needed, even with the use of a CO2-rich atmosphere and fertilizers. Because of the psychological and ecological need to grow a wide range of crops, we estimate that 150-170 square kilometers of farmland will be needed to comfortably feed a city of 1 million people[16].

The range of uses of farmland across the city are:
- Carbohydrates like potatoes, wheat, and corn
- Protein-rich plants such as legumes and soybeans
- A range of mixed nutrition fruits and vegetables
- Oil-producing plants (e.g. rapeseed, sunflowers)
- Providing open space for both views and exercise for colonists, as well as auxiliary atmosphere regeneration
- Water-filled or partially flooded farm structures for fish farming. Fish form a key part of the food system; they digest plant waste and are an excellent low-energy protein source.

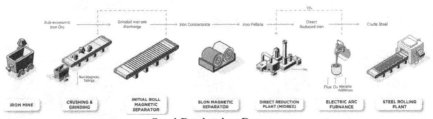

Steel Production Process

Steel has been used for a case study in metal processing, as it is both a critical material for building purposes and for industry. Ore from mines and quarries is mechanically crushed, separated magnetically to increase purity, and then reduced to iron in a direct-reduction plant. This DRI plant would initially use hydrogen, as this is the most energy-economical of the various reducing gases proposed in the literature[17], but may switch to methane if it is more available from biochemical sources. The iron from this plant will be alloyed to various steels as required in electric arc furnaces[18], and then shaped in a rolling mill. Cold-rolled steel is preferable to machined or hot-rolled steel in almost all cases, due to superior strength, manufacturing speed, and wastage.

One of the staples of construction materials is basalt fiber, which is used extensively in BFRP due to its high specific strength of 1790 kNm/kg and exceptionally low manufacturing energy of just 1.1 kWh/kg[19]. It is produced through simple melting and extrusion of mined basalt, using rejected process heat to reduce power consumption. Another material used widely across the colony is sulfur concrete, which can be cured on the surface without climate control - a significant benefit for exterior structures.

One of the biggest industrial challenges on Mars is the total lack of petroleum. It may be possible to build a city using no or only imported hydrocarbons, but it is far more promising to consider how such chemicals can be produced.

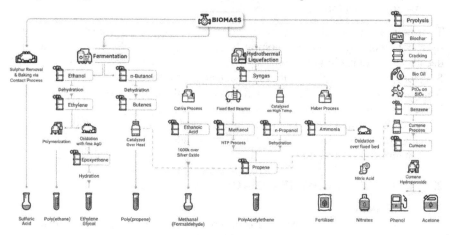

The usual method of producing base hydrocarbons is chemical synthesis from atmospheric CO_2, most commonly the Sabatier process, producing methane. This will be used for the production of many simple olefins and carbonyls, but larger hydrocarbons are generally more complex. A promising route is biochemistry; in particular, the processing of biomass to produce various targeted compounds. The biomass input is provided by waste plant matter and algae. Algae can be grown in photobioreactors with extremely high growth rates (0.38 kg/m3/day), making it ideal for this purpose. This biomass is liquefied under high pressure to produce oil that can be processed very similarly to petroleum. The benefit of biopetrochemistry is the ability to use modified terrestrial methods for processing, rather than developing entire new synthesis routes. With this feedstock, a chemical industry capable of producing many invaluable consumables, such as rocket propellant, fertilizer, explosives and surgical anesthetics.

The most important product of this bio- and electro-petrochemical industry is the output of large volumes of energetically cheap plastics. This is the key to producing large structures in BFRP with low energy demands and no imports from Earth, and thus grow Nexus Aurora at the rate needed to reach a population of 1 million.

1.5   Logistics

While the vast majority of transport in Nexus Aurora occurs indoors, large surface rovers are needed to move cargo, equipment, and occasionally people to installation sites on the surface. As with robotics, these rovers will be outfitted with a wide range of modular tools for specialist construction (crane, bulldozer), extraction (drill, backhoe), and transport roles[20] (flatbed cargo, pressurized transport for crew). A rough analysis estimates that we will need approximately 90 rovers within the first 10 years of the colony, with corresponding equipment. The rover platform is sized for transporting very large loads, including whole habitation modules or tens of tons of raw regolith for industry. To prevent excessive wear (a problem that has plagued Mars rovers for decades), a shape-memory $Fe-Mn-Si$[21] alloy will be used for the wheels of rovers, as well as the wheels and exposed moving parts of other heavy-capacity outdoor equipment.

Transport of equipment inside and outside the base will be handled almost entirely by robots working either autonomously or remotely controlled. During the first decade of colonization, we expect to unload over 8,000 tons of material from Starships, which will require a dedicated cargo handling infrastructure. Autonomous rovers in flatbed configuration will be used to carry equipment between storage yards and installation sites. Dedicated cargo airlocks will allow multiple tons of material to be moved in and out of the base, with air jets removing contaminating dust. Inside the colony, small autonomous carts[22] will transport equipment from storage and production to consumers.

2.   ENGINEERING NEXUS AURORA

2.1   Structures

Nexus Aurora will far outgrow the 6 m modular housing that is designed for the early years, but the design principles will be maintained in the cavernous habitation spaces. BFRP will remain a heavily-utilized material, used for pressure vessels across the city. However, the single short tubes will be superseded by clustered cylinders hundreds of meters long, building up vast interior spaces (in which houses, shops, and recreation spaces are situated) to create soaring expanses and bustling city blocks that transport the citizens far from the reality of a totally enclosed existence.

The smallest commonly-used structure of this "clustered cylinder" architecture is the 570Hab, a structure composed of four 18 m diameter, 150 m long BFRP cylinders, providing a total usable land area of 8000 m2 at a minimum height of 8 m - sufficient for three-story buildings. The hab can house 570 people at a high population density[23] (living in compact apartments with narrow streets), but we have specified a maximum of 400 to allow for spacious streets and sufficient natural light.

Radiation shielding is incorporated into structural elements of the 570Hab, and it is taken to extremes. All four sides of the hab are enclosed with a 5 m thick wall of rammed soil blocks (ingress and egress provided by steel tunnels of various sizes). The roof is covered with a 3 m layer of water (kept liquid by a low-pressure layer above) that provides necessary radiation shielding[24] while allowing natural light inside. \

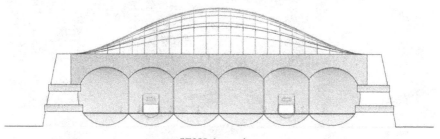

570Hab section

The interior of the 570Hab will take a number of forms, the most common being a city block. However, uses such as multi-story manufacturing facilities and recreational open space have also been identified. The exact nature of the cityscape that can be built within the confines of such a structure is detailed later in this document.

Despite their size, 570Habs only form the smallest and least dense sections of mass housing in Nexus Aurora - analogous to the suburbs of a terrestrial city. For the buildings that make up the urban core, the principle of clustered cylinders is taken a step further with the Air Mattress concept. This still uses joined BFRP cylinders with overhead water for shielding, but on a significantly larger scale and in a less standardized fashion. Each Air Mattress would be individually designed with consideration of both the available land area and local ground features, to avoid having urban cores be too homogenous.

Due to the specialized design nature, it isn't possible to give detailed information about specific instances of the Air Mattress. Plans are dependent on survey maps in greater detail than is available from orbit. However, various concepts for district centers involve Air Mattresses with a land area of over 150,000 m2, equal to many city blocks. Such a large internal volume will contain not just homes for thousands of people but wide tree-lined boulevards, urban parks, and shopping districts. Air Mattress structures provide the space to create the environment most like a terrestrial city on Mars, and we intend to make full use of this.

And yet, even Air Mattresses do not provide enough space for some activities. They will have an exceptionally large footprint but the limits of the BFRP cylinder model constrain maximum height, and will require steel tensile rods to fill the space. The final piece of commonplace architecture is the centerpiece of neighborhoods across Nexus Aurora, providing soaring volumes and open public spaces.

A 200 m diameter dome, formed of a flattened sphere of transparent BFRP half-buried inside an appropriately sized crater. Radiation shielding is minimal, but the risk is deemed acceptable given the psychological and sociological advantages of having large open community spaces and the short times that people will spend in the domes. 200 m is close to the practical limit of the possibilities of transparent BFRP domes - larger structures may be possible with stronger plastics (such as Dyneema[25]) but these have not been extensively considered on grounds of energy to construct and manufacturing complexity.

The interior space will be entirely given to community use - considered uses include playing fields, forested parks, plazas for social gatherings, and amphitheaters. These domes also act as central transport hubs, where pedestrian arteries for the city connect.

Beyond the relatively regular, repeated buildings that provide housing (as well as the simpler structures for farms and industry, discussed in other sections of this report) we have identified the need for a number of landmarks to be built around Nexus Aurora. These serve a number of purposes - a few are functional but the majority serve as cultural and social nexuses for the population of Nexus Aurora. We have considered dozens of ideas (from radio telescopes to skyscrapers to mausoleums) but a few have been developed in more detail, and one has proven such a compelling concept amongst the community that detailed 3D models have been produced. A few of these landmarks are:

Monument to the Fallen Wanderers, a structure that serves as both a resting place for the remains of those who die in the early stages of the colony, and a memorial to all those who made incredible sacrifices to make colonization of Mars possible.

Another proposal is large infrared telescopes[26] for astronomical research. The Martian atmosphere absorbs virtually no infrared, especially in MIR and FIR bands, and 50 m collectors would be able to carry out observations of protoplanetary discs and galaxies in the early universe, impossible to do through Earth's wet atmosphere.

The most developed landmark is called the Tree of Life. The giant redwood tree, planted on the first day of the founding of the colony, will require a structure that can be expanded around it for support and protection from the environment. The interior will contain a spiraling walkway stretching from the ground to the canopy of the tree, along which people can walk or sit.

The Tree of Life is intended to embody the growth of the colony, from humble sapling to an ecosystem towering over the landscape in a structure built entirely from Martian materials. It will outlive the entire founding era of the colony and may even live to the day when Mars is terraformed to the degree where a pressure vessel is not needed.

All of the structures in Nexus Aurora, be they vast Air Mattress habitats or tiny modules holding industrial equipment, will be placed with careful regard to both human experience and the demands of the surface. Urbanism on Earth in the past century has seen cities spread across the landscape without regard for the nature that existed before, and grow in directions led by commercialism and short-sighted thinking that have rendered the lives of many city-dwellers a miserable concrete monotony. We cannot afford to repeat either of these mistakes on Mars.

Human experience is the driving force behind our architectural decisions. We have chosen to design walkable streets and wide-open spaces to prevent a glum tunnel city. Light and fresh air are precious and the colony will flounder without them, even though they are difficult to design and build. Every habitat is designed with a view, either over communal spaces or the Martian landscape, to ensure that people do not feel boxed in by the city.

Simultaneously, we are placing our buildings carefully with due regard for the natural landscape of Mars. Habitats will lie along contours and follow riverbeds, and we will not flatten the landscape with bulldozers. Through this philosophy, we can build a city better suited for habitation than perhaps even Earth cities.

2.2   Infrastructure and Transport

To feed, build, and supply a colony of a million people, infrastructure projects on a colossal scale are required. In particular, to transport the tens of millions of tons of material per year (mostly raw ore, but also food, water and manufactured material) around the city, vehicles running on roadways are far from adequate. The energy requirements of operating such a system are enormous, as are the material demands of hundreds of kilometers of concrete roadway. Instead, rail is the far superior option.

Rail Car Configurations

A Martian 3 m rail gauge has been developed with the capacity to carry hundreds of tons per electric locomotive at 250 km/hr across the surface. This capacity is dictated by the millions of tons of ore that need to be transported from quarries to furnaces. Cargo cars are set up to use standard containers, and they are interchangeable with passenger modules.

The material cost of producing railways is significant (160 kg of steel and 625 kg of sulfur concrete per meter), but analysis shows that if a given rail route has a lifetime throughput of over 2 million tons, energy will be saved compared to vehicles running on compacted regolith. The minimum projected system size is 300 km of rail, which is easily achievable with steelmaking and mining capacity in the latter stages of the project. Rail lines will form the transport backbones of Nexus Aurora, with dense urban areas situated nearby to rail lines.

While Nexus Aurora will become significantly less dependent on imported equipment as the city population approaches 1 million, the cadence of ship landings will increase as the population swells. In order to meet this goal, over 5000 Starships (or equivalent) will be landing per transfer window by the year 2100. This requires significant ground-based infrastructure to receive, process, and re-launch so many vehicles. For this purpose, we have picked 4 crater-based launch complex sites in the vicinity of the city, sufficient to process all landings in just 100 sols.

The design of these complexes is modelled on Baikonur Cosmodrome, with trenches to funnel heat and debris away from vehicles[27]. All launch complexes are connected to the colony by rail and have extensive storage depots to process the cargo and arrivals. To mitigate risk for the colony in the event of launch or landing failure, all complexes are located in relatively high-walled craters and exclusion zones have been drawn around them to prevent damage to buildings in the event of an accident.

As Nexus Aurora grows, the complexity and scale of the supply chains required grow proportionally. The beauty of building a city from the ground up is that this system can be unified and controlled centrally. This allows both comprehensive control of strategic stockpiles[28], optimization of automated delivery of goods, and prevention of waste.

Transportation routes across the colony have been designed with spaces for automated robotic systems to carry cargo. A standardized container design has been developed to allow maximum interoperability of cargo handling systems. This standard has been implemented for a variety of equipment outside containers to allow for modular production lines with automated operation.

The ultimate goal of indoor logistics is for all transport of goods or equipment further than around 25 m to be carried out by autonomous robots. This represents a significant investment for Nexus Aurora, as many of the complex robot systems will need to be imported from Earth. However, it also allows for significant improvements in efficiency, "24.6/7" operation, and removes a significant burden of menial labor (either in manual moving of such cargo, or remote control of such robots) from the inhabitants of the city. Massively automated systems such as this have been proven to be incredibly effective on Earth[29] and Nexus Aurora intends to expand this framework beyond the warehouse[30] to cover the entire city footprint.

These transport robots will be the most visible part of the city's robotic infrastructure, but they will be only a small part of it. The demands of the Martian economy and practicalities of operation on the surface necessitate the removal of human operators wherever possible; they will be replaced by a semi-autonomous system controlled by an overseer. As such, the vast majority of construction, mining and industry will be carried out without direct human involvement. Instead, supervisors will give top-level guidance to a semi-autonomous system and work to further automate menial or repetitive tasks.

We foresee robots and remote-controlled systems carrying out almost all tasks on the surface: mining operations (particularly blasting); control of trains and rovers transporting equipment; installation of buildings and equipment; and moving equipment between the surface and the city through intermediate zones. These zones are intended to remove toxic perchlorate-containing dust and other contaminants. Many of the jobs traditionally carried out by entry-level human labourers will also be automated or robotically-aided[31] to some degree.

2.3   Production

Power generation in the completed Nexus Aurora colony is not significantly different from that in the early stages of colonization. The notable difference is the diversification of the power generation and storage infrastructure to safeguard against failure. We have also varied the number of technologies considered as a precautionary measure against any one energy production method becoming significantly more or less viable in the 80 years from today to the completion of the project. In practice, Nexus Aurora will use some blend of these generation techniques with exact numbers determined by the colonists. The total power demand of the base is estimated to peak at 10 GW, with the vast majority of that needed for the production of materials for construction.

Nuclear power is our primary plan for power generation in the later years of the colony, as it removes the need for storage. The most suitable architecture for our purposes is CANDU, as it operates with low-purity uranium, has many passive safety features and does not require large pressure vessels to be manufactured. Providing 10 GW using purely CANDU reactors would require approximately 1400 tons of unenriched uranium per year, which can be imported or mined in-situ. The primary concern of a nuclear power system on this scale is cooling - radiator arrays would have to span around 20 km2, including buildings radiating heat.

Ground-based solar energy remains highly effective, scalable, and resilient against failure. Manufacturing lightweight, high-efficiency photovoltaic cells on Mars is not practical due to the complex production process, so heavier and less efficient arrays will have to be deployed in greater numbers. One-axis tracking is also very promising for increasing output. To fully power the city in this way, roughly 300 km2 of solar farms would be required.

Space-based solar systems with power beaming are very promising, as they provide power almost around the clock with virtually no storage required. This technology is still in its infancy, so it is hard to design for and thus we have not planned to rely on it.

Power storage needs can be reduced by running power-hungry industrial processes only during daylight hours, but around 100 MWh of capacity will still be needed as backup for essential systems. Sodium-sulfur is the most promising battery chemistry, as it has high density and the electrolytic chemicals are both abundant on Mars. Mechanical storage, such as flywheels or hydroelectric, does not have anywhere near the energy density required.

Farm buildings take largely the same form as habitats (utilizing the BFRP tubes) but the major difference is that much of the later-stage farming complexes have been moved north, to a massive farming outpost, in order to increase received light throughout the year. More detail is given later in this report.

Apart from gradual upscaling as the colony grows, production processes of water and air are largely unchanged, although water production demands have increased with both industry and habitation. The glaciers in the region of Hellas Planitia are of such great scale that they will practically never be exhausted. Current estimates for the size of the glacier that exists within kilometers of our landing site is 2-5 km3, an order of magnitude greater than the requirements for Nexus Aurora.

The biopetrochemical and electropetrochemical industry remains a linchpin of the economy of Nexus Aurora. In addition to the colossal demands of tens of thousands of tons of polymer for the construction industry, polymers provide a major export. As the production scales, larger and more efficient chemical processes become practical. This reduction in energy cost is the primary driving factor that allows the colony the luxury of large open domes and towering landmark buildings.

Mining in Nexus Aurora, as with all other industries, takes place on a colossal scale to keep up with the pace of expansion of the city. Raw ore and basalt demand is estimated at 2 million tons per year. Given the scale of this demand and the hardness of the rock in our region (basalt is typically 6-7 on the Mohs scale), industrial explosives are required. Two quarry sites north of the city have been identified for producing basalt (needed for basalt fiber and iron) and plain regolith (needed for soil for farming, rammed soil blocks, among other uses).

A major industrial zone for Nexus Aurora has been planned inside Geim Crater. This area is shielded from both the colony and mining areas and is situated at an intersection in the rail network, ideal for both importing and exporting material. This holds the production facilities for almost all heavy industry. Steel, basalt fiber, polymers, sulfur concrete, cement composite and many other less-used materials are required. They will be processed with rolling mills, composite layup machines and casting equipment. Light manufacturing of consumer goods and other small-volume chemical products is largely completed inside Nodes to enable people to work close to their homes.

2.4   Medical

Nexus Aurora will model a medical system that provides accessible healthcare for all citizens. We can derive a large portion of the Mars-based medical system specifications from what is accepted to be the medical standard on Earth, and from medical knowledge of living in low Earth orbit. The details of medical standards are largely omitted from this document, which instead focuses on facets of health and medicine unique to Mars.

Life on Mars carries a number of health risks: hypo-gravity may decrease bone density and immune function; solar radiation exposure can cause cancer at higher rates of incidence and severity; isolated lifestyles can degrade mental health; risk of depressurization events threaten barotrauma and "the hurls" (depressurization sickness); and several others. A heavy focus will be placed on these Mars-specific ailments, each with a dedicated outpatient center for both treatment and research.

The means of travel will often be limited on Mars, necessitating an increase in the widespread accessibility of medical services. To remedy this, we have designed a network of conferencing equipment dedicated to medical triage (referred to as tele-triage). Patients can securely video conference with an on-call triage nurse from a public tele-triage station, and the nurse will typically provide instruction to the caller, such as which health care provider to visit. Patients can call in with their personal devices or access public access stations, which will be strategically placed throughout the colony, especially in remote and high-risk environments such as mining quarries. While telemedicine is possible with this system, it is not intended to replace in-person medical visits. Rather, it exists primarily to reduce unnecessary hospital traffic and provide triage staff with a powerful logistics tool.

To serve the medical demands of Nexus Aurora, we have prepared facility designs for 3 large hospitals (each with a roughly 750-bed capacity), several buildings dedicated to outpatient specialties, and around 100 clinics of varying sizes. All clinics will be public, and all specialty beds found in clinics and outpatient buildings will be counted as ICU/specialty beds made available for hospital overflow purposes. With these facilities, we target a total medical capacity of 4000 beds.

The potential for depressurization events imply some likelihood of hermetic lockdowns in which people can become stuck in certain buildings or corridors. As the requirement of hermetically sealing doors nullifies the logic of egress regulations, we design each hermetic segment in the city to include standardized emergency medical kits so that trapped people will never be without access to medicine. Near each medical kit will be a tele-triage conferencing system to reach emergency responders and medical staff.

Modular Medicine Production Line

Manufacturing pharmaceuticals on Mars will be critical to the sustainability of the colony. While importing pharmaceuticals from Earth is possible and will remain a viable source, there is no financial substitute to manufacturing these drugs. Due to the lack of chemical feedstock, we have deemed a biomanufacturing approach to be superior. This requires less energy and can produce a wider variety of products than a chemical production line[32]. To this end, a modular medicine production platform has been designed to allow a scalable, flexible production line.

This system was illustrated with a production system for salicylic acid, a widely used drug and chemical precursor. Genetically engineered E. coli[33] and perfusion cultivation[34] are used to produce high yields with little equipment and waste. This system has been modelled to fit multiple production lines in a single 570Hab, along with support facilities such as quality control and water purification through reverse osmosis. Logistics robots carry out much of the work of moving material and equipment around the facility. A single production line with a footprint of around 1200 m2 employing 40 technicians and scientists[35] can produce salicylic acid at an average rate of 55 kilograms per day. 36 of these production lines are sufficient to produce the 58 most-consumed drugs from WHO's Essential Medicine list[36].

2.5   Beyond the City

Beyond the dense urban center of Nexus Aurora and the outer ring of quarries and solar farms, human presence in Hellas Planitia reaches far across the landscape. A number of small scientific outposts, at various points of interest identified from orbit, are planned, as well as small mining projects near expected mineral deposits. We anticipate the placement of many more of these mining projects as in-situ analysis detects deposits of valuable minerals such as lithium and copper. We also envisage a limited number of outposts far from the city for purely tourism or sightseeing purposes, although these are harder to justify economically.

The planned scientific outposts are located in sites of particular geologic or geographic interest. For example, the confluence point of Niger and Dao Vallis is an excellent point to find deposited sediment, and Nidavellir Point, which is located at the lowest point on Mars, ideal for deep-crust geology and potentially even astrobiology.

Mining outposts near Cañas crater and Terby Crater have been identified, as orbital data suggest that concentrations of hydrogenous minerals, sulfates, and ferric oxides are significantly higher in the regions close to the city.

However, the scale of these specialist mining and science outposts pale in comparison to Calorie Outpost, a sub-city located 1030 km north of Nexus Aurora at a significantly higher altitude in Tyrrhena Terra. Moving the majority of the farming far north is advantageous. Increasing light during winter is a massive boon for farming, as it reduces the amount of food that needs to be stockpiled for winter when productivity is lower, unless we resort to the power-hungry method of artificially lighting hundreds of square kilometers of land. This will reduce the number of farms that need to be built.

Calculating this reduction factor is multivariate, but our best estimates are that moving farms to the north reduces the land area needed by around 45%. This represents a colossal decrease in energy demands for population growth, even when considering the additional cost of building and operating a rail line to connect Calorie Outpost and the main sections of Nexus Aurora. For this reason, the decision to move farming north comes easily.

Part of the economic role of Nexus Aurora will be to support a significant manned presence in low Mars orbit. At a minimum, this orbital outpost would just be a refueling station for outbound Starships. It will use solar power to convert water and CO_2 from asteroids and the Martian moons into fuel with the Sabatier process. However, this fuel station could well expand to a fully-fledged orbital city with habitation for thousands of people, orbital manufacturing, and provisioning for spacecraft travelling between Mars and the outer solar system.

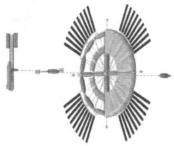

Starship re-fueling space station

It is hard to predict exactly how much orbital infra-structure will be needed, because it is heavily dependent on the amount and nature of deep-space development elsewhere in the solar system. We have made plans for a basic habitation and refueling station, which can house a few hundred people in a rotating ring, protected from radiation by a layer of asteroid regolith and ice enclosed in a Kevlar bag. If and when the orbital station expands, it could either do so by enlarging this station (by adding additional counter-rotating rings, or increasing the radius of the existing ring) or building additional stations in similar orbits.

In addition to large stations in Martian orbits, the connectivity needs of the colony are addressed by an array of communication satellites. Earth-Mars links are provided by a number of high-orbit satellites with 20 m dish EHF links, each providing 9 Gb/s. Communication to the Martian surface is done with X-band radio to large ground stations, with fiber connections connecting to ground-based data centers and consumers.

3   LIVING IN NEXUS AURORA

We have explored many creative methods of engineering a city for 1 million people. However, it is equally important to create a city that 1 million people will actually be comfortable living in. Since the city is designed for permanent residence, it must be appealing enough to motivate people to leave their belongings and loved ones behind forever and begin a new life on Mars. This requires titanic efforts in architecture, sociology, urban planning and macroeconomics to make Mars not only a place where people can live, but where people want to spend their lives.

3.1   Urban Planning
The challenges of urban planning on Mars are extreme. It is difficult to create a living system where inhabitants can flourish given the necessary physical restraints of Martian life. The solutions to these problems come from innovative applications of existing concepts in urban design theory, married with creative uses of the structures available.

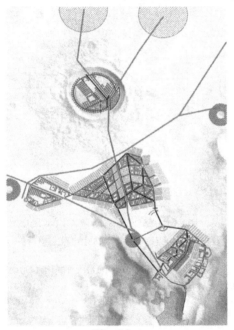

City Map

The top-level layout of the city is modelled on Howard's Garden City concept[37]. It avoids sprawling urban cores in favor of distributed small towns, with employment scattered both in these towns and in dedicated industrial hubs such as the Geim Crater complex and launch complexes. Each of these "Nodes" is organized around a central open space, generally a 200 m dome embedded in a natural crater. However, long, thin spaces built into lava tubes are also an option. Large areas of housing, both in smaller 570Habs and Air Mattresses, provide neighborhoods of 10-15 thousand residents, interspersed with zones for employment. The Nodes differ in size depending on the differences in land value between more and less desirable areas, without creating low-value segregated areas.

Intra-town transport is provided depending on traffic levels. The shortest or quietest routes are composed of service tubes which are 6 m in diameter and contain multiple levels: a pedestrian walkway, service tubes for robots, and passages for water, power, and air distribution. This type of tunnel can also be fitted for indoor light rails, which are used for higher-traffic transport along longer distance routes.

Higher-traffic, longer-distance routes will be for more than just transportation. Larger roads, collectively termed "boulevards," have been developed to contain multiple features. These will include: pavements for pedestrian traffic; wide roads for autonomous cargo transport, bicycles and bus-style vehicles; and in the busiest areas, ample shopfronts. The intention is that almost all routes over 50 m long will be boulevards, making the city supremely walkable and bikeable. Unlike car-focused American suburbs[38], the transport areas will be more like European high streets that act as areas for congregation.

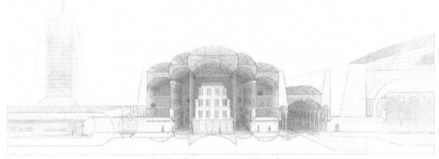

Typical urban space section

On a larger scale, the zoning of the city is focused on following major transport lines to allow easy movement of people and equipment. The roles of Nodes vary across the city, adapted to the various industrial regions. For example, we have mapped out a large industrial zone in Geim Crater, and a region dense with nuclear power generation near Leeuwenhoek Glacier. The exact locations of these Nodes are determined entirely by the lie of the land - the domes at the center of each Node are placed in naturally occurring craters and depressions.

The character, culture and architectural style of each Node varies according to the preferences of the inhabitants. Nexus Aurora will provide the tools for citizens to shape their habitat into a home. The available areas for personalization include the architectural styles of buildings within habitats, street furniture (including trees, which will be a feature of almost all streets), and the form and function of communal buildings.

A multi-purpose NFC card can be used for many purposes such as access, ID, and contactless payments. It allows for streamlined access and services across the city. Within the city, all restricted access, private or industrial, is controlled by NFC card authorization. This allows for anonymous monitoring of travel across the entire colony, which is a valuable tool for optimizing traffic flow and emergency response. NFC card access is also useful for ensuring security and peace of mind. However, the implications for the creation of a surveillance state are severe, and a number of anonymous protocols have been developed for preventing unscrupulous governments or bad actors from harvesting personal data.

3.2   Character of the City

City Block Section

The cultural character of Nexus Aurora is shaped by its people. By 2100, we have estimated that 95% of the city's population will be Earth-born. We have attempted to identify which Earth regions will send the most migrants to Mars based on socioeconomic factors. We predict that the largest populations will be from the Anglosphere (around 30% of immigrants), East Asia (20%), and South Asia (15%). We will have no nationality-based selection criteria for immigrants to Nexus Aurora, so the city will be a cultural melting pot.

Great effort will be made to allow citizens to preserve their cultural identity. Each Node will have a designated second official language, alongside one of the three main languages of the colony which will be determined by census. Citizens will be encouraged to style their homes and streets according to familiar design practices from Earth. This will allow communities to shape their communal spaces, determining the form, usage and nature of parks, recreation areas, and places of worship. The form and function of street furniture like statues, benches, and street markets would also be open to the discretion of communities, with common rules to prevent offensive statues or art being installed.

The nature of urban design in Nexus Aurora produces natural communities of all scales. The city has pressure-tight dividing barriers, creating blocks of a few hundred houses, each with small amounts of shared open space. Clusters of a few dozen 570Habs and houses within Air Mattresses are formed between boulevards, with populations of a few thousand sharing amenities like shops, parks and community halls. These then fit into neighborhoods (sharing schools and places of worship) and Nodes (with communal open spaces). This emphasis on fractal communities is to avoid feelings of isolation and to provide citizens with a real sense of belonging in their homes.

A vital part of community engagement is sport and recreation. This is key for increasing inter-generational and inter-household belonging in the entire city, while promoting physical and mental health. The recreational needs of a city of 1 million will be functionally equivalent to that of an Earth city. Bars, community centers with nightly activities, plays, cinemas and sports arenas have all been considered essential parts of life. Any indoor spaces with no specialist structural needs (such as bars and cinemas) can be built into the Air Mattress with no difficulty. These buildings will be scattered throughout the city on "high streets" within the residential areas and along Boulevards to create gathering points near homes rather than isolated in retail parks.

There are a few activities limited by the size of pressurized volumes - arenas for pitch sports like football and rugby are difficult (although possible in Node domes), so court and track sports like tennis, basketball and running should be emphasized for live spectators. The Mars-specific needs of the community are near-impossible to predict; open spaces to remind the city dwellers of Earth are strong candidates, so urban planning has emphasized the inclusion of parks. Additionally, the long distances of urban farmland make running, walking and cycling possible, so radiation-shielded walkways have been added to all farms for this purpose.

The family unit is at the core of the community, no matter the demographic or type of family. The architecture of the city provides private apartments for each family with space for bedrooms, a kitchen, bathroom and sitting room. These "standard-issue" apartments are comparable in size to city apartments on Earth. While they are not overly spacious, they are a comfortable place for a family to call home. Apartments will be owned, not rented, so citizens can decorate and modify the spaces within building regulations.

While diversity is celebrated, we also recognize that unwanted prejudices, biases and bigotry will travel to Mars as well. The first method for mitigating such issues is education. As part of their permanent induction to Nexus Aurora, new citizens will attend a small-group style teaching programme which will contain content on directly tackling and subtly subverting existing prejudices. Secondly, community action and coordinated enrichment aims to prevent the formation of exclusionary communities. This includes cultural exchanges between Nodes with large cultural groups, inclusionary festivals and all-ages education programmes to proactively counteract tensions. An array of measures will also be taken to prevent active discrimination. This includes anonymized recruitment and performance review practices across Nexus Aurora, colony-sponsored immigration for highly skilled individuals without the funds to travel from Earth, and active engagement between community advocacy groups and government to allow any issues to be raised.

3.3   Economics
Building a strong economy for Nexus Aurora carries many broad-stroke similarities to projects in contemporary international politics and macroeconomics. It is

important to distribute the economic benefits of growth among both societal and corporational interests while minimizing the chance and effects of economic depression. The overarching goal of the fiscal plan of Nexus Aurora is to achieve self-sufficiency from terrestrial economics, which is to minimize investment tying Martian policy to the interests of groups on Earth, and a neutral or even positive trade balance.

There are no precedents for building an economic system for Mars that can be simply adapted for Nexus Aurora. Rigid shipping times, a hostile environment requiring high welfare, long payback time on investments and a deep political involvement in the economy means that a historic economic model cannot be simply adapted.

The early stages of Nexus Aurora will be in deficit - a colony of tens of thousands can produce almost nothing to offset the large cost of shipping production equipment to Mars. Small income streams may exist, such as high-value tourism or scientific research, but the colony will run at an enormous loss for decades.
To fill this gap, investment from private and public groups on Earth will be needed. This will primarily include national governments, large firms and private individuals. In this early stage, all investment will be quantified and unified into a national bond system. This minimizes the undesirable political entanglement from non-monetary investment, a strategy commonly employed in neo-colonialist enterprises[39]. The incentive for national governments providing this investment is the fact that almost all of the initial funds will be spent on manufacturing and engineering on Earth, essentially turning investment in Nexus Aurora into a national jobs programme.

During later stages of Nexus Aurora, the national government will aim to buy back these bonds from terrestrial investors. This holds multiple benefits: investors willingly hand in holdings on Nexus Aurora, and the potential for significant returns will further incentivize investment in the project. It also provides a "carrot and stick" for groups not adhering to government policy, by freezing buyback of stocks controlled by such groups.

Financial import-export charts

In the long term, a solution must be provided to account for the huge initial trade (and thus economic) deficit. We feel the optimal solution to this is a long-term strategic policy of currency devaluation. Having a slow reduction in the effective value of Martian currency allows cheaper imports in early years while making exports competitive in the later stages. This scheme has diminishing returns as other colonies use the same strategy, which makes Nexus Aurora competitive with other colonies as other organizations develop cities of their own, which will also serve to make our export strategy significantly more lucrative.

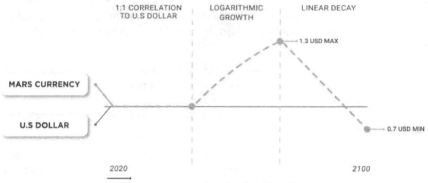

Mars currency devaluation timeline

In the long term, the trade balance of Nexus Aurora will be the following. Imported products will primarily be vital engineering components, luxury, and heritage goods. The bulk of exports will be to other Martian colonies[40], with some exports also to outposts elsewhere in the Solar System (Mars' decreased gravity and thin atmosphere make launching material much easier than on Earth).

We have come to the conclusion that it is neither possible nor beneficial to prevent other major powers setting up Mars colonies. On the contrary, we should encourage this. Trade with other locations on Mars does not carry the prohibitively expensive launch costs of Mars-Earth trade, and the same launch costs mean that materials can be sold with a significant markup (which are slightly less than shipping costs) and a significant market will still exist. As such, the most exported goods will be those needed in the largest quantities by other colonies, and those that Nexus Aurora can produce most effectively: food, polymer composites, and steel. Secondary physical exports will include products that Nexus Aurora can produce in greater quantities and at lower prices than other colonies: pharmaceuticals and biochemical products.

However, the most lucrative economic sector of Nexus Aurora is the services industry, particularly intellectual property and quaternary-sector services. The concentration of both highly educated workers and technological development will, as with Silicon Valley before it, be a fertile breeding ground for development of technologies and products with markets across the solar system. Tertiary- and quaternary-sector exports have the additional advantage of not requiring transport to Earth.

The internal economics of Nexus Aurora will also shift as it grows from an industrial and research outpost to a full-fledged city. To facilitate innovation and attract immigration and investment to the colony, we intend to encourage partnerships with both Earth-based and Martian companies to create a free market for goods and services for the citizens of Nexus Aurora. However, this market will require new regulatory structures to adapt to the peculiarities of the Martian economy.

The most prominent such structure is the formation of vertically-integrated consortiums of similar businesses, which serve to stabilize the prices of goods and services against demand shocks and provide a streamlined method of regulation. To deter cartel formation, the Martian government will have tight regulatory control over the foundation, role and internal hierarchical structure of consortiums. This creates de-facto nationalized industries while still allowing free market competition between firms within each consortium. One example of this regulation is sterilization of many products, especially those shipped from Earth, to ensure smooth delivery and support of all equipment.

Another difference is the existence of a level of state-provided welfare considered radical for Earth, to ensure the indefinite welfare of the citizens and to prevent various dystopian "oxygen fees apply" scenarios. This takes the form of a system of pseudo-UBI, where the basic needs of citizens (life support, basic food) are provided by the state rather than with cash. This is backed up with a price ceiling on certain goods, to prevent profiteering by private firms.

Beyond these economic activities, there is one lucrative source of income of note: the influx of new colonists. It is assumed that the vast majority of colonists either pay for their own passage to Mars (either directly, in the much touted "sell your house, move to space" model, or via sponsorship by an employer or nation). However, it takes a small number of individuals paying for a premium experience on Mars - either short-stay tourists[41] or wealthy individuals hoping to buy a more luxurious trip and home in Nexus Aurora - to completely offset the diminishing cost of shipping and technology procurement.

Managing this entire system of welfare is the Martian Sovereign Wealth Fund. Modelled on similar funds in countries like Norway and Singapore, this organization is funded by exports from centrally-owned industries such as food and basic construction materials. It is used not only to procure these necessities of life, but to perform the bond buyback described above.

For the incomes and outgoings of the government of Nexus Aurora, costs remain high in initial years, driven by the massive import of industrial equipment. However, this is increasingly offset by profit from export, tourism income, and taxation. Assuming that it began in 2020, Nexus Aurora is profitable by 2100, but will take several further decades to balance the colony's debt, with bond buyback leading to true fiscal independence.

It is important to note that economic and pseudo-economic factors have been of utmost importance in almost all decision making in Nexus Aurora. The drive to reduce mass and energy consumption has been critical in our structural designs, with the ultimate goal of reducing mass shipped from Earth.

3.4   Administration
In the broadest sense, a system of government is a bureaucratic structure that serves the inhabitants of the community. Nexus Aurora represents an opportunity to develop a more empathetic, coherent framework around the lives of citizens. We

aspire to be optimistic but not utopian, and acknowledge the need to have a mutable system that can be improved by the changing needs of the population. As such, we have set out the goals and specifications of political, education and judicial systems according to the overarching ideals of Nexus Aurora.

The ultimate goal of the government of Nexus Aurora is the preservation of the rights of Mars inhabitants, both to their personal rights and the right to collectively shape the future of the community. While we cannot foresee every future challenge facing Nexus Aurora, we have outlined several potential pitfalls and made provisions in the governmental structure to avoid them. In brief, it is an open, representative democracy backed by technical advisory groups providing stability and informed decision making for the good of the city.

One particular risk is the formation of collusive oligopoly: corporations on which Nexus Aurora is dependent, particularly Earth-Mars shipping, can form cartels that can strongarm the elected government into developing financial interests above the interests of citizens. To prevent this, it is vital to ensure that companies are competing in a fair and open market for the considerable business that Nexus Aurora can offer.

The keywords of the government structure are accountability, transparency and cooperation. These are all essential elements when any power structure is in possession of life-or-death power over citizens. Citizens must be able to trust that elected officials are working in the best interests of the community. The following methods are in place to assure this: frequent referenda; career term limits for politicians; accessible meetings; and a completely open public record of decision-making. The following precautionary methods are in place should a particularly divisive leader take power: inflexible limits on the conduct and spending of elected officials; required approval of certain decisions from subject-expert advisory groups; and a multi-stage system of checks and balances on all significant acts.

The education system is of prime importance. The foundations of Martian society are built upon research, a desire to explore the unknown, and the need to build future generations with the skills needed to thrive in such a unique environment. The education system aims to import the most effective parts of Earth theories allied with specialized developments to fit the needs of Mars.

The educational system for citizens aged 3 to mid-20s aims to break from established doctrine of linear, test-based progression[42] between educational stages with little emphasis on practical life skills. Instead, students learn on mixed-age campuses in purpose-designed buildings, with classes organized into broad developmental tiers. Progression occurs not on a fixed-time basis but when students demonstrate mastery of relevant skills, via exams, coursework or capstone projects. Mixed-age groups encourage cooperation and respect and make use of student mentoring schemes to develop the skills of both learner and educator[43].

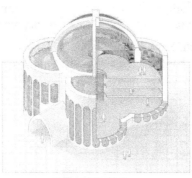

School Building

Within schools, there is an increased emphasis on practical as well as academic subjects. This includes technical skills (programming, mechanical maintenance), interpersonal skills (presentations, debating, teamwork) and practical skills (emergency preparedness, sign language for use in EVA suits).

Schools will provide ample work-life experiences to smooth the transition from education to profession. The primary method is internships (especially in high demand professions[44] to fill these gaps without resorting to forced career routes), with school alumni encouraged to open their workspaces for current students.

The socio-economic and political environment of Nexus Aurora will not be perfect, but we expect it to mitigate many of the systemic conditions that lead to crime on Earth. Our environment will include a quasi-UBI system providing basic housing and food, a more comprehensive social security net, and increased regulation of imports of psychoactive substances. While this is not comprehensive, it is the beginning of an improved system.

The emphasis on criminal justice in Nexus Aurora is on restorative justice. This is both ethically preferable to punitive justice and imprisonment models, and removes the burden of a non-productive prisoner group on the colony. House arrest and community service models for all but extreme crimes facilitate reintegration into society and reduces recidivism[45,46]. When harsher prisons are required, a Scandinavian model[47,48] of less punitive environments will be employed.

Outside of criminal justice, civil disputes[49] will be handled by community mediation and dispute panels wherever possible - courts will only intervene where absolutely necessary. Lawmaking will be handled primarily by the elected representatives, but approval from panels of judges, sociologists and psychologists will be required to evaluate the effect of new laws.

Life on Mars is fundamentally dangerous: dust storms, fires, depressurisation, and chemical leaks are but a few of the potential emergencies that may arise. Due to these dangers, proper disaster relief is essential to life in Nexus Aurora. Live risk assessments will be carried out by Joint Emergency Response Centres in all urban centres, which also coordinate mitigation, preparedness and response across various departments. Drills for various emergencies will occur semi-regularly and training will be a mandatory part of schooling to ensure citizen preparedness.

Diplomatic relations with Earth, other Martian colonies, and the rest of the solar system are a priority to ensure smooth flow of migration and trade. Immigration will be handled by a multi-tier visa system that includes both temporary and permanent residents. Cooperation between terrestrial nations, key stakeholders (private and industrial) and supranational organizations will occur at the highest level of Martian elected government.

4   OUR NEXUS AURORA

We hope this summary, while brief, properly conveys our vision for a bold future and the path for building it. These 20 pages barely scratch the surface of the work carried out by our phenomenal international team. Behind this report are over 100 pages of technical reports, proposals for economic and political systems, evaluations of existing technologies and brainstorms of new ones. We have created over 5 GB of 3D models, diagrams and renders of our civilization, and all the images in this application have been rendered by our team. Planning and discussion stretches over 60,000 messages from 50+ team members in a multitude of time zones. We have no intention of halting progress now. We were brought together by this competition, but development of these concepts will continue after it is over. We will continue building until the day humans stand on Mars, in a city that defies every force of nature standing before it, until we can truthfully say, "Mars is our home."

4.1   Contributors List

- **Coordinators**: Adrian Moisa (SpaceInstructor), David W. Noble (Marsorbust), Henry Manelski, Koen Kegel (koenikova), Kim von Däniken, jakobberg, Felix M. Müllner, Vikas, NLDukey, Elias Malak (EliTheDude), Sean Wessels, Mitchell Farnsworth, Mae Elizabeth, Sitanshu Pani, Sam Ross (document editor)
- **Administration**: Felix M. Müllner, Hyeonseo Daniel Kim (tankat), HoospitalBall, einarhg, CertifiedDefaultToday, thefiendhitman, Zam, David Zühlke
- **Buildings**: Jakobberg, Sean Wessels, Koen Kegel (koenikova), SK9, Sitanshu Pani, Apollo, Kosaki Onoder, Dr. Orion Lawlor
- **Economics**: Hyeonseo Daniel Kim (tankat), Felix M. Müllner, Arxivist
- **Equipment**: ukudaly, Will Coulson, Henry Manelski, Marco, CaliDToday, Ignis Cogitare, Sitanshu Pan, Max Jagiella, Cameron Rough, PiTim, Bryce Bornhorn, Il Ho Cho
- **Infrastructure**: Arxivist, Tankat, Sean Wessels, Arxvisit, blakemw, Sitanshu Pani, Dr. Orion Lawlor, Hyeonseo Daniel Kim
- **Lifestyle**: Sitanshu Pani, Henry Manelski, Sean Wessels

- **Logistics**: Mitchell Farnsworth, Sam Ross, Henry Manelski, Zam, fredfredb3rg, questionablyhuman, agustin99, shalrath, thephatpeach, Pranav Kishan, Cameron Rough
- **Medical**: Kim von Däniken, b.funkt, Sean Wessels, Mae Elizabeth, Apollo, Julian, Brooke Shepard, Liam Berry
- **Production**: Peteware7, Felix, Sean Wessels, MikeGo, Mark Watt Ne, Mae Elizabeth, Henry Manelski, CertifiedDefaultToday, Dr. Orion Lawlor, Julian, deadlifts12, Kim von Däniken
- **Transport**: Cameron Rough, Bilal, Jakobberg, Martin Eizinger
- **Urban Planning**: Koen Kegel (koenikova), Sean Wessels, Olivier Rinfret, zen, JWFJWF, Sitanshu Pani, Il Ho Cho, Daniela Castro
- **Utilities**: Jakobberg, Koen Kegel (koenikova), Michael Dealbert, blakemw, idyllicchimp
- **Arts and 3D Design**: Vikas, yury2bcu, Sitanshu Pani, splinter_vx, DrQuandryPHDToday, Elias Malak(EliTheDude), Mihai Haceanu, Framps, somepoint_sound_derm, felix, NLDukey, Bobbbay, Sebash, Natron, Calvin, THEWILDOFFICIAL, Felix M. Müllner, Rory Hyland, Alex Doyi

4.2   Compiled References

1) https://mepag.jpl.nasa.gov/topten.cfm?topten=10
2) https://ntrs.nasa.gov/archive/nasa/casi.ntrs.nasa.gov/19650016474.pdf
3) https://www.energy.gov/sites/prod/files/2017/12/f46/2016_Occupational_Radiation_Expos ure_Report.pdf
4) https://www.fosterandpartners.com/projects/mars-habitat/
5) B. Soares, R. Preto, L. Sousa, and L. Reis, "Mechanical behavior of basalt fibers in a basalt-UP composite," Procedia Structural Integrity, vol. 1, pp. 82–89, 2016.
6) https://www.livescience.com/radiation-mars-safe-lava-tubes.html
7) I. Repins et al., "19·9%-efficient ZnO/CdS/CuInGaSe2solar cell with 81·2% fill factor," Progress in Photovoltaics: Research and Applications, vol. 16, no. 3, pp. 235–239, May 2008.
8) https://www.westinghousenuclear.com/Portals/0/new%20plants/evincitm/eVinci%20Micro %20Reactor%20NPJ%20M-A%202019.pdf?ver=2019-04-30-211410-367
9) https://www.mhps.com/products/gasturbines/lineup/ft8mp/index.html
10) https://www.nasa.gov/mission_pages/station/research/long_duration_sorbent_testbed
11) https://ntrs.nasa.gov/archive/nasa/casi.ntrs.nasa.gov/20180007365.pdf
12) https://www.nasa.gov/pdf/683215main_DressingAltitude-ebook.pdf
13) J. P. Grotzinger, "Analysis of Surface Materials by the Curiosity Mars Rover," Science, vol. 341, no. 6153, pp. 1475–1475, Sep. 2013.
14) Schilt et al., Perchloric acid and perchlorates 2003
15) Salisbury, F.B., Growing Crops for Space Explorers on the Moon, Mars, or in Space,Advances in Space Biology and Medicine, Volume 7, pages 131-162, 1999, JAI Press Inc.
16) K. M. Cannon and D. T. Britt, "Feeding One Million People on Mars," New Space, vol. 7, no. 4, pp. 245–254, Dec. 2019
17) https://www.eceee.org/library/conference_proceedings/eceee_Industrial_Summer_Study/20 18/2-sustainable-production-towards-a-circular-economy/rethinking-steelmaking-zero-emissions-and-flexibility-with-hydrogen-direct-reduction/
18) https://new.abb.com/news/detail/54933/next-generation-electric-arc-furnace-stirrer-improves-steel-production-output

19) V. G. Luk'yashchenko et al., "Technology of Electric Melting of Basalt for Obtaining Mineral Fiber," Journal of Engineering Physics and Thermophysics, vol. 92, no. 1, pp. 263–270, Jan. 2019.

20) https://electrek.co/2018/08/30/volvo-new-electric-mining-vehicle-prototypes/

21) Y. H. Wen, H. B. Peng, D. Raabe, I. Gutierrez-Urrutia, J. Chen, and Y. Y. Du, "Large recovery strain in Fe-Mn-Si-based shape memory steels obtained by engineering annealing twin boundaries," Nature Communications, vol. 5, no. 1, Sep. 2014, doi: 10.1038/ncomms5964.

22) https://www.wired.com/story/amazon-warehouse-robots/

23) https://space.nss.org/settlement/nasa/75SummerStudy/3appendB.html

24) C. E. Hellweg and C. Baumstark-Khan, "Getting ready for the manned mission to Mars: the astronauts' risk from space radiation.," Die Naturwissenschaften, vol. 94, no. 7, pp. 517–526, Jul. 2007.

25) https://www.dsm.com/dyneema/en_GB/our-products/dyneema-fiber/dyneema-xbo-technology.html

26) Lhttps://www.nasa.gov/directorates/spacetech/niac/2020_Phase_I_Phase_II/lunar_crater_radio_telescope

27) Hall, R. and Shayler, D., 2001. The Rocket Men. Chichester: Springer, pp.49-51.

28) http://www.apics.org/mediaarchive//omnow/Crack%20the%20Code.pdf

29) https://www.ocadotechnology.com/blog/2019/1/14/experimenting-with-robots-for-grocery-picking-and-packing

30) https://www.ocadotechnology.com/blog/2019/1/14/experimenting-with-robots-for-grocery-picking-and-packing

31) K. Junge, J. Hughes, T. G. Thuruthel, and F. Iida, "Improving Robotic Cooking Using Batch Bayesian Optimization," IEEE Robotics and Automation Letters, vol. 5, no. 2, pp. 760–765, Apr. 2020.

32) M. Gavrilescu and Y. Chisti, "Biotechnology — a sustainable alternative for chemical industry," Biotechnology Advances, vol. 23, no. 7–8, pp. 471–499, Nov. 2005,

33) Metabolic design of a platform Escherichia coli strain producing various chorismate derivatives" I S. Noda, et. al., Metabolic Engineering 33, p. 119-129, 2016

34) E. Heinzelmann, "BioTech 2019 – ZHAW Waedenswil, 2 – 3 July 2019: Part 1 From Innovation to Technology Breakthrough," CHIMIA International Journal for Chemistry, vol. 73, no. 9, pp. 763–766, Sep. 2019

35) https://biotechnet.ch/sites/biotechnet.ch/files/aktuell/dateien/chimia_2019-9_singleuse-biotech_2019.pdf

36) World Health Organization Model List of Essential Medicines, 21st List, 2019. Geneva: World Health Organization; 2019. Licence: CC BY-NC-SA 3.0 IGO

37) E. Howard, Garden cities of to-morrow, 1898

38) J. Jacobs, The death and life of great American cities, 1961

39) https://www.ft.com/content/9f5736d8-14e1-11e9-a581-4ff78404524e

40) https://forum.nasaspaceflight.com/index.php?topic=44411.0

41) https://planete-mars.com/an-economic-model-for-a-martian-colony-of-a-thousand-people/

42) https://www.researchgate.net/profile/Nissa_Dahlin-Brown/publication/255685603_University_Community_Schools/links/02e7e5202812dedd65000000.pdf#page=7

43) http://citeseerx.ist.psu.edu/viewdoc/download?doi=10.1.1.157.2967&rep=rep1&type=pdf

44) https://eric.ed.gov/?id=ED477535

45) https://bja.ojp.gov/sites/g/files/xyckuh186/files/Publications/RAND_Correctional-Education-Meta-Analysis.pdf

46) https://www.unodc.org/documents/justice-and-prison-reform/18-02303_ebook.pdf

47) https://worldjusticeproject.org/rule-of-law-index/global/2020/Norway/table

48) https://www.kriminalomsorgen.no/information-in-english.265199.no.html

49) https://via.library.depaul.edu/cgi/viewcontent.cgi?referer=https://www.google.com/&httpsredir=1&article=3268&context=law-review